FORMULA NUMBER	DESCRIPTION	FORMULA
5.4	T score	$T = 10z + 50$
5.5	Formula for transforming a percentile rank to a raw score	Raw score $= (\sigma)(z) + \mu$
	Linear Correlation	
6.1	z-score formula for r	$r = \dfrac{\Sigma(z_X \cdot z_Y)}{N}$
6.2	Raw-score formula for r	$r = \dfrac{CP_{XY}}{\sqrt{(SS_X)(SS_Y)}}$
	Linear Regression	
7.1	Unrealistic z-score regression model	$z'_Y = z_X$
7.2	Realistic z-score regression model	$z'_Y = rz_X$
7.3	Raw-score regression model	$Y' = bX + a$
7.4	Slope of a regression line	$b = r\left(\dfrac{S_Y}{S_X}\right)$
7.5	Intercept of a regression line	$a = \bar{Y} = b\bar{X}$
7.6	Standard error of estimate	$S_{est.Y} = \sqrt{\dfrac{\Sigma(Y - Y')^2}{N - 2}}$
7.7	Standard error of estimate	$S_{est.Y} = \left(\sqrt{S_Y^2(1 - r^2)}\right)\left(\sqrt{\dfrac{N}{N-2}}\right)$
7.8	Formula to partition variance	$\left(Y - \bar{Y}\right) = \left(Y' - \bar{Y}\right) + (Y - Y')$
7.9	Coefficient of determination	$r^2 = S_{explained}^2 / S_Y^2$
7.10	Coefficient of nondetermination	$k = S_{pred.\ errors}^2 / S_Y^2$
7.11	Coefficient of nondetermination	$k = (1 - r^2)$
	Sampling Distributions	
8.1	Variance of the sampling distribution of the mean	$\sigma_{\bar{X}}^2 = \dfrac{\sigma^2}{N}$
8.2	Standard error of the mean	$\sigma_{\bar{X}} = \sqrt{\sigma_{\bar{X}}^2}$
8.3	Standard error of the mean	$\sigma_{\bar{x}} = \sqrt{\dfrac{\sigma^2}{N}}$
8.4	Standard error of the mean	$\sigma_{\bar{X}} = \dfrac{\sigma}{\sqrt{N}}$
8.5	z score for the sampling distribution of the mean	$z = \dfrac{\bar{X} - \mu}{\sigma_{\bar{X}}}$
8.6	Confidence interval of the mean (based on a z statistic)	$\bar{X} \pm z_c \sigma_{\bar{X}}$
8.7	Confidence interval of the proportion (based on a z statistic)	$P \pm z_c \sigma_P$
8.8	Standard error of the proportion	$\sigma_P = \sqrt{\dfrac{pq}{N}}$

STRAIGHTFORWARD STATISTICS

STRAIGHTFORWARD STATISTICS

FOR THE BEHAVIORAL SCIENCES

James D. Evans
Lindenwood College

Brooks/Cole Publishing Company

I(T)P™ *An International Thomson Publishing Company*

Pacific Grove · Albany · Bonn · Boston · Cincinnati · Detroit · London
Madrid · Melbourne · Mexico City · New York · Paris · San Francisco
Singapore · Tokyo · Toronto · Washington

To Lois

SPONSORING EDITOR: JIM BRACE-THOMPSON
MARKETING TEAM: JEAN THOMPSON AND GAY MEIXEL
MARKETING REPRESENTATIVE: SUSAN HAYS
EDITORIAL ASSISTANT: JODY HERMANS
PRODUCTION COORDINATOR: FIORELLA LJUNGGREN
PRODUCTION: GREG HUBIT BOOKWORKS
INTERIOR DESIGN: CHRISTY BUTTERFIELD
INTERIOR ILLUSTRATION: LOTUS ART
COVER DESIGN: LISA BERMAN
COVER PHOTO: P. R. PRODUCTION/SUPERSTOCK, INC.
MANUSCRIPT EDITOR: CAROL REITZ
ART COORDINATOR: GREG HUBIT
TYPESETTING: ETP/HARRISON
COVER PRINTING: COLOR DOT GRAPHICS, INC.
PRINTING AND BINDING: R. R. DONNELLEY & SONS
  COMPANY, CRAWFORDSVILLE

Printed in the United States of America
10 9 8 7 6 5 4 3 2 1

Library of Congress Cataloging-in-Publication Data

Evans, James D.
  Straightforward statistics for the behavioral sciences / James D. Evans.
    p. cm.
  Includes bibliographical references and index.
  ISBN 0-534-23100-4
  1. Social sciences—Statistical methods. I. Title.
HA29.E79 1996
519.5—dc20   95-18736
                                                CIP

FOR MORE INFORMATION, CONTACT:

BROOKS/COLE PUBLISHING COMPANY
511 Forest Lodge Road
Pacific Grove, CA 93950
USA

International Thomson Publishing Europe
Berkshire House 168-173
High Holborn
London WC1V 7AA
England

Thomas Nelson Australia
102 Dodds Street
South Melbourne, 3205
Victoria, Australia

Nelson Canada
1120 Birchmount Road
Scarborough, Ontario
Canada M1K 5G4

International Thomson Editores
Campos Eliseos 385, Piso 7
Col. Polanco
11560 México D. F. México

International Thomson Publishing GmbH
Königswinterer Strasse 418
53227 Bonn
Germany

International Thomson Publishing Asia
221 Henderson Road
#05-10 Henderson Building
Singapore 0315

International Thomson Publishing Japan
Hirakawacho Kyowa Building, 3F
2-2-1 Hirakawacho
Chiyoda-ku, Tokyo 102
Japan

BRIEF CONTENTS

CONTENTS

PART 4   **METHODS OF INFERENTIAL STATISTICS**   **275**

Chapter 10   **One-Sample t Statistic**   **276**

Chapter 11   **Two-Sample t Tests**   **305**

Chapter 15   **Nonparametric Tests**   **443**

Chapter 16   **Bringing It All Together**   **483**

PREFACE

This book is a relatively concise introduction to behavioral science statistics for undergraduates. It uses realistic data sets and research problems to teach essential concepts and procedures. A unique opening chapter is designed to engage the student's interest and reduce anxiety by (a) illustrating everyday statistical thinking and (b) establishing the relevance of the topic to careers in the behavioral sciences. The second chapter complements the first by carefully linking both descriptive and inferential statistics to the process of doing behavioral research. The text distinguishes itself from similar books by using the following classroom-teaching tactics (versus textbook expository methods) to convey the material:

1. Maintaining a focus on the fundamental topics
2. Using "pointing" techniques
3. Inserting "reminders" at critical points as more advanced concepts are introduced

My intent in writing this book was to improve both the information-processing aspects and the motivational dimension of behavioral statistics courses. My need to strive for these improvements is based on my belief that this course is the most important one that our undergraduates take.

Rationale for Developing the Text

During the last two decades of teaching statistics, I have been continually disturbed by the inability of most students to benefit adequately from reading their statistics textbooks. What appeared to me to be clearly written and organized books seemed to thoroughly confuse and frustrate the majority of my undergraduates. In contrast, some of the same students who alleged that the textbook was unintelligible often remarked (spontaneously and, in most cases, without a conspicuous ulterior motive) that the lectures made it all seem so simple and obvious.[1] Such comments caused me to wonder what magic the classroom lectures possessed that generally was not present in otherwise excellent texts on the topic. So, for a

---

[1] I would have felt like a miracle worker had several colleagues not also reported hearing the same comments in their statistics classes.

while, I got into the habit of making mental notes on what I was doing at precisely those times when students exhibited the "Aha!" reaction.

My observations revealed that, in contrast to most statistics books, the effective statistics teacher (a) uses physical movement to "point," emphasize, and clarify; (b) repeats crucial ideas to consolidate them; (c) tends to address only the most fundamental and central aspects of applied statistics; (d) describes mathematical operations as the formulas are used; (e) focuses on the essential forms and applications of probability (eschewing the marbles-in-the-urn approach); (f) *reminds* students of the meaning and use of elementary concepts and formulas when they are reintroduced in the context of advanced procedures; and (g) anticipates the student's mental set at critical points.

With the aim of developing a statistics book that teaches as well as informs, I have attempted to implement these instructional devices in *Straightforward Statistics*.

Features of the Text

The conceptual level of *Straightforward Statistics* is geared to undergraduates who are taking their first course in behavioral statistics. The book is more conceptual than mathematical and presumes only that the student has had the equivalent of high school algebra. I have tried to present a good balance of concepts, applications, and "how to" descriptions.

To use the teaching devices described earlier, the book:

Points and emphasizes through:

1. An "internal glossary" of key terms, in boldface type, followed by their italicized definitions, as they are introduced
2. A list of key terms and a summary at the end of each chapter
3. Boxes that highlight important points and summarize basic procedures
4. Plentiful, clear illustrations throughout
5. Arrows that visually tie labels or explanations to numbers, formulas, or critical elements of graphs (an especially significant characteristic of the book)

Reviews and reminds through:

1. Review questions at the end of each chapter
2. Distributed repetition of major ideas both within and between chapters
3. "Asides" inserted at crucial junctures within the body of the text

The "asides" feature merits some elaboration because it is what renders this book most teacher-like. Asides are sidebar comments and miniboxes that are strategically positioned in the exposition itself. They do what the classroom teacher does as he or she lectures: refresh the student's memory, translate math operations and symbols, and call attention to matters and ideas that cannot be overlooked because they are so fundamental to mastery of the material.

Following are *additional features* of the text.

Emphasis on the most fundamental material The early chapters carefully and methodically present the pivotal concepts, rationales, formulas, and definitions of applied statistics. The second half of the book describes and illustrates specific techniques and applications within inferential statistics. Throughout the book, peripheral issues and esoteric formulas are shunned in the service of clearly conveying and illustrating the essential elements of behavioral statistics. This focus on essentials at the sacrifice of tangentials is, of course, the reason for the book's title.

Integrated math Explanations of simple mathematical operations are not shoved into some appendix for possible use by exceptionally conscientious students. Rather, any such operations are clearly annotated as they appear within the main chapters.

Integrated probability theory and applications There is no separate chapter on probability. Instead, the most generally useful aspects of probability are introduced or reviewed and applied in the context of common statistical procedures and statistical interpretation. Emphasis is placed on "probability as percentage of area under a curve" and "probability as long-run percentage."

Continuity of examples Although each chapter contains new data sets, a particular data set might appear in two (or even three) successive chapters. Such substantive continuity is valuable for two reasons: It conveys the realistic idea that a particular sample of observations might need to be analyzed in several ways, and it directly illustrates the relationships among different statistical concepts and procedures.

Anticipation of the student's perspective One of the most difficult lessons to learn as a professor is that the student often gets off track relative to the well-planned sequence of information one is presenting. I have attempted to use my 20+ years of experience in teaching statistics to predict when the student might have a question or be confused about something, and meet the situation with an answer to the likely question or a second example that clears up the confusion.

An informal writing style that simulates an engaging discussion between teacher and student   This feature was incorporated to further the basic objective of developing a text that presents the subject matter of statistics in a way that the concerned instructor might.

Organization of the Text

Straightforward Statistics is organized in a conventional fashion. Chapters 1 and 2 provide an introduction to the course and background definitions. Chapters 3–7 present the fundamentals of descriptive statistics—from measures of central tendency through linear regression. Chapters 8 and 9 introduce the basic concepts of sampling distributions and hypothesis testing, which become the foundation for specific statistical tests and procedures. Chapters 10–15 cover the most essential significance tests, starting with *t* tests (Chapters 10 and 11), proceeding through analysis of variance (Chapters 12–14), and ending with nonparametric tests (Chapter 15). Chapter 16 ("Bringing It All Together") is a summary that provides flowcharts to help students choose the appropriate statistical procedures for specific kinds of data sets.

Computer Analysis

Using computers to carry out statistical analyses is hardly an option anymore. The reality is that today few researchers do statistics "by hand." Yet learning to use software does not substitute for comprehending the concepts of statistics. Accordingly, an introductory book should do two things in regard to computerized statistical analysis. It should encourage the student to learn how to use statistical software to do the grunt work in data analysis. But, at the same time, it should avoid overwhelming the student with computer printouts. That is, the student should not be distracted from learning about the logical system of statistics by an intrusive emphasis on computer analyses.

This textbook introduces the reader to the advantages and limitations of computerized analysis in the very first chapter. Instructions and examples illustrating the use of MINITAB and MYSTAT are given in Appendix C. Although computer printouts do not appear within the chapters themselves, I have placed computer icons in the margins whenever a data set in the text is also analyzed by MINITAB or MYSTAT in Appendix C. Since these two packages are so widely available, it should be easy for adopters of this text to include computer assignments in their courses.

Solutions to Problems

Each chapter ends with review questions that ask the student to apply concepts covered in that chapter. Students will find answers to the odd-numbered questions in Appendix B. I have also included solutions to all the review questions in the *Instructor's Manual* that accompanies the book.

Supplements and Ancillaries

Professors who adopt this text as required reading in their courses will be provided with the *Instructor's Manual* upon request. In addition to the solutions to review questions, the manual provides numerous test items for each chapter. These test items are also available on computer diskette, and they come with a test-building program. The computerized test bank is available in both IBM-compatible and Macintosh versions.

Students who wish to strengthen their grasp of behavioral statistics can purchase the *Student Study Guide to Accompany Straightforward Statistics for the Behavior Sciences.* This excellent learning aid contains a glossary of concepts for each chapter, procedure summaries, practice quizzes with answers, and drill questions.

ACKNOWLEDGMENTS

I am grateful to many people for their support, expertise, and wise counsel in the development and production of this text. I wish to express my appreciation to the reviewers of the manuscript for their valuable comments and suggestions. They are Robert Allan, Lafayette College; Mary Allen, California State University–Bakersfield; Paul J. Ansfield, University of Wisconsin–Oshkosh; Bryan C. Auday, Gordon College; Madison Dengler, Luther College; Dana S. Dunn, Moravian College; Marvin Dunn, Florida International University; Jane Halpert, DePaul University; Marle Kelley, Western Oregon State College; Gerald Peterson, Saginaw Valley State University; Mark Shatz, Ohio University; Allen Shoemaker, Calvin College; and Philip Tolin, Central Washington University.

I have worked with a number of publishers on a variety of projects. In my opinion, Brooks/Cole Publishing Company is tops. I would like to thank the following members of the Brooks/Cole team: Jim Brace-Thompson (Psychology Editor, friend, and guide), Fiorella Ljunggren (Production Services Manager), Kelly Shoemaker (Senior Designer), Faith Stoddard (Supplements Editor), Jodi Hermans (Editorial Assistant), and the many other players working behind the scene.

Special acknowledgment is due copyeditor Carol Reitz, who has the eyes of an eagle and the wisdom of Solomon; production editor Greg Hubit, whose tireless attention to detail and timelines made everything flow so smoothly toward the publication date; Laurel Technical Services for their thorough problem checking; and ETP/Harrison for their compositorial excellence. In the same spirit, kudos is due Professor Shirley Hensch for writing one of the most useful student study guides I've seen.

I would like to thank Lindenwood College for granting the sabbatical leave that made it possible for me to begin work on this textbook. Gratitude is also owed my Lindenwood students, who provided feedback on many of the examples and explanations that appear in this book.

Finally, I want to express my sincerest appreciation for the patience and support extended by my family as I pursued yet another textbook project. Well, the book is finished now. I'm back to normal for a while. Thanks.

TO THE STUDENT

I wrote this book with you in mind. I have tried to make behavioral statistics interesting and understandable, and clarify its relevance to the profession that you're grooming for. I trust that your course instructor will do the same. But successful classroom learning is the result of a team effort. Here's what you can do to get the greatest benefit from this book:

1. Before reading each chapter, page through it to get some general idea of what it is about, and read the summary.
2. Next, read the chapter slowly and carefully. If you don't understand something, go over it again until it sinks in. Pay special attention to the "little arrows," boxed notes, and italicized points. Statistics has to be digested a little at a time; it can't be gobbled up like some other subject matters.
3. When you finish the chapter, reread the summary and do as many review questions as your schedule permits. I developed these questions for the express purpose of sharpening your statistical knowledge and skills. You will find answers to the odd-numbered review questions in Appendix B.
4. Review the material at least twice before taking an exam on it. Statistical ideas can be fairly likened to a wet watermelon seed: A firm grasp requires several tries.

I hope you find the book interesting and as easy to learn from as I have tried to make it.

James D. Evans
Internet address: evans@lc.lindenwood.edu

PART I GETTING STARTED

You are about to begin studying a field that is very practical—so practical, in fact, that almost no area of science can get along without it. Indeed, the application of statistical methods is bound to play a very prominent role in your career in the behavioral sciences. As you read this book, I hope you will not only see the practicality of statistics but also experience the personal rewards that come from understanding the logic behind the methods. The purpose of the first section is to give you a good sendoff on this challenging intellectual adventure. Chapter 1 addresses the questions of *why* and *wherefore*: What benefits to you and your field result from studying and using statistics? Chapter 2 examines how statistics relates to and supports the general process of research in the behavioral sciences. The material in Chapter 2 also provides the conceptual foundation for ideas presented throughout the rest of the book.

CHAPTER 1   WHY STUDY STATISTICS?

This book teaches you how to apply a set of logical procedures to the practical and theoretical questions studied in the behavioral sciences, which include fields such as psychology, sociology, political science, criminal justice, and management. You will find that none of these disciplines could exist in its present form without the regular application of the methods described here.

Should you be bored or intimidated by the term *statistics*? All I ask is that you reserve judgment for the time being. Although I can't guarantee that you will be swept off your feet with excitement, I can say there is a strong probability that you will find the material challenging in a positive sense, and an even higher probability that you will discover many uses for it in the years ahead.

Students learn better when they have a sense of orientation and purpose at the outset. In statistics courses, people often attempt to get their bearings by asking the questions: Why study statistics? What good will it do if I know about this stuff?—certainly reasonable things to ask. I believe there are three excellent reasons for seeking a basic mastery of applied statistics:

1. Studying statistics and related courses can improve your effectiveness in everyday problem solving.
2. A knowledge of the fundamentals of statistical methods will enhance your career and employment prospects.
3. Applying statistical analysis is the only way the behavioral sciences can make significant advancements.

Let's consider how each of these reasons pertains to you and this course.

STATISTICAL TRAINING AND EVERYDAY PROBLEM SOLVING

Sir Ronald A. Fisher, well known for his early contributions to the field of statistics, long ago portrayed the work of the statistician as merely a refinement of the layperson's native tendency to learn through first observing real-world events and then drawing appropriate conclusions (Fisher, 1966). (Also see Argyris, 1980.) At its base, statistics is simply a sensible model for making decisions in a rational way. Viewed from this vantage point, courses in statistics can be expected to enhance our general thinking ability. Is there any evidence to support this idea?

Each of the following problems represents a practical question that you might encounter in everyday life.[1] Carefully consider each

---

[1] The first two of these examples are taken directly from Lehman, Lempert, and Nisbett (1988).

problem and try to select the best answer. (The correct answers are given at the end of this section.)

"A high school student has to choose between two colleges. The student has several friends, who are similar to himself in values and abilities, at each school. All of his friends at school A liked it on both educational and social grounds; all of them at school B had deep reservations on both grounds. The student visited both schools for a day, and his impressions were the reverse." That is, the student personally preferred school B. In your opinion, which school should the student choose?

(a) School A   (b) School B

"After the first two weeks of the major league baseball season, newspapers begin to print the top ten batting averages. Typically, after two weeks, the leading batter has an average of about .450. Yet no batter in major league history has ever averaged .450 at the end of a season. Why do you think this is?

(a) A player's high average at the beginning of the season may be just a lucky fluke.
(b) A batter who has such a hot streak at the beginning of the season is under a lot of stress to maintain his performance record. Such stress adversely affects his playing.
(c) Pitchers tend to get better over the course of the season, as they get more in shape. As pitchers improve, they are more likely to strike out batters, so batters' averages go down.
(d) When a batter is known to be hitting for a high average, pitchers bear down more when they pitch to him.
(e) When a batter is known to be hitting for a high average, he stops getting good pitches to hit. Instead, pitchers 'play the corners' of the plate because they don't mind walking him."

Suppose that you've made a habit of investing $100 in stocks each month. For the most part, your investments have tended to give you average to slightly above average returns. But the stocks you've chosen in each of the past seven months have taken nose dives. As you consider your investment options for this month, which strategy would be the wisest?

(a) Invest the same amount as you usually do.
(b) Invest less than $100, in view of the fact that you have been on a losing streak.
(c) Invest more than $100, since you are due for a "lucky break" in the market.
(d) Stop investing until your luck changes.

As you can see, these kinds of questions represent real-world decision-making situations. Although they might not seem to be directly related to the discipline of statistics, a familiarity with statistics appears to improve students' ability to answer them correctly. Specifically, a series of well-controlled investigations by Lehman, Lempert, and Nisbett (1988) has revealed that

> people's solutions of everyday-life problems using statistical rules are greatly enhanced not only by instruction in college statistics courses but even by relatively brief training sessions. These training sessions are effective even when the training is highly abstract and formal and does not make any reference to everyday-life content. (p. 433)

Moreover, this line of research suggests that by learning about and applying the basic statistical principles covered in this text, you will:

- Develop, improve, and elaborate on fundamental problem-solving processes that are adaptive
- Become a more logical thinker
- Become a more effective decision maker

In my opinion, the enhancement of your ability to think logically is the most significant benefit of studying statistics. But, as mentioned at the start of this chapter, there are two additional benefits to such an undertaking. Let's now examine them more closely.

NOTE: The answer to each of the three practical problems is (a).

STATISTICS AS A PROFITABLE ENTERPRISE

As a part of your general education, some degree of statistical expertise will help you to better comprehend the data-oriented reports that now populate both the journals in your field and periodicals published for the general public. The hard fact is that all branches of the social sciences are becoming increasingly quantitative. Regardless of your particular major, you probably will be at a great disadvantage if you are not able to decipher company reports and journal articles that contain statistical analyses.

On the positive side, numerous career benefits are associated with knowing something about statistics. Such knowledge potentially can help you deal with some difficult job-related problems, enhance your employment potential, and make you more promotable. I will support these claims by citing the job-related experiences of some of my former students. The names have been changed, but all the accounts are true.

Turning Down the Heat

Bob is a supervisor in the bottling department of a large, nationally known brewery. He developed a temporary problem with his general manager when it was discovered that his department was pumping too much beer into some bottles and not enough into others. Since the actual bottle filling process was automated, Bob could not very well blame his employees. And a check of his equipment operation revealed no mechanical glitches.

It so happened that Bob was enrolled in a behavioral statistics course at the time. Something he had recently learned in the course prompted him to take a random sample of the *bottles themselves* and check their volume variations. He then applied some simple statistical tests to his bottle-volume measurements, and the tests confirmed that the source of the problem was an unusual amount of fluctuation in the dimensions of the bottles. The brewery switched to a new bottle vendor, the general manager was again happy, and the heat was off Bob.

Getting That Job

Mary was unable to find a suitable job in her major (accounting), but, partly because her college transcript showed a good grade in behavioral statistics, she did land a well-paid position as a statistician for the U.S. General Accounting Office.

Emily, a psychology major, was having a hard time making ends meet on her entry-level social worker's pay. Fortunately, she was able to use her knowledge of statistics to get a lucrative job as a researcher with a large pharmaceutical firm.

Movin' on Up

A bright sociology major named Nancy became disillusioned when she found that her only job opportunity immediately after graduation was as a floor clerk in a children's clothing chain. Within a year, however, her excellent work record and her undergraduate course in statistics qualified her to fill a new professional position in the store—director of marketing research.

Employment Prospects: Some Data

The examples are merely a handful of the dozens of similar ones I can recall. They illustrate that real people can, and sometimes do, derive tangible benefits from their familiarity with statistical methods—often without anticipating or seeking such benefits. I chose these particular former students because in each case the person had vigorously protested the college's statistics requirement on the grounds that the material was irrelevant to his or her career objectives.

Although no one can guarantee that you will realize monetary gain or career enhancement as a direct result of studying statistics, there appears to be a strong market for people with ability in this discipline. And the pay isn't bad. Data collected by the U.S.

Bureau of Labor Statistics and published in the *Occupational Outlook Handbook* (1990) indicate that

employment opportunities for persons who combine training in statistics with knowledge of computer science or a field of application—such as biology, economics, or engineering—generally are expected to be favorable through the year 2000....

Private industry will...require increasing numbers of statisticians to monitor productivity and quality in the manufacture of various products.... Business firms will rely more heavily than in the past on statisticians to forecast sales, analyze business conditions, modernize accounting procedures, and help solve management problems. In addition, sophisticated statistical services will increasingly be contracted out to consulting firms. (p. 72)

What's more, "psychologists with extensive training in quantitative research methods and computer science may have a competitive edge over applicants without this background" (*Outlook,* 1994, p. 126). And sociologists "well trained in quantitative research methods—including survey techniques, advanced statistics, and computer science—will have the widest choice of jobs" (*Outlook,* 1994, p. 128).

On the matter of remuneration for statistical types, *Outlook* (1994) reported that

the average salary for statisticians in the Federal Government in non-supervisory, supervisory, and managerial positions was $51,893 in 1993.... According to a 1992 American Statistical Association salary survey of statisticians in departments with statistics programs the median starting salary for assistant professors was $40,000...and for professors $54,000. (p. 99)

You may also find it interesting that many behavioral science professionals who are not full-time statisticians do part-time statistical consulting to handsomely supplement their regular incomes. At the time of this writing, statistical consultants in my neck of the woods (St. Louis, Missouri) are typically paid between $60 and $200 per hour for their technical knowledge. Most of their clients are medium to large business firms.

STATISTICS AS THE KEY TO KNOWLEDGE ADVANCEMENT

The third general reason for learning how to apply statistical methods to your field is that using a statistical approach is the only way we can gain an accurate understanding of human behavior and hence be able to predict it. The goals of understanding and predicting behavior are paramount in the behavioral sciences.

Consider the following sample questions that have been posed by behavioral scientists:

Does psychotherapy really help people? If so, how much does it help? Are some kinds of therapy more effective than others?

Does drinking coffee help, hurt, or have no effect on students' ability to retain textbook material?

If we send 50 cents along with a marketing questionnaire, are more people likely to complete and return it? Will sending $1.00 make an even greater difference in the rate of participation?

Is our manager-training program cost effective?

Is there a relationship between enrollment levels in community colleges and the unemployment rate in the surrounding region?

Do personalized comments on returned tests cause students to improve their performance on subsequent tests?

Without systematically collecting data on these kinds of questions and analyzing the results statistically, we cannot answer them with any degree of precision or confidence. In fact, in the absence of a statistical approach to phenomena, about all we are able to do is develop logical arguments to support one position or another on each such question. Since almost any position can be supported through argument alone, our disciplines would constantly be at impasses. We just would not know whose argument was most correct. In contrast, a statistical approach requires that (1) facts about behavior be systematically collected under controlled conditions, (2) these facts be converted to numbers, and (3) theories about behavior be evaluated and modified in the context of these data. In short, statistical models and procedures allow us to separate useful theories from worthless ones by applying objective decision-making criteria to observations of real-world events. And that is how scientific fields make progress rather than remaining stuck in argument.

As an example, take the last question from the list. Social reinforcement theory predicts that giving students positive written feedback on their exam papers will cause them to increase their efforts on subsequent exams and thus obtain better scores. To test this prediction, a few years ago I conducted a little experiment with a large class of students. Beginning with exam 1 in the course, I wrote encouraging, personalized comments on some of the students' returned exams, and I gave the remaining students nothing but a grade. The average scores on three exams for the two groups of students are shown in Table 1.1.

Table 1.1

AVERAGE TEST SCORES ON
THREE EXAMS FOR TWO
GROUPS OF STUDENTS

	EXAM AVERAGES		
GROUP	EXAM 1	EXAM 2	EXAM 3
Personal feedback	35.39	32.93	34.33
No feedback (control)	33.67	32.46	31.30
Difference	1.72	0.47	3.03

Since I had made sure that the two sets of students were matched in ability, their scores should not have differed on the first exam.[2] If social reinforcement theory is correct, the group with personal feedback should have gotten higher average scores on exams 2 and 3.

What do the test score averages in Table 1.1 suggest to you? My conclusion was that the special feedback did increase the students' performance on exam 3 and, hence, that social reinforcement theory was supported by the outcome. A colleague of mine disagreed, saying that none of the differences between the two groups appeared large enough to support the claim that personal feedback had a positive effect.

Without a statistical analysis, the two of us could have argued endlessly, and perhaps fruitlessly, about the meaning of these results and the merits and shortcomings of social reinforcement theory. As you will see later in this book, however, the statistical analysis that I carried out on the data provided some resolution for us: The 1.72 and 0.47 group differences for exams 1 and 2 were simply "chance" differences stemming from what statisticians call "sampling error." But the statistical test indicated that the exam 3 difference of 3.03 was very likely a "real effect" of positive feedback. So the difference in the group averages on the third test did support social reinforcement theory after all.

Not all findings in the behavioral sciences require statistical analysis for their evaluation and application. But most of the results we get are somewhat ambiguous, like those presented here.

---

[2] I ensured that the experimental and control groups were "statistically equivalent" in average academic ability by randomly assigning students to one condition or the other. As is explained in Chapter 2, assigning research participants to treatment conditions on the basis of chance (i.e., randomly) usually does accomplish the goal of initial equivalence between the experimental and control groups. That way, if the two groups show performance differences later on, you know that the experimental procedures had an effect.

It is in these more typical cases that the application of statistics makes the difference between productive interpretations of the findings and the interminable deadlock of conflicting opinions.

TAKING STOCK

This chapter has presented some good reasons for learning about statistics and (I hope) provided an orientation that will help make further study of the topic more meaningful for you. In the next chapter you will encounter some ideas and definitions that will strengthen the foundation of statistical thinking you have begun to build. In the process of studying those topics, you will become aware of the crucial role that statistics plays in behavioral research.

COMPUTERS AND STATISTICS

In our electronic society computers take the tedium out of many tasks while adding enjoyment to others. As you either know or likely have surmised, most behavioral scientists do statistical analyses of their data via computer programs, occasionally supplementing the computer-generated results with calculator-based analyses. (I do virtually all of my statistical work on a computer.) And it is likely that your instructor will want you to become familiar with computerized statistics as a component of your statistics education. Accordingly, this text provides applications of two well-known software packages—**MINITAB** and **MYSTAT**—to problems presented in each chapter. You will find these examples in Appendix C.

Important: As you read about certain research problems and data sets in each chapter, you will see the icon in the margin of the page. This icon signifies that the corresponding chapter in Appendix C illustrates how to analyze the same data set using the MINITAB and MYSTAT programs.

I suggest that each time you see the computer icon, you run the associated data set through MINITAB or MYSTAT in order to get practical experience in using the software. (See the icon in the margin of this page? Check out Chapter 1 of Appendix C to find out how to create and save data files.) This practice will pay benefits early in your career or, if you are heading to graduate school, early in your postgraduate work. For the present, you will find it convenient and helpful to utilize the programs for checking your solutions to problems given at the end of each chapter. Your instructor might assign additional computer analysis problems.

Before going on, I would like to make four points about computerized statistical analyses:

1. *Knowing how to use statistical software is not sufficient for understanding statistics.* Learning how to run data through computer programs is not the same as learning how to select an appropriate statistical procedure to use with a particular research problem and how to interpret the outcome of that procedure. To actually understand statistics, you need to study this book, listen to your instructor, and practice solving statistics problems "by hand." In short, you won't master statistics simply by becoming adept at running MINITAB and MYSTAT.

2. *Knowing how to use statistical software is not necessary for understanding statistics.* You don't need to learn about computerized statistical analysis in order to gain a fundamental grasp of behavioral statistics. Indeed, depending on time constraints and personal philosophy, many instructors who require this textbook will choose to not assign any computer analysis problems—with little or no sacrifice in the quality of the learning experience.

3. *Knowing how to use statistical software can make the application of statistics to your discipline much easier and more enjoyable.* Once you are sure that you comprehend the logic, appropriate use, and proper interpretation of a procedure, delegating the actual computational process to a computer can liberate you from the only aspect of statistics that is a drudgery: repetitive number crunching. This will enable you to spend more time contemplating the meaning of your findings in the context of statistical logic, since you will be devoting almost no time to the mechanics of doing the calculations. Another bonus is that the computer is considerably less error prone than we humans are.

4. *To be up to date in your discipline, almost regardless of what it is, you have to know how to do statistics on a computer.* Thus it is ideal for you to take up this challenge in the context of this course. If you can't work it in now, however, you should take it up as soon as possible in your college experience.

SUMMARY

1. Statistics can be thought of as a set of logical procedures that help us to address basic questions posed by the behavioral sciences.

2. Reasons for learning about statistics include enhancing of our

thinking and decision-making abilities, improving our professional competence and career opportunities, and advancing
knowledge in our disciplines.

3. Recent research has shown that studying statistics can improve
   our ability to make good judgments about everyday-life situations.

4. A knowledge of basic statistics can bolster one's performance
   on the job, one's promotability, and one's chances of finding sat-
   isfactory employment.

5. Most research findings in the behavioral sciences are not clear
   cut and thus require statistical analysis if they are to be pro-
   ductively interpreted.

6. Computers facilitate statistical analysis by reducing the
   amount of time that you must devote to performing numerical
   calculations. By itself, however, computational software can't
   teach you how to apply and interpret statistics.

REVIEW QUESTIONS

1. According to the text, what are three major reasons that it is important for behavioral science students to study statistics?

2. The textbook asserts that without applying statistical methods
   to behavioral science questions, "we cannot answer them with
   any degree of precision or confidence." In what specific ways
   does a statistical approach to the behavioral sciences enhance
   the accuracy of our understanding, compared to a rational-
   argument approach.

3. Summarize the advantages and limitations of learning how to
   do statistical analyses on computer.

CHAPTER 2   BASIC CONCEPTS AND IDEAS

Chapter 1 gave a rationale for learning about and using statistics in the behavioral sciences. This chapter will begin to build your statistical vocabulary while illustrating how statistics fits into the overall scheme of behavioral research. Unless you digest these terms and concepts first, the procedures and formulas that appear later will not make much sense.

The preceding sentence may puzzle you if you have begun reading this book with the preconception that statistics is "math." True, theoretical statistics is almost purely a branch of mathematics. And even applied statistics (the kind used by behavioral scientists) borrows a lot from mathematical theory. However, to the researcher or other professional who works with statistics daily, **statistics is** viewed not primarily as math but *as a set of logical procedures used to:*

1. *Summarize data*
2. *Test hypotheses*
3. *Make inferences and predictions about behavior*

Applied statistics does rely on numbers, formulas, and, most important, probability theory. But, strictly speaking, a mathematical system does not have to make contact with the real world so long as it is internally consistent and can be defended logically. Theoretical calculus is a good example. In contrast, applied statistics is almost always tied to the real world through its data, which, in turn, have been generated by scientific observations of behavior. What's more, your successful use of this subject matter will depend more on your grasp of when to apply particular statistical procedures and how to interpret the results than on your aptitude for mathematics. It is important to remember that behavioral statistics does not exist for its own sake but rather as one component of the process of behavioral research.

Consistent with that view of the field, this chapter will present statistical concepts within the context of studies that illustrate the usual stages of a behavioral science investigation. First let's distinguish between two kinds of "groups" or "sets" that are involved in statistical research.

POPULATIONS AND SAMPLES

A researcher usually is not primarily interested in the behavior of the group of people directly observed in an investigation—that is, the sample. Rather, the goal often is to use the sample as the means of understanding or predicting the characteristics of some larger, but unobserved, group called the population.

Since a behavioral scientist sometimes investigates things and events as well as people, a **population** should be thought of as *the entire set of people, things, or events that the researcher wishes to study*. The members of any well-defined population can be clearly identified on the basis of some trait or set of traits that they share. It is this common trait that defines the population in question.

For example, a political scientist might survey 1000 registered voters across the United States in order to obtain some statistics on their political beliefs. In this case the researcher is not merely interested in the opinions of these particular respondents. The goal of such a study is to make a general statement about the entire set of 100 million or so registered voters in the United States—the population targeted by the study.

Any well-defined group of people, events, or things can be designated a population for research purposes. It all depends on the researcher's aim in conducting the investigation. Your statistics class could be designated a population if that is the set of people that the researcher wants to examine. Likewise, a population might be defined as all college students in North America, or all people in your hometown who are over the age of 65. The main consideration is that the trait or set of traits that identifies the population be carefully specified in advance of the study (and, of course, that it reflects the goals of the study).

A **sample** is simply a *subset of a population*, some portion of the larger group of people, events, or things targeted by the study. The people, events, or things that the researcher actually observes and measures are the sample used in the investigation. In the political scientist's survey, the 1000 people actually interviewed are the sample. Researchers observe samples because it would be too expensive, too time consuming, or just plain impossible to gather facts on all members of most populations.

However, there's a catch to the practice of investigating a population by observing a subset of it: The sample must be representative of the population; that is, it must yield results similar to those that would be obtained if the whole population were observed. And this is likely to happen only if the sample accurately reflects most of the essential characteristics of the population.

Unfortunately, many samples used in the social sciences are biased; they do not accurately represent the population of interest. This problem usually is a result of using an inappropriate method of selecting the participants in a sample. In general, the best way to draw an unbiased sample is to use **simple random sampling**: *The sample is selected in such a way that each member of the population has an equal chance of being chosen for inclusion in the sample, and each selection is made independently of all others*. In practice, this definition means that people are selected for sample membership on the basis of chance alone.

You probably are familiar with simple random sampling through your experience with raffles and lotteries. The lottery tickets are the population. Their numbers are placed into a large box, urn, or rotating drum. After the tokens or stubs are scrambled unsystematically, the winning numbers—that is, the sample—are drawn out of the container in an unsystematic fashion. Note that each ticket's chance of winning is approximately the same (it is roughly the ratio of prizes to tickets) and that each winner is selected from the population independently of other selections. Much more will be said about sampling and sampling techniques in later chapters.

Box 2.1

STATISTICAL PROVERBS

A researcher's sample receives most of his or her immediate attention, but the aim of this attention is to get to know a population better.

A good sample is unbiased. It is very much a chip off the ol' block and reflects the essential traits of the parent population.

The best way to get a representative sample is to let chance do the selecting.

STATISTICS AND THE RESEARCH PROCESS

A research project in the behavioral sciences normally consists of six stages:[1]

1. Formulating questions
2. Choosing a research design
3. Observing and measuring
4. Finding relationships
5. Interpreting relationships
6. Generalizing

Although statistical concerns exist in all phases of a research project, they are most prominent in phases 4 and 6, finding relationships and generalizing. Let's now examine the statistical aspects of the research process within the framework of two investigations.

Stage 1: Formulating Questions

A scientific study begins when the researcher poses a question about nature that can be addressed by making systematic observations under controlled conditions. Consider these questions:

---

[1] The six stages of a research project discussed here are adapted from Evans's (1985) "Five Phases of Scientific Inquiry."

Will giving college students personalized, positive comments on their returned exams cause them to perform better on subsequent exams?

Are jurors who believe in capital punishment more likely than other jurors to convict any defendant regardless of the nature of the alleged crime?

These questions pertain to significant problems in human behavior. And, as you will see, each of them can be converted into an observation-based study. The word *observation* is the key here because the behavioral sciences are **empirical sciences**; that is, the *theories are founded upon, evaluated by, and modified on the basis of observations* of behavior.

Empirical: Based on or pertaining to sensory experience and observation.

Where do research questions come from? Actually, social and behavioral scientists draw from a large variety of sources in developing their investigations (McGuire, 1973). Very often, brand new research questions grow out of the researcher's casual observations of how people function and adapt in a certain situation. The researcher's initial investigations of a phenomenon involve converting an informal hunch—based on preliminary observations—to a formal prediction about behavior. In turn, a research design is developed to test this prediction under controlled circumstances (see below). In other cases, the research questions are suggested by prior investigations or existing theories about a particular problem. Practical problems, such as drug use among law enforcement officers, can also inspire research projects.

Identifying variables The "things" or "events" that research questions refer to are called **variables**: *the characteristics of people, objects, and events that can take on different values.* In investigating the relationship between advocacy of the death penalty and tendency to convict, for example, "belief in capital punishment" is one variable; it *varies* between people who advocate capital punishment (the first value of the variable) and people who do not (the second value of the variable). "Tendency to convict" is the other variable that can take on at least two values: "convict the defendant" versus "don't convict the defendant."

NOTE: It is important not to confuse the **values of a variable**—that is, *the different states or magnitudes that a characteristic can assume*—with the variable (characteristic) itself. In the other research question being considered here, for instance, "type of feedback" on exam papers is one variable, and *personal feedback versus no personal feedback are its possible values.* Likewise, "performance on classroom exam" is the second variable, and *its values are the individual test scores* that people make on the exam.

When you think about it, almost everything studied by behavioral scientists—from political beliefs to consumer preferences, migration patterns, and intelligence—is a variable. And most variables can be quantified—that is, converted to numbers for purposes of statistical analysis. For example, we quantify intelligence by giving IQ tests. The resulting "intelligence quotient" is a number between 1 and 160 (or so) that reflects one's level of general intelligence.

Hypotheses and relationships   After a researcher refines the research question and identifies the critical variables within it, the question is likely to be rephrased as a research hypothesis. A hypothesis is *a proposed relationship between two variables*. It succinctly expresses what kind of relationship the researcher *expects* to observe between two variables. In the first research question considered here, the hypothesis might take this form:

Research hypothesis: Personalized feedback on returned exams will increase college students' performance on their subsequent exams in the same course.

The hypothesis for the second question might look like this:

Research hypothesis: When placed in the role of jurors, people who advocate capital punishment will be more likely to convict a defendant than will people who do not believe in capital punishment.

Notice that each research hypothesis predicts a relationship between two variables, where a **relationship** is defined as *a systematic going together of, or association between, two variables*. Thus, if the first hypothesis is correct, then students' exam scores will be associated with the level of personalized feedback they get: Those who get personalized feedback will tend to obtain relatively higher scores, and those who receive no such feedback will tend to make relatively lower scores. A relationship is said to exist to the extent that one variable is *predictable* from another variable. Discovering relationships between variables and interpreting those relationships within a theoretical context are what the behavioral sciences are all about. Statistics is a tool that we use to efficiently pursue these goals.

Some relationships are of the cause-and-effect kind, whereas others are merely predictive. A **causal relationship** exists if *it can be demonstrated that a change in one variable contributes to or causes a change in the other variable.* Note that our first research hypothesis implies a causal association between type of feedback on exams and performance on subsequent exams.

Two variables have only a **predictive relationship** if *changes in one of them can be forecast from changes in the other, but there is insufficient evidence to claim that one variable causes changes in the other.* Our second research hypothesis implies only that tendency to convict can be predicted from beliefs about capital punishment; it does not state that jurors' beliefs about capital punishment directly cause them to make a conviction decision.

What determines whether a relationship between variables allows only predictions or also provides information about cause and effect? Usually this determination is based on the kind of research design used to gather the data.

Box 2.2

SUMMARY OF HOW TO FORMULATE RESEARCH QUESTIONS

The question-generation stage of research is the most complicated part of the process. The steps in formulating a research problem include:

1. Developing the initial question
2. Identifying the main variables addressed by the question
3. Converting the initial question to a testable hypothesis that asserts a particular relationship between the variables

---

Stage 2: Choosing a Research Design

Types of research design   A **research design** is *the specific plan and set of procedures that a researcher uses to gather observations about behavior.* Generally speaking, research designs fall into three basic categories:

1. **Correlational designs:**   A large class of procedures that enable the researcher to discover *predictive relationships* between variables but that do not provide information on causality. Most nonexperimental research studies are correlational.
2. **Experimental designs:**   Research procedures that involve systematically manipulating one variable while holding all other variables constant, thereby providing information on the *causal relationship* between the manipulated variable and some behavior of interest.
3. **Parameter-estimation designs:**   Research procedures that simply identify the degree or level of a characteristic in a population—as in political, economic, or social surveys.

Although a discussion of parameter-estimation research will be largely postponed until later chapters, this chapter will illustrate the other two types of design.

An experimental design study   One of our hypotheses asserts that giving students personalized positive comments on their returned tests will cause them to improve their performance on subsequent exams. Such a causal hypothesis requires that a well-controlled experiment be performed to assess the validity of the hypothesis. Let's first consider the simple logic of the experimental method and then see how it is applied to test the personalized feedback hypothesis.

1. First the experimenter develops a hypothesis that describes a cause-and-effect relationship between an **independent variable** (the hypothesized "cause") and a **dependent variable** (the hypothesized "effect"). Note that the dependent variable is *the behavior of interest* that the experimenter observes and measures.

2. The research conditions and procedures are set up so that all factors that could influence the dependent variable are held constant.

3. One variable, the independent variable, is then *systematically manipulated,* or varied, by the experimenter while all else continues to be held constant.

4. If the dependent variable (i.e., the target behavior being measured) changes after the independent variable has been deliberately varied, the researcher concludes that the independent variable caused changes in the dependent variable. The crux here is that *since all other potential influences were kept constant,* only the one thing that was varied—the independent variable—could have produced the change in the dependent variable.

The investigation of the effects of personalized feedback used this logic (Evans & Peeler, 1979). The independent variable that was systematically manipulated was feedback: personalized feedback on exams versus no personalized feedback. Students who received the personalized feedback were the experimental group, and the remaining students were the control group. The differential treatment given to these two groups was the manipulation of the independent variable. The dependent variable was the students' performance on exams in the course; this was the behavior of interest that was observed and measured.

How were all other variables controlled in this experiment? Through **random assignment** of students to the two conditions. *On the basis of a purely chance process*—the flip of a coin, to be precise—46 of 92 students were randomly assigned to the experimental group (personalized feedback), and the remaining 46 students were assigned to the control group (no personalized feedback). Assigning research subjects[2] to the treatment conditions on the basis of chance alone makes the different conditions **statistically**

---

[2] *Subjects* is the term that is conventionally used to label the people who serve as participants in behavioral research projects.

equivalent. Statistically equivalent means that *at the start of the study, the two groups of subjects were likely to have only small chance differences between them on the dependent variable and everything that could influence the dependent variable.* That way, if the two groups showed a sizable difference in exam scores at the *end* of the study, we would be confident that the difference was produced by the one thing that was intentionally varied between the two groups: the type of feedback that they received on their returned exams.

Three exams were given in the course, and students in the experimental group received positive personalized comments on all of the exams at the time they were returned. All exam scores were carefully recorded for later statistical analysis.

A correlational study[3]   Since the hypothesis concerning the relationship between tendency to convict and advocacy of capital punishment implied only a predictive relationship between the two variables, a correlational research design was used to test this question. The researchers first *randomly selected* 25 adult subjects from a listing of residents in the local community. The selected subjects were contacted by telephone and asked to participate in "a special kind of legal survey." All of the subjects agreed to complete and return the survey materials that the researchers promised to mail to them.

What the subjects actually received was a summary of a court transcript describing some ambiguous evidence against a robbery suspect. After reading the transcript, the subjects rated the degree to which they would probably vote to convict the suspect if they were sitting on the jury in this case. Finally, they responded yes or no to a question asking whether they believed in the use of the death penalty in cases of capital murder. The subjects then mailed the materials back to the researchers.

A major difference between this investigation and the first one is that subjects were not randomly assigned to the treatment groups. Rather, the research subjects assigned themselves to one of two "groups" by indicating whether they believed in the use of capital punishment (group 1) or not (group 2). Since random assignment was not used, the two groups of respondents may have differed in a number of ways—education, religion, political preferences, and so on—in addition to holding different beliefs about capital punishment. Therefore, if the groups also differed in their tendency to convict, it would be impossible to say whether that result had been produced by education, religious factors, political

---

[3] This investigation was conducted by one of my research methods classes. Special thanks to all the students involved in this project.

leanings, or beliefs about capital punishment. This lack of control over variables was acceptable, however, because the researchers were merely interested in finding out whether there was a predictive relationship between two variables. They were not attempting to find out whether advocacy of the death penalty directly causes the tendency to vote for conviction.

> NOTE: A principal distinction between experiments and correlational studies is that subjects are randomly assigned to treatment conditions in experiments but not in correlational research. Random assignment is what gives the experiment its control over irrelevant variables and hence enables the experimenter to determine what is causing behavior to change.

Recall, however, that even though the researchers did not randomly assign subjects to treatment conditions, they did *randomly select* their sample of people. They did so for the purpose of obtaining a sample that would be representative of the population of adults in their community.

The importance of randomness The words *random* and *randomly* appear frequently in statistics textbooks. The reason is that both the logic of research designs and the mathematical theories underlying statistical procedures assume that either random assignment or random selection has been implemented at some point in the data-generation process. This is an important concept to remember because conclusions and decisions that stem from statistical analysis are valid only if the randomness assumption has been satisfied by the research procedure.

Randomness always *implies a lack of predictability between events,* as in the behavior of a pair of dice or a roulette wheel. It means that outcomes are purely a function of chance.

Another important point to remember is that random selection and random assignment serve different purposes in research. Random selection of a sample of people from a population is done to maximize the sample's chance of accurately representing the essential characteristics of the population. This is done so that results and conclusions derived from the sample can be validly generalized to the population. Random selection is most often used in surveys and correlational studies.

In contrast, random assignment is conducted in order to make different treatment groups in an experiment statistically equal on all variables that could affect the dependent variable. Figure 2.1 illustrates these two functions of randomness.

> NOTE: Random selection is used to maximize the chance that a sample will represent its population, whereas random assignment is used for the purpose of experimental control.

Figure 2.1 *The Separate Functions of Random Selection and Random Assignment*

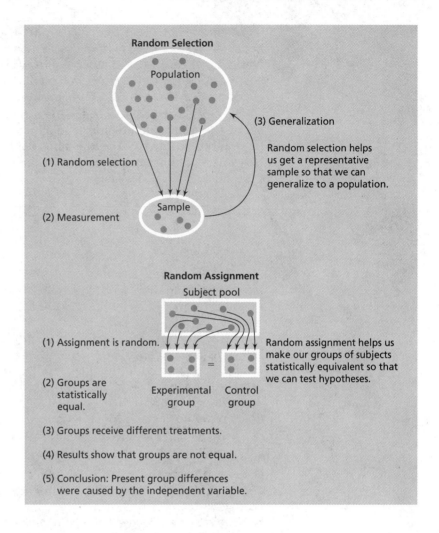

Stage 3: Observing and Measuring

In the third stage of a scientific investigation, the planned research procedure is carried out to generate data related to the question being asked. The researcher makes observations of the subjects' behaviors and, in the process, measures them on the dependent variable. **Measurement** is defined as *converting observations to numbers according to a rule.* Statistical research is quantitative and therefore always involves the conversion of observations to numbers.

In the personalized feedback study, the students' test-taking behavior was observed (within the context of regular classroom examinations), and those observations were transformed to test scores according to this rule: For each student, add up the number of correct answers on a 50-point multiple-choice test.

Two measurements were taken in the investigation of beliefs about the death penalty and tendency to convict. First, the subjects were asked to rate their likelihood of voting to convict the robbery suspect by placing a check mark at the appropriate point on the following ten-point scale:

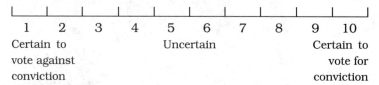

Measurement in this case followed the rule: The score on tendency to convict equals the number located below the subject's check mark.

The second measurement in this study concerned the subjects' beliefs regarding capital punishment. Each subject was instructed to answer yes or no to the question, Do you believe in using the death penalty if a defendant is convicted in a case of capital murder? The subject was then assigned a "death-qualified" score of 1 if he answered yes or a "death-qualified" score of 2 if he answered no. Essentially, this measurement operation simply *categorized* subjects into group 1 or group 2.

Quantitative versus qualitative measurement   In the research on beliefs about capital punishment, tendency to convict is a **quantitative variable**. It was *measured with a number scale that is capable of distinguishing differing amounts of some property*. In contrast, belief in capital punishment is a **qualitative variable**. It was measured by simply *identifying differences in kind*; that is, people were categorized as either death-qualified jurors or non-death-qualified jurors. Unlike quantitative variables, qualitative variables don't have true numerical status. Nonetheless, qualitative variables are often given arbitrary number codes, such as 1 and 2, for purposes of statistical analysis. All fields within the behavioral sciences investigate both quantitative and qualitative variables, and statistical procedures exist for analyzing the results of both types of measurement.

Data and raw scores   The product of behavior measurement procedures is referred to as **data**: *the numbers that the researcher statistically analyzes*. **Data** is the plural form of the word, and **datum**, defined as *a numerical record of a single observation*, is the singular form. The *data that you start out with*, prior to doing any statistical analyses, are called **raw scores**. For example, the raw score data from the investigation of beliefs about capital punishment are shown in Table 2.1. Notice that the raw scores are just the ratings that the 25 subjects gave in response to the question about their likelihood of voting to convict the suspect. Also note how the ratings are listed according to the subjects' beliefs concerning the death penalty. Any such *list of data* is called a **data set**.

Stage 4: Finding Relationships

Once the data have been collected and listed, the real fun begins. It is at this point that researchers apply statistical analyses to the data in order to find out whether the relationship predicted by the research hypothesis does in fact exist.

Descriptive versus inferential statistics   There are two major branches within the field of applied statistics: descriptive statistics and inferential statistics. Typically both branches are used in the analysis of data, with the descriptive calculations preceding the inferential procedures.

Descriptive statistics consists of very elementary *procedures for describing and summarizing* a data set. Such statistics can be applied to either sample data or data from an entire population (if it is a small population); they produce a simplified picture of general

Table 2.1

RAW SCORES OF DEATH-QUALIFIED AND NON-DEATH-QUALIFIED JURORS ON A TEN-POINT RATING SCALE OF TENDENCY TO CONVICT	(1) Death Qualified (N = 15)	(2) Non-Death Qualified (N = 10)
	9	3
	6	6
	8	2
	5	4
	7	5
	7	4
	9	4
	5	4
	8	5
	10	3
	6	
	7	
	4	
	7	
	7	

Note: 60% of the respondents were "death qualified"—that is, believed in using the death penalty in cases of capital murder.

trends and characteristics in the data. A frequently used descriptive statistic is the arithmetic average, or *mean*, which is calculated by summing all the raw scores in a list and dividing the sum by the number of scores. The arithmetic averages found in the two studies being considered here are shown in Tables 2.2 and 2.3.

Inferential statistics consists of more advanced *procedures that enable us to generalize from sample data to population characteristics*—in other words, to infer something about a large set of people or situations on the basis of a small subset of people or situations. One function of inferential statistics is to assess the **statistical significance** of relationships found in sample data—that is, *to determine whether the relationship shown by the sample is reliable* or merely a chance outcome unique to that small subset of data. More will be said about inferential statistics in later parts of this chapter.

Seeing relationships in data   By computing some descriptive statistics on a set of data, you can obtain at least a preliminary indication of whether there is a relationship between two variables within a sample of data. For example, consider the average (mean) scores on tendency to convict of the death-qualified and non-death-qualified subjects that appear in Table 2.3. Since the average tendency to convict was higher for the death-qualified subjects (7.00) than for the non-death-qualified subjects (4.00), these descriptive statistics suggest that there is a relationship between believing in capital punishment and the tendency to convict a suspect regardless of his crime—just as was hypothesized.

The value of inferential statistics   The average test scores from the personalized feedback investigation, shown in Table 2.2, are harder to understand. Clearly, there was not much difference between the feedback and control groups on the second exam, but there was some average difference between them on exams 1 and 3. The exam 3 difference was predicted by the research hypothesis; however, the groups should have been equal on exam 1. Why? Because the groups were not treated differently until *after* the first test (when they first got their graded exams back), and the groups had been "equated" at the beginning of the course through a random-assignment procedure.

Table 2.2

AVERAGE SCORES ON THREE EXAMS AS A FUNCTION OF FEEDBACK CONDITION

Feedback Condition	EXAM AVERAGES		
	Exam 1	Exam 2	Exam 3
Personalized feedback	35.39	32.93	34.33
Control condition	33.67	32.46	31.30

Table 2.3

AVERAGE RATINGS ON CONVICTION SCALE FOR DEATH-QUALIFIED AND NON-DEATH-QUALIFIED SUBJECTS	(1) Death Qualified	(2) Non-Death Qualified
Average (mean)	7.00	4.00

Fortunately, inferential statistical tests clarified the situation by showing that the group difference on exam 1 was a *chance* difference that resulted from random variation in the data. This means that even though the groups were not numerically equivalent on the first test, they were "statistically equivalent." The same type of inferential test also revealed that the exam 3 difference of 3.03 points, on the average, was statistically reliable and not just a chance difference. Therefore, the predicted relationship between personalized feedback and performance on subsequent exams did occur by the time the third exam was administered.

You should note that finding a relationship in sample data requires both descriptive and inferential statistics. Descriptive statistics reveals general trends in the data, and inferential statistics indicates whether or not the relationships are due to chance. We will study methods for discriminating between real and chance relationships in Chapters 9–15.

Stage 5: Interpreting Relationships

After a researcher has found a reliable relationship between the variables in a study, he or she proceeds with an interpretation of that finding. Interpreting a relationship consists of (1) determining the extent to which cause and effect can be asserted and (2) drawing conclusions about what the finding means within a broader theoretical context.

For reasons considered earlier, we know that statements about causal relationships can be applied to the results of the personalized feedback project (an investigation that used the experimental

Box 2.3

IMPORTANT POINTS ABOUT STATISTICAL SIGNIFICANCE (STATISTICAL RELIABILITY)	If the result of an inferential statistical test indicates that a relationship between two variables is statistically significant, then it is likely (but not certain) that the patterns or trends shown by the data in the sample:

1. Are not just chance outcomes
2. Will be repeated if the same study is conducted again
3. Are not unique to the particular sample used and therefore can be tentatively generalized to other people, places, and times

method) but not to the investigation of beliefs about capital punishment (a correlational study). A more complete interpretation of those findings would also involve relating the results of the personalized feedback project to general theories of learning. Similarly, we would want to relate the outcome of the capital punishment study to the social psychology of attitudes.

Stage 6: Generalizing

The final step in doing a scientific investigation is to make statements about the applicability of the findings to situations or populations outside the investigation itself, to the extent that such statements are warranted. This is called **generalizing** the results.

Testing for statistical significance   When you carry out an inferential statistical test and determine that the relationship shown by your sample is reliable, and not merely a chance outcome, you are already generalizing in a sense. A "statistically significant" or "statistically reliable" finding is one that is likely to recur if the original study is repeated. This means that your findings probably are applicable to other persons and situations similar to those observed in your study.

Consider the study of personalized feedback, in which a statistical test showed that the experimental group scored "significantly" higher than the control group on the third exam in the course. Such a result means that the positive effect of personalized feedback in that study is likely to occur in similar samples of students at other times, in other places. Remember that the function of all inferential statistics is to assess or establish the generality of results obtained from samples.

Parameter estimation   In addition to testing for the reliability of a relationship, inferential statistics can be used directly to *estimate the characteristics of a population on the basis of a sample drawn from that population.* This procedure, which is illustrated in Figure 2.2, is called **parameter estimation.**

Parameter Estimation: Estimating a numerical characteristic of a population on the basis of sample statistics.

A **parameter** is *a numerical characteristic of a population.* An example is the percentage of adults in a local community who believe in using the death penalty in cases of capital murder. The population in this example is defined as all persons over the age of 17 who live in the community of interest. If a representative sample is drawn from that population using simple random sampling, then a **statistic**—that is, *a numerical characteristic of a sample*—can be computed and used to infer the corresponding population parameter. But note that parameter estimation can be successfully conducted only if the sample used to make the inference is randomly selected.

Recall that the 25 subjects in the study of beliefs about capital

Figure 2.2 *Schematic of Parameter Estimation*

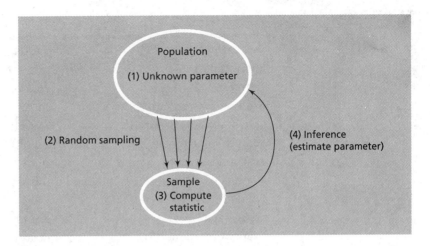

punishment were randomly selected from a list of adult residents in a local community. Therefore, the percentage of respondents who were "death qualified" in the sample (60%; see Table 2.1) can be used as a statistic to infer the percentage of respondents in the community population who advocate the use of the death penalty. Our estimate is that the population parameter is approximately 60%. (Parameter estimation is actually a little more complicated than this. See Chapters 8 and 10 for the details.)

> NOTE: A **statistic** is a *numerical characteristic of a sample*, and a **parameter** is a *numerical characteristic of a population*. If the sample is representative of the population, then the sample's statistic can be used to estimate the corresponding parameter in the population.

AN ENDLESS CYCLE

The six stages of scientific inquiry—formulating questions, choosing a research design, observing and measuring, finding relationships, interpreting relationships, and generalizing—have been presented as a sequential activity that has a definite beginning and a clear end. But I need to add that, in practice, this sequence is a self-sustaining cycle; that is, each time a research project "ends" with the researcher's statement of general conclusions from the most recent investigation of a topic, new research questions emerge from certain theoretical implications and unanticipated findings present in the researcher's data. When these new questions have been formulated, the whole research cycle is repeated, which leads to still more researchable questions, and so on. Behavioral science is truly humankind's perpetual motion machine, and statistics is the set of tools that keeps the machine humming efficiently and anchored to the real world.

KEY TERMS

statistics

population

sample

simple random sampling

six stages of research

empirical sciences

variables

values of a variable

hypothesis

relationship

causal relationship

predictive relationship

research design

correlational design

experimental design

parameter-estimation design

independent variable

dependent variable

random assignment

statistically equivalent

randomness

measurement

quantitative variable

qualitative variable

data

datum

raw scores

data set

descriptive statistics

inferential statistics

statistical significance

generalizing

parameter estimation

parameter

statistic

SUMMARY

1. Statistics is a set of logical procedures used to summarize data, test hypotheses, and make inferences and predictions about behavior.

2. A population is the entire set of people, things, or events that the researcher wishes to study.

3. A sample is a subset of a population. Samples are used to make inferences about a population, to the extent that they are representative of the population of interest.

4. Simple random sampling involves selecting a sample in such a way that each member of the population has an equal chance of being chosen—a good way to obtain representative samples.

5. A behavioral science research project consists of six stages: formulating questions, choosing a research design, observing and measuring, finding relationships, interpreting relationships, and generalizing.

6. Stage 1 of research—formulating questions—entails developing a research question, identifying the variables, and stating a hypothesis (a proposed relationship between variables).

7. Stage 2 of research involves selecting a research design. Experimental research designs test causal hypotheses, correlated designs test predictive hypotheses, and parameter-estimation designs estimate characteristics of populations.

8. Randomness, in the form of either random selection or random assignment, is an important assumption of the mathematical theories underlying statistical procedures.

9. Stage 3 of research—observing and measuring—consists of collecting data that bear on the research hypothesis. Variables may be measured quantitatively or qualitatively. Most statistical data are in the form of raw scores.

10. Stage 4 of research—finding relationships—includes two steps: (a) using descriptive statistics to summarize the overall trend in the data and (b) using inferential statistics to assess the reliability (significance) and generality of the results.

11. Stage 5 of research—interpreting relationships—also has two parts: (a) determining the extent to which cause and effect can be asserted and (b) placing the findings in a broader theoretical context.

12. Stage 6 of research—generalizing—involves making statements about the applicability of the findings to situations or populations outside the investigation. Parameter estimation is a special procedure for generalizing, whereby the level of a population characteristic is estimated from a sample statistic.

REVIEW QUESTIONS

1. Briefly compare and contrast applied statistics and mathematics, noting both the overlap and the difference between the two fields.

2. A behavioral scientist is interested in determining the level of achievement motivation in female executives in the United States. She compiles a list of more than 7000 names and addresses from the membership rosters of various associations of professional managers. She then contacts 500 of the female executives, 400 of whom agree to take a test that (unbeknownst to them) assesses the extent of each woman's achievement motivation. The researcher finds that her sample of 400 executives obtains an average score of 82 points (of a possible 100) on the test. She then draws a conclusion about female executives in general. Within the context of this example, define the population and identify the sample. How well do you think the sample represented the target population? What events or problems might have reduced the representativeness of the sample in this project?

3. Define parameter and statistic, and identify each in the example described in Question 2.

4. Suppose that, while sitting in statistics class, you hear one person tell another that the students in the class may be thought of as either a sample or a population, depending on the purpose of the particular investigation that uses the students. Is this assertion correct? If so, why? If not, why not?

5. Define or describe: population, sample, and simple random sampling.

6. Which of the following are variables, and which are values of variables?

   (a) An IQ score
   (b) Men
   (c) Intelligence
   (d) Gender
   (e) Creativity
   (f) Sophomore

7. Distinguish between a relationship and a hypothesis, and make up two examples of each.

8. Research has shown that college students who smoke a lot tend to get lower grades in their courses than students who smoke very little; also, nonsmokers tend to get the best grades overall. There is a relationship in these findings. Identify the variables, and give some examples of values of those variables.

9. Is the relationship described in Question 8 a predictive relationship, a causal relationship, or both? Defend your answer with definitions and justification within the context of the example.

10. Distinguish between random assignment and random selection, and describe the primary research function fulfilled by each.

11. In your own words, tell why random assignment of subjects to treatment conditions is usually considered an essential ingredient in well-controlled behavioral experiments.

12. Define or describe: statistically equivalent, correlational study, experimental design, independent variable, and dependent variable.

13. A research-minded educator wishes to conduct an experiment to compare the relative effectiveness of two approaches to instruction, conventional lecture and discovery learning, within the context of a plane geometry course. In a large school district the geometry teachers who have agreed to participate in the research are randomly assigned to one instructional method or the other. At the end of the semester, all geometry students in the study are given the same standardized examination, and the average test scores of the respective instruc-

tional methods are compared. The conventional-lecture average (89%) is slightly higher than the discovery-learning average (85%). What are the independent and dependent variables in this study? Does this investigation qualify as a "true experiment," or is it basically just a correlational study? Explain your answer.

14. In Question 13, is the dependent variable a qualitative or a quantitative variable? Explain your choice.

15. In Question 13, identify the descriptive statistics used. Why is it necessary to use inferential statistics in such an investigation? Be specific.

16. Define and contrast the following: independent vs. dependent variables; qualitative vs. quantitative variables; descriptive vs. inferential statistics; and experimental vs. correlational research.

17. In what two stages of a research project is inferential statistics most likely to be used? For what specific purposes?

18. In what sense is the determination of the "statistical significance" of a finding an act of generalizing research results?

19. For the purpose of doing parameter estimation, which is more important: random selection or random assignment? Explain your answer.

20. Define or describe: statistical significance, parameter, statistic, parameter estimation, and generalizing results.

21. Using your own words, describe the relationship between testing research results for statistical significance and "generalizing" those results.

22. Develop an original example of a study that uses a true experimental design. Identify the independent and dependent variables. Tell why you consider the study to be an experiment rather than a correlational study.

23. Develop an original example of a behavioral investigation that has only a correlational design—not an experimental design. Identify the variables involved, and state your hypothesis. Tell why this investigation may not be considered a true experiment.

24. Create an original example of a parameter estimation study. Identify the population, the sample, and the parameter in question. What kind of sampling procedure would you use in this investigation? Why?

25. State what *randomness* implies, and tell why the assumption of randomness is essential to the application of behavioral statistics.

PART 2

DESCRIPTIVE STATISTICS

The next five chapters introduce you to the fundamentals of summarizing and describing data from behavioral science research. This introduction includes organizing and graphing data (Chapter 3), computing simple numerical indexes to represent a data set (Chapter 4), interpreting individual values in a distribution of data (Chapter 5), and finding and applying relationships between variables (Chapters 6 and 7). It is important to understand how to compute and interpret descriptive statistics because describing your data set is a prerequisite to using the more powerful (and more interesting) inferential procedures described in later parts of the book.

CHAPTER 3 FREQUENCY DISTRIBUTIONS AND GRAPHS

The first step in statistically analyzing a set of data is to tally the numbers in an organized fashion and perhaps show that tally on some kind of graph. The chief purpose of graphs, tallies, and the simple statistical indexes that go along with them is to summarize and describe data sets. Such basic procedures fall into the general category of **descriptive statistics** (see Chapter 2).

The most elementary way of organizing raw scores is to arrange them in what is called a "frequency distribution." A graph of a frequency distribution helps the researcher to efficiently communicate the results of a study to others. Accordingly, this chapter will present some very basic concepts and techniques pertaining to the organization and communication of data.

SCALES OF MEASUREMENT

Scale of measurement: refers to the set of mathematical properties that may be assumed to exist in a set of data.

Often the specific method that you can use to organize and graph a set of data depends on how the variables in your study were measured. Different measurement techniques yield raw scores with *different mathematical properties*. These properties are conceptualized as four "levels" of measurement called **scales of measurement**. From the lowest to the highest, the levels are referred to as nominal, ordinal, interval, and ratio scales. The lowest scale (nominal) has no true mathematical properties, and the highest scale (ratio) has the greatest number of properties, allowing the widest range of permissible mathematical operations (for example, multiplying and dividing numbers).

Nominal Measurement

Qualitative variables, such as gender (female versus male), are on a **nominal scale**. A nominal scale, then, allows you to specify *differences in kind* but not differences in amount; that is, you can say that being a female is different from being a male but not more or less than being a male. Since nominal "measurement" merely classifies or categorizes observations, any numbers one assigns to these observations are arbitrary and are used chiefly to label them (for example, male = 1; female = 2). However, you can count the number of observations in a sample that fall into each nominal category. And, as you will see later in this book, some statistical methods can be applied to simple frequency counts.

Though lowly by mathematical standards, nominal variables are often studied by behavioral scientists. Some examples are "treatment groups" in an experiment (that is, the experimental group versus the control group), consumer preferences (prefer versus not prefer a product), cultural categories, personality types, marital status, and religious affiliation.

Ordinal Measurement

The next three scales of measurement—ordinal, interval, and ratio—possess some true mathematical properties. They are used to measure quantitative variables—that is, to numerically represent differences in the amount of some characteristic.

An **ordinal scale** of measurement allows you to *determine the relative ranks* of observations within a data set. The only permissible mathematical operation is determining inequalities. You can say that one observation is "greater than" or "less than" another along the measured dimension, but *you can't tell how much difference* there is between those observations. For example, a teacher with a small class can rank students on creativity, from most creative to least. He can even assign meaningful numbers to represent the students' ranks: 1 for the most creative (Jane), 2 for the second most creative (Dick), and so on. However, if you look at the teacher's ranked list of names, it is impossible to determine whether Jane is twice as creative as Dick or just slightly more creative than Dick.

Letter grades (A, B, C, D, F) in courses are also on an ordinal scale of measurement. The grades essentially place students into ranked levels of course mastery. But is the difference in course mastery between A and B students equal to the difference between B and C students? That would be difficult to say in most cases. We know that A students have more course-relevant knowledge than B students, and B students more than C students, but we do not know how much more. Again, the only quantitative information available is "greater than" and "less than."

Several statistical indexes and tests can be applied to ordinal data (ranks). We will study two of them in Chapter 15.

Interval Measurement

An **interval scale** not only provides information on inequalities among observations but also indicates the *amount of distance between different observations.* This is because on an interval scale, *every unit of measurement is equal to every other unit* along the entire range of the characteristic being measured. Hence, scale values can be meaningfully added and subtracted from one another (for example, $4 - 2 = 6 - 4$), and *differences* between scale values can be meaningfully multiplied and divided [$(4 - 2)/2 = (6 - 4)/2$]. These important mathematical properties permit a wide range of statistical tests to be conducted on data measured on the interval scale.

What an interval scale lacks is a true—or absolute—zero point that represents the complete absence of the characteristic being measured. The Fahrenheit thermometer is the classic example of an interval scale because each degree of temperature is equal to every other degree, but 0°F does not represent the total absence of heat. In the northern half of the United States, the winter temperature regularly plummets below 0°. Thus, the 0° value is an **arbitrary zero point**. And therein lies the limitation of interval scales. Since the zero value is not a true zero, *it is not valid to form ratios of individual scale points*. On the Fahrenheit thermometer, for instance, 80° is not twice as warm as 40°. In this case, 80/40 does not equal 2—at least not in terms of the underlying temperature property that the numbers represent.[1]

A great many measures used in the behavioral sciences are considered to be on an interval scale: measures such as intelligence and aptitude test scores, (some) rating scales, and various attitude scales. It is to your advantage to work with data at this level of measurement because most of the more powerful statistical techniques you'll be learning about assume that variables are measured on scales with equal intervals.

Ratio Measurement

Variables measured on a **ratio scale** have both *equal intervals and a true zero point*. Examples are most physical variables, such as the height of developing children (inches or centimeters), body weight (pounds or kilograms), money (dollars and cents), and reaction times in perceptual and psychomotor tasks. If your data are measured on this kind of scale, then it is legitimate to form ratios of scale values: 1000 milliseconds is twice as long as 500 milliseconds, or (1000/500) = 2, because 0 is absolute and represents a complete absence of the characteristic in question (time). Likewise, if a child was 18 inches long at birth and is 54 inches tall at age 10, then that child has literally tripled her birth height. Because 0 inches means the absence of length, it is meaningful to say that $\frac{54}{18} = 3$.

Many "behavioral" variables, such as attitudes and aptitudes, cannot be measured on a ratio scale. For example, is IQ a ratio-scale variable? Not quite. It might be reasonable to assume that each unit on the IQ scale equals every other unit; hence, 120 – 119 = 40 – 39. But since psychologists cannot define or measure "zero intelligence," it is not reasonable to assert that a person with an IQ of 120 is three times as intelligent as someone with an IQ of 40. An individual with a 120 IQ might be less than twice as intelligent as the person with the IQ of 40, or she might be several times more intelligent. Without being able to define a true zero point for

---

[1] Since the total absence of heat occurs at –459.67° on the Fahrenheit scale, the ratio of 80°F to 40°F is actually [80 – (459.67)]/[40 – (–459.67)] = 539.67/499.67 = 1.08.

intelligence, we simply cannot form valid ratios of scale points. IQ satisfies the equal-interval criterion, but not the absolute-zero criterion, of ratio measurement. Consequently, it is measured on an interval scale but not a ratio scale.

Fortunately, *the most important statistical procedures assume only that the data conform to an interval scale*, and many behavioral variables satisfy that assumption.

General Points Concerning Measurement Scales

Higher scales "include" lower scales   As you may have noticed, a higher scale of measurement, such as a ratio scale, has all the properties of the lower scales plus some additional ones. In fact, some variables can be represented at all four levels of measurement. This point is illustrated in Figure 3.1, which shows how the variable of height can be measured with each type of scale. The figure also serves as a convenient summary and comparison of the four measurement scales.

Figure 3.1 shows that any group of adult males can be divided into arbitrary classifications (that is, nominal categories) labeled "short men" and "tall men." Such crude "measurement" is not very informative, but it is a starting point in the analysis of a variable such as height. We can also rank people on height, from tallest to shortest, without specifying the number of inches that separate persons at adjacent ranks. This is ordinal measurement. It provides more information about each person than does nominal measurement, and yet it retains the information available in the nominal categories. If we want to, we can use the ranks to categorize the group into short and tall men.

With nothing more sophisticated than a ruler, we can specify the number of inches by which any two of the men differ in height— that is, the "distance" between people on the measurement dimension. Since all inches on the ruler are equal, and since we can accurately determine the relative distance of each height from some arbitrary zero point, this measurement operation represents an interval scale. Observe, however, that we retain ordinal, or rank, information even when we use an interval scale. Thus, an interval scale implicitly contains an ordinal scale. Notice in Figure 3.1 that Pablo's interval-scale height is 3 inches and Patrick's is 2 inches. But since we haven't gauged height against a true zero point, it doesn't make sense to say that Pablo is 1.5 times as tall as Patrick.

If we use a measuring tape to gauge each man's absolute height relative to ground *zero* (the floor), we are specifying height on a ratio scale. Figure 3.1 shows that Pablo is 5' 11" (71 inches) and Patrick is 5' 10" (70 inches) tall. Now that all measurements have been made relative to a true zero point, it is accurate to say that Pablo is 71/70 times, or 1.014 times, as tall as Patrick. Note that the ratio

Figure 3.1 *Measuring Height with Four Scales of Measurement*

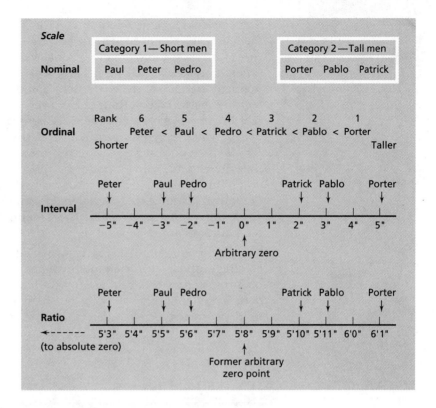

scale includes all the information on classification, rankings, and relative distances provided by the three lower scales of measurement.

Scales confused are scales misused Interpretation problems can arise when ordinal and interval measurements are treated as ratio-scale data. You're already familiar with the futility of trying to conclude that one person is twice as smart as another on the basis of interval-scale IQ scores.

Now consider an example that involves ordinal measurement. A clinical psychologist is treating four clients who have multiple personality disorder. After six months of therapy, she ranks the four patients in terms of the extent to which each has "integrated" his or her various personalities. The results, from most integrated to least, are listed here:

Client	Integration Rankings
Maggie	1
Tom	2
Larry	3
Harriet	4

Since there is no true zero point on the "personality integration" dimension, it is nonsensical to say that Larry's personalities are three times less integrated than Maggie's, or that Maggie's are three times more integrated than Larry's. And although one might be tempted to claim, at the very least, that Maggie has much better integrated personalities than Larry, even that statement is not correct in view of the level of measurement used. Since the clinician has provided only crude rankings of the patients, Maggie might very well be only slightly better off than both Tom and Larry. On the other hand, Harriet might be *much* less integrated than the other three. Again, ordinal measurement does not provide information on distances between persons on the dimension of interest. Thus, any statements about the amount of difference between magnitudes cannot be valid.

The measurement scale determines the type of analysis   Upcoming sections of this chapter will show that nominal data must be grouped somewhat differently than ordinal, interval, and ratio data. Later chapters will point out how the scale of measurement influences the type of statistical analysis that you can perform on your data.

CONSTRUCTING FREQUENCY DISTRIBUTIONS

Consider the data set shown below. It contains scores from a test of human relations skills that was taken by 30 managers in a suburban industrial firm. The test consisted of 40 multiple-choice questions, so it was possible to score as low as 0 or as high as 40. The scores are on an interval scale of measurement.

17	23	21	28	23	28
26	25	22	34	23	27
29	21	22	33	33	26
24	25	28	31	20	23
25	35	31	23	24	30

A set of raw data, even one as comparatively small as this, is not too informative in its original "raw state." If you had collected these data to find out something about the managers or at least their aptitude for dealing with problems in human relations, you probably would begin by asking the following kinds of questions:

1. How can I go about interpreting individual test scores? For example, how common or unusual is a score of 24?
2. Is there much variation in the scores?

3. How can I organize and communicate these results to other people?
4. Are the scores distributed in a typical way, or is their distribution unusual in some respect?

These questions pertain to using descriptive statistics to extract information from the raw data. The last two questions, in particular, can be addressed by organizing the scores in some logical way and subsequently constructing a picture of the data, a *graph*, that emphasizes the statistical characteristics of the data set.

The Frequency Distribution

The simplest way to organize a set of data is to construct a **frequency distribution**, which is *a statistical table that shows:*

- *What responses* (scores or values) the research subjects made
- *How many subjects made each of the responses* (the **frequency**)

The basic frequency distribution for these 30 scores appears as Table 3.1. First, notice that the first column in the table lists the entire range of raw scores attained by the managers. That range

Table 3.1

FREQUENCY DISTRIBUTION OF 30 SCORES ON A TEST OF HUMAN RELATIONS SKILLS	RAW SCORES	FREQUENCY	RELATIVE FREQUENCY (%)
	35	1	3.33
	34	1	3.33
	33	2	6.67
	32	0	0.00
	31	2	6.67
	30	1	3.33
	29	1	3.33
	28	3	10.00
	27	1	3.33
	26	2	6.67
	25	3	10.00
	24	2	6.67
	23	5	16.67
	22	2	6.67
	21	2	6.67
	20	1	3.33
	19	0	0.00
	18	0	0.00
	17	1	3.33
		Sum: 30	

includes all the test scores that actually occurred as well as three values that did not appear in the raw data (18, 19, 32). These values were added to the table so there would be no numerical gaps in the frequency distribution.

Second, observe that the number of times each value occurs in the data set is listed in the frequency column. This information reveals which scores are relatively common and which ones are relatively rare; it also indicates at what part of the score range the scores tend to bunch up or cluster.

The third column in the table gives the **relative frequency** of each score, which simply *converts each frequency count to a percent.* Converting frequencies to percents helps you to more easily interpret and describe the findings. To get a relative frequency, you first convert the frequency to a **proportion**:

$$\text{Proportion} = \frac{\text{frequency of a score}}{\text{total number of scores}} \qquad (3.1)$$

Then make the proportion a percent by multiplying it by 100.

For example, to transform a frequency of 5 to a relative frequency, first divide 5 by the total number of scores (30):

number of test scores that equal 23
$$\downarrow$$
$$\text{Proportion} = \frac{5}{30} = .1667$$
$$\uparrow$$
total number of test scores

Now, to make the proportion a percent:

multiply
$$\downarrow$$
$$\text{Percent} = \text{proportion} \cdot 100 \qquad (3.2)$$
$$= (.1667) \cdot 100$$
$$= 16.67$$

NOTE: In this book, a dot lying midway between two values means that you are to multiply the first quantity by the second. For example, $(.1667) \cdot 100$ means $(.1667)$ times 100.

You should be aware that both a proportion and a percent are considered relative frequencies, but in this chapter the term *relative frequency* consistently refers to a percent.

If *a frequency distribution has only the relative frequencies of scores*, it is called a **relative frequency distribution**. Therefore, Table 3.1 shows both a basic frequency distribution, consisting of the raw scores and the raw frequencies, and a relative frequency distribution, consisting of the raw scores and the percent associated with each.

A **relative frequency distribution** displays the percent of subjects who obtained each raw score.

The Grouped Frequency Distribution

Sometimes it is advantageous to display the frequencies of raw scores as a function of ranges of values on the variable measured in a study. *Listing frequencies by successive score ranges* yields a **grouped frequency distribution** (so called because each range is a "group" of values). Why do this? There are two reasons:

1. To simplify the graphing of a data set
2. To make certain characteristics of the data's distribution stand out more clearly

Table 3.2 displays a grouped frequency distribution based on the 30 scores from the test of human relations skills. You can see that the original 18-unit range—from 17 to 35—has been reduced to seven *three-unit* ranges. The lowest range includes the values 17, 18, and 19; the second lowest range includes the values 20, 21, and 22; and so on. *Each range of values in a frequency distribution* is called a **class interval**. So 17–19 is the first class interval, 20–22 is the second class interval, and so on.

Some conventional practices The rules for constructing a grouped frequency distribution are not etched in stone, but some traditional guidelines have been passed down through generations of behavioral scientists. Most of these guidelines aim at making the resulting distribution as informative and interpretable as possible. Following are some major procedural conventions:

1. The number of class intervals you use is up to you, but it should be somewhere between 6 and 20. (This figure varies somewhat from one statistician to another.) You should not use so few intervals that the characteristics of the data distribution (e.g., its basic shape) are obscured or distorted; you should not use so many intervals that you defeat the purpose of grouping values of the variable (see the reasons given earlier). In general, the more raw scores you have, the more class intervals you can use without producing intervals with zero frequencies. The more intervals you use, the narrower the intervals will be.

Table 3.2

GROUPED FREQUENCY DISTRIBUTION OF 30 SCORES ON A TEST OF HUMAN RELATIONS SKILLS	CLASS INTERVALS	FREQUENCY
	35–37	1
	32–34	3
	29–31	4
	26–28	6
	23–25	10
	20–22	5
	17–19	1

2. All class intervals should be the same size, or width (for example, all three units wide or four units wide, but not a mixture of three-unit and four-unit intervals).
3. The set of class intervals should be **mutually exclusive**; that is, *no raw score should lie in more than one interval.* Intervals should not overlap.
4. The set of class intervals should be **all inclusive**; that is, *every raw score should belong to an interval.* None should fall outside the interval structure.

Constructing the class intervals   The main idea behind class intervals is that you are dividing a fixed range of values into a fixed number of equal-size segments. The first step in setting up the intervals is to find the **range** of the distribution, which equals the highest value in the distribution minus the lowest value. Looking back at Table 3.1, you can see that the highest value *(H)* is 35 and the lowest *(L)* is 17. Therefore,

$$\text{Range} = H - L \qquad\qquad (3.3)$$
$$= 35 - 17$$
$$= 18$$

Now add 1 to the range:

$$\text{Range} + 1 = 18 + 1 = 19$$

This is called the **exact range** of the distribution. Any time your raw data are in the form of whole numbers (integers), the exact range can be *computed by adding 1 to the range of the distribution.*

To determine the size of the intervals, divide the exact range by the number of intervals you want to have (seven in this example):

$$\text{Interval size} = \frac{\text{exact range}}{\text{number of intervals}} \qquad\qquad (3.4)$$

$$\text{Interval size} = \frac{19}{7} = 2.71$$

When the result of this division is not in the same unit of measurement as the raw scores (whole numbers in this case), you should *round up* to be sure that the topmost scores are included in the highest interval. Rounding 2.71 up to a whole number yields:

$$\text{Interval size} = 3$$

Next, arrange the class intervals in a table with the lowest interval at the bottom of the table. Notice that each interval is defined by a low value, called the **stated lower limit**, and a high value, called the **stated upper limit**. If the raw data are whole numbers, the stated upper limit of any interval can be found this way:

$$\text{Stated upper limit} = \text{stated lower limit} + (\text{interval size} - 1) \quad (3.5)$$

Therefore, the stated upper limit of the first interval in this example is $17 + (3 - 1) = 19$.

To get the next class interval, use the next higher data value for the stated lower limit and find the stated upper limit using formula (3.5). In the example, the second lowest interval begins with 20. The upper limit of that interval is 20 + (3 − 1) = 22.

Proceed in this fashion until all intervals have been constructed and listed in the table. Finally, count the number of raw scores that lie in each interval and list those frequency counts in the "frequency" column. The general steps used in constructing group frequency distributions are summarized in Box 3.1.

Box 3.1

RECOMMENDED STEPS IN CONSTRUCTING GROUPED FREQUENCY DISTRIBUTIONS

Note: These steps should be used exactly as they appear if the raw data are in the form of whole numbers.

1. Find the *range* of scores plus 1: Subtract the lowest value in the distribution from the highest value and add 1 to that difference.

2. Determine the class *interval size*: Divide the result of step 1 by the number of class intervals to be used. In the behavioral sciences, use 6 to 20 intervals. If the result of the division is not a whole number, *round up* so that all scores will be within the range of intervals.

3. Set up the *list of intervals*: The lowest score in the distribution is the lower limit of the first interval.* To get the upper limit, add the class interval size to the lower limit and subtract 1 from the result. Therefore, upper limit = lower limit + (interval size −1). The lower limit of the next interval equals the lower limit of the first interval plus the interval size: lower limit of next interval = lower limit of present interval + interval size. The upper limit of this second interval is found in the same fashion as before. Continue this process until all intervals are listed from lowest (at the bottom of the table) to highest (at the top of the table).

4. Determine the *frequency of scores* for each class interval by counting the number of raw scores that lie in that interval, and list the frequencies in their own column.

5. Check your results:

   a. All class intervals should be the same size.

   b. No two class intervals should overlap.

   c. Every raw score should "fit into" one of the class intervals.

---

*There are exceptions to the rule of starting with the lowest score in the distribution. Some variables have a fixed top score. An example is college grade point average, which cannot exceed 4.00. When defining class intervals along such a constrained dimension, you should start at the fixed top score and form the intervals in a backward direction. This strategy ensures that the upper limit of the highest class interval will not exceed the natural maximum score of the variable. It wouldn't make sense for a grade point interval to extend above 4.00.

In still other situations, you might wish to make the class intervals perceptually simpler by starting not with the lowest score but with a score that is some multiple of the interval width. If the interval width is 3 and the lowest score is 17, for example, then the intervals might look like this: 15–17, 18–20, 21–23, and so on.

Figure 3.2 *Illustration of Exact Limits on a Portion of the Scale of Human Relations Skills*

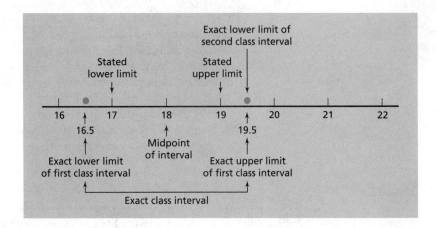

Finding Exact Limits

According to mathematical theory, the actual boundaries (that is, "limits") of a class interval lie just beyond the interval's stated limits. That way, no raw score rests on any interval boundary. These true boundaries are called exact limits of class intervals.

An **exact limit**, be it a lower or an upper limit, *lies precisely halfway between the stated upper limit of one interval and the stated lower limit of the next higher interval.* Figure 3.2 illustrates this concept for the first class interval of the present frequency distribution. Notice that the stated interval is 17–19 but the exact interval is 16.5–19.5. Adding the interval size (3) to the exact lower limit gives the exact upper limit.

Table 3.3 lists the exact limits for all the class intervals. Two facts should be apparent as you view the table:

1. The exact limits are found by subtracting 0.5 from each stated lower limit and by adding 0.5 to each stated upper limit.
2. The exact upper limit of any interval equals the exact lower limit of the next higher interval.

On the practical side, exact limits are useful (and sometimes essential) for drawing the graph of a frequency distribution.

Table 3.3

GROUPED FREQUENCY DISTRIBUTION SHOWING EXACT LIMITS ON A SCALE OF HUMAN RELATIONS SKILLS

CLASS INTERVALS	EXACT LIMITS	FREQUENCY
35–37	34.5–37.5	1
32–34	31.5–34.5	3
29–31	28.5–31.5	4
26–28	25.5–28.5	6
23–25	22.5–25.5	10
20–22	19.5–22.5	5
17–19	16.5–19.5	1

Figure 3.3 *General Two-Axis Structure for Graphing Statistical Data*

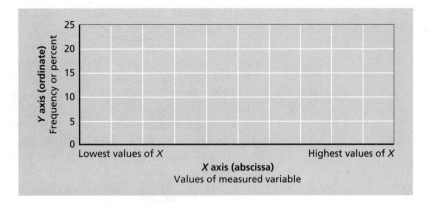

GRAPHING A FREQUENCY DISTRIBUTION

A **graph** is *a pictorial representation of either a distribution of data or the descriptive statistics derived from that distribution.* Graphs often facilitate the communication of findings to others. Graphs also help to clarify overall trends in data.

Most statistical graphs are displayed on the general two-axis structure shown in Figure 3.3. The *horizontal axis* is called the **X axis**, or **abscissa**, and it usually represents either the individual values or the class intervals of the variable being graphed. The lowest values are at the left portion of the *X* axis, and the high values are at the right. The *vertical axis* is called the **Y axis**, or **ordinate**, and it normally represents the frequencies or relative frequencies associated with the respective values of the variable being graphed.

 Figure 3.4 shows one type of graph that can depict a frequency distribution (either ungrouped or grouped) when the raw data are

Figure 3.4 *Histogram of Scores on a Test of Human Relations Skills*

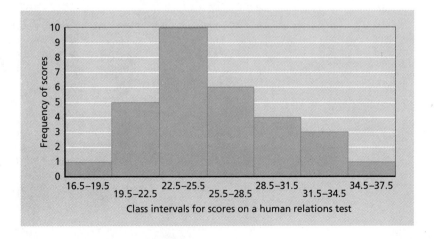

on any scale of measurement except a nominal scale. This kind of graph is called a **histogram**. It *uses "bars" or "columns" of various heights to represent the frequencies associated with the class intervals* in Table 3.3. Each bar has the same width, and the vertical edges of the bars originate at the exact limits of the class intervals. The bars share boundaries to indicate that the variable was measured on a quantitative scale (not qualitative or nominal).

Some conventional guidelines for the construction of graphs are given in Box 3.2. You may wish to take a look at them now, since they will help you to understand the remaining graphs shown in this chapter.

RECKONING WITH OUR QUESTIONS

Now that you have read about the fundamentals of setting up frequency distributions and developing graphs from those distributions, you have the necessary knowledge to intelligently address

---

Box 3.2

RULES OF CONVENTION IN THE CONSTRUCTION OF A STATISTICAL GRAPH

1. Both the vertical axis and the horizontal axis should be accurately labeled to indicate what each axis represents. In most graphs, the horizontal (*X*) axis represents the values, levels, or amounts of the variable being graphed, and the *Y* axis represents the frequency or percent of data associated with each level or value on the *X* axis.

2. The quantity marks along the vertical axis should be laid out in equal intervals, going from the lowest value (usually 0) at the bottom to the highest value at the top. Each quantity mark should be labeled with a number.

3. If the variable on the horizontal axis is measured in an interval scale or a ratio scale, then the marks along the axis should be laid out in equal intervals. Each increment mark should be labeled with a number.

4. If the variable on the horizontal axis is measured in an ordinal scale, the increment marks should be arranged from the lowest rank on the left to the highest rank on the right, with equal spaces between them. All marks should have appropriate labels (numbers representing ranks).

5. If the variable on the horizontal axis is in a nominal scale, the category labels along the axis should be arranged alphabetically from left to right, with equal spaces between them. *Only bar graphs* (not frequency polygons or histograms) *should be used to depict the distribution of nominal variables.* (See the later section on "Graphing Nominal Data.")

the basic kinds of questions about descriptive statistics that were posed early in this chapter. Recall that these queries concerned the set of 30 human relations test scores that we have been working with. Let's look back at each question in the light of what you have just learned.

1. *How can I go about interpreting individual test scores? For example, how common or unusual is a score of 24?* Frequency distributions, such as those in Tables 3.1 and 3.2, provide a convenient context for evaluating individual values in a set. Table 3.2 shows that a value of 24 is in the most populated class interval; hence, a test score of 24 is a "typical" outcome. In contrast, a score as low as 17 or as high as 35 is very unusual and unlikely in this distribution. A score of 27 is slightly higher than the typical score, and so on.

2. *Is there much variation in the scores?* The ungrouped frequency distribution shown in Table 3.1 clearly reveals a range of 18 points in the test scores. Figure 3.4 provides a visual expression of the variation in values. The table and figure could be used to compare the variation in test scores for the present sample of managers to the amount of variation in other groups of managers from different business firms.

3. *How can I organize and communicate these results to other people?* Grouped frequency distributions, such as that shown in Table 3.2, and histograms, such as the one in Figure 3.4, are efficient and easy-to-grasp methods of conveying the overall trends of a study to other interested parties.

4. *Are the scores distributed in a typical way, or is their distribution unusual in some respect?* Figure 3.4 indicates that, in one sense, the distribution of test scores is not typical: It deviates from the balanced "bell shape" of the so-called normal distribution. Statisticians would say the distribution shown in Figure 3.4 is "skewed, or unbalanced from left to right."

ANOTHER EXAMPLE OF GRAPHING

I have a new set of data to illustrate several additional aspects of graph construction. The data are shown in Table 3.4 in a basic (ungrouped) frequency distribution. The data values are the response times required by 85 telephone operators to provide directory assistance to customers. The times are in seconds, and they are measured to the nearest *tenth of a second*. Since the dimension of time has equal intervals and an absolute zero point, the values are on a ratio scale of measurement.

Important: *Because the accuracy of measurement was taken to tenths of a second, the basic unit of measurement here is not the whole number 1 but the decimal number 0.1.* This will affect some of the computations used in tabulating and graphing the data.

Effect of the Unit of Measurement

The grouped frequency distribution of the data on operators' response times is shown in Table 3.5, along with a few additional quantities that will be discussed shortly. Working with decimal numbers rather than whole numbers requires slight modifications in the procedures used to compute the exact range, the interval size, and the exact limits of the distribution.

Exact range   The exact range in this example is found by adding 0.1 (one-tenth)—not 1—to the range of values: Exact range = (6.2 – 4.1) + 0.1, or (highest value – lowest value) + *unit of measurement*. The result is (2.1) + 0.1 = 2.2.

Interval size   It is decided that the grouped frequency distribution will have 11 class intervals. From formula (3.4), the interval size is computed as follows:

Table 3.4

	TIME (SECONDS)	FREQUENCY (SECONDS)
UNGROUPED FREQUENCY DISTRIBUTION OF THE AMOUNT OF TIME (IN SECONDS) REQUIRED BY LONG-DISTANCE OPERATORS TO HANDLE DIRECTORY ASSISTANCE REQUESTS	6.2	2
	6.1	2
	6.0	4
	5.9	5
	5.8	6
	5.7	7
	5.6	10
	5.5	12
	5.4	8
	5.3	5
	5.2	5
	5.1	3
	5.0	3
	4.9	2
	4.8	2
	4.7	2
	4.6	0
	4.5	3
	4.4	1
	4.3	1
	4.2	1
	4.1	1

$$\text{Interval size} = \frac{\text{exact range}}{\text{number of intervals}}$$

$$= \frac{2.2}{11}$$

Because of measuring response time to the nearest tenth of a second, rather than in whole seconds, the class interval size is a decimal number ($\frac{2}{10}$, or 0.2) rather than a whole number.

The lowest class interval shown in column 1 of Table 3.5 is set up by using the lowest value (4.1) as the first stated lower limit and adding (interval size – 0.1) to it to obtain the first stated upper limit: 4.1 + (0.2 – 0.1) = 4.2. Then the remaining ten intervals are constructed according to the same method.

Exact limits   Column 2 of Table 3.5 lists the exact limits of the class intervals. Figure 3.5 illustrates that, as defined previously, the exact limit lies precisely halfway between two stated limits. However, note that because the basic unit of measurement in this example is 0.1, the constant subtracted from the lower limits and added to the upper limits is 0.05—or half of 0.1. Contrast that with the earlier example, where the data were in whole numbers and the constant used to compute exact limits was 0.5 (or half of 1). When exact limits are set up, then, the *general rule* is to add or subtract *half the unit of measurement* that exists in the original raw scores.

The cumulative percent distribution   It is sometimes useful to know what percent of scores in a distribution lie at and below a particular class interval. To obtain this information, you must first

Table 3.5

GROUPED FREQUENCY DISTRIBUTION OF THE AMOUNT OF TIME (IN SECONDS) REQUIRED BY LONG-DISTANCE OPERATORS TO HANDLE DIRECTORY ASSISTANCE REQUESTS	(1) CLASS INTERVAL	(2) EXACT LIMITS	(3) MIDPOINTS	(4) FREQUENCY	(5) CUMULATIVE FREQUENCY	(6) RELATIVE CUMULATIVE FREQUENCY (%)
	6.1–6.2	6.05–6.25	6.15	4	85	100.00
	5.9–6.0	5.85–6.05	5.95	9	81	95.29
	5.7–5.8	5.65–5.85	5.75	13	72	84.71
	5.5–5.6	5.45–5.65	5.55	22	59	69.41
	5.3–5.4	5.25–5.45	5.35	13	37	43.53
	5.1–5.2	5.05–5.25	5.15	8	24	28.24
	4.9–5.0	4.85–5.05	4.95	5	16	18.82
	4.7–4.8	4.65–4.85	4.75	4	11	12.94
	4.5–4.6	4.45–4.65	4.55	3	7	8.24
	4.3–4.4	4.25–4.45	4.35	2	4	4.71
	4.1–4.2	4.05–4.25	4.15	2	2	2.35

Figure 3.5 *Illustration of Exact Limits on a Portion of the Scale of Long-Distance Operators' Response Times*

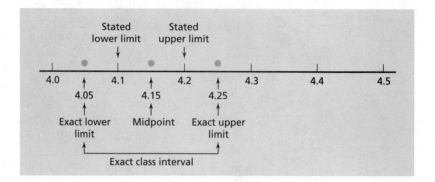

find the **cumulative frequency** of each interval, which is defined as *the number of values in that interval plus the number of values that fall below that interval.* The cumulative frequencies for the distribution of operators' response times are listed in column 5 of Table 3.5.

Next the cumulative frequencies are converted to **cumulative relative frequencies** (also called **cumulative percents**) by *dividing each cumulative frequency by the total number of scores* (85) *and multiplying the result by 100:*

$$\text{Cumulative percent} = \frac{\text{cumulative frequency}}{\text{total number of values}} \cdot 100 \tag{3.7}$$

Column 6 of Table 3.5 shows the results of these calculations. Each of those cumulative relative frequencies represents the percent of scores that fall below the exact upper limit of a particular class interval. For example, we could say that about 85% of the telephone operators in this study responded to directory-assistance requests in less than 5.85 seconds.

Graphing the Distribution

A new kind of graph will be used to depict the distribution of operators' response times: the frequency polygon. A frequency polygon serves the same purpose as a histogram. It is an alternative way of graphing distributions when the raw data were measured on an ordinal, interval, or ratio scale.

To construct a frequency polygon, you must first find the **midpoint** of each class interval, which is *the halfway point between the lower and upper limits of the interval:*

$$\text{Midpoint} = \frac{\text{upper limit} - \text{lower limit}}{2} \tag{3.6}$$

The midpoints of the present class intervals are listed in column 3 of Table 3.5.

As shown in Figure 3.6, a frequency polygon is constructed through the following steps:

Figure 3.6 *Frequency Polygon of Directory Assistance Operators' Response Times*

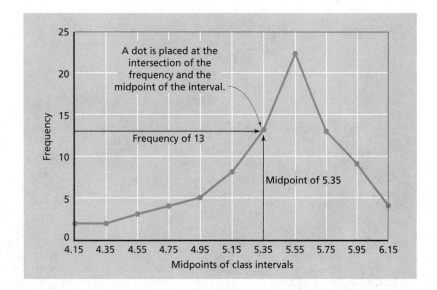

1. For each class interval, place a dot (or other marker) at the point in the graph where the interval frequency intersects the interval's midpoint.
2. Connect the dots with straight lines to produce a many-sided figure that charts the distribution of values.

GRAPHING NOMINAL DATA

Frequencies and relative frequencies associated with nominal categories (also known as qualitative measurement) *are usually depicted in a* **bar graph**, such as that shown in Figure 3.7. A bar graph

Figure 3.7 *Bar Graph of the Status of People over Age 64*
(Based on data from the U.S. Bureau of the Census. The figure illustrates how a nominal-scale variable is graphed.)

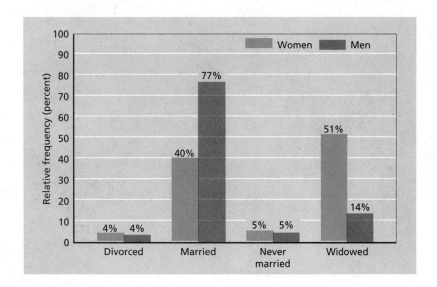

Box 3.3

ON THE MISUSE OF GRAPHS

When the rules and conventions of graphing are violated, you can bet that the graph's author either is naive about constructing statistical graphs or is misusing a graph to "lie with statistics."

This point is never illustrated better than in courtroom testimony that utilizes distorted pictorial representations of data to sway the jury. For instance, a court case concerning sex discrimination in the laying off of employees featured two graphs based on the same data. Each figure showed the percent of males and females terminated by a company during a large reduction in force. But the respective pictures conveyed very different impressions.

The prosecution's graph is presented here. Notice that the vertical axis is much longer than the horizontal axis. Also, the prosecution's statistician cut off the bottom of the vertical axis, so that the "Percent terminated" begins at 10 rather than at 0. Both of these bad practices are designed to exaggerate the apparent male–female difference in termination rate, to support the charge that the employer's layoff practices discriminate against women.

In contrast, the defense's statistician supplied the next graph, *which is based on the same termination percents.* Obviously, this figure creates the impression that there is only a minimal difference in the termination rates for males and females. This feat is accomplished by manipulating axis length in the opposite way. The vertical axis in this case is only about two-thirds as long as the horizontal axis. In addition, notice that the defense's statistician uses larger units to represent the vertical axis (5 points per tick, rather than 1 point per tick). The effect of this is to make any difference in the height of the two vertical bars small in comparison to the apparent scale.

These examples of lying with pictures make two points worth remembering:

1. Always evaluate a graph with a skeptical eye, lest its creator mislead you via sleight of pen.

2. It is important to follow conventional rules and procedures when constructing your own graphs, lest you inadvertently mislead the gullible (or look foolish to the knowledgeable).

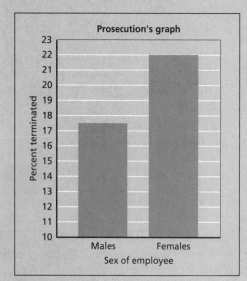

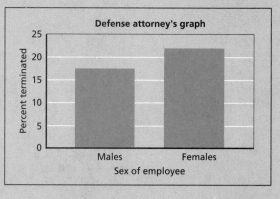

is put together in the same way as a histogram, except for two differences:

1. The "bars" in the graph do not share boundaries or touch one another. This reflects the fact that the variable is one of qualitative differences (i.e., differences in kind), not quantitative differences (i.e., differences in amount).
2. Relatedly, the *X* axis (abscissa) is not marked off on a quantitative scale but instead is labeled with the names of nominal categories being studied.

STATISTICAL CURVES

Symmetrical: The left side of a figure or form is the mirror image of the right side.

Table 3.6 shows a large relative frequency distribution of the final exam scores of 200 statistics students. The frequency polygon of the distribution is graphed in Figure 3.8. Notice that the polygon is **symmetrical** and shaped somewhat like a bell. It is also "smoother" looking—that is, less choppy—than the frequency polygon in Figure 3.6.

As a general rule, frequency polygons tend to become smoother, more "curved," and less angular as the number of observations in the data set increases. As a result, *the relative frequency polygons of larger populations of values*, both real and theoretical, are referred to as **statistical curves**. And many of those curves closely approximate the theoretical curve shown in Figure 3.9.

Figure 3.9 is the so-called **normal curve**, which is *the chief theoretical model underlying inferential statistics*. You will encounter it and its many applications throughout this text. It may be helpful to bear in mind that the normal curve is just a special kind of relative frequency polygon based on an infinitely large set of observations and graphed by plotting percent as a function of data values. It is interesting and important that many natural phenomena, including a wide range of human behaviors, tend to be distributed in a way that approximates the normal curve.

Unfortunately, however, not all distributions are like the normal curve. Some other possible statistical curves are displayed in Figure 3.10.

Skewed distributions are not symmetrical but *asymmetrical*. In a **negatively skewed distribution**, *most of the data "pile up" over the high values of the measured variable*. Scores like this might occur, for example, if 12th-grade students were given a 6th-grade math test.

In a **positively skewed distribution**, *most of the data lie at the low end of the scale*, a few scores lie in the middle range, and just a trickle of values can be found over the high end. A very difficult exam in your statistics class might produce scores on this kind of curve.

Table 3.6

	RAW SCORES	FREQUENCY	RELATIVE FREQUENCY (%)
UNGROUPED RELATIVE FREQUENCY DISTRIBUTION OF FINAL EXAM SCORES IN A BEHAVIORAL SCIENCE STATISTICS COURSE (N = 200)	92	1	0.5
	91	1	0.5
	90	2	1.0
	89	3	1.5
	88	4	2.0
	87	6	3.0
	86	9	4.5
	85	12	6.0
	84	15	7.5
	83	18	9.0
	82	19	9.5
	81	20	10.0
	80	19	9.5
	79	18	9.0
	78	15	7.5
	77	12	6.0
	76	9	4.5
	75	6	3.0
	74	4	2.0
	73	3	1.5
	72	2	1.0
	71	1	0.5
	70	1	0.5
Sums:		200	100.0

Figure 3.8 *Percent (Relative Frequency) Polygon Showing the Distribution of Final Exam Scores in a Behavioral Science Statistics Course*

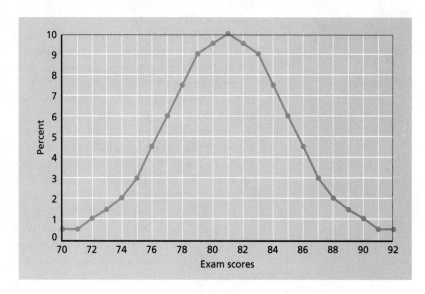

Figure 3.9 *The Normal Curve, also Known As the Normal Distribution*

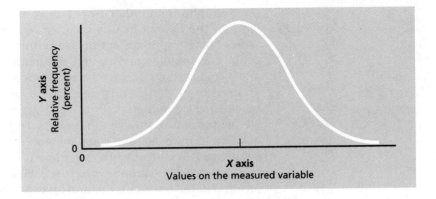

Figure 3.10 *Statistical Curves for Various Types of Distributions Found in the Behavioral Sciences*

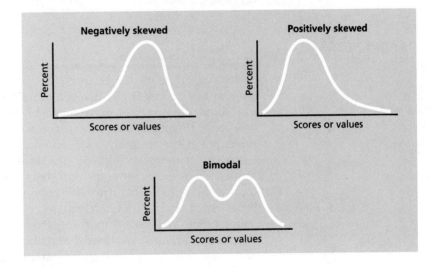

A **bimodal distribution** *has two "peaks"* and tends to occur when data from two different populations are combined in the same graph. Such a curve would result from plotting the heights of all males and females at your college. There would be much overlap, but the heights of the female population would tend to cluster at a different place in the graph than those of the male population.

COMPUTER AUTOMATION OF GRAPHING

I've shown you how to construct statistical graphs "by hand," so that you will understand the conventions and logic underlying such representations of data. In everyday statistical work, however, most behavioral scientists use computer programs to automatically build graphs from either raw data or frequency distributions. In fact, all the figures in this chapter were originally drawn by my Lotus 1-2-3®* spreadsheet program. (A spreadsheet program specializes in rapidly carrying out mathematical operations on data and summarizing the data for presentation purposes.)

In view of the blazing speed and nearly perfect accuracy of spreadsheet graphing, it is to your great advantage to learn the simple procedure involved in using software in this way. Not only will this knowledge save you time and frustration, but it will also enable you to conveniently experiment with different numbers of class intervals and graph types to see what configuration best communicates the data. You will find an example of spreadsheet graphing in Appendix C.

TAKING STOCK

At the beginning of this chapter you began with an unorganized block of test scores that conveyed almost no meaningful information about the group of managers who took the human relations skills test. Having thought about the methods and interpretations presented in the chapter, however, you now have the ability to organize and summarize almost any set of behavioral data to communicate statistical findings to others. Henceforth you will also know what textbooks and articles in your field mean when they use terms such as *distribution, range, class interval, normal,* and *skewed.* Perhaps most important, this chapter has explained the nature and origin of statistical distributions and "curves," which play huge roles in upcoming chapters. Look for them!

KEY TERMS

scales of measurement

nominal scale

ordinal scale

interval scale

arbitrary zero

ratio scale

frequency distribution

frequency

relative frequency

proportion

relative frequency distribution

grouped frequency distribution

class interval

mutually exclusive

all inclusive

range

exact range

interval size

---

*Lotus 1-2-3 is a registered trademark of Lotus Development Corporation.

stated lower limit
stated upper limit
exact limits
graph
X axis, abscissa
Y axis, ordinate
histogram
midpoint

cumulative frequency
cumulative relative frequency
bar graph
statistical curve
normal curve
negatively skewed
positively skewed
bimodal distribution

SUMMARY

1. The first step in statistically analyzing a set of data is to tally the data in an organized framework called the frequency distribution and then graph the results of the study.
2. The specific method used to organize and graph a data set can be affected by the scale of measurement: nominal, ordinal, interval, or ratio.
3. Nominal measurement categorizes, ordinal measurement ranks, interval measurement provides information on the distance between ranks, and ratio measurement combines an interval scale with a true zero point.
4. A frequency distribution shows what values the data take on and how many times each value occurs in the data set.
5. Relative frequency distributions show the percent of the distribution associated with each value of the measured variable.
6. A grouped frequency distribution divides the range of a distribution into equal-size class intervals and displays the number of values that fall into each interval.
7. Distributions are graphed on a two-axis structure defined by a horizontal X axis (the abscissa), which represents the data values, and a vertical Y axis (the ordinate), which represents the frequency of each data value.
8. Histograms and frequency polygons are used to graph ordinal, interval, and ratio data. Bar graphs are used to graph nominal data.
9. A cumulative relative frequency distribution shows the percent of values lying below each of the exact upper limits in a frequency distribution.
10. When very large data sets are graphed, the resulting frequency polygon takes the form of a statistical curve.
11. The most useful statistical curve is the normal curve, which is bell shaped and symmetrical. Other types of curves result from skewed and bimodal distributions.

REVIEW QUESTIONS

1. Define or describe: scale of measurement, nominal scale, ordinal scale, interval scale, and ratio scale. Use your own words.
2. Identify the "highest" possible scale of measurement associated with each of the following (where nominal scales are the lowest level of measurement and ratio scales are the highest):
   (a) Reaction time
   (b) Social class
   (c) Scores on the Scholastic Aptitude Test (SAT)
   (d) Treatment groups in an experiment
   (e) The national rankings of college football teams
3. If Paula finished second and Penny fourth in a talent contest, is it accurate to assert that Paula is twice as talented as Penny? Justify your answer in the context of scales of measurement.
4. For each of the four scales of measurement, list two variables in your major field that are measured at that level. Also, state your reasons for saying that each of the listed variables is measured at a particular level.
5. Define or describe: frequency distribution, relative frequency, proportion, class interval, mutually exclusive, and exact range. Use your own words.
6. For what main reasons do behavioral researchers construct frequency distributions? That is, what functions do those devices serve in the research process?
7. The following set of data is the IQ scores of 50 children. Construct a basic (ungrouped) frequency distribution and a basic relative frequency distribution for these data.

87	103	113	97	100	100	85	118	115	82
90	92	107	103	102	98	110	93	97	108
100	101	100	95	98	99	97	103	102	105
99	101	96	94	100	102	101	98	99	99
98	104	106	101	100	100	99	101	83	117

8. Use the data from Question 7 to construct a grouped frequency distribution and a grouped relative frequency distribution with ten class intervals. (You may work with the stated limits of the class intervals.) In what ways, if any, is the grouped frequency distribution more or less informative than the ungrouped distribution? Does the grouped frequency distribution appear to be basically symmetrical or asymmetrical?
9. Construct a histogram for the data in Question 7.

10. The following data are reaction times (in seconds) from a face recognition experiment. Construct a relative frequency distribution using the exact limits to define class intervals. Try using seven intervals.

0.5 1.2 1.9 1.2 2.4 0.8 0.9 0.3 2.1 1.3

1.4 2.1 0.9 0.7 0.5 2.0 1.3 1.1 2.2 1.4

1.0 0.7 0.6 0.9 1.0 1.1 1.2 1.2 1.3 1.2

11. Create a cumulative relative frequency distribution for the data in Question 10.

12. Construct a relative frequency polygon for the data in Question 10. Would you describe the distribution as mainly symmetrical, negatively skewed, positively skewed, or bimodal? Give specific reasons for choosing a particular label.

13. If the highest score in a set of data is 542 and the lowest score is 71, compute the following quantities in connection with constructing a frequency distribution that has 18 class intervals: range, exact range, interval size, exact limits of the lowest interval, and midpoint of the highest interval.

14. In a 100-student psychology class, 45 students are psychology majors, 25 are business administration majors, 10 are physical science majors, 15 are sociology majors, and 5 are humanities majors. Draw an appropriate graph to represent the distribution of these data. State a specific reason for choosing to use the particular kind of graph you created.

15. What is wrong with the following set of class intervals? (*Hint:* There are at least three problems with the intervals per se.)
8–10, 6–8, 3–6, 0–3

16. Briefly define or describe mutually exclusive, all inclusive, stated lower limit, exact lower limit, and cumulative frequency.

17. Draw a frequency polygon for the data in Table 3.3.

18. Construct a relative frequency histogram for the data in Table 3.5.

19. Define or describe: bar graph, statistical curve, normal curve, negatively skewed, positively skewed, and bimodal distribution.

20. Do the following matching exercise:

___ Figure 3.4   a. Approximately normal
___ Figure 3.6   b. Bimodal
___ Figure 3.8   c. Negatively skewed
                   d. Positively skewed

21. Here are 20 scores on the extroversion scale of the Myers-Briggs Type Indicator. Construct a relative frequency distribution

using the exact limits to define class intervals. Try using three intervals and then seven intervals. Which frequency distribution do you prefer? Why?

12, 13, 12, 14, 23, 17, 16, 13, 9, 15, 7, 16, 19, 17, 6, 20, 20, 17, 10, 2

22. Create a cumulative relative frequency distribution for the data in Question 21.

23. Construct a relative frequency polygon for the data in Question 21. Would you describe the distribution as mainly symmetrical, negatively skewed, positively skewed, or bimodal? Give specific reasons for choosing a particular label.

24. Here are 23 scores from a final exam in a psychotherapy course. Construct a relative frequency distribution using the exact limits to define class intervals. Try using 11 intervals and then seven intervals. Which frequency distribution do you prefer? Why?

74, 87, 93, 83, 97, 90, 74, 57, 43, 70, 47, 53, 60, 80,

77, 57, 77, 77, 77, 67, 50, 93, 53

25. Construct a relative frequency polygon for the data in Question 24. Would you describe the distribution as mainly symmetrical, negatively skewed, positively skewed, or bimodal? Give specific reasons for choosing a particular label.

CHAPTER 4   SUMMARY MEASURES

This chapter will cover the most fundamental kinds of descriptive statistics. Many of these descriptive indexes simply summarize the major trends or characteristics in a sample or population of data. Others have both descriptive and inferential functions. They not only summarize the characteristics of a sample but also are the first step in making inferences about the population represented by that sample.

Descriptively speaking, distributions of data possess two chief characteristics: (1) an *average* or typical value (called a measure of central tendency) and (2) a certain amount of *dispersion* of values around that average (called a measure of variation). Accordingly, I divide the topics conceptually into measures of central tendency and, later, measures of variation.

Important: These topics and those in the remainder of the textbook will involve the mathematical operation of division. Since dividing quantities often results in a decimal number, it may be necessary to "round off" the answer. The rules for rounding used in this book are presented in Box 4.1. Because your method of rounding affects the accuracy of the results you get, you should study Box 4.1 before proceeding.

MEASURES OF CENTRAL TENDENCY

Figure 4.1 *Two Distributions with Different Central Tendencies*

Consider the two hypothetical distributions (A and B) of data shown in Figure 4.1. Each represents the relative frequencies (i.e., percents) of values on some variable we'll call X. Note that the values of X in this example range from about 5 to about 25 and that *the height of either curve at any particular point indicates how frequently a particular value of X occurs in the data set.*

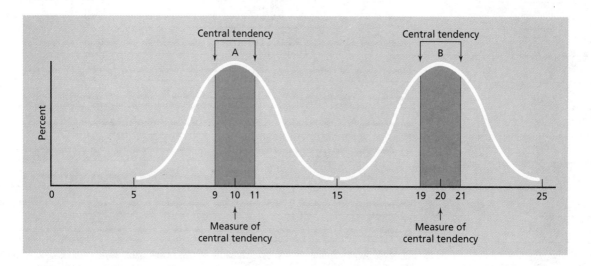

Box 4.1

SUGGESTED RULES FOR ROUNDING DECIMAL NUMBERS

Computing statistical indexes, such as means, often results in decimal numbers—example; $x = \frac{68}{12} = 5.6666. \ldots$ In these cases, you need to decide how to "round off" the decimal portion of the number.

Rounding poses two questions:

1. How many decimal places should be kept in the number?
2. Should the digit in the last decimal place kept be increased or allowed to stay the same?

Different statistics instructors differ somewhat in their answers to these questions, but the following suggestions are acceptable in most instances:

Rule 1. Round to two decimal places more than there were in the original data used in the calculation.

Example: If the original data were whole numbers and they produced a mean of 5.66666...., then round to 5.67.

Example: If the original data produced a mean of 5.66666. ... and were measured to the first decimal place (that is to the tenths level of decimation), then round to 5.667.

Rule 2. If the digit just beyond the last digit to be kept is 5 or greater, then round up; if the digit just beyond the last digit to be kept is less than 5, then the last digit remains the same.

Example: Assuming that the original data were whole numbers:

last digit to be kept
↓
4.26666 rounds to 4.27
↑
digit just beyond last digit to be kept

4.26499 rounds to 4.26

4.26500 rounds to 4.27

The next thing to note is that even though the values in distribution A range from 5 to 15, its values tend to "pile up" or cluster between 9 and 11, with the most common value being 10. *The tendency of scores to cluster around some value in a distribution* is called the distribution's **central tendency**. *A particular value that is used to typify or characterize a distribution's central tendency* is referred to as a **measure of central tendency**.

Distribution B's values tend to cluster between 19 and 21, and 20 is identified as the measure of central tendency. Clearly, distribution B is located in a different place than distribution A. It should be clear, then, that a measure of central tendency serves the important function of telling us where a distribution of data tends to be located on a variable's measurement scale (which is why measures of central tendency are occasionally called "measures of location"). It follows that we can use measures of central tendency to do at least two things:

1. Describe the typical, or average, value that represents any particular distribution
2. Compare two or more distributions in terms of their typical, or average, values

Later sections of this book will demonstrate that a measure of central tendency can also be used as a standard for interpreting and assessing individual values in a data set.

Table 4.1

	X
RAW SCORES OF 30 MANAGERS ON A TEST OF HUMAN RELATIONS SKILLS	17
	23
	21
	28
	23
	28
	26
	25
	22
	34
	23
	27
	29
	21
	22
	33
	33
	26
	24
	25
	28
	31
	20
	23
	25
	35
	31
	23
	24
	30
	780 = sum of raw scores

There are three types of central tendency measure, each representing a different way of finding the average value in a distribution: the mode, the median, and the mean. Each of these statistics will be discussed within the context of some familiar data: the raw scores made by 30 managers on a test of human relations skills. Recall that this data set—-shown in Table 4.1—was used in the last chapter to construct a frequency distribution.

In Table 4.1, observe that the column of raw scores is labeled with a capital *X*. This is the conventional way of symbolizing raw scores in the field of statistics.

> NOTE: Whenever you see a capital *X* symbol in a statistical formula, you know that the formula requires you to do something with raw scores. *X* is the universal symbol for raw scores or a set of raw scores.

Mode

The simplest measure of central tendency is the **mode**, defined as *the most frequently occurring value* in a distribution. To determine the mode, you simply tally the number of times each value occurs; then you identify the most frequent value as the mode. The mode is routinely found in the process of constructing a basic, ungrouped frequency distribution.

Ranking the values in a data set from highest to lowest makes the mode stand out. For example, the ordered list of data in Table 4.2 shows that the modal value on the human relations test is 23. It occurs five times—more often than any other score.

Median

A second kind of statistical average is called the **median**. It is *the middle score in a distribution*. Theoretically, and often in fact, 50% of the values in a distribution lie above the median and 50% lie below it. This property makes the median the 50th **percentile** (see Chapter 5).

To determine the median, follow these steps:

Percentile: A point or value on the *X* axis of a distribution below which a particular percent of a distribution lies. The 50th percentile of a distribution is the value that is higher than the bottom 50% of the scores and lower than the top 50% of the scores; in short, it is the median.

1. Arrange the data in a hierarchy that orders the values from highest to lowest, and assign a numerical rank to each value. The highest score gets a rank of 1, the second highest score a rank of 2, and so on. The lowest value in the data set should have a rank of *N*, where **N** equals *the number of values in the data set.*

   > NOTE: The *N* of a sample is also referred to as the "sample size."

2. Determine the rank of the median value by this formula:

$$\text{Median rank} = \frac{N + 1}{2}$$ *MeDian formula*   (4.1)

NOTE: Formula (4.1) gives only the middle rank, not the median itself.

3. Determine the median by identifying the value that holds the median rank.

 Table 4.2 illustrates how the median is found for the set of human relations test scores.

Table 4.2

AN ORDERED LIST OF 30 HUMAN RELATIONS TEST SCORES SHOWING THE MODE, THE MEDIAN, AND THE MEAN OF THE DISTRIBUTION	RANK	X	
	1	35	
	2	34	
	3	33	
	4	33	
	5	31	
	6	31	
	7	30	
	8	29	
	9	28	
	10	28	
	11	28	
	12	27	
	13	26	
	14	26	
	15	25	
	16	25	←Median = 25
	17	25	
	18	24	
	19	24	
	20	23	
	21	23	
	22	23	Mode = 23
	23	23	
	24	23	
	25	22	
	26	22	
	27	21	
	28	21	
	29	20	
	30	17	

$\Sigma X = 780$ = sum of raw scores

$\Sigma X/N = 780/30 = 26.00$ = mean

1. The values are ordered from highest to lowest, and each value is given a numerical rank. N is 30 because there are 30 values.
2. The median rank is:

$$\frac{N+1}{2} = \frac{30+1}{2} = \frac{31}{2} = 15.5$$

3. The median is identified as the value that holds the rank of 15.5—in other words, the value halfway between the 15th score (25) and the 16th score (also 25) in the ordered list. The value halfway between 25 and 25 is (25 + 25)/2, or 50/2, or 25. In this example, then, the median is 25.

Notice that <u>when your data set has an even number of values, the median is the average of the two middlemost scores. When there is an odd number of values, the median is the middle score,</u> as in the following simple example, where $N = 5$:

10

9

8

7

6

Mean

The most useful measure of central tendency is the **mean**, or _arithmetic average of a set of values._ To obtain a mean:

1. Add up all of the values in a data set.
2. Divide the total by the number of values.

The formula for the _mean of a sample_ is:

$$\overset{①}{\underset{②}{\overline{X} = \frac{\Sigma X}{N}}}$$

$\hfill$ (4.2)

> **FORMULA GUIDE**
>
> ① Add up all the raw scores.
> ② Then divide by the number of scores (i.e., by the sample size).

NOTE: The Greek symbol Σ, uppercase sigma, means "sum the values that follow." Whenever you see Σ in formulas, you are being instructed to perform addition. Thus, ΣX means add the raw scores in a data set.

Applying the formula to the sample of data shown in Tables 4.1 and 4.2, we get

$$\overline{X} = \frac{\Sigma X}{N}$$

or

sum of raw scores in Table 4.2
$$\downarrow$$
$$\overline{X} = \frac{780}{30}$$
$$\uparrow$$
number of raw scores in Table 4.2

$$= 26.00$$

$\overline{X}$, pronounced "*X-bar*," always symbolizes the sample mean. (A different symbol is used to represent the mean of the population.)

The mean is referred to as the "center of gravity" of a distribution. This label reflects the following algebraic property: <u>If you sum all deviations of raw scores from their mean, the result will be 0.</u> A deviation from the sample mean is symbolized as $(X - \overline{X})$, which says "subtract the mean from a raw score." <u>The *result produced by subtracting the mean from a raw score*</u> is called a **deviation score**. For example, subtracting the mean of 26.00 from the highest raw score in Table 4.2—that is, 35 – 26.00—produces the deviation score of 9. Since some deviation scores are negative—for instance, 17 – 26.00 = –9—the sum of the deviation scores will be 0: $\Sigma (X - \overline{X}) = 0$. Hence, the mean is the algebraic "balance point" of a distribution.

Another property of the mean is that it is based on the **least squares principle:** *The sum of the squared deviations from the mean is less than the sum of the squared deviations from any other value.* To "square" a deviation score, you multiply it by itself. Thus, $(X - \overline{X})^2 = (X - \overline{X})(X - \overline{X})$; for example, $(17 - 26.00)^2 = (-9)(-9) = 81$. Box 4.2 demonstrates that, compared with the mode and the median, the mean yields the "least" sum of squared deviations. Try this demonstration with any value in the data set, if you like. The mean will always win the contest of least squares. (The only exception occurs when the distribution is perfectly symmetrical. In that case, the mode and median have the same value as the mean.)

✸ NOTE: It is important to remember the least squares principle. It will be used again when you study linear regression in Chapter 7.

Other Uses of the Mean

In addition to describing the typical score in a data set, the sample mean is frequently used to:

1. Compare the average values of two or more distributions
2. <u>Estimate a population mean</u>

Box 4.2

DEMONSTRATION OF LEAST SQUARES PRINCIPLE FOR THE MEAN

Data set: $\{2, 4, 4, 6, 7, 9, 17\}$; $N = 7$

Mode = 4,   median = 6   mean = 7

Note: *SS* stands for the sum of squared deviations, or simply the sum of squares.

Sum of squares based on the mode:

$$
\begin{aligned}
SS_{\text{mode}} &= \Sigma(X - \text{mode})^2 \\
&= (2 - 4)^2 + (4 - 4)^2 + (4 - 4)^2 + (6 - 4)^2 + (7 - 4)^2 + (9 - 4)^2 + (17 - 4)^2 \\
&= \ \ -2^2 \ + \ \ 0^2 \ \ + \ \ 0^2 \ \ + \ \ 2^2 \ \ + \ \ 3^2 \ \ + \ \ 5^2 \ \ + \ \ 13^2 \\
&= \ \ \ 4 \ \ + \ \ 0 \ \ + \ \ 0 \ \ + \ \ 4 \ \ + \ \ 9 \ \ + \ \ 25 \ \ + \ \ 169 \\
&= \ \ 211
\end{aligned}
$$

Sum of squares based on the median:

$$
\begin{aligned}
SS_{\text{median}} &= \Sigma(X - \text{median})^2 \\
&= (2 - 6)^2 + (4 - 6)^2 + (4 - 6)^2 + (6 - 6)^2 + (7 - 6)^2 + (9 - 6)^2 + (17 - 6)^2 \\
&= \ \ -4^2 \ + \ -2^2 \ + \ -2^2 \ + \ \ 0^2 \ + \ \ 1^2 \ + \ \ 3^2 \ + \ \ 11^2 \\
&= \ \ 16 \ \ + \ \ 4 \ \ + \ \ 4 \ \ + \ \ 0 \ \ + \ \ 1 \ \ + \ \ 9 \ \ + \ \ 121 \\
&= \ \ 155
\end{aligned}
$$

Sum of squares based on the mean:

$$
\begin{aligned}
SS_{\text{mean}} &= \Sigma(X - \overline{X}^2) \\
&= (2 - 7)^2 + (4 - 7)^2 + (4 - 7)^2 + (6 - 7)^2 + (7 - 7)^2 + (9 - 7)^2 + (17 - 7)^2 \\
&= \ \ -5^2 \ + \ -3^2 \ + \ -3^2 \ + \ -1^2 \ + \ \ 0^2 \ + \ \ 2^2 \ + \ \ 10^2 \\
&= \ \ 25 \ \ + \ \ 9 \ \ + \ \ 9 \ \ + \ \ 1 \ \ + \ \ 0 \ \ + \ \ 4 \ \ + \ \ 100 \\
&= \ \ 148
\end{aligned}
$$

Conclusion: The sum of the squared deviations from the mean is less than the sum of the squared deviations from any other measure of central tendency. (*Exception*: When the distribution is perfectly symmetrical, the mean, median, and mode produce the same sum of squares.)

Comparing distribution averages   An experiment in the behavioral sciences normally uses two or more samples, where each sample represents one level, or value, of the independent variable. If the independent variable has an influence on the measured, or dependent, variable, then the means of the respective samples should be different from one another.

CONCEPT RECAP

The *independent variable* in an experiment is the variable that the researcher systematically manipulates (see Chapter 2). The *dependent variable* is the behavior that the researcher observes and measures.

Box 4.3

COMBINING AVERAGES: THE WEIGHTED MEAN

There will be times when you want to combine the means of two or more groups of observations to compute an overall mean of all the groups. If the groups are of equal sizes—that is, if they have the same number of observations—then computing the overall mean is simply a matter of adding the individual means and dividing by the number of means.

The situation is slightly more complicated, however, if you wish to combine the means of groups that contain different numbers of observations. In that case you need to compute a weighted mean using the formula:

$$\text{Weighted mean} = \frac{\Sigma(N \cdot \text{mean})}{\Sigma N}$$

where mean = any group mean, and N = the number of observations in that group of data.

This formula tells you to:

1. Multiply each group mean by the number of observations in the group; that is, "weight" each mean by the number of observations that it is based on.

2. Sum the products generated in step 1.

3. Divide the sum of the products by the sum of the group sizes.

The rationale of this procedure is that *each group mean should influence the overall mean in proportion to the number of observations that it represents.* In a sense, each mean "votes" in proportion to the number of constituents it represents.

Take the following example: A physical education teacher has 17 girls and 10 boys "military-press" as much weight as they can, using a barbell. The girls lift a mean weight of 90 pounds, and the boys lift a mean weight of 100 pounds. What is the average amount of weight pressed by the two groups combined? It is tempting to say 95 pounds, but that is incorrect. Since the group of girls is larger than the group of boys, the girls' mean must be given more weight in the overall average. Using the above formula, we get:

$$
\begin{aligned}
\text{Weighted mean} &= \frac{(17 \cdot 90) + (10 \cdot 100)}{17 + 10} \\
&= \frac{1530 + 1000}{27} \\
&= \frac{2530}{27} \\
&= 93.70 \text{ pounds}
\end{aligned}
$$

Table 4.3 displays the results of an experiment that investigated the effect of "distributed practice" on college students' ability to remember words from a long list of nouns. Each student studied a list of 120 words in which each word occurred twice. The three levels of the independent variable, distribution of practice, were established in the following way:

Table 4.3

RECALL SCORES AND MEANS FOR THREE LEVELS OF SPACING IN A STUDY OF DISTRIBUTED PRACTICE EFFECTS	SPACING BETWEEN WORD PRESENTATIONS		
	SPACING = 0 X_1	SPACING = 5 X_2	SPACING = 10 X_3
	10	21	24
	18	23	23
	13	18	18
	12	22	21
	12	17	25
	18	19	22
	17	20	20
	20	19	24
	15	22	19
	15	19	24
Sums:	$\Sigma X_1 = 150$	$\Sigma X_2 = 200$	$\Sigma X_3 = 220$
N's:	$N_1 = 10$	$N_2 = 10$	$N_3 = 10$
Means:	$\overline{X}_1 = 15$	$\overline{X}_2 = 20$	$\overline{X}_3 = 22$

Level 1. Forty of the words were repeated immediately after their initial occurrences. This was the "Spacing = 0," or "massed practice," condition.

Level 2. For the second set of 40 words, the two occurrences of each word were separated from each other by five intervening words. That was the "Spacing = 5," or "distributed practice 5," condition.

Level 3. For the third set of 40 words, the two occurrences of each word were separated from each other by ten intervening words. This was the "Spacing = 10," or "distributed practice 10," condition.

The words were presented visually, one at a time. Following their viewing of the list, the students attempted to write down as many of the words as they could from memory.

As the means given in Table 4.3 indicate, the average number of words recalled increases as a function of the distribution, or spacing, of word repetitions in the study list:

Spacing = 0: $\overline{X}_1 = 15$

Spacing = 5: $\overline{X}_2 = 20$

Spacing = 10: $\overline{X}_3 = 22$

Notice that when sample means are compared, a numerical subscript is used to identify the different means with their respective samples—$\overline{X}_1$ versus $\overline{X}_2$ versus $\overline{X}_3$. In Table 4.3, the various sample sizes *(N)* and sums *(ΣX)* are also distinguished by sub-scripts.[1]

In studies that compare several means, it is conventional to graph the sample means as a function of the levels of the independent variable. The purpose is to clearly illustrate the major trend or effect produced by the independent variable.

As Figure 4.2 shows, a **line graph** of means is constructed somewhat differently from the frequency polygons that you learned about in Chapter 3. Specifically, *the Y axis (i.e., vertical axis) represents the dependent-variable means of the various samples in the experiment.* It does not represent frequencies. The X axis represents the levels of the independent variable (as in Chapter 3). A dot or other marker is placed at the point where each sample mean intersects with the corresponding level of the independent variable, and then the dots are connected by straight lines.

Topics covered in Chapters 11–14 will show you how to test the statistical reliability of the differences between means obtained in a behavioral experiment.

Parameter estimation   For every sample mean $\overline{X}$, there is a corresponding population mean that is represented by the sample mean. A population mean is symbolized with μ—pronounced "myoo"—the lowercase "mu" in the Greek alphabet. Recall that a numerical characteristic of a sample, such as $\overline{X}$, is called a *statistic* and that the corresponding characteristic of a population (μ in this case) is referred to as a **parameter**.

Figure 4.2 *Line Graph of Means from an Experiment on Distributed Practice Effects*

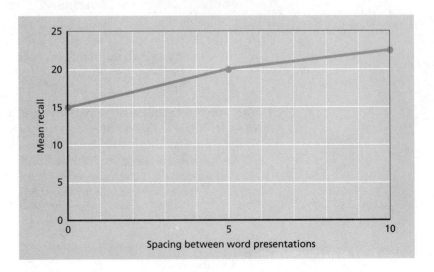

---

[1] Throughout this text, subscripts are used to identify different sample statistics that are being compared.

NOTE: It is conventional in the field of statistics to use Greek letters to symbolize parameters (population indexes) and Roman letters to symbolize statistics (sample indexes).

If a population is **finite**—that is, *small enough to be countable*—then it is possible to directly calculate the population mean using the formula:

$$\text{Population mean} = \mu = \frac{\Sigma X}{N} \tag{4.3}$$

where N is the number of observations in the population.

Hypothetical populations in statistical theory, as well as large populations actually studied by behavioral scientists, are considered **infinite**: *The number of observations is so great as to be indefinable and "uncountable."* In the case of an infinite population, μ cannot be calculated (since N is either undefined or indeterminable); rather it must be estimated. It can be shown algebraically that $\overline{X}$ is an **unbiased estimator** of μ (Hays, 1988). This means that if all possible samples of a given size are drawn from a population, *the average sample mean will be exactly equal to* μ. Another way to express this fact is:

Expected value of $\overline{X} = \mu$

where the **expected value** is the *average value of a statistic* ($\overline{X}$ in this case) that would be found in the long run. It is for this reason that $\overline{X}$ can be used to estimate the mean of an infinite population. Techniques for estimating μ will be addressed in Chapters 8 and 10.

RELATIVE LOCATIONS OF CENTRAL TENDENCY MEASURES

Figure 4.3 illustrates the effects of a distribution's shape on the comparative magnitudes of the mode, median, and mean. If the distribution is perfectly symmetrical, with the left half of the distribution precisely mirroring the right half, the mode, median, and mean have the same numerical value—unless the distribution is bimodal.

In a skewed distribution, the median is always somewhere between the mode and the mean. Of the three measures, the mean has the lowest value in a negatively skewed distribution and the highest value in a positively skewed distribution because it is influenced by extreme scores.

These statements are not applicable to bimodal or multimodal distributions.

EVALUATING MEASURES OF CENTRAL TENDENCY

All three measures of central tendency have their respective uses in behavioral statistics. Let's now take a brief look at the advantages and limitations of each of these indexes of "average."

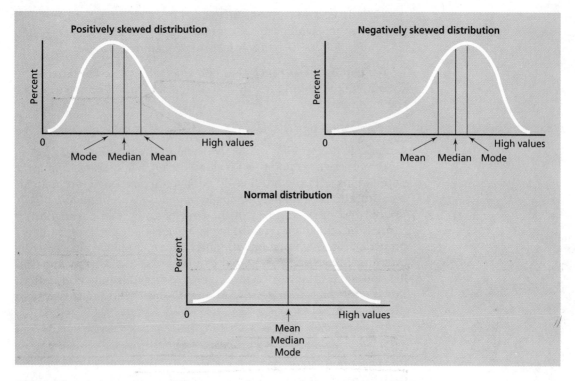

Figure 4.3 *Relative Locations of the Mean, Median, and Mode in Skewed Distributions*

Evaluating the Mode

Advantages of the mode   For small samples, the mode can be quickly and easily determined. Also, if a variable is on a nominal scale of measurement, the mode is often the only measure of central tendency that can be meaningfully applied to the data. For instance, although it would not be meaningful to calculate the mean or median college major at your school, you could determine the "modal major." At many colleges, the business administration major is the mode; that is, there are more students in that nominal category than in any other.

Disadvantages of the mode   It is not unusual for small data sets to have more than one mode. This outcome makes the mode an ambiguous index of central tendency. A second disadvantage is that the mode is a "statistical dead end"; that is, it cannot be used for more advanced statistical purposes, such as making inferences about populations. The reason is that there is no practical statistical theory that relates sample modes to population modes in the way that sample means can be linked to population means.

Evaluating the Median

Advantages of the median   If your data are on an ordinal scale of measurement, where observations can be ranked but the measurement scale does not have equal intervals, the median can nonetheless be used as an average value. (The mean could not be used, because its calculation assumes an interval scale.) The median is a valid measure of central tendency in that case because the ability to order the values is all that is required to find the middle score.

When a distribution is severely skewed, with a few very extreme values at either the high end or the low end of the X axis, the median is usually a better descriptive index of central tendency than the mean. This is because the median value is not influenced by the extreme values. As you will see shortly, the mean can be greatly distorted by the outlying values in a skewed distribution—to the extent that the mean is not an accurate description of the average value of the distribution. It is for this reason that the U.S. Bureau of Labor Statistics always reports the average individual income as a median rather than a mean. The handful of multimillionaires and billionaires in our society pulls the mean income so far up the pay scale that it does not typify earnings of the average American. Developing an accurate description of the average price of new homes presents a similar problem. Again the existence of a skewed distribution makes the median the preferred descriptive statistic, as indicated by this January 1990 newspaper clipping:

The average [i.e., mean] price for a new home in November rose to $155,900, from $147,600 in October. The median price in November rose to $127,000 from $123,000 the previous month. (*St. Louis Post-Dispatch*, January 4)

Disadvantages of the median   Like the mode, the median is not amenable to most of the more advanced and powerful techniques of inferential statistics. It is predominantly a descriptive statistic. As with the mode, the reason for this limitation on the median is that there is no practical statistical theory that relates sample medians to population medians in the way that sample means can be linked to population means.

Evaluating the Mean

Advantages of the mean   Of the three types of average covered here, the mean is by far the most practical—in the sense that it is used in dozens of advanced statistical procedures and is particularly valuable as an inferential statistic. In fact, the majority of techniques discussed in the remainder of this book will utilize the mean in some fashion.

An additional advantage of the mean is that it is calculated by combining all of the scores in the data set, whereas the median and mode are based on just one score each. This fact makes the mean a more stable measure of central tendency. If several random samples are drawn from a given population, the various sample means usually will vary less from one another than will the modes and medians of the samples. Stability is a valuable characteristic for a statistic to have for the following reason: *The more stable the sample statistic, the more accurately it estimates the corresponding population parameter.* This point will become more important when you study parameter estimation in Chapters 8 and 10.

Disadvantages of the mean   The mean cannot be calculated meaningfully unless the scale of measurement has equal intervals; there is no valid mean for nominal and ordinal data. In addition, the mean may not be the most informative measure of central tendency to use if the distribution of values is severely skewed. In that case, the median is usually preferable for summarizing the typical score. Why? Consider the mean and median for the two small data sets shown here. Notice how making one score more extreme affects the mean but has no effect on the median. In the distribution on the right, which measure of central tendency seems to more accurately describe the typical score, 7 or 11?

$\dfrac{X}{}$	$\dfrac{X}{}$	
3	3	
5	5	
7	7	
9	9	
11	31	← **One score is extreme.**
$\Sigma X = 35$	$\Sigma X = 55$	
$\overline{X} = 7$	$\overline{X} = 11$	← **The mean is distorted.**
Median = 7	Median = 7	← **The median is not affected.**

Is the Measure of Central Tendency Enough Information?

Now you know about three ways to represent the "average" outcome in a set of data. You'll learn about a wide variety of ways to use this knowledge in the remaining chapters of this text. But is it usually sufficient to calculate only a measure of central tendency when you wish to analyze research results? Although the central tendency of a data set is a very important piece of information, it does not tell the whole story about the data's distribution. In order to get a more complete picture of a distribution and to fully utilize the statistical information within it, you also need to compute an index of its variation.

THE CONCEPT OF VARIATION (DISPERSION)

Statistical **variation** is defined as *the degree to which the values in a distribution tend to depart from the distribution's average value.* Variation is also referred to as the **dispersion** of values away from their average.

To better appreciate the need to consider variation in a distribution, take a look at the two samples of raw scores that are shown in Table 4.4. The data are the reaction times of student drivers who were required to perceive and respond to a simulated automobile accident situation under two different drug states. On a random basis, one-half of the students were administered a moderately high dose of an amphetamine (a stimulant) and the other half of the students were given a placebo, a capsule that had the same appearance as the amphetamine but contained no active ingredient.

Table 4.4

REACTION TIMES (IN TENTHS OF A SECOND) OF STUDENT DRIVERS UNDER TWO DRUG STATES

	DRUG CONDITION	
	PLACEBO	AMPHETAMINE
	19	20
	18	15
	21	25
	20	19
	23	22
	19	20
	20	21
	17	17
	19	20
	21	24
	20	20
	20	17
	18	18
	22	18
	21	23
	22	21
	20	16
	21	22
	20	23
	19	19
Sums:	$\Sigma X = 400$	$\Sigma X = 400$
N's:	$N_1 = 20$	$N_2 = 20$
Means:	$\overline{X}_1 = 20$	$\overline{X}_2 = 20$

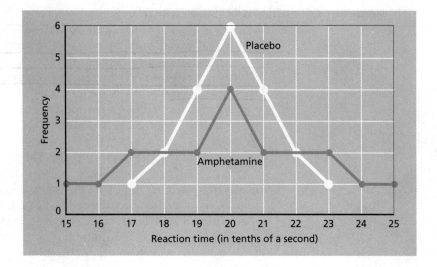

The reaction times are presented as integers but actually represent the number of *tenths of a second* required for each response. A score of 19 represents 1.9 seconds.

The summary statistics at the bottom of Table 4.4 show that the two samples are exactly equal in terms of (1) their sums of raw scores, (2) their sample sizes ($N_1 = N_2 = 20$), and (3) their means ($\overline{X}_1 = \overline{X}_2 = 20$). But are the two sample distributions equivalent?

The frequency polygons of the distributions, displayed in Figure 4.4, indicate that the distributions definitely are not the same. By viewing those graphs, you can see that even though the two samples have the same central tendency, the amphetamine sample had more *variation* than the placebo group. The scores in the amphetamine group are more dispersed, or spread out, relative to the mean of 20. It appears that the stimulant drug makes some people more efficient reactors while making others less efficient, with the overall result being a larger average difference among the subjects' scores.

MEASURES OF VARIATION

Range

The simplest numerical index of variation is the **range** (or **simple range**), defined as *the difference between the highest and lowest scores in a distribution:*

$$\text{Range} = X_{max} - X_{min} \tag{4.4}$$

Recall that we used the range in Chapter 3 to calculate the class interval size. From formula (4.4), the range for the placebo sample is 23 – 17 = 6. The range for the amphetamine sample is 25 – 15 = 10. In general, then, *the greater the dispersion of values in a distribution, the larger is its range.*

Though easy and quick to calculate, the range is a crude descriptive statistic that has no additional statistical applications. It is considered to be an unstable measure of variation because a drastic change in either one of its defining values (X_{max} or X_{min}) causes an equally drastic change in the index of variation—even though all other scores remain the same. Typically it is used only as a preliminary indicator of variation.

Interquartile Range and Semi-Interquartile Range

The **interquartile range** is *the range of values that includes the middle 50% of the distribution.* To determine the interquartile range, you first need to find the following values:

- The 75th percentile, which is also called the "third quartile" and is symbolized as Q_3. The 75th percentile is just above the bottom 75% of the raw scores and just below the upper 25% of the raw scores.

- The 25th percentile, which is also called the "first quartile" and is symbolized as Q_1. The 25th percentile is just above the bottom 25% of the raw scores and just below the upper 75% of the raw scores.

Since 25% of the raw scores are above Q_3 and 25% of the raw scores are below Q_1, the scores between Q_3 and Q_1 make up the middle 50% of the distribution. Once you have determined which value lies just below the top 25% of the distribution and which value lies just above the bottom 25% of the distribution, you can compute the interquartile range:

$$\text{Interquartile range} = Q_3 - Q_1 \tag{4.5}$$

A related measure of variation is the **semi-interquartile range**, defined as *one-half of the interquartile range.* You can obtain the semi-interquartile range with the formula:

$$\text{Semi–interquartile range} = \frac{Q_3 - Q_1}{2} \tag{4.6}$$

The greater the variation among values in a distribution, the larger are interquartile and semi-interquartile ranges. You already know, for example, that the scores in the amphetamine group have more variation than the scores in the placebo group. So you would expect the interquartile and semi-interquartile ranges of the amphetamine subjects to be greater than those of the placebo subjects. This expectation is confirmed by the calculations shown in Table 4.5, which displays these two ranges. As you inspect Table 4.5, notice that *to calculate the interquartile range, you must first sort the scores from highest to lowest.* This allows you to easily identify the lowest 25% and highest 25% of the scores.

The interquartile and semi-interquartile ranges are alternatives to the simple range as measures of variation in data sets. They are considered more stable than the simple range because they are based on less extreme scores than the latter. But, alas, the interquartile and semi-interquartile ranges are less useful than we might hope. They can be criticized on the basis of their main advantage. Because they ignore the most extreme 50% of the scores in a distribution, they fail to fairly represent the total variation in

Table 4.5

COMPUTATION OF THE INTERQUARTILE AND SEMI-INTERQUARTILE RANGES USING DATA FROM THE AMPHETAMINE EXPERIMENT	PLACEBO GROUP X	AMPHETAMINE GROUP X
	23	25
	22	24
	22	23
	21	23
	21 ←Q_3 = 75th percentile = 21	22 ←Q_3 = 75th percentile = 22
	21	22
	21	21
	20	21
	20	20
	20	20
	20	20
	20	20
	20	19
	19	19
	19 ←Q_1 = 25th percentile = 19	18 ←Q_1 = 25th percentile = 18
	19	18
	19	17
	18	17
	18	16
	17	15

Interquartile range = $Q_3 - Q_1$ = 21 − 19 = 2   Interquartile range = $Q_3 - Q_1$ = 22 − 18 = 4

Semi-interquartile range = $\dfrac{Q_3 - Q_1}{2} = \dfrac{2}{2} = 1$ Semi-interquartile range = $\dfrac{Q_3 - Q_1}{2} = \dfrac{4}{2} = 2$

the data set. Also, the system of statistical analysis that was built around these measures of variation has been replaced by a more powerful system based on a statistic called the *variance.*

Variance

Relative to the range and the interquartile range, the variance is a much more sophisticated measure of variation that is used in a diversity of descriptive and inferential statistical procedures. It is also more stable than those other indexes because it is based on all the observations in a distribution—not just two scores. The **variance** is defined as *the average of the squared deviations from the mean.* The formula associated with this definition is:

$$\text{Sample variance} = S^2 = \frac{\overset{③}{\downarrow} \overset{①}{\downarrow} \overset{②}{\downarrow}}{\underset{\overset{\uparrow}{④}}{N}}\Sigma(X - \overline{X})^2 \qquad (4.7)$$

FORMULA GUIDE

① Subtract the mean from each raw score to get *N* deviation scores.
② Square each deviation score.
③ Sum the squared deviations.
④ Divide that sum by the sample size.

Since $\Sigma(X - \overline{X})^2$ is called the **sum of squares (SS)**—meaning *the sum of squared deviations from the mean*—another way to express the formula for the sample variance is:

$$S^2 = \frac{SS}{N} \qquad (4.8)$$

where $SS = \Sigma(X - \overline{X})^2$.

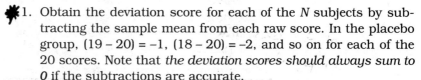

Table 4.6 shows how the variances of the placebo sample and the amphetamine sample are calculated using formula (4.7). These steps are used in calculating each variance:

1. Obtain the deviation score for each of the *N* subjects by subtracting the sample mean from each raw score. In the placebo group, $(19 - 20) = -1$, $(18 - 20) = -2$, and so on for each of the 20 scores. Note that *the deviation scores should always sum to 0* if the subtractions are accurate.

Table 4.6

CALCULATION OF SAMPLE
VARIANCES FOR TWO GROUPS
OF STUDENT DRIVERS UNDER
DIFFERENT DRUG STATES

	PLACEBO GROUP			AMPHETAMINE GROUP		
RAW SCORES X_1	DEVIATION SCORES $X - \bar{X}_1$	SQUARED DEVIATIONS $(X - \bar{X}_1)^2$	RAW SCORES X_2	DEVIATION SCORES $X - \bar{X}_2$	SQUARED DEVIATIONS $(X - \bar{X}_2)^2$
19	−1	1	20	0	0
18	−2	4	15	−5	25
21	1	1	25	5	25
20	0	0	19	−1	1
23	3	9	22	2	4
19	−1	1	20	0	0
20	0	0	21	1	1
17	−3	9	17	−3	9
19	−1	1	20	0	0
21	1	1	24	4	16
20	0	0	20	0	0
20	0	0	17	−3	9
18	−2	4	18	−2	4
22	2	4	18	−2	4
21	1	1	23	3	9
22	2	4	21	1	1
20	0	0	16	−4	16
21	1	1	22	2	4
20	0	0	23	3	9
19	−1	1	19	−1	1
Sums: 400	0	42	400	0	138

$$S_1^2 = \frac{\Sigma(X - \bar{X}_1)^2}{N_1} = \frac{42}{20} = 2.1 \qquad S_2^2 = \frac{\Sigma(X - \bar{X}_2)^2}{N_2} = \frac{138}{20} = 6.9$$

2. Square each of the N deviation scores; for example, $(-1)^2 = (-1)(-1) = 1$, $(-2)^2 = (-2)(-2) = 4$, and so on for each of the 20 deviation scores.
3. Sum the squared deviation scores. For the placebo group, $\Sigma(X - \bar{X})^2 = 42 = SS_1$.
4. Obtain S_1^2 by dividing the sum of the squared deviations by the number of observations (sample size): $S_1^2 = SS_1/N_1 = 42/20 = 2.1$.

The results of the calculations show that the variance of the placebo group (S_1^2) is 2.1, whereas the variance of the amphetamine group (S_2^2) is 6.9. In general, then, the greater the variation of scores in a distribution, the larger is the value of the variance.

Computational formula for S^2 Formula (4.7) for the sample variance is a *definitional formula*; it defines the variance as the average of the squared deviations from the mean. Such definitional formulas will be presented throughout this book when a statistical method is introduced and explained. However, definitional formulas often are not the simplest devices for actually computing a statistic; also, many of them are subject to substantial rounding error. Consequently, you will also be given *computational formulas* for most of the statistical methods covered in this text. Although computational formulas frequently look complex and formidable, they are much easier to use—and often more accurate—than the corresponding definitional formulas. Computational formulas usually permit you to work directly with raw scores, rather than slogging your way through deviation scores and the like.

The **computational formula** for the sample variance is

$$S^2 = \frac{\Sigma X^2 - (\Sigma X)^2 / N}{N} \leftarrow \text{⑤} \tag{4.9}$$

FORMULA GUIDE

① Square each raw score.
② Sum the squared raw scores.
③ Sum all raw scores.
④ Square the sum of raw scores.
⑤ Divide the squared sum of raw scores (step 4) by N.
⑥ Subtract the result of step 5 from the result of step 2.
⑦ Divide the result of step 6 by N, the number of raw scores.

Table 4.7 illustrates how the computational formula (4.9) is used to calculate the sample variances for the placebo group and the amphetamine group. Observe that the computational formula produces exactly the same answer as the definitional formula (4.7) but without requiring one to work with deviation scores. This advantage exists because of the following algebraic equivalences:

$$\text{Sum of squares} = SS = \Sigma(X - \overline{X})^2 = \Sigma X^2 - \frac{(\Sigma X)^2}{N} \tag{4.10}$$

The population variance   The parameter that corresponds to the sample variance is σ^2 (lowercase Greek *sigma* squared), known as the **population variance.** If the population is finite, then σ^2 is calculated:

$$\text{Population variance} = \sigma^2 = \frac{\Sigma(X - \mu)^2}{N} \qquad (4.11)$$

where X represents the population raw scores, μ is the population mean, and N is the size of the finite population (that is, the number of observations in it).

If the population is infinite, σ^2 cannot be computed because N is undefined. Therefore, σ^2 must be estimated. In Chapter 10, you will learn how to estimate the population variance by using a slight modification of the sample variance.

The Standard Deviation

Still another measure of variation is the **standard deviation**, defined as *the square root of the variance.* For a sample:

$$\text{Sample standard deviation} = S = \sqrt{S^2} = \sqrt{\frac{SS}{N}} \qquad (4.12)$$

NOTE: By logical deduction, if the standard deviation is the square root of the variance, then the variance is the square of the standard deviation: $S^2 = (S)^2 = S \cdot S$.

Referring to Table 4.7, you can see that the placebo group's standard deviation is $S_1 = \sqrt{S_1^2} = \sqrt{2.1} = 1.45$. The amphetamine group's standard deviation is $S_2 = \sqrt{S_2^2} = \sqrt{6.9} = 2.63$.
For a population:

$$\text{Population standard deviation} = \sigma = \sqrt{\sigma^2} \qquad (4.13)$$

If the standard deviation is simply the square root of the variance, why do we bother calculating it? Part of the answer is convenience. Most often, the sample standard deviation is used for the same purpose as the sample variance: to describe the dispersion of values within a sample, compare variation in two or more samples, and make inferences about a population. In some of these situations, it is simply more computationally convenient to use the standard deviation in a formula than to use its square, the variance.

As you will see in the next chapter, the standard deviation has a unique relationship to the normal curve that makes it a valuable tool for (1) interpreting individual values within a distribution and (2) converting several different distributions to a standard distribution for purposes of comparison. When you use the standard deviation to interpret individual values in a distribution, *it is helpful to think of it as the standard, or typical, amount by which raw scores deviate from their mean.* Then each raw score can be thought of in terms of how many standard units of variation lie between it and the mean (see Chapter 5). You can't use the variance in this way.

Table 4.7

CALCULATION OF SAMPLE
VARIANCES USING THE
COMPUTATIONAL FORMULA

PLACEBO GROUP		AMPHETAMINE GROUP	
RAW SCORES X_1	SQUARED SCORES X_1^2	RAW SCORES X_2	SQUARED SCORES X_2^2
19	361	20	400
18	324	15	225
21	441	25	625
20	400	19	361
23	529	22	484
19	361	20	400
20	400	21	441
17	289	17	289
19	361	20	400
21	441	24	576
20	400	20	400
20	400	17	289
18	324	18	324
22	484	18	324
21	441	23	529
22	484	21	441
20	400	16	256
21	441	22	484
20	400	23	529
19	361	19	361

$\Sigma X_1 = 400 \qquad \Sigma X_1^2 = 8042 \qquad\qquad \Sigma X_2 = 400 \qquad \Sigma X_2^2 = 8138$

$$SS_1 = \Sigma X_1^2 - [(\Sigma X_1)^2/N_1]$$
$$= 8042 - [(400)^2/20]$$
$$= 8042 - [160{,}000/20]$$
$$= 8042 - 8000 = 42$$
$$S_1^2 = SS_1/N_1 = 42/20 = 2.1$$

$$SS_2 = \Sigma X_2^2 - [(\Sigma X_2)^2/N_2]$$
$$= 8138 - [(400)^2/20]$$
$$= 8138 - [160{,}000/20]$$
$$= 8138 - 8000 = 138$$
$$S_2^2 = SS_2/N_2 = 138/20 = 6.9$$

Why? The variance is expressed in "squared units" relative to the raw scores, whereas the standard deviation is always in the original units of measurement. If the raw data represent dollar amounts, for instance, then the standard deviation reflects variation in dollar units, whereas the variance represents variation in "squared dollar amounts." Thus, the standard deviation is the statistic of choice whenever the researcher wants the measure of variation to be in the same units of measurement as the raw data.

Measures of Variation and Scales of Measurement

A restrictive feature of the variance and standard deviation is that you can use them only if you are working with data that are on an interval or ratio scale. To calculate the variance, for example, you must square deviation scores [i.e., $(X - \bar{X})^2$] and add them. These operations make mathematical sense only if the scale has equal units. In contrast, it is legitimate to apply the interquartile range and semi-interquartile range to ordinal data as well as to interval and ratio data.

Notice how these measurement-scale requirements parallel those for measures of central tendency. Since the median works well with ordinal data, you can expect to see the interquartile range used as a description of data variation whenever the median is the measure of central tendency. Similarly, either the variance or the standard deviation is normally used in tandem with the mean, which requires interval or ratio data.

There is no conventional measure of variation that works with nominal-scale data. But the number of nominal categories can serve as a very crude index of variation. Data sets that represent many categories can be considered to have more variation than data sets that represent just two or three categories.

TAKING STOCK

The average and the variation about the average are the most important characteristics of any distribution of data. In a sense, most of this book is about interpreting specific outcomes in the context of behavioral averages and behavioral variation. Having read this chapter, you are now better equipped to comprehend those precise methods of interpreting findings. In fact, in the next chapter you will use your new knowledge to evaluate a person's test score by determining how many standard deviations the score departs from the mean of the test score distribution.

KEY TERMS

central tendency	mean
measure of central tendency	uppercase sigma (Σ)
mode	deviation score
median	least squares principle
percentile	line graph
N	parameter

finite population   interquartile range

infinite population   semi-interquartile range

unbiased estimator   variance

expected value   sum of squares

variation   computational formula for S^2

dispersion   population variance

range   standard deviation

SUMMARY

1. Distributions of data possess two chief characteristics: an average value, known as the measure of central tendency, and the dispersion of values around that average, known as the variation.
2. There are three different measures of central tendency. The mode is the most frequent value, the median is the middle value in a distribution, and the mean is the arithmetic average of all the values in a distribution.
3. The main advantage of the mode is that it is the only type of average that can be used with nominal-scale data. The median is the best description of central tendency when the distribution is badly skewed. The mean is superior to the mode and median, in that it can be used in a large number of advanced statistical procedures.
4. The sample mean is an unbiased estimator of the population mean. It is also used to compare different samples in an experiment.
5. In a skewed distribution, the mode is always located at the highest point in the curve, the mean is always "pulled out" closest to the longer tail of the curve, and the median is always between the mean and the mode.
6. The range of a distribution, defined as the difference between the highest and lowest scores, is the simplest measure of variation. A second measure of variation is the interquartile range, defined as the difference between the two values that bracket the middle 50% of scores in a distribution. A third measure, the semi-interquartile range, is one-half the interquartile range. None of the "range" measures of variation is very amenable to inferential statistics
7. The variance, defined as the average of the squared deviations from the mean, is a sophisticated measure of variation that is used in a variety of inferential procedures.

8. The standard deviation is the square root of the variance. For convenience, it sometimes replaces the variance in statistical formulas. The standard deviation also is used to interpret individual values in a distribution.

REVIEW QUESTIONS

1. Briefly define or describe: central tendency, measure of central tendency, mode, median, percentile, N, mean, and least squares principle.
2. Briefly define or describe: line graph, parameter, finite population, infinite population, unbiased estimator, and expected value.
3. Using appropriate rounding rules, round off the following mean values:

   7.81500, when the raw scores are whole numbers

   7.79500, when the raw scores are whole numbers

   100.7664, when the raw scores are decimal numbers taken out one decimal place (to the tenths level)

   0.239949, when the raw scores are decimal numbers taken out to the second decimal place (to the hundredths level).

4. The following final exam scores are obtained by college seniors taking an honors seminar in sociology: 20, 27, 35, 26, 33, 22, 22, 39. Find the mode, mean, and median for this data set.
5. In your own words, summarize the advantages and disadvantages of the mode, mean, and median.
6. Within the electronics department of a large aerospace firm, an anonymous in-house survey reveals that 355 of the employees consider themselves to be Democrats, 292 claim to be Republicans, 49 label themselves as Independents, and the remaining 9 say that they belong to the American Socialist Party. Use one of the three measures of central tendency to statistically summarize the "typical" political allegiance within the department, and state the main reason or reasons that you chose that particular measure rather than one of the remaining two.
7. The following IQs belong to a class of kindergartners. Compute the mode, median, and mean IQs for these pupils. Based on the results of your calculations, is the distribution basically symmetrical, positively skewed, or negatively skewed?

   X: 81, 106, 84, 127, 80, 100, 116, 91, 112, 113, 96, 106

8. Calculate the mean of the following sample of values of original scores: 10, 14, 7, 21, 15, 17, 11, 16, 7, 12. Now add a con-

stant of 3 to each raw score and recalculate the mean. Finally, subtract a constant of 7 from each of the original raw scores and recalculate the mean. What do you conclude about the effect on the mean of changing each raw score by a fixed amount?

9. Compute the mean of the following sample of original scores: 13, 18, 16, 11, 11, 23, 14, 24, 8, 22. Now multiply each of the raw scores by 3 and recalculate the mean. Finally, divide each of the original raw scores by 2 and recalculate the mean. What do you conclude about the effects on the mean of multiplying or dividing raw scores by a constant?

10. An educational psychologist investigated the effects of caffeine on high school students' ability to solve anagram (scrambled word) problems. Each of four randomly assigned groups of students ingested a different amount of caffeine one-half hour prior to attempting the problems in timed trials. Group 1 received the lowest dose, and groups 2, 3, and 4 ingested progressively larger amounts. The results, in terms of the number of problems solved, are given in the accompanying table. Compute the mean for each condition and construct a line graph of the means, using proper graphing techniques. On the basis of the graph, what conclusion do you draw about the effect of caffeine on performance in this task?

Caffeine Dose

500 mg	1000 mg	1500 mg	2000 mg
3	15	11	5
3	5	15	3
7	10	5	5
7	8	9	5

11. For each of the three samples of data listed here, demonstrate, through appropriate calculations, that the sum of the squared deviations from the mean is less than the sum of the squared deviations from either the median or the mode.

X_1: 24, 18, 22, 14, 22, 10, 24, 22, 20, 24

X_2: 10, 14, 7, 21, 15, 17, 11, 16, 7, 12

X_3: 13, 18, 16, 11, 11, 23, 14, 24, 8, 22

12. The sample mean, $\overline{X}$, is often used to estimate the mean of an infinite population. Why does the population mean have to be estimated in this case instead of calculated? What is the rationale or justification behind using $\overline{X}$, rather than some other index, to estimate μ? Be as specific and detailed as possible.

13. If there are 167 values in a distribution, what is the rank of the median value? If a distribution has 1002 scores in it, what is the rank of the median value? Describe the general procedure for determining the rank of the median value.

14. Eight trainees for electronic circuit assembly jobs report the following numbers of errors in their first training module: 20, 22, 34, 35, 23, 39, 37, 23. Suppose that a final check of the errors reveals that the 39 in the data set was actually a transposition error and should have been 93. How does this correction in the data affect the mode, median, and mean, respectively? On the basis of this example, what conclusion do you draw about the relative stability of the three measures of central tendency as descriptions of the typical score in a distribution?

15. Twenty disadvantaged orphans of preschool age were adopted into well-to-do homes. The children's IQs were measured at the time of the adoption and again 3 years later. The pretest and posttest scores are shown in the table. Compute (a) the mean of the pretest scores, (b) the mean of the posttest scores, (c) the posttest – pretest difference for each child, and (d) the mean of the posttest – pretest differences. Is the mean of the difference scores equal to, less than, or greater than the difference between the posttest and pretest means?

Child	Pretest	Posttest
1	102	102
2	87	107
3	94	100
4	85	80
5	100	96
6	70	110
7	80	92
8	64	79
9	78	98
10	92	90
11	54	88
12	97	103
13	89	88
14	84	96
15	77	92
16	87	105
17	94	112
18	90	84
19	90	94
20	88	107

16. Briefly define or describe: variation, dispersion, range, interquartile range, semi-interquartile range, variance, sum of squares, σ^2, and standard deviation.

17. Verify through appropriate calculations that the following two sets of data have identical means but different variances, ranges, and interquartile ranges.

    X_1: 99, 84, 127, 99, 117, 87, 89, 107, 109, 108, 86, 100
    X_2: 81, 106, 84, 127, 80, 100, 116, 91, 112, 113, 96, 106

18. Add 10 to each of the scores in X_1 and X_2 of Question 17. Recalculate the mean and variance of each group of scores. Based on your results, how do you describe the effect that adding a constant to scores has on the mean and the variance, respectively?

19. Compute the variance and standard deviation for the following sample of raw scores: 25, 21, 38, 20, 30, 31, 34, 33. Now add 2 to each raw score and recalculate the variance and standard deviation. How does adding a constant value to each raw score affect the variance and standard deviation, respectively?

20. Compute the variance and standard deviation for the raw scores listed in Question 19. Now multiply each raw score by 2 and recalculate S^2 and S. How does multiplying raw scores by a constant affect their variance and standard deviation, respectively?

21. A sociologist studying human migration habits surveys ten urbanized adult Navajos about the number of different cities they've lived in. Her data are 18, 23, 6, 13, 20, 9, 14, 13, 16, 11. Calculate the mean, median, and mode for these data.

22. Compute the range, variance, and standard deviation of the data in Question 21. By how many standard deviations does the most extreme datum differ from the mean of the distribution? (*Hint*: The most extreme datum is the one with the largest absolute deviation score.)

23. A sociologist is studying the lifestyles of inner city children who are "drug runners." The sociologist discovers that his 16 subjects began their hazardous work at the following ages: 10, 12, 5, 14, 7, 9, 9, 11, 10, 9, 8, 8, 15, 9, 7, 11. Calculate the mean, median, and mode of this data set. Judging from the relative sizes of the three measures of central tendency, does the distribution of scores tend to be positively or negatively skewed?

24. Compute the range, variance, and standard deviation of the data in Question 23. By how many standard deviations does the most extreme datum differ from the mean of the distribution? (*Hint*: The most extreme datum is the one with the largest absolute deviation score.)

25. For the distribution shown in Question 23, what percent of the scores lie within two standard deviations of the mean? What percent lie within three standard deviations of the mean?

CHAPTER 5   RELATIVE MEASURES AND THE NORMAL CURVE

Suppose a college friend of yours named Glenda tells you that she scored 88 points on her final exam in behavioral statistics and 47 points on her humanities final. Should you conclude that Glenda knows more about statistics than about the humanities? Your first impulse might be to say yes. But since you are studying statistics, you might respond in a more cautious and reasoned fashion: Eighty-eight points out of how many possible points? A score of 47 points relative to what?

Relative to what? is a pertinent question because comparisons and interpretations of statistical results are almost always relative matters. Few statistical quantities or indexes are meaningful by themselves. Most of their usefulness derives from comparing them to expected standards or statistical **norms**—putting them in context, if you will.

Norm: An average value or some other standard statistical characteristic of a population of observations.

This chapter will introduce you to **relative measures**, defined as *mathematical transformations of raw scores to standard number scales.* One advantage of relative measures is that by converting raw scores from different distributions to the same standard number scale, we can more accurately compare values from different distributions. Later chapters will also show how certain relative measures enable us to compare the means of different distributions with one another.

COMMON RELATIVE MEASURES

From previous chapters in this book as well as your past experience with simple arithmetic, you are familiar with two kinds of relative measures: proportions and percents. In Chapter 3 you read that proportions and percents are *relative* frequencies.

Proportion

A **proportion** is *a ratio of the number of target events to the total number of events in a set.* A proportion is calculated by dividing the number of times an outcome of interest (i.e., target event) occurs by the total number of outcomes (i.e., target events + nontarget events).

For example, you could ask what proportions of possible points Glenda got in her statistics class and humanities class. To answer this question, you need the following additional information: There were 100 possible points on the statistics exam and 50 possible points on the humanities exam. Now you can convert Glenda's raw scores to relative measures—that is, place the scores on a standard number scale for purposes of comparison. Since Glenda's raw score in statistics was 88,

$$\text{Proportion correct in statistics} = \frac{\text{target events}}{\text{possible events}} = \frac{88}{100} = .88$$

And since Glenda's raw score in humanities was 47,

$$\text{Proportion correct in humanities} = \frac{\text{target events}}{\text{possible events}} = \frac{47}{50} = .94$$

Now both test outcomes have been converted to a standard number scale that always ranges between .00 and 1.00. And it appears that Glenda knew relatively more about the specific test content in humanities (proportion = .94) than in statistics (proportion = .88). (But did she? As you will see shortly, there are other ways of comparing two test scores that might lead us to a different conclusion.)

Proportions are used frequently in statistics, and they are often abbreviated with **lowercase** *p*. I will follow this practice throughout this chapter.

> **NOTE:** It is conventional to use the letter *p* to stand for a proportion. The letter *p* is also used to represent probability. This double usage of *p* is appropriate because in the field of statistics, probability is best interpreted as a proportion.

Percent

Making a proportion a percent   A **percent** is simply *a proportion multiplied by 100*. Converting raw scores to percents places them on a standard number scale where 0% is the lowest possible figure and 100% is the highest possible. Glenda's percent correct in statistics was $(p)(100) = (.88)(100) = 88\%$; in humanities her percent correct was $(p)(100) = (.94)(100) = 94\%$.

Is percent the best way to go?   Using the percent of possible points as a relative measure is only one way of comparing values from different distributions. Often it is not the best way to make such comparisons. For example, what if Glenda's statistics exam was a lot harder that her humanities exam? Suppose half of Glenda's class scored 94% or above in humanities, but only a couple of students scored above 88% in statistics. Such facts might cause you to change your assessment of Glenda's relative knowledge in the two disciplines.

Thus, a second way to calculate relative measures is to convert raw scores to what are called **measures of relative standing** in a distribution. Measures of relative standing are *mathematical transformations that locate a raw score's rank or position in a sample or population.* In effect, measures of relative standing compare a particular person's score to other scores in the distribution rather than to an external standard such as "percent of possible points." Because different measuring procedures (for example, academic tests) have different levels of "difficulty," a particular percent correct figure does not mean the same thing from one set of scores to another. Accordingly, statisticians generally prefer to use measures of relative standing, rather than percent correct, in comparing values from different distributions.

We will study two measures of relative standing in this chapter: percentile ranks and standard scores.

PERCENTILES AND PERCENTILE RANKS

A **percentile** is defined as *a value below which a certain percent of a distribution lies.*[1] A **percentile rank** is *the percent of a distribution lying below a certain value.*[2] In all uses of these concepts, percentile helps define percentile rank, and vice versa. When one has been determined, so has the other.

For example, the median of a distribution is the 50th percentile. It is the value below which 50% of the distribution lies. Therefore, we can say that a person in a population whose raw score equals the median has a percentile rank of 50; that is, 50% of the values in that distribution are lower than that person's score.

How to Figure a Percentile Rank from Cumulative Percents

 Let's apply these measures of relative standing to Glenda. Tables 5.1 and 5.2 display the ungrouped cumulative percent distributions of final exam scores in Glenda's statistics and humanities classes. Recall that a cumulative percent distribution (also known as a cumulative relative frequency distribution) shows the percent of observations that lie below the true upper limit of each score in a distribution. So this kind of distribution is well suited for determining percentile ranks.

Note the following points:

1. There were 200 people in Glenda's statistics class and 40 people in her humanities class.

---

[1] In contrast, some statistics textbooks define a *percentile* as "a value *at or* below which a particular percent of the distribution lies." The latter definition is technically more accurate than the one used here, but it is more difficult to grasp and discuss. Therefore, I decided to use the simpler definition in order to more easily introduce the basic idea. Moreover, in most practical applications, either definition adequately serves the intended purpose.

[2] See footnote 1.

2. Glenda's raw score is identified in each table.

There are two steps in computing a percentile rank from data in a cumulative percent distribution.

Step 1: Determine the percent of scores below the target value
Column 5 of Table 5.1 shows that 96.5% of the statistics scores were below the true upper limit (88.5) of Glenda's raw score (88). But note that Glenda's percentile rank is the *percent of people who scored below her*. So drop down one row in column 5 of the table. The cumulative percent there shows that 94.5%, or approximately 95%, of the test scores are lower than Glenda's 88.

Step 2: Add one-half of the percent of scores *at* the target value
Because of the theory underlying the calculation of percentile ranks in frequency distributions, *we must also assume that one-half of the*

Table 5.1

UNGROUPED CUMULATIVE PERCENT DISTRIBUTION OF FINAL EXAM SCORES IN A BEHAVIORAL STATISTICS CLASS (N = 200)	(1) RAW SCORES	(2) TRUE UPPER LIMIT	(3) FREQUENCY	(4) CUMULATIVE FREQUENCY	(5) CUMULATIVE PERCENT
	92	92.5	1	200	100.0
	91	91.5	1	199	99.5
	90	90.5	2	198	99.0
	89	89.5	3	196	98.0
	88 ←Glenda 88.5		4	193	96.5
	87	87.5	6	189	94.5
	86	86.5	9	183	91.5
	85	85.5	12	174	87.0
	84	84.5	15	162	81.0
	83	83.5	18	147	73.5
	82	82.5	19	129	64.5
	81	81.5	20	110	55.0
	80	80.5	19	90	45.0
	79	79.5	18	71	35.5
	78	78.5	15	53	26.5
	77	77.5	12	38	19.0
	76	76.5	9	26	13.0
	75	75.5	6	17	8.5
	74	74.5	4	11	5.5
	73	73.5	3	7	3.5
	72	72.5	2	4	2.0
	71	71.5	1	2	1.0
	70	70.5	1	1	0.5

Sample mean: $\overline{X} = 81.00$ Sample standard deviation: $S = 4.0$

scores that equal Glenda's are actually below hers.[3] So two additional values (half of the four 88's) are considered to be below Glenda's score. These two scores amount to 1% of the distribution. How's that? Since there were $N = 200$ students in Glenda's statistics class, 2 scores make up $\frac{2}{200} \times 100$, or 1%, of the data set. Now, to arrive at Glenda's exact percentile rank, we must add this 1% to the 94.5% of the values that fall below Glenda's test score. Therefore, Glenda's percentile rank is 94.5 + 1 = 95.5, or approximately 96. Accordingly, a score of 88 is approximately the 96th percentile.

In the same fashion, Table 5.2 indicates that Glenda's percentile rank in humanities is 90 (85% + 5%). Her raw score of 47, then, is the 90th percentile. The 5% added to the cumulative percent below Glenda's score in this case represents one-half of the humanities students with a score of 47. Since there were $N = 40$ students in the humanities class, 2 scores amount to $\frac{2}{40} \times 100$, or 5%, of the observations.

Now, when you consider Glenda's relative standing in the two groups of students, you would conclude that Glenda did comparatively better on the statistics test, where she topped 96% of her peers, than on the humanities test, where she exceeded 90% of her fellow students.

Table 5.2

	(1)	(2) TRUE UPPER	(3)	(4) CUMULATIVE	(5) CUMULATIVE
UNGROUPED CUMULATIVE PERCENT DISTRIBUTION OF FINAL EXAM SCORES IN A HUMANITIES CLASS ($N = 40$)	RAW SCORES	LIMIT	FREQUENCY	FREQUENCY	PERCENT
	48	48.5	2	40	100
	47 ←Glenda	47.5	4	38	95
	46	46.5	6	34	85
	45	45.5	16	28	70
	44	44.5	6	12	30
	43	43.5	4	6	15
	42	42.5	2	2	5

Sample mean: $\overline{X} = 45.00$ Sample standard deviation: $S = 1.41$

---

[3] The reason for this step is that percentiles assume that the trait or ability measured by the test is on a continuous, unbroken scale, even though scores on the test are expressed in discrete whole numbers. Thus, if four people have a score of 88 on the test, it is assumed that these individuals actually possess minutely different levels of the ability being measured and that half of them are at or slightly below the whole-number score of 88, while the others are at or slightly above that exact point on the scale.

General Applications of Percentile Ranks

The percentile rank is the relative measure that is most widely used in the interpretation of people's performance on standardized tests, such as IQ tests, the Scholastic Aptitude Test, the Graduate Record Exam (GRE), and many others. Every year thousands of college seniors take the GRE to gain admission to graduate school. That test measures academic aptitude in three general areas: verbal ability, quantitative ability, and analytical ability. The actual percentile-rank norms of the GRE for the years 1983 to 1986 are shown in Table 5.3. The three aptitudes measured have different means and standard deviations, and thus they are on different numerical scales. Nonetheless, note how the use of percentile ranks allows the test taker to compare his or her relative strengths across all three aptitudes.

THE STANDARD SCORE

Another measure of relative standing that is extremely useful in statistical analysis is called the *standard* score, or *z* score. Standard scores are used not only to assess the relative standings of individual scores in a distribution but also to demonstrate (or actually carry out) a variety of more advanced statistical procedures. Consequently, you will encounter it repeatedly in later sections of this book. Standard scores are based on the standard deviation.

The Standard Deviation As a Unit of Measurement

In the last chapter, you were introduced to a versatile measure of raw score variation called the *standard deviation*. As Figure 5.1 suggests, the standard deviation of a distribution can be used as a unit of measurement along the horizontal axis (i.e., abscissa) of a statistical curve. In Figure 5.1 (p. 104), the distribution is normal, and the parameter symbol for the standard deviation (σ) indicates that this is a population distribution.

> **CONCEPT RECAP**
>
> The population standard deviation, σ, is the square root of the population variance, σ^2. Each of these parameters is a measure of the extent to which values in a distribution tend to vary from one another and hence from the mean. In the case of a finite population, the standard deviation is calculated this way:
>
> $$\sigma = \sqrt{\frac{\Sigma(X - \mu)^2}{N}}$$
>
> If the population is considered infinite, then σ and σ^2 cannot be calculated; they must be estimated. (See Chapter 10 for details.)

Table 5.3

GRADUATE RECORD EXAMINATION NORMATIVE DATA: PERCENTILE RANKS USED ON SCORE REPORTS (1983–1986). (Source: GRE materials selected from Guide to the Use of the Graduate Record Examinations Program, Educational Testing Service, (1994). Reprinted by permission of Educational Testing Service. Permission to reprint GRE materials does not constitute review or endorsement by Educational Testing Service of this publication as a whole or of any other testing information it may contain.)	SCALED SCORE	PERCENT OF EXAMINEES SCORING LOWER THAN SELECTED SCALED SCORES		
		VERBAL ABILITY	QUANTITATIVE ABILITY	ANALYTICAL ABILITY
	800	99	98	99
	780	99	97	98
	760	99	94	97
	740	98	90	95
	720	96	87	93
	700	95	83	91
	680	93	79	88
	660	91	74	84
	640	88	71	81
	620	85	65	77
	600	82	61	72
	580	78	56	67
	560	73	52	61
	540	68	46	56
	520	63	42	50
	500	57	37	44
	480	52	32	38
	460	45	27	33
	440	40	23	27
	420	34	19	23
	400	28	16	18
	380	24	13	15
	360	18	10	11
	340	15	8	9
	320	11	6	6
	300	8	4	4
	280	6	3	3
	260	3	2	2
	240	2	1	1
	220	1	1	1
Mean:		475	546	516
Standard deviation:		127	140	129
Number of examinees	816,621			
Percent women	49			
Percent men	51			

The first thing to notice about Figure 5.1 is that particular points along the abscissa of the normal curve can be described as follows:

One standard deviation above the mean: $+1\sigma$

One standard deviation below the mean: -1σ

Two standard deviations above the mean: $+2\sigma$

And so on

The standard deviation serves not only as an index of variation but also as *a constant unit of distance along the abscissa* of a statistical curve. Because of this, it is both meaningful and useful to describe particular values as being "1 standard deviation below the mean," "2.5 standard deviations away from the mean," and so on.

The second thing to notice is that *if a distribution is normal, fixed percentages of the curve always lie between the standard deviation markers.*

34.13% of the area under the curve lies between $+1\sigma$ and μ.

34.13% of the area under the curve lies between -1σ and μ.

Therefore, 68.26% of the curve lies between -1σ and $+1\sigma$.

Through a similar analysis:

95.44% of a normal distribution lies between -2σ and $+2\sigma$.

99.74% of a normal distribution lies between -3σ and $+3\sigma$.

The constant relationships between the standard deviation and areas under the normal curve always hold true *if (and only if) the distribution in question is normal.* As you will see, the existence of this relationship is of tremendous value in numerous statistical applications.

Figure 5.1 *Percent of Area under the Normal Curve As a Function of Standard Deviation Units*

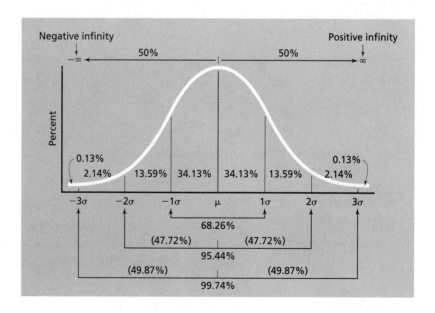

From μ to Infinity

At this point it is natural to ask, How many standard deviation units are needed to include 100% of the normal curve? The answer is an infinite number. As is suggested by the top portion of Figure 5.1, exactly 50% of the curve's area is below the mean, and exactly 50% is above the mean. (A certain point, such as μ, takes up no space in a theoretical distribution.) But since the tails of the normal curve are assumed to extend outward to both positive and negative infinity, no number of standard deviation units—no matter how big that number might be—can completely include all of the curve. For practical purposes, however, –3σ and +3σ are usually treated as the outer boundaries of a normal distribution.

The *z* (Standard) Score and σ

Statisticians have formalized a way to express raw scores in terms of how many standard deviations they are away from their mean. The device for doing this is called a **standard score**, or **z score**, defined as *a deviation score expressed in standard deviation units.* You probably remember that, for population values, a deviation score is $X - \mu$. To express this deviation score in standard deviation units, you simply divide it by the standard deviation (σ). Thus, a *z* score (or standard score) for population values is computed as:

$$z = \frac{\overset{\textcircled{1}}{\overset{\downarrow}{X - \mu}}}{\underset{\underset{\textcircled{2}}{\uparrow}}{\sigma}} \qquad (5.1)$$

FORMULA GUIDE

① Subtract the mean from a raw score to obtain the deviation score.

② Divide the deviation score by the standard deviation.

Since a *z* score always represents the number of standard deviation units that a particular value deviates from its mean, a value that converts to a *z* score of –2 is 2 standard deviations *below* the mean, a value that converts to a *z* score of +1.9 is 1.9 standard deviations *above* the mean, and so on. Since *z* scores are equivalent to standard deviation units, *z* has the same relationship as σ to areas under the normal curve. Figure 5.2 illustrates this fact. *A normal distribution of z scores,* such as the one shown, is referred to as the **standard normal distribution** (that is, a normal distribution of standard scores); the segments within it are laid out in terms of various proportions of its total area. I will describe some of its uses in this chapter.

Important Points About Working with *z* Scores
- Even though some parts of this chapter show you how to work with *z* scores in a normal distribution, *z* scores can be computed for any distribution of data that are on an interval or ratio scale of measurement, *whether the distribution is normal or nonnormal.*

- Even when a distribution of *z* scores is not normal, the *z* scores still express deviation scores in standard deviation units.

- Converting a nonnormal distribution of raw scores to *z* scores does not make the distribution normal. In fact, converting raw scores to *z* scores does not change the shape of the distribution in any way (see the discussion below).

---

z scores in a sample   Standard scores are commonly computed for values in a sample distribution. But since samples are not normally distributed, the z scores derived from them do not have the exact relationship to area of the distribution shown in Figure 5.2 (although z scores from some very large samples might approximate that relationship). The formula for sample z scores is:

Figure 5.2 *The Standard Normal Distribution: Proportion of Area under the Normal Curve As a Function of z Scores*

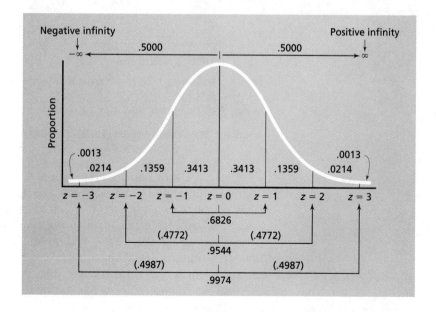

$$z = \frac{X - \overline{X}}{S} \tag{5.2}$$

NOTE: The S is the sample standard deviation. It is computed using:

$$S = \sqrt{\frac{\Sigma(X - \overline{X})^2}{N}}$$

(Refer to Chapter 4.)

Comparing values across distributions   For practice, let's compute Glenda's sample z score on her statistics exam, where her score (X) is 88, the mean $(\overline{X})$ is 81, and the standard deviation (S) is approximately 4:

$$z = \frac{X - \overline{X}}{S}$$

$$= \frac{88 - 81}{4}$$

$$= \frac{7}{4}$$

$$= 1.75$$

So Glenda's score is 1.75 standard deviations above the mean of her class.

How does she compare in relative standing on her humanities exam, where her raw score (X) is 47, the mean $(\overline{X})$ is 45, and the standard deviation (S) is 1.41?

The answer is given by

$$z = \frac{47 - 45}{1.41}$$

$$= \frac{2}{1.41}$$

$$= 1.42$$

In terms of standard deviations above the mean, Glenda's performance was slightly lower in humanities than in statistics.

Properties of z Scores

Note how easy it was in the last example to compare values for two different distributions when those values are converted to standard scores. Values may be compared across all standard-score distributions because the conversion of raw scores to z scores gives all of the distributions involved the same mean and variance.

The mean of a distribution of z scores is always 0 (because the z scores sum to 0), and *the standard deviation of the distribution is always 1.* Since the variance is the square of the standard deviation, it too equals 1 in any z-score distribution. These statements are always true, regardless of whether the distribution in question is normal. You can find brief algebraic proofs of these points in Hays (1988), but here's a logical approach to the matter that should be easy to digest.

The conversion of raw (X) scores to z scores is a *point-for-point* "remapping" of the data. This means that *the transformation from X scores to z scores does not in any way change the relationship between scores within the distribution.* An observation that was equal to the mean in the raw-score distribution is equal to the mean in the z-score distribution; an observation that was exactly one standard deviation above the mean as a raw score is exactly one standard deviation above the mean as a z score; and so on.

There are three implications of this point-for-point remapping idea:

1. Converting raw scores to z scores in no way alters the shape of the distribution. A skewed raw-score distribution becomes a skewed z-score distribution. And it is skewed—point for point— in exactly the same way as the raw-score distribution. *It is important to remember that transforming X scores to z scores does not make a nonnormal distribution normal.*
2. It follows from the first implication that a raw score that is equal to its mean converts exactly to the z-score mean. Now consider that when $X = \mu$, the corresponding z score is $(\mu - \mu)/\sigma = 0/\sigma = 0$. Therefore, a raw score mean of μ converts to a z-score mean of 0.
3. With the same logic, a raw score that is one standard deviation above (or below) its mean converts to a z score that is one standard deviation away from its mean. If X is one standard deviation above μ, then $X = \mu + \sigma$. So in this situation, $(X - \mu)$ actually is $([\mu + \sigma] - \mu)$. The corresponding z score is $([\mu + \sigma] - \mu)/\sigma$, or σ/σ, or 1. Hence, a raw-score standard deviation of σ converts to a z-score standard deviation of 1.

Transformed Scores

One of the most important applications of the z score is its use to transform raw scores from one number scale to a second one that has a different mean and standard deviation. Such transformations often are carried out to directly compare values from two different distributions. Alternatively, the researcher who carries out such a transformation might simply want the data distribution to have a particular mean or standard deviation in order to facilitate the interpretation of individual scores.

Consider this example: When the Graduate Record Examination (GRE) was first developed, its creators intended each of the scales to have a mean of 500 and a standard deviation of 100. Yet, if you recheck Table 5.3, you'll see that as of 1986 the mean of the GRE-Verbal scale was 475, and its standard deviation was 127. To transform GRE-Verbal scores to a new scale with the desired mean and standard deviation, the following formula can be applied:

$$①\ ②$$
$$\downarrow\ \downarrow$$

Transformed score = (new standard deviation)(z) + new mean

(5.3)

FORMULA GUIDE

① Multiply any *z* score on the old scale times the new standard deviation.

② Add the product obtained in step 1 to the new mean.

In this particular example, the formula is expressed as:

Transformed score = 100z + 500

To transform the highest scaled score ($X = 800$) in Table 5.3, you first convert it to a z score using the original mean (475) and standard deviation (127):

$$z = \frac{800 - 475}{127} = \frac{375}{127} = 2.56$$

Then apply formula (5.3), using the new mean and standard deviation:

Transformed score = (100)(2.56) + 500

= 256 + 500

= 756

This result means, very simply, that if the GRE-Verbal test scores had a mean of 500 and a standard deviation of 100 (instead of 475 and 127, respectively), a student who achieved a score of 800 on the present number scale would have a score of 756 on the new number scale. Table 5.4 shows the results of converting all of the listed "scaled scores" on the verbal ability test to the new distribution with $\mu = 500$ and $\sigma = 100$.

Note that score transformations of this type do not affect either the shape of the distribution or the percentile ranks of the values within the distribution.

Over the years, psychologists and educators have come to prefer a particular version of the transformed score, called a **T score**, where

$$T = 10z + 50 \tag{5.4}$$

You can see that this formula converts z values to a distribution that has a mean of 50 and a standard deviation of 10.

Table 5.4

ORIGINAL AND TRANSFORMED SCORES ON THE GRE VERBAL ABILITY SCALE

ORIGINAL SCORES	TRANSFORMED SCORES
800	756
780	740
760	724
740	709
720	693
700	677
680	661
660	646
640	630
620	614
600	599
580	583
560	567
540	551
520	535
500	520
480	504
460	488
440	472
420	457
400	441
380	426
360	409
340	394
320	378
300	362
280	347
260	331
240	315
220	299

	(Original)	(New)
Mean:	475	500
Standard deviation:	127	100

You may be interested to know that, for most intelligence tests, the Intelligence Quotient (IQ) is a transformed score based on

$$IQ = 15z + 100$$

This formula permits IQs derived from different intelligence tests to be conveniently compared with one another, regardless of differences in the raw-score means and standard deviations among the various tests in use.

USING THE STANDARD NORMAL DISTRIBUTION

Recall that the standard normal distribution is a normal distribution of z scores with a mean of 0 and a standard deviation (and variance) equal to 1.[4] Also recall that the z-score units along the abscissa of the standard normal distribution have a constant relationship to a proportion of the area under the normal curve. Take a second look at Figure 5.2 to help you visualize that relationship.

The constant relationship of the z statistic to areas under the curve is an important concept for you to both remember and investigate further. It is the main theoretical model for the procedures of inferential statistics covered in later sections of this book. In the remainder of this chapter, you will learn how to use the standard normal distribution to:

- Find the percentage of the area between any two values along the abscissa of the normal curve
- Find the percentile rank of any score in a normal distribution
- Estimate the percentile rank of any raw score in a distribution of actual data that approximates a normal distribution

One Distribution, an Infinity of Values

The standard normal distribution in Figure 5.2 shows only seven z values, which should be thought of as benchmarks. Actually the standard normal distribution is an infinite distribution; it has so many values in it that it is impossible to count them all. Furthermore, the abscissa of this distribution represents a **continuous variable**. *The variable can be measured or expressed in infinitely smaller units—integers, tenths, hundredths, thousandths, and so on—to any degree of precision.*

The upshot of these statements is that, given an appropriately calibrated table of z values, you can determine the proportion of

---

[4] Note that the mean of 0 and variance of 1 are not unique to the standard normal distribution, but rather characterize all distributions of z scores. What does set the standard normal distribution apart from other z-score distributions are its "normal" shape and "normal" properties (see the discussion in this section).

the normal curve that lies between any two values in the distribution—not just between $z = -1$ and $z = +1$ but between, say, $z = -1.96$ and $z = +2.58$. Table Z of Appendix A is used for this very purpose.

Understanding Table Z

Table Z is so central to an understanding of statistics that it has been reproduced in Box 5.1 for your immediate use.

General table structure   Observe that each page of Table Z contains three "blocks" and that each block contains three columns of figures: columns 1, 2, and 3.

Columns in Table Z   Within each block of Table Z, the leftmost column contains z-score values. The range of these values from the beginning to the end of the table is $z = 0.00$ to $z = 4.00$ (i.e., four standard deviations above or below the mean).

Column 2 of each block in Table Z shows the proportion of the area of the normal curve that lies between a z score and the mean of the distribution. For example, find a z score of 1.00 in Block C of the table. Notice that the number in column 2 (next to a z score of 1.00) is .3413. This means that 34.13% of the normal curve lies between a z score of 1 and the mean ($z = 0$)—something you probably already knew from examining earlier figures that show the standard normal distribution.

> NOTE: In Table Z, the area under the curve is given in terms of proportions. To convert these proportions to percents, simply move the decimal point two positions to the right. This amounts to multiplying each proportion by 100.

Column 3 in each block of Table Z gives the proportion of the area beyond a particular z value—that is, between the z value and infinity. For a z score of 1.00 (see Block C, column 3), .1587, or 15.87%, of the area lies beyond a z score of 1.

Important points   At the top of Table Z, the shaded areas in the schematic pictures of the normal curve tell you what general area of the distribution the proportions in that column pertain to. This relationship is summarized in Figure 5.3. Note in the schematic pictures:

- Since the normal curve is perfectly symmetrical, the proportion of the area that applies to a positive z score also applies to a negative z score of equal magnitude. For example, the proportion of the area between $z = -1.00$ and the mean is .3413, the same as that associated with $z = +1.00$.
- The total area under the normal curve is 1.0000 (or 100%), with each half of the distribution equaling .5000 (or 50%) of the area.

Box 5.1 Table Z

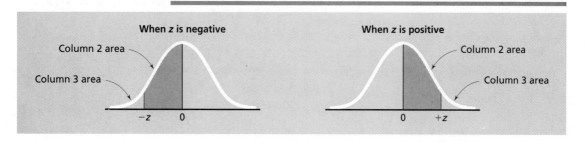

When z is negative — Column 2 area, Column 3 area, −z 0

When z is positive — Column 2 area, Column 3 area, 0 +z

	BLOCK A			BLOCK B			BLOCK C	
1	**2** AREA BETWEEN z AND THE MEAN	**3** AREA BEYOND z	**1**	**2** AREA BETWEEN z AND THE MEAN	**3** AREA BEYOND z	**1**	**2** AREA BETWEEN z AND THE MEAN	**3** AREA BEYOND z
z			z			z		
0.00	.0000	.5000	0.40	.1554	.3446	0.80	.2881	.2119
0.01	.0040	.4960	0.41	.1591	.3409	0.81	.2910	.2090
0.02	.0080	.4920	0.42	.1628	.3372	0.82	.2939	.2061
0.03	.0120	.4880	0.43	.1664	.3336	0.83	.2967	.2033
0.04	.0160	.4840	0.44	.1700	.3300	0.84	.2995	.2005
0.05	.0199	.4801	0.45	.1736	.3264	0.85	.3023	.1977
0.06	.0239	.4761	0.46	.1772	.3228	0.86	.3051	.1949
0.07	.0279	.4721	0.47	.1808	.3192	0.87	.3078	.1922
0.08	.0319	.4681	0.48	.1844	.3156	0.88	.3106	.1894
0.09	.0359	.4641	0.49	.1879	.3121	0.89	.3133	.1867
0.10	.0398	.4602	0.50	.1915	.3085	0.90	.3159	.1841
0.11	.0438	.4562	0.51	.1950	.3050	0.91	.3186	.1814
0.12	.0478	.4522	0.52	.1985	.3015	0.92	.3212	.1788
0.13	.0517	.4483	0.53	.2019	.2981	0.93	.3238	.1762
0.14	.0557	.4443	0.54	.2054	.2946	0.94	.3264	.1736
0.15	.0596	.4404	0.55	.2088	.2912	0.95	.3289	.1711
0.16	.0636	.4364	0.56	.2123	.2877	0.96	.3315	.1685
0.17	.0675	.4325	0.57	.2157	.2843	0.97	.3340	.1660
0.18	.0714	.4286	0.58	.2190	.2810	0.98	.3365	.1635
0.19	.0753	.4247	0.59	.2224	.2776	0.99	.3389	.1611
0.20	.0793	.4207	0.60	.2257	.2743	1.00	.3413	.1587
0.21	.0832	.4168	0.61	.2291	.2709	1.01	.3438	.1562
0.22	.0871	.4129	0.62	.2324	.2676	1.02	.3461	.1539
0.23	.0910	.4090	0.63	.2357	.2643	1.03	.3485	.1515
0.24	.0948	.4052	0.64	.2389	.2611	1.04	.3508	.1492
0.25	.0987	.4013	0.65	.2422	.2578	1.05	.3531	.1469
0.26	.1026	.3974	0.66	.2454	.2546	1.06	.3554	.1446
0.27	.1064	.3936	0.67	.2486	.2514	1.07	.3577	.1423
0.28	.1103	.3897	0.68	.2517	.2483	1.08	.3599	.1401
0.29	.1141	.3859	0.69	.2549	.2451	1.09	.3621	.1379
0.30	.1179	.3821	0.70	.2580	.2420	1.10	.3643	.1357
0.31	.1217	.3783	0.71	.2611	.2389	1.11	.3665	.1335
0.32	.1255	.3745	0.72	.2642	.2358	1.12	.3686	.1314
0.33	.1293	.3707	0.73	.2673	.2327	1.13	.3708	.1292
0.34	.1331	.3669	0.74	.2704	.2296	1.14	.3729	.1271
0.35	.1368	.3632	0.75	.2734	.2266	1.15	.3749	.1251
0.36	.1406	.3594	0.76	.2764	.2236	1.16	.3770	.1230
0.37	.1443	.3557	0.77	.2794	.2206	1.17	.3790	.1210
0.38	.1480	.3520	0.78	.2823	.2177	1.18	.3810	.1190
0.39	.1517	.3483	0.79	.2852	.2148	1.19	.3830	.1170

(continued)

Box 5.1 Table Z (cont.)

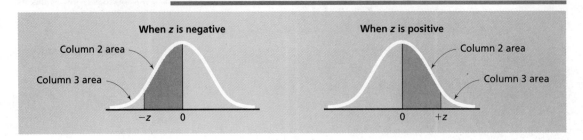

When *z* is negative — Column 2 area, Column 3 area, −*z*, 0

When *z* is positive — Column 2 area, Column 3 area, 0, +*z*

	BLOCK A			BLOCK B			BLOCK C	
1	**2**	**3**	**1**	**2**	**3**	**1**	**2**	**3**
	AREA BETWEEN z AND	**AREA BEYOND**		**AREA BETWEEN z AND**	**AREA BEYOND**		**AREA BETWEEN z AND**	**AREA BEYOND**
z	**THE MEAN**	**z**	**z**	**THE MEAN**	**z**	**z**	**THE MEAN**	**z**
1.20	.3849	.1151	1.60	.4452	.0548	2.00	.4772	.0228
1.21	.3869	.1131	1.61	.4463	.0537	2.01	.4778	.0222
1.22	.3888	.1112	1.62	.4474	.0526	2.02	.4783	.0217
1.23	.3907	.1093	1.63	.4484	.0516	2.03	.4788	.0212
1.24	.3925	.1075	1.64	.4495	.0505	2.04	.4793	.0207
1.25	.3944	.1056	1.65	.4505	.0495	2.05	.4798	.0202
1.26	.3962	.1038	1.66	.4515	.0485	2.06	.4803	.0197
1.27	.3980	.1020	1.67	.4525	.0475	2.07	.4808	.0192
1.28	.3997	.1003	1.68	.4535	.0465	2.08	.4812	.0188
1.29	.4015	.0985	1.69	.4545	.0455	2.09	.4817	.0183
1.30	.4032	.0968	1.70	.4554	.0446	2.10	.4821	.0179
1.31	.4049	.0951	1.71	.4564	.0436	2.11	.4826	.0174
1.32	.4066	.0934	1.72	.4573	.0427	2.12	.4830	.0170
1.33	.4082	.0918	1.73	.4582	.0418	2.13	.4834	.0166
1.34	.4099	.0901	1.74	.4591	.0409	2.14	.4838	.0162
1.35	.4115	.0885	1.75	.4599	.0401	2.15	.4842	.0158
1.36	.4131	.0869	1.76	.4608	.0392	2.16	.4846	.0154
1.37	.4147	.0853	1.77	.4616	.0384	2.17	.4850	.0150
1.38	.4162	.0838	1.78	.4625	.0375	2.18	.4854	.0146
1.39	.4177	.0823	1.79	.4633	.0367	2.19	.4857	.0143
1.40	.4192	.0808	1.80	.4641	.0359	2.20	.4861	.0139
1.41	.4207	.0793	1.81	.4649	.0351	2.21	.4864	.0136
1.42	.4222	.0778	1.82	.4656	.0344	2.22	.4868	.0132
1.43	.4236	.0764	1.83	.4664	.0336	2.23	.4871	.0129
1.44	.4251	.0749	1.84	.4671	.0329	2.24	.4875	.0125
1.45	.4265	.0735	1.85	.4678	.0322	2.25	.4878	.0122
1.46	.4279	.0721	1.86	.4686	.0314	2.26	.4881	.0119
1.47	.4292	.0708	1.87	.4693	.0307	2.27	.4884	.0116
1.48	.4306	.0694	1.88	.4699	.0301	2.28	.4887	.0113
1.49	.4319	.0681	1.89	.4706	.0294	2.29	.4890	.0110
1.50	.4332	.0668	1.90	.4713	.0287	2.30	.4893	.0107
1.51	.4345	.0655	1.91	.4719	.0281	2.31	.4896	.0104
1.52	.4357	.0643	1.92	.4726	.0274	2.32	.4898	.0102
1.53	.4370	.0630	1.93	.4732	.0268	2.33	.4901	.0099
1.54	.4382	.0618	1.94	.4738	.0262	2.34	.4904	.0096
1.55	.4394	.0606	1.95	.4744	.0256	2.35	.4906	.0094
1.56	.4406	.0594	1.96	.4750	.0250	2.36	.4909	.0091
1.57	.4418	.0582	1.97	.4756	.0244	2.37	.4911	.0089
1.58	.4429	.0571	1.98	.4761	.0239	2.38	.4913	.0087
1.59	.4441	.0559	1.99	.4767	.0233	2.39	.4916	.0084

Box 5.1 Table Z (cont.)

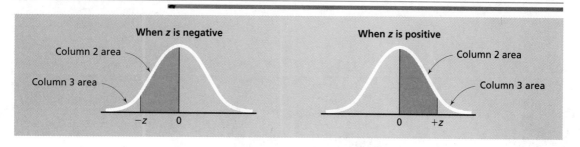

When z is negative
Column 2 area
Column 3 area
−z   0

When z is positive
Column 2 area
Column 3 area
0   +z

BLOCK A			BLOCK B			BLOCK C		
1	2	3	1	2	3	1	2	3
z	AREA BETWEEN z AND THE MEAN	AREA BEYOND z	z	AREA BETWEEN z AND THE MEAN	AREA BEYOND z	z	AREA BETWEEN z AND THE MEAN	AREA BEYOND z
2.40	.4918	.0082	2.72	.4967	.0033	3.04	.4988	.0012
2.41	.4920	.0080	2.73	.4968	.0032	3.05	.4989	.0011
2.42	.4922	.0078	2.74	.4969	.0031	3.06	.4989	.0011
2.43	.4925	.0075	2.75	.4970	.0030	3.07	.4989	.0011
2.44	.4927	.0073	2.76	.4971	.0029	3.08	.4990	.0010
2.45	.4929	.0071	2.77	.4972	.0028	3.09	.4990	.0010
2.46	.4931	.0069	2.78	.4973	.0027	3.10	.4990	.0010
2.47	.4932	.0068	2.79	.4974	.0026	3.11	.4991	.0009
2.48	.4934	.0066	2.80	.4974	.0026	3.12	.4991	.0009
2.49	.4936	.0064	2.81	.4975	.0025	3.13	.4991	.0009
2.50	.4938	.0062	2.82	.4976	.0024	3.14	.4992	.0008
2.51	.4940	.0060	2.83	.4977	.0023	3.15	.4992	.0008
2.52	.4941	.0059	2.84	.4977	.0023	3.16	.4992	.0008
2.53	.4943	.0057	2.85	.4978	.0022	3.17	.4992	.0008
2.54	.4945	.0055	2.86	.4979	.0021	3.18	.4993	.0007
2.55	.4946	.0054	2.87	.4979	.0021	3.19	.4993	.0007
2.56	.4948	.0052	2.88	.4980	.0020	3.20	.4993	.0007
2.57	.4949	.0051	2.89	.4981	.0019	3.21	.4993	.0007
2.58	.4951	.0049	2.90	.4981	.0019	3.22	.4994	.0006
2.59	.4952	.0048	2.91	.4982	.0018	3.23	.4994	.0006
2.60	.4953	.0047	2.92	.4982	.0018	3.24	.4994	.0006
2.61	.4955	.0045	2.93	.4983	.0017	3.25	.4994	.0006
2.62	.4956	.0044	2.94	.4984	.0016	3.30	.4995	.0005
2.63	.4957	.0043	2.95	.4984	.0016	3.35	.4996	.0004
2.64	.4959	.0041	2.96	.4985	.0015	3.40	.4997	.0003
2.65	.4960	.0040	2.97	.4985	.0015	3.45	.4997	.0003
2.66	.4961	.0039	2.98	.4986	.0014	3.50	.4998	.0002
2.67	.4962	.0038	2.99	.4986	.0014	3.60	.4998	.0002
2.68	.4963	.0037	3.00	.4987	.0013	3.70	.4999	.0001
2.69	.4964	.0036	3.01	.4987	.0013	3.80	.4999	.0001
2.70	.4965	.0035	3.02	.4987	.0013	3.90	.49995	.00005
2.71	.4966	.0034	3.03	.4988	.0012	4.00	.49997	.00003

- For a given z value, the sum of the column 2 proportion (area between the z score and the mean) and the column 3 proportion (area beyond the z score) always equals .5000 (or 50% of the curve).

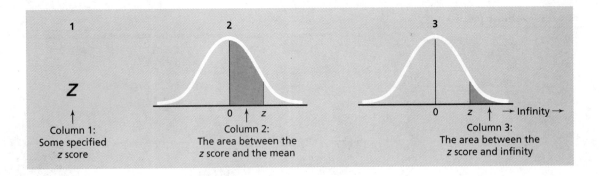

Figure 5.3 *How to Interpret Table Z*

Finding Areas under the Normal Curve

I will now acquaint you with several standard situations that involve using Table Z to find the proportion of a normal curve associated with particular z scores.

> Important Recommendation: To most easily and accurately find areas under the normal curve, *always draw a picture of the normal curve first.* Then shade in the approximate area of the curve that you want to work with.

Area between a z score and the mean To determine the proportion of the normal curve that lies between any z value and the mean, consult Table Z. Then follow these steps:

1. Locate the z score in question in column 1 of one of the "blocks" in Table Z.
2. Look to the immediate right of that z value; the proportion of the area between that z value and the mean is shown in column 2.

Panel (a) of Figure 5.4 illustrates such a problem using a z of +1.5. Once you locate a z value of 1.5 in Block A, column 1 of Table Z, moving to column 2 (immediately to the right of the z value) reveals that the proportion of the area between z = 1.5 and the mean (z = 0) is .4332. In other words, 43.32% of the distribution is included between those two points.

Figure 5.4 *Area Between a Particular z Score and the Mean*

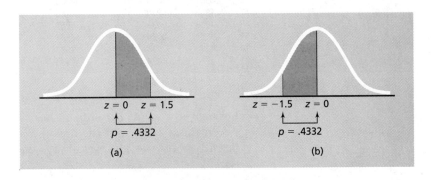

Panel (b) of Figure 5.4 illustrates that the proportion is the same for a z score of -1.5. Thus, negative z values are always associated with the same proportion of the area under the normal curve as positive z values of equal size. The left (negative) side of a normal curve is the "mirror image" of the right side in all respects.

Area beyond a particular z score   Panel (a) of Figure 5.5 addresses this question: What proportion of the area under a normal curve lies *beyond* a z value of -1.5—that is, between $z = -1.5$ and negative infinity? To answer that question, go to Block A, column 1 of Table Z and locate $z = -1.5$ (attach the negative sign mentally to make it -1.5, if you like). Then move to column 3, and you see that the proportion of the area *beyond* $z = -1.5$ is .0668. This means that 6.68% of the curve falls between $z = -1.5$ and negative infinity.

Panel (b) of Figure 5.5 illustrates a fact referred to earlier. Each half of a normal distribution contains exactly 50% of the area under the statistical curve. In this case, the sum of the area between $z = -1.5$ and the mean plus the area beyond $z = -1.5$ is equal to .4332 plus .0668, or .5000.

Summing positive and negative areas   What proportion of a normal distribution falls between a (positive) z value of 1.33 and a (negative) z value of -1? The solution appears in Figure 5.6. The procedure for solving this type of problem is given here:

1. Use column 2 of Table Z to determine the proportion of the area associated with the positive z value ($p = .4082$ in this example).
2. Use column 2 of Table Z to determine the negative z value ($p = .3413$).
3. Sum the two proportions: $.4082 + .3413 = .7495$.

This procedure works whenever the two z values in question are on opposite sides of the mean in a normal distribution.

Subtracting areas   When you wish to calculate the proportion of the area between any two z values on the *same side of the mean*, a

Figure 5.5 *Area Beyond a Particular z Value*

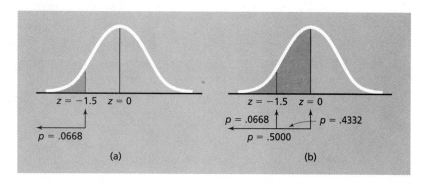

(a)   (b)

Figure 5.6 *Area Between Two z Scores on Opposite Sides of the Mean*

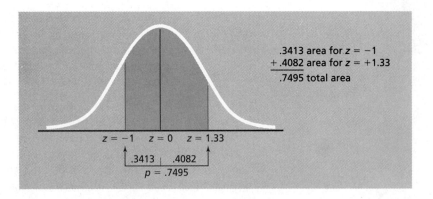

.3413 area for *z* = −1
+ .4082 area for *z* = +1.33
.7495 total area

z = −1   *z* = 0   *z* = 1.33

.3413 | .4082

p = .7495

simple subtraction procedure works. Suppose, for example, that your aim is to determine what proportion of a normal distribution is contained between *z* = −1.6 and *z* = −0.5. The correct procedure is given here and illustrated in Figure 5.7.

1. Use Table Z to find the area between the larger *z* value and the mean. In the example, the proportion of the curve between *z* = −1.6 and *z* = 0 (the mean) is .4452.
2. Use Table Z to find the area between the smaller *z* value and the mean. The relevant proportion (.1915) lies between *z* = −0.5 and *z* = 0.
3. Subtract the smaller proportion from the larger one: .4452 − .1915 = .2537. This shows that 25.37% of a normal distribution falls between *z* = −1.6 and *z* = −0.5.

For your reference, Box 5.2 provides several guidelines for finding areas under the normal curve.

Converting *z* Scores to Percentile Ranks

In a normal distribution it is easy to determine any *z* score's percentile rank (that is, the percent of a distribution lying below the *z* score).

Figure 5.7 *Area Between Two z Scores on the Same Side of the Mean*

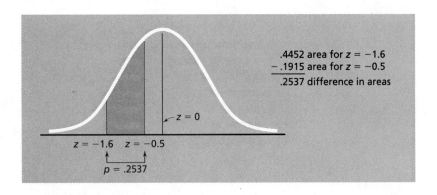

.4452 area for *z* = −1.6
− .1915 area for *z* = −0.5
.2537 difference in areas

z = 0

z = −1.6   *z* = −0.5

p = .2537

Box 5.2

GUIDELINES FOR FINDING THE PROPORTION OF THE AREA BETWEEN ANY TWO VALUES IN A NORMAL DISTRIBUTION

1. Always carefully note the precise values and signs (positive or negative) of the two z scores in the problem.

2. **Important:** *Draw a picture* of the standard normal distribution, place the two values at appropriate points along the X axis, and shade in the area to be determined (as in Figures 5.4 to 5.7).

3. Turn to Table Z in Appendix A of this book, and locate the z values in column 1.

4. To determine the proportion of the area between any two z values:

   a. If the z values are on opposite sides of the mean, add the two areas (example: Figure 5.6).

   b. If the z values are on the same side of the mean, subtract the smaller area from the larger one (example: Figure 5.7).

z score at the mean   If the z value is at the mean of the distribution (z = 0), then the corresponding percentile rank is 50 because 50% of a normal distribution lies below its mean.[5]

z scores above the mean   If the z value is above the mean, then its percentile rank is found in the following way:

1. Using column 2 of the appropriate block in Table Z, determine the proportion of the area between the z score and its mean.
2. Add .5000 (the area of the lower half of the curve) to the result of step 1. This sum is the proportion of values in the distribution that are below the z scores in question.
3. Multiply the result of step 2 by 100 to make the proportion a percent.

These steps are illustrated in Figure 5.8, which shows that a z value of 1.33 represents the 91st percentile.

z scores below the mean   If a z value is below the mean, the following steps apply:

1. Use column 3 of the appropriate block of Table Z to find the proportion of values that are below the z score in question (that is, the area between a negative z and negative infinity).
2. The z score's percentile rank equals 100 times the result of step 1.

---

[5] Bear in mind that this is true because a normal distribution is perfectly symmetrical; otherwise, either more or less than 50% of the distribution would fall below the mean.

Figure 5.8 *Converting a z Score to a Percentile Rank in a Normal Distribution*

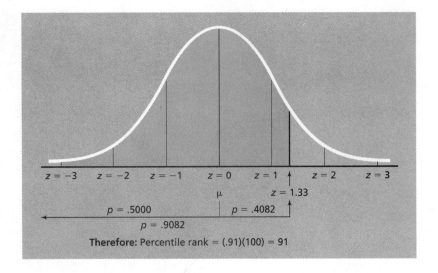

If you use this approach with a *z* score of –1, you get:

Percentile rank = (.1587)(100) = 15.87, or 16

Thus, a person who is 1 standard deviation below the mean in a normal distribution has a percentile rank of 16.

Estimating Percentile Ranks in Data Sets

Now let's leave the strictly theoretical realm and take a look at how to apply the principles discussed above to a distribution of real data.

If a large set of data *closely approximates a normal curve* in its shape and area distribution, then there is a simple way to estimate any raw score's percentile rank. Since the distribution of GRE scores in the 1983–1986 norm group is nearly normal, I'll use it in an illustration of this estimation procedure. (See Table 5.3.)

Consider a raw score of 680 on the GRE scale of verbal ability. Since the verbal scale's mean is 475 and its standard deviation is 127, the *z* score that corresponds to a raw score of 680 is

$$z = \frac{680 - 475}{127} = \frac{205}{127} = 1.61$$

Since $z = 1.61$ is above the mean, the next step is to look up the proportion of the area between it and the mean ($z = 0$) of the standard normal distribution, using column 2 of Table Z. Adding .5000 to that proportion yields the total proportion of the area below the original raw score. Finally, the percentile rank is found by multiplying 100 by the total proportion of the curve below the score. Figure 5.9 shows that the estimated percentile is 95—a reasonably accurate approximation of the actual percentile rank of 93 for a score of 680 (see Table 5.3).

Figure 5.9 *Finding the Percentile Rank Corresponding to a Score of 680 on the GRE Scale of Verbal Ability*

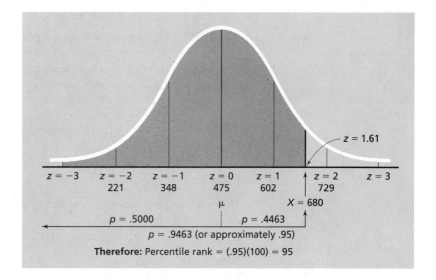

$z = -3$	$z = -2$	$z = -1$	$z = 0$	$z = 1$	$z = 2$	$z = 3$
	221	348	475	602	729	

$z = 1.61$

μ $X = 680$

$p = .5000$ $p = .4463$

$p = .9463$ (or approximately .95)

Therefore: Percentile rank $= (.95)(100) = 95$

NOTE: The procedures described in this section and the next section produce accurate results only to the extent that the distribution of values you are working with closely approaches the shape and properties of the normal curve. The procedures should never be used when the data distribution is clearly non-normal. They are not intended for use with small sample distributions.

Box 5.3 summarizes the procedures for estimating the percentile ranks of raw scores in empirical data distributions.

Box 5.3

GUIDELINES FOR CONVERTING RAW SCORES TO PERCENTILE RANKS IN A NORMAL DISTRIBUTION

1. Draw a picture of the normal curve and shade in the area below the raw score.

2. Using the raw score (X), the mean (μ), and the standard deviation (σ) of the population, convert the raw score to a z score: $z = (X - \mu)/\sigma$.

3. Go to Table Z in Appendix A and locate the z score in column 1.

4. Determine the proportion of the area below the raw score.

    a. If the raw score is higher than the mean (i.e., z is positive), then add the proportion in column 2 of Table Z to .5000.

    b. If the raw score is lower than the mean (i.e., z is negative), then find the proportion in column 3 of Table Z.

5. Round the proportion found in step 4 to a two-digit proportion, and multiply that figure by 100. The result is the percentile rank of the raw score.

Converting Percentile Ranks to Raw Scores

You have learned how to estimate a percentile rank from knowing a raw score and the mean and standard deviation of the distribution that contains that score. The same procedure can also be used in reverse. Given a person's percentile rank and the mean and standard deviation of the distribution, you can estimate that person's raw score. This procedure not only is practical but also improves your understanding of and ability to work with the standard normal distribution.

There are two steps involved in transforming a percentile rank to a raw score:

1. Determine the z score that corresponds to the individual's percentile rank using Table Z.
2. Apply the following formula to that z score:

$$\text{Raw score} = (\sigma)(z) + \mu \tag{5.5}$$

where σ is the standard deviation of the raw score distribution and μ is the mean of that distribution.

To illustrate this procedure, let's suppose Hanna's percentile rank on the GRE-Verbal scale is 40. We know that a percentile of 40 is *below the mean* of the test score distribution and, therefore, that *we will be working with a negative z score.* Since the 40% of scores below Hanna's lie between some negative z score and negative infinity, we need to consult column 3 of Table Z to find the proportion of the "area beyond z." There we can see that the z score that corresponds to a proportion of .4013 (or approximately the lowest 40% of the normal distribution) is –0.25. Next we insert $z = -0.25$ into formula (5.5):

$$\text{Raw score} = (\sigma)(z) + \mu$$

$$= (127)(-0.25) + 475$$

$$= -31.75 + 475$$

$$= 443.25$$

Thus, Hanna's score on the GRE-Verbal scale is approximately 443.

TAKING STOCK: PROPORTION, PERCENT, AND PROBABILITY

As you read this chapter, you learned:

- There is a fixed relationship between z scores and areas under the normal curve.

- Any particular area under the normal curve is expressed as a proportion of the total area.
- Any particular proportion of the area under the normal curve can be expressed as a percent by multiplying the proportion by 100.

The third point correctly implies that "proportion of area" and "percent of area" provide essentially the same information. Indeed, I use both terms in the context of the normal curve throughout this book. But what you also need to realize and remember is that proportion and percent of area also express the probability of randomly selecting values from a normal distribution. Thus, the probability of randomly selecting a z value between $z = -1.00$ and $z = +1.00$ is exactly equal to the proportion of the distribution that lies between those two values: probability $= .3413 + .3413 = .6826$. In other words, if you repeatedly select values at random from a normal distribution, approximately 68% of the random draws will yield z values between -1.00 and $+1.00$ (inclusively). Therefore, values in that range have a 68% probability of being drawn.

This is a small point now, but it will become increasingly important as you proceed through the text: probability = proportion = percent/100. **Please be sure to remember it!**

NOTE WELL: Throughout this textbook, **probability** is defined as *some proportion of the total area under a statistical curve* (such as the normal curve). This is only one of several ways of defining probability, but it is the most practical definition in the context of applied statistics.

KEY TERMS

norm	standard score
relative measures	z score
proportion	standard normal distribution
percent	T score
measures of relative standing	continuous variable
percentile	probability
percentile rank	

SUMMARY

1. Comparisons and interpretations of statistical results are almost always relative matters.

2. Statistical comparisons use relative measures, defined as mathematical transformations of raw scores to a standard number scale.

3. Commonly used relative measures include proportion, defined as a ratio of the number of target events to the total number of events in a set, and percent, defined as a proportion multiplied by 100.

4. Measures of relative standing, such as percentile ranks and standard scores, are mathematical transformations that express a value's relative position in a sample or population.

5. A percentile rank is the percent of a distribution lying below a certain value.

6. A standard score, or z score, expresses the distance between a raw score and the mean in standard deviation units; z scores below the mean are negative values.

7. Regardless of whether a set of z scores is normally distributed, the mean of the z scores is always 0, and the standard deviation is always 1.

8. A transformed score is a special version of a standard score in which the original data are converted to a new number scale that has a preselected mean and standard deviation; the distribution of transformed scores has the same shape as the original distribution.

9. The standard normal distribution is a normal distribution of z scores.

10. In the standard normal distribution, z scores have a constant relationship to proportions of the area under the normal curve. This fact allows you to find the proportion of the distribution lying between any two z scores, convert z scores to percentile ranks, and estimate percentile ranks from z scores in empirical data distributions.

11. In the context of a normal distribution, probability = proportion = percent/100.

REVIEW QUESTIONS

1. Define or describe: relative measures, proportion, percent, measures of relative standing, percentile, and percentile rank.

2. Describe the main advantage of using relative measures, and give two examples of the use of relative measures in everyday life.

3. In a class of 400 students, the mean number of daily absences during the school term was 30. What proportion of the class was missing on a typical day? What percent?

4. A small business had a total sales volume of $1,200,000 last year. After she paid for all of her overhead, the proprietor was left with $240,000. What proportion of total revenues represented profit? What percent?

5. The test results for a class of 20 showed that Paul was tied with one other person and was lower than 17 others. What was Paul's percentile rank in the group?

6. In Table 5.1, what value represents the 50th percentile?

7. In Table 5.1, what percentile rank is held by a raw score of 85?

8. For what practical reason might you elect to use a measure of relative standing (e.g., percentile rank) rather than "percent correct" to interpret or evaluate a person's score on a test or other assessment device?

9. Using your own words, describe the relationship between the z score and the standard deviation.

10. Convert the following raw scores to z scores. Then calculate the mean and variance of the set of z scores: 21, 38, 20, 30, 31, 34, 33.

11. Robert scored 370 on the SAT-Quantitative scale. Assuming that the test has a mean of 500 and a standard deviation of 100, what was Robert's z score?

12. A test of manual dexterity has a mean of 266 and a standard deviation of 16. Saleh's raw score on the test is 306. Convert his score to a T score that has a mean of 50 and a standard deviation of 10.

13. Construct a frequency polygon from the data in Table 5.2. (If necessary, refer to Chapter 3 for a review of how to do this.) Next convert the raw scores in Table 5.2 to the same number of z scores. Finally, construct a frequency polygon of the z scores. What characteristics of the frequency polygon changed as a result of the conversion to standard scores? Did the shape of the polygon change in any way?

14. Using Table Z in Appendix A, determine the proportion of the normal curve that is between $z = 1.96$ and the mean ($z = 0$). Also, find the proportion of the area between $z = -1.96$ and $z = +1.96$.

15. Using Table Z in Appendix A, find the percent of the normal curve between $z = -2.575$ and $z = -1.96$.

16. Amy scored 70 on a standardized IQ test that has a mean of 100 and a standard deviation of 15. Assuming that IQs are normally distributed, estimate Amy's percentile rank in the population.

17. Bob is a member of a large national sales force that specializes in the marketing and distribution of mainframe computers. His personal sales dollar volume was $12,670,000 during the past year, which was the 90th percentile in the whole population of

individual sales volumes. Assuming that the salespersons' gross sales figures are normally distributed, what was Bob's approximate z score in that distribution?

18. What is the probability of randomly selecting values that are between $z = -1.96$ and $z = +1.96$ in a normal distribution? What is the probability of randomly selecting values that are at least 1.96 standard deviations away from the mean (in either a positive or a negative direction)?

19. What is the probability of randomly selecting values that are at least as extreme as $z = 2.575$ in a normal distribution? (*Hint:* Carefully examine the wording of this problem. Values can be extreme in two ways.)

20. When values are randomly drawn from a normal distribution, what is the probability of getting values that are within 1.00 standard deviation of the mean?

21. Assume that an IQ test has a mean of 100 and a standard deviation of 16. Also assume that the scores based on the test are normally distributed. If Norman's percentile rank on the test is 20, what score did he get on the IQ test?

22. Assume that an IQ test has a mean of 100 and a standard deviation of 16. Also assume that the scores based on the test are normally distributed. If Nancy's percentile rank on the test is 60, what score did she get on the IQ test?

23. Assume that an IQ test has a mean of 100 and a standard deviation of 16. Also assume that the scores based on the test are normally distributed. What percent of the IQ distribution lies between an IQ of 68 and an IQ of 132?

24. Assume that the national mean on the ACT (composite) college admissions test is 18 and that the standard deviation is 3.5. On a single random draw, what is the probability of selecting a student with an ACT score of at least 25?

25. A self-concept test yields scores that are normally distributed with a mean of 30 and a standard deviation of 6.5. If a "poor self-concept" is defined as being in the bottom 20% of the distribution, how low would one have to score to fall into the poor-self-concept category?

CHAPTER 6   LINEAR CORRELATION

All empirical sciences base their theories, predictions, and practical applications on relationships between variables. Indeed, it is fair to say that *science is*, in the simplest sense, *a search for regular patterns of "connectedness" between variables.* In the behavioral sciences, we are interested in such things as:

- How well scores on a personnel selection test relate to employees' performance on the job
- To what extent personal adjustment and happiness are related to people's self-esteem
- The degree of association between a society's level of economic frustration and its tendency to engage in aggression against other societies
- The relationship between the number of hours per week that elderly people devote to physical fitness training and their reported frequency of sexual intercourse
- The extent of association between achievement motivation and grades in college

The list of behavioral science questions could go on and on interminably. And almost every question would address a possible relationship between two variables, where **relationship** is defined as:

- "A 'going together' of two variables: it is what the two variables have in common" (Kerlinger, 1979, p. 22)
- "A patterned mutual change between variables. That is, as *X* changes, *Y* changes in a patterned way" (Walizer & Wiener, 1978, p. 60)
- "A connection or association between two variables, such that a value of variable *Y* is at least partially predictable from the corresponding value of variable *X*" (Evans, 1985)

In the field of statistics, *a relationship between two variables* is referred to as a **correlation**. In this chapter you will learn when and how to calculate and interpret a statistical index called the **linear correlation coefficient**, which is the device most widely used by behavioral researchers to discover and summarize relationships between variables.

Don't let the simplicity of the correlation coefficient lead you to underestimate its importance: This index is the statistical essence of science in general.

Linear: Pertaining to or forming a straight line. Two variables that have a linear relationship tend to be related to each other in a straight-line fashion. That is, when you graphically plot changes in one of the variables as a function of increases in the other variable, the resulting pattern of plot points tends to form a straight line.

THE PEARSON *r*

The coefficient of correlation I have been referring to is known as the Pearson *r*. This index was developed by British mathematician Karl Pearson, and the *r* symbol is used simply to identify the specific type of correlation coefficient being referred to. (There are many other types of correlations, but none is as widely used as the Pearson *r*.)

The **Pearson r** is defined as *a measure of the strength and direction of linear association between two variables*. Let's take a detailed look at what is meant by this definition.

Assumptions of *r*

Depending on the specific use made of *r*, several assumptions underlie its computation and interpretation. The two most elementary assumptions are that (1) the data of the study are in the form of "pairs" of observations and (2) the two variables being correlated have an essentially linear relationship with each other.[1]

Pairs A correlation can be calculated only if each person in your investigation has been measured on each of two variables, variable *X* and variable *Y*, and each person's value on the *X* variable is aligned (i.e., paired) with her value on the *Y* variable. Table 6.1 shows pairs of scores for five hypothetical research subjects. Note how each subject's *X* score is paired with her *Y* score.

By convention, within the context of correlation, the *X* variable is referred to as the **independent**, or **predictor**, variable and the *Y* variable is called the **dependent**, or **criterion**, variable. The researcher views the *X* and *Y* variables as changing together (i.e., covarying), with changes in one typically preceding changes in the other. The variable thought to change first under ordinary circumstances is labeled the *X* variable, and the other is called the *Y* variable. Within some types of research design, it can be assumed that the independent (*X*) variable causes the dependent (*Y*) variable to change, but this assumption is not necessary. In fact, in most

Table 6.1

PAIRS OF SCORES FOR FIVE HYPOTHETICAL RESEARCH SUBJECTS

| | PAIRS OF SCORES | |
SUBJECT	VARIABLE *X* SCORE	VARIABLE *Y* SCORE	
Person 1	1	2	←one pair of scores
Person 2	2	3	
Person 3	3	4	
Person 4	4	5	
Person 5	5	6	

[1] In Chapter 7 you will encounter two additional assumptions made about *r* in the context of linear regression: the assumptions of homoscedasticity and bivariate normality.

uses of *r*, no assumption of causality can be made. What's more, *you should studiously avoid the temptation to conclude that a correlation between two variables implies that one variable causes the other to change, unless the correlation in question was computed on data from a well-controlled experiment.* Assertions about causality must be based on the type of research design used, not on the existence of some statistical result. More on this later.

Linear relationship   When you calculate a linear correlation coefficient, the resulting number is an accurate index of the strength of the association between variables *X* and *Y* to the extent that Y *values are a straight-line function of the X values that they are paired with.* A perfect straight-line relationship between *Y* and *X* is graphed in Figure 6.1. That figure was drawn by plotting the *Y* values in Table 6.1 as a function of the corresponding *X* values.

Perfect linear relationships, such as the one in Figure 6.1, are almost never found in the behavioral sciences. But many relationships that we study tend to approximate a linear function. Therefore, they essentially satisfy the crucial assumption made by the Pearson *r* correlation coefficient.

NOTE: Pairs of data do not have to form a perfectly linear relationship in order for us to use the Pearson *r* with them; rather, the relationship need only be a reasonable approximation of a straight line.

Uses of the Pearson *r*

The linear correlation coefficient has both scientific and applied uses. The behavioral sciences assume that human nature is governed by an inherent lawfulness, which is represented in reliable associations between variables. The scientific function of *r* is to reveal the laws of human behavior by providing a numerical index of relationships present in research data.

Figure 6.1 *Plot of Y Values As a Function of X Values (Note: The plot forms a straight line, thus meeting the most important assumption of the Pearson r correlation.)*

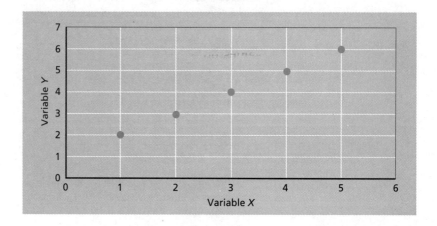

The applied function of r consists of using the linear correlation index to predict future values of Y from present values of X. An example is using job candidates' aptitude test scores to predict which one of several applicants is most likely to succeed at the job. The test scores in this case are variable X, and future performance on the job is variable Y. To the extent that Y is *strongly correlated* with X, such predictions can be made accurately and, hence, enable an employer to make better hiring decisions. The applied use of the Pearson r is known as "linear regression analysis," a topic we will study in Chapter 7.

Properties of r

The linear correlation coefficient has two major properties: magnitude and direction.

Magnitude of association   The absolute value of r (that is, disregarding its negative or positive sign) can range from .00 to 1.00, inclusively.[2] The closer r is to .00, the weaker is the relationship between the two variables being analyzed; the closer r is to 1.00, the stronger is the relationship. A correlation coefficient of 1.00 means that values of Y are perfectly associated with—and hence perfectly predictable from—corresponding values of X. A correlation coefficient of .00 means that Y values vary randomly as a function of X values, and there is no predictability from one of the variables to the other.

In the behavioral sciences, most correlations range between .00 and .50 in absolute value (that is, ignoring the negative sign). Very rarely is r right at 1.00. The important point to remember is that the strength of the association between the variables is greater at higher values of r: $r = .75$ represents a stronger relationship than $r = .50$, which represents a stronger relationship than $r = .25$.

Direction of relationship   A linear correlation coefficient can be **positive** (reflecting a *direct relationship*) or **negative** (reflecting an *inverse relationship*). If two variables have a positive correlation, then higher values on variable Y tend to be associated with higher values on X, and lower values on Y tend to be associated with lower values on X. In short, the two variables increase or decrease together. For example, arousal level (Y) gets higher as one increases one's caffeine ingestion (X) and lower as one decreases the caffeine ingestion. Positive, or **direct**, correlations are expressed with either a + sign (e.g., $r = +.40$) or no sign at all (e.g., $r = .40$).

Negative, or **inverse**, correlations exist between variables that tend to change in opposite directions. For example, as the supply of a commodity increases, the demand for it tends to decrease. For

---

[2] The number itself is the "coefficient" part of the correlation coefficient concept.

another example, higher levels of self-esteem tend to be associated with lower levels of maladjustment, and lower levels of self-esteem tend to be associated with higher levels of maladjustment. Negative correlations are expressed with a – sign (e.g., $r = -.40$).

It is *important* to remember the following, perhaps surprising, fact about correlation: *The size (magnitude) of a correlation is interpreted independently of its sign (positive or negative).* Thus, a relationship represented by a correlation coefficient of –.80 is just as strong as a relationship represented by a correlation coefficient of +.80. A negative correlation coefficient is not the same thing as a negative number. The negative sign simply tells about the direction of the relationship between X and Y. The strength of that relationship is entirely indicated by the size of the correlation, irrespective of the sign.

Is $r = -.80$ a stronger or weaker correlation than $r = +.50$? The coefficient of –.80 is the stronger one. Strength is independent of direction and is always indicated by the size of the number irrespective of the sign.

The direction of a correlation can often be determined by looking at a special type of graph called a scatter diagram.

THE SCATTER DIAGRAM

Before you compute and interpret a correlation coefficient, it is always wise to first *plot the Y values against their corresponding X values,* to form a graph called a **scatter diagram**. A scatter diagram, also referred to as a **scatterplot**, provides preliminary answers to the following critical questions about variables X and Y:

1. Are Y values systematically related to X values?
2. Is the relationship essentially linear, or is it basically nonlinear?
3. Is the relationship a positive (direct) or a negative (inverse) one?
4. Does the relationship appear to be a strong, moderate, or weak one?

Actually, Figure 6.1 is a simple example of a scatter diagram. But let's consider a realistic example. First, take a look at the data in Table 6.2. The data are from a small study conducted by the personnel department in an industrial plant. Each of the ten employees in the study took an electronics aptitude test just before being hired to work in the circuit assembly department of XYZ Company (fictitious name). Their scores on the so-called selection test are the values of variable X. Variable Y is represented by the ratings of the employees' job performance some six months after being hired. The basic question in this investigation is: To what extent is job performance related to performance on the selection test, if at all?

Table 6.2

PAIRS OF SCORES FOR TEN EMPLOYEES OF XYZ COMPANY		PAIRS OF SCORES	
	EMPLOYEE	VARIABLE X SCORE ON SELECTION TEST	VARIABLE Y PERFORMANCE EVALUATION
	M. Higgins	40	90	←pair of scores for Higgins
	T. Purdue	32	74
	R. Gutwald	46	96
	J. Gomez	50	90
	O. Danis	22	44
	E. Pompe	10	50
	C. Clover	18	56
	Y. Arnel	30	76
	D. Evers	26	42
	L. Kluver	30	64

The scatter diagram that addresses this question is constructed by plotting each person's Y value as a function of his or her X value within the framework of the two-axis structure shown in Figure 6.2. Note that the *vertical (Y) axis* in the figure is called the **ordinate**, and the *horizontal (X) axis* is referred to as the **abscissa**.

When specific values are attached to the X and Y axes, a marker (usually a dot or some other small symbol) can be placed on the graph *precisely where a person's Y score intersects with the corresponding X score*. The result of this type of plotting for the data in Table 6.2 is the scatter diagram in Figure 6.3. Note how M. Higgins is represented by the dot that appears in the figure at the intersection of her Y value (90) with her X value (40). Can you see that each marker in the scatter diagram represents one pair of values?

Figure 6.2 *General Two-Axis Structure Used in the Graphing of Scatter Diagrams*

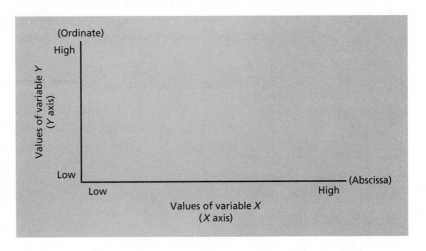

Figure 6.3 *Scatter Diagram Showing a Plot of Job Performance Scores As a Function of Selection Test Scores*

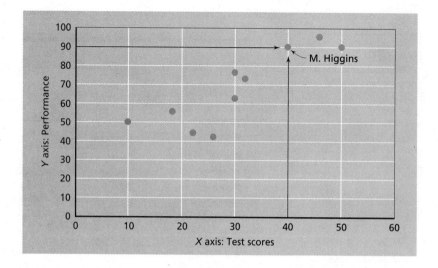

The plot in Figure 6.3 answers the first question that a scatter diagram should address. There definitely is a relationship between the variables *X* and *Y*. As scores on the selection test increase, there is a systematic tendency for job performance to improve as well. Now let's consider the remaining types of information that are provided by scatterplots.

The Linearity Question

One important function of the scatter diagram is to give a preliminary verification of the assumption that the *X* and *Y* variables are related in an essentially straight-line fashion. Remember that if the assumption of linearity is not met, then the Pearson *r* correlation is not an appropriate—or accurate—index of the relationship in question.

Figure 6.4 shows a second plot of the data in Table 6.2. This time a **line of best fit** has been drawn *through the middle of the dots* to show that the scatter pattern is basically linear. The line of best fit, also known as the **regression line**, was mathematically determined by procedures that you will learn about in Chapter 7. Right now, let's just consider it to be a graphical summary of the dot pattern in a scatter diagram. It suggests that the present relationship is predominantly linear.

The Question of Direction

Another important piece of information that we can glean from the scatter diagram is the overall direction of the association between variables *X* and *Y*—positive (direct) or negative (inverse).

The plot in Figures 6.3 and 6.4 shows that job performance (*Y*) is positively related to scores on the personnel selection test (*X*); that is, as the *X* values go from low to high, there is a clear tendency

Figure 6.4 *Scatter Diagram Showing That the Relationship Between Job Performance and Test Scores Is Essentially Linear*

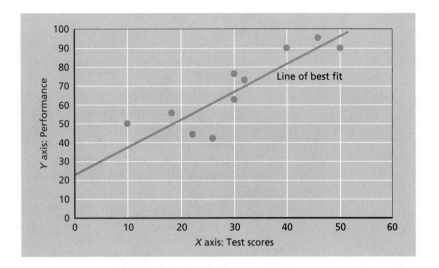

for the *Y* values to do the same. Note that the overall orientation of the dot pattern (and hence the line of best fit) is upward to the right. The pattern "runs" from the lower left-hand corner to the upper right-hand corner of the graph.

Now consider a second data set that exhibits the opposite direction in a linear relationship. The data in Table 6.3 show that

Table 6.3

SELF-ESTEEM AND NEUROTICISM: PAIRS OF RAW SCORES

PERSON*	SCORES ON SELF-ESTEEM TEST X	SCORES ON NEUROTICISM TEST Y
a	27	25
b	32	18
c	40	16
d	33	20
e	24	24
f	35	21
g	20	28
h	21	29
i	36	17
j	26	26
k	30	22
l	28	24
m	26	22
n	27	31
o	29	20

*Each of the 15 people in this sample has a pair of scores; each person's X value is aligned with his or her Y value. To calculate a linear correlation, you must tabulate scores in pairs of this type.

people who get low scores on a test of self-esteem tend to get high scores on a test of "neuroticism." There is an *inverse* relationship—a *negative* correlation—between the two variables. Persons who have low opinions of themselves tend to develop many symptoms of "neurosis," whereas those who possess high self-opinions tend to have few such symptoms.

Figure 6.5 shows what that inverse relationship looks like in a scatter diagram. Note that the orientation of the dot pattern is exactly the opposite of that shown in Figures 6.3 and 6.4. This time, the pattern runs from the upper left-hand corner to the lower right-hand corner of the graph. As X values increase, Y values tend to decrease.

It is important to note that the different linear relationships shown in Figures 6.4 and 6.5 are approximately equal in strength ($r = .86$ and $r = -.83$, respectively). The relationships differ only in direction. This fact makes the point, once again, that the *size (strength) of a correlation is independent of its sign (direction).*

The Question of Strength

The final information provided by the scatter diagram is the strength, or magnitude, of the correlation. In general, the more strongly variables X and Y are associated with (and therefore *predictive* of) each other, the more narrow and elongated the scatterplot is. Conversely, the weaker the covariation (relationship) between the variables, the "shorter and fatter" the dot pattern is. Notice that the *slope*—that is, the degree of slant—in the dot pattern tends to be less in weaker relationships than in stronger ones. If the correlation is about .00, then the scatterplot will appear to have no upward or downward trend at all. Figure 6.6 illustrates

Figure 6.5 *Scatter Diagram Showing a Negative (Inverse) Linear Relationship Between Self-Esteem and Neuroticism*

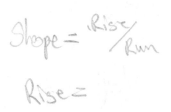

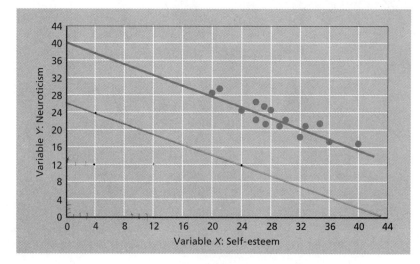

Figure 6.6 *Scatter Diagrams Showing How the Plot Pattern Changes As a Function of the Strength of the Correlation Coefficient (Note: Weaker r's are associated with a broader scatter of dots.)*

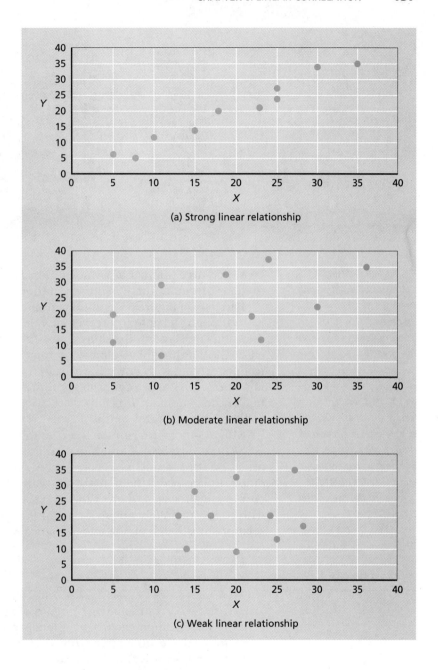

(a) Strong linear relationship

(b) Moderate linear relationship

(c) Weak linear relationship

what happens to the scatter diagram as the correlation coefficient goes from $r = .98$, panel (a), to $r = .50$, panel (b), to $r = .15$, panel (c).

The two panels in Figure 6.7 present scatter diagrams that result from **null relationships**—that is, from *zero correlations*. Panel (a) shows that a zero correlation can stem from too wide a scatter of dots. This situation indicates that for any given *X* value, there can

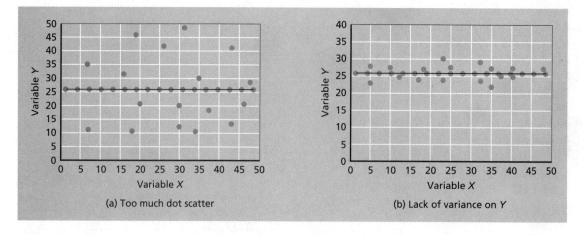

(a) Too much dot scatter

(b) Lack of variance on Y

Figure 6.7 *Scatter Diagrams of Null (r = .00) Relationships*

[In panel (a) r = .00 because values of Y vary too greatly around the respective values of X. Panel (b) shows a case of r = .00 that results from a lack of adequate variation in Y values (restriction of range). In each case, the dot pattern lacks slope, and the line of best fit is parallel to the horizontal axis (abscissa).]

be a very large number of Y values. Hence, it is nearly impossible to predict Y from X. There is no association between Y and X values.

Panel (b) of Figure 6.7 exhibits another type of null relationship. Even though the dot pattern is long and narrow (similar to that of a strong correlation), the line of best fit has no slope to it. In this case, the Y values have almost no variance, which prevents them from covarying with any other variable, including variable X. This is a special instance of a problem that sometimes surfaces in the use of the Pearson r. It is called "restriction of range," and we will consider it again later in this chapter.

Still another problem for the Pearson r is illustrated in Figure 6.8. There variables X and Y are, in fact, strongly related to each other, but the actual linear correlation in this case is –.02. This example underscores the noteworthy point that the Pearson r is accurate only in cases where Y has an essentially linear relationship with X. It is definitely not accurate for gauging **curvilinear** (curved) relationships, such as the one in Figure 6.8.

COMPUTING *r*

There are two ways to start with pairs of scores and calculate the Pearson r correlation coefficient: the z-score method and the raw-score method. To understand the z-score approach, you need to first consider what r, the numerical index, basically reflects.

The Meaning of *r* As a Number

The r coefficient is *a numerical index of the extent to which people's ranks (or relative positions) on variable Y correspond in a systematic way to their ranks (or relative positions) on variable X.* To the extent that r is a large number (approaching 1.00), you can predict a per-

Figure 6.8 *Scatter Diagram Showing a Curvilinear (Curved) Relationship (Note: Although Y is substantially and systematically related to X, r = .00 because the Pearson r correlation is accurate only to the extent that the relationship in question conforms to a straight line.)*

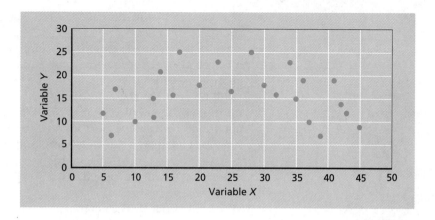

son's position on one variable from a knowledge of his or her position on another variable. The regular correspondence of positions from X to Y can be either direct, representing a positive correlation, or inverse, representing a negative correlation.

Given that the Pearson r as a numerical index of correlation is mainly sensitive to the correspondence of people's positions on two variables, you ought to be able to use some statistical measure of the relative positions on two variables to compute r. You can, with a measure of position already familiar to you—the z score.

The z-Score Formula

The definitional formula for r, developed by Karl Pearson himself, is as follows:

$$r = \frac{\overset{②}{\downarrow}\ \overset{①}{\downarrow}}{\underset{\underset{③}{\uparrow}}{N}}\Sigma(z_X \cdot z_Y)$$

(6.1)

FORMULA GUIDE

① Multiply each person's z score on variable X by his or her z score on Y to obain the cross-product of the z scores.

② Sum all of the z-score cross-products in the data set.

③ Divide the result of step 2 by N, the number of pairs of observations.

This formula represents the mathematical concept of correlation. It tells you that:

- The size of r will increase as the sum of the z-score cross-products increases.
- r will be large to the extent that large z scores on X tend to be multiplied by large z scores on Y (i.e., there is a good correspondence of relative positions from one variable to the next).
- r will be small to the extent that large z scores on X tend to be multiplied by small z scores on Y (i.e., there is a poor correspondence of relative positions from one variable to the next).
- r will be based on a negative sum, and hence inverse, if positive z scores on X tend to be paired with negative z scores on Y; otherwise, r will be based on a positive sum, representing a direct relationship

To use formula (6.1) to compute r, you must first convert each raw score on X and Y to a standard score (see the Concept Recap).

CONCEPT RECAP

Recall that a z score is calculated by dividing a raw score's deviation from its mean by the standard deviation of the data set. Hence, $z_X = (X - \overline{X})/S_X$ and $z_Y = (Y - \overline{Y})/S_Y$ for each pair of scores.

The interesting relationship of linear correlation to the relative positions—and hence the z scores—of paired X and Y values is illustrated in Figures 6.9 through 6.11.

Figure 6.9 shows a perfect positive relationship between variables X and Y ($r = 1.00$). The figure presents raw scores and the associated z scores of five hypothetical persons on vertical X and Y dimensions. Note that, for each person, the z score on Y is entirely predictable from a knowledge of the corresponding z score on X. Also note that positive z scores on X are paired with positive z scores on Y—and negatives with negatives—*so that all the z-score cross-products are either positive or 0* (no negatives). Because the sum of the z-score cross-products is positive, r is positive; because that sum is large, r is large. The lower portion of Figure 6.9 presents the details of the z-score calculation of r.

Figure 6.10 shows a perfect negative relationship between variables X and Y ($r = -1.00$). Again, once you know that the correlation is perfect, you can easily and accurately predict a given Y value from a knowledge of the corresponding X value. This time the pattern of correspondence is inverse: Large positive z scores on X are associated with large negative z scores on Y, and large negative z's on X with large positive z's on Y. Thus, the sum of z-score cross-products is negative, and so is r.

Figure 6.11 lacks the predictability present in Figures 6.9 and 6.10. Notice the absence of patterned correspondence between the

Figure 6.9 *Z scores and r: Example of a Perfect Positive Correlation.* (Source: *Figure adapted from* Invitation to Psychological Research *by James D. Evans, copyright © 1985 by Holt, Rinehart and Winston, Inc., reproduced by permission of the publisher.)*

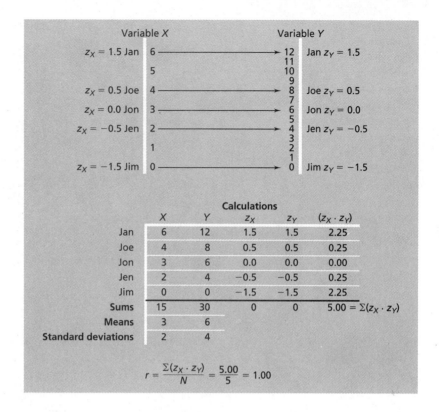

hypothetical subjects' positions on X and their positions on Y. There is little, if any, predictability from one variable to the other. The sum of the z-score cross-products is nearly 0, which yields a negligible correlation ($r = -.05$).

Although almost all linear correlations in the behavioral sciences are smaller than 1.00 and larger than $-.05$, the general rule illustrated by these figures always holds true. The better the patterned correspondence between paired positions on the X and Y dimensions, the greater the sum of the z-score cross-products, and the higher the resulting r coefficient.

Table 6.4 shows how the z-score formula (6.1) is used to calculate the correlation between selection test scores and job performance (from the data in Table 6.2). Notice that *the calculated r of .86 is simply the mean of the z-score cross-products.*

The Raw-Score Formula: *r* the Easy Way

The z-score method of computing r is simple to understand, and it conveys the meaning of linear correlation. But, because of all the decimal numbers involved in its use, formula (6.1) is not considered a convenient formula for linear correlation. It is much easier to

Figure 6.10 *Z scores and r: Example of a Perfect Negative Correlation.* (Source: *Figure adapted from* Invitation to Psychological Research *by James D. Evans, copyright © 1985 by Holt, Rinehart and Winston, Inc., reproduced by permission of the publisher.*)

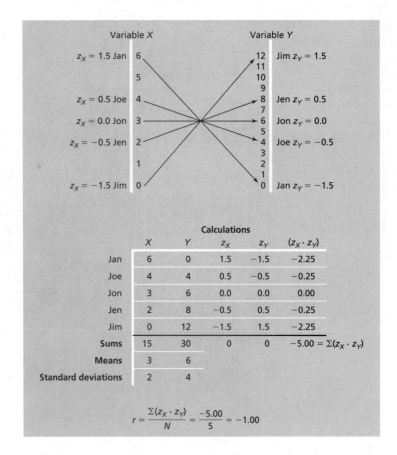

1344.142

~27718~
39675.57

Table 6.4

CORRELATION CALCULATIONS FOR PERFORMANCE VERSUS TEST SCORES: Z-SCORE METHOD

PERSON	X	Y	z_X	z_Y	$z_X z_Y$
M. Higgins	40	90	0.816	1.151	0.940
T. Purdue	32	74	0.136	0.306	0.042
R. Gutwald	46	96	1.327	1.468	1.947
J. Gomez	50	90	1.667	1.151	1.919
O. Danis	22	44	−0.714	−1.278	0.913
E. Pompe	10	50	−1.735	−0.961	1.667
C. Clover	18	56	−1.055	−0.644	0.679
Y. Arnel	30	76	−0.034	0.412	−0.014
D. Evers	26	42	−0.374	−1.383	0.518
L. Kluver	30	64	−0.034	−0.222	0.008
Sums:	304	682	0.000	0.000	8.618
N = 10					
Means:	30.40	68.20	0.000	0.000	0.8618
S's:	11.76	18.94	1.000	1.000	

$$r = \Sigma(z \cdot z_Y)/N = 8.618/10 = .8618, \text{ or } .86$$

Figure 6.11 *Z scores and r: Example of a Negligible (Near-Zero) Correlation. (Source: Figure adapted from* Invitation to Psychological Research *by James D. Evans, copyright © 1985 by Holt, Rinehart and Winston, Inc., reproduced by permission of the publisher.)*

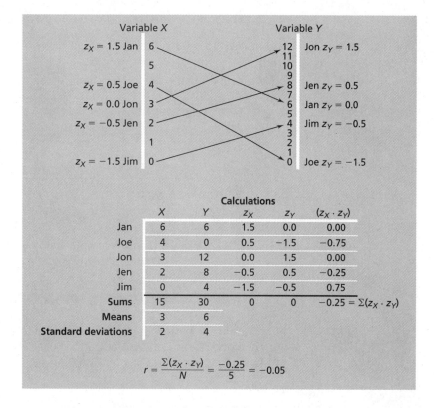

compute *r* with the raw-score computational formula. This formula consists of three parts:

1. The adjusted sum of the cross-products of *X* and *Y*:

$$CP_{XY} = \Sigma(XY) - (\Sigma X \cdot \Sigma Y)/N$$

$$\begin{array}{ccccc} \uparrow & \uparrow & \uparrow & \uparrow & \uparrow \\ ② & ① & ⑤ & ③ & ④ \end{array}$$

> **FORMULA GUIDE**
> ① Multiply each *X* raw score by the corresponding *Y* raw score.
> ② Sum the results of step 1.
> ③ Multiply the sum of the *X* raw scores by the sum of the *Y* raw scores.
> ④ Divide the product of the sums from step 3 by *N*, the sample size.
> ⑤ Subtract the result of step 4 from the result of step 2.

2. The sum of squares on variable *X*, formula (4.10) from Chapter 4:

$$SS_X = \Sigma X^2 - \frac{(\Sigma X)^2}{N}$$

3. The sum of squares on variable Y (same logic as step 2):

$$SS_Y = \Sigma Y^2 - \frac{(\Sigma Y)^2}{N}$$

Putting all three parts together yields:

$$r = \frac{CP_{XY}}{\sqrt{SS_X \cdot SS_Y}} \leftarrow ③ \qquad\qquad (6.2)$$

$$\underset{②}{\uparrow} \quad \underset{①}{\uparrow}$$

FORMULA GUIDE

① Multiply the sum of squares on variable X by the sum of squares on variable Y.
② Take the square root of the result of step 1.
③ Divide the result of step 2 into the adjusted sum of the cross-products.

Table 6.5 shows how the raw-score formula is used to compute the correlation between selection test scores and job performance [note that the value of r obtained agrees with that produced by the

Table 6.5

PERFORMANCE VERSUS
SELECTION TEST SCORES:
RAW-SCORE CALCULATION OF
r (N = 10, NUMBER OF PAIRS
OF SCORES, OR SAMPLE SIZE)

PERSON	X	Y	X^2	Y^2	XY
M. Higgins	40	90	1,600	8,100	3,600
T. Purdue	32	74	1,024	5,476	2,368
R. Gutwald	46	96	2,116	9,216	4,416
J. Gomez	50	90	2,500	8,100	4,500
O. Danis	22	44	484	1,936	968
E. Pompe	10	50	100	2,500	500
C. Clover	18	56	324	3,136	1,008
Y. Arnel	30	76	900	5,776	2,280
D. Evers	26	42	676	1,764	1,092
L. Kluver	30	64	900	4,096	1,920
Sums:	304	682	10,624	50,100	22,652

$$SS_X \rightarrow 1382.40 = \Sigma X^2 - (\Sigma X)^2/N = 10,624 - (304)^2/10$$

$$SS_Y \rightarrow 3587.60 = \Sigma Y^2 - (\Sigma Y)^2/N = 50,100 - (682)^2/10$$

$$CP_{XY} \rightarrow 1919.20 = \Sigma XY - (\Sigma X \cdot \Sigma Y)/N = 22,652 - (304 \cdot 682)/10$$

$$r \rightarrow .862 = \frac{CP_{XY}}{\sqrt{SS_X \cdot SS_Y}} = \frac{1919.20}{\sqrt{1382.40 \cdot 3587.60}} = \frac{1919.20}{2226.99}$$

z-score formula (6.1)]. Table 6.6 presents the raw-score calculation of the correlation between self-esteem and neuroticism. Even though the raw-score formula (6.2) might seem more complex and intimidating than the z-score approach, it is much simpler to use because it allows you to work entirely with whole numbers. Formula (6.2) was algebraically derived from the z-score formula (6.1). See Hays (1988) for the proof.

INTERPRETING *r*

In general, *r* is a numerical index of the direction and strength of a linear relationship between two variables. But once *r* is calculated, how are you to interpret it in specific ways when you are comparing two or more correlations? The three standard ways of interpreting linear correlations are in terms of (1) absolute magnitude, (2) statistical reliability, and (3) variance accounted for.

Table 6.6

NEUROTICISM VERSUS SELF-ESTEEM: RAW-SCORE CALCULATION OF *r* (*N* = 15, NUMBER OF PAIRS OF SCORES, OR SAMPLE SIZE)

PERSON	X	Y	X^2	Y^2	XY
a	27	25	729	625	675
b	32	18	1,024	324	576
c	40	16	1,600	256	640
d	33	20	1,089	400	660
e	24	24	576	576	576
f	35	21	1,225	441	735
g	20	28	400	784	560
h	21	29	441	841	609
i	36	17	1,296	289	612
j	26	26	676	676	676
k	30	22	900	484	660
l	28	24	784	576	672
m	26	22	676	484	572
n	27	31	729	961	837
o	29	20	841	400	580
Sums:	434	343	12,986	8117	9640

$$SS_X \rightarrow 428.93 = \Sigma X^2 - (\Sigma X)^2/N = 12,986 - (434)^2/15$$

$$SS_Y \rightarrow 273.73 = \Sigma Y^2 - (\Sigma Y)^2/N = 8117 - (343)^2/15$$

$$CP_{XY} \rightarrow -284.13 = \Sigma XY - (\Sigma X \cdot \Sigma Y)/N = 9640 - (434 \cdot 343)/15$$

$$r \rightarrow -.829 = \frac{CP_{XY}}{\sqrt{SS_X \cdot SS_Y}} = \frac{-284.13}{\sqrt{428.93 \cdot 273.73}} = \frac{-284.13}{342.65}$$

Magnitude Interpretation

The most intuitively obvious way to interpret a linear correlation is to attach labels to arbitrary ranges of r values, as follows:

If r is in this range:	Give r this label:
.80 to 1.00	Very strong
.60 to .79	Strong
.40 to .59	Moderate
.20 to .39	Weak
.00 to .19	Negligible to very weak

Although such a framework often serves as a basis for a discussion of the meaning of correlations, it does have two short-comings. First, the particular magnitude intervals must be arbitrarily selected; there's nothing wrong, for example, with defining the "Very strong" interval as .75 to 1.00 (rather than .80 to 1.00). The subjective aspect of interval selection and labeling makes this interpretation method somewhat ambiguous.

A second, and perhaps more serious, problem with the magnitude interpretation is that it ignores the fact that random error in sample data sometimes produces a fairly large numerical correlation that is not reliable—that is, *it will not be reproducible in additional studies of the same relationship.* Such a **spurious r** is most likely to appear when very small samples ($N = 3$ to 15) are used in a study, although even large samples are not immune to the possibility of a large r resulting from random error. Hence, to avoid treating a random correlation as a real association between X and Y, behavioral scientists typically check the "statistical reliability" of their correlations before applying any sort of magnitude interpretation.

Statistical Reliability Interpretation

The topic of statistical reliability, or "statistical significance," will be discussed at length in Chapter 9. Here, a brief overview of the concept will have to suffice.

Interpretation of r in the context of statistical reliability leads to one of two possible conclusions: either the r is deemed "significantly different from 0" or the r is labeled "nonsignificant," meaning unreliable and not likely to be reproducible in replications of the research study. Nonsignificant correlations normally receive no further attention, whereas significant correlations are given more examination and interpretation.

To be declared statistically reliable, a correlation coefficient must be compared with a large set of random (chance) r values and found to be a very unlikely event among such random outcomes (see Chapter 9). The test of statistical reliability always takes sample size into account. The general rule is that the larger the number of paired scores in your sample, the more confidence you should have that an *r* of a given size is reliable. For example, *r* = .40 is not statistically significant (i.e., not considered reliable) if *N* = 10, but it is significant (i.e., considered reliable) if *N* = 20. When *N* = 10 the linear correlation must exceed .63 in order to be "significant." Therefore, the *r* calculated in Table 6.5 is considered reliable, because .86 exceeds .63.[3]

Don't let these mysterious numbers discombobulate you. Their origin and meaning will be revealed in later portions of this book. For the time being, just try to get the general meaning of statistical reliability as it applies to interpreting correlations.

Variance Accounted for Interpretation

The most abstract—but perhaps most useful—way to discuss the meaning of a correlation is to refer to the **proportion of variance accounted for** by the correlation coefficient. The proportion of variance in *Y* "accounted for" by variance in *X* is equal to the square of *r*—that is, $r \cdot r$, or r^2. In the case of the association of selection test scores with job performance, the correlation is .86. Therefore, the proportion of variation in job performance scores accounted for by variation in selection test scores is $.86^2$, or .7396. Because of the way proportion underlies percent, it is accurate to say that about 74% of the variance in job performance is predictable from a knowledge of selection test scores. This, in turn, means that about 26% of the variation in job performance is associated with variables other than the ability measured by the selection test.

In the vocabulary of statisticians, r^2 is called the **coefficient of determination**. The complement of r^2, $(1 - r^2)$, is referred to as the **coefficient of nondetermination** (equal to .26 in the above example).

If you're a typical first-time reader of this chapter, the concept of "variance accounted for" probably is not too clear yet. If so, don't feel alone. This idea is a difficult one to grasp. You will find that "variance accounted for" progressively accumulates meaning as you become more knowledgeable about and more experienced with statistics. Right now, let's try to add a little more meaning to the concept by considering it from three vantage points: information analysis, overlapping variance, and comparative strength of correlations.

Percent of information conveyed All the different job performance scores shown in Table 6.2 are 100% of the information

---

[3] See Chapter 9 if you absolutely can't wait to get all the facts on the topic of statistical reliability.

available about the employees' job-related competence. Since the square of the correlation between test scores and performance is approximately .74, it is accurate to say that *74% of the information about the employees' job-related competence is known on the basis of their test scores alone*—prior to the appraisal of their work behavior. This indicates that the selection test is a valid instrument to use in determining which applicants for that type of job should be hired. After all, on an average you will have about 74% of the information on their future job proficiency at the time they take the test.

Overlapping variance Percent of variance accounted for also refers to the percent of statistical overlap between the variance of X and the variance of Y—in other words, the percent of variation in Y that is covariation with X.

This way of looking at "variance accounted for" is illustrated in Table 6.7, which shows the scores of four hypothetical subjects on three variables: X_1, X_2, and Y. The unique feature of the data set in Table 6.7 is that each value of variable Y is equal to the sum of the corresponding values of X_1 and X_2; that is, for any one of the four subjects, $Y = X_1 + X_2$. Furthermore, X_1 and X_2 are completely uncorrelated with each other ($r_{X1X2} = .00$), and each X variable is correlated with Y: $r_{X1Y} = .447$ and $r_{X2Y} = .894$. If you square these two correlations, you find that X_1 accounts for 20% of the variance in Y and X_2 accounts for 80% of the variance in Y. It is not surprising, then, that 100% of the variation in these Y values is accounted for by the two X values. This is an unrealistic data set, but it does

Table 6.7

PROPORTION OF VARIANCE IN Y ACCOUNTED FOR BY X_1 AND X_2	PERSON	X_1	X_2	Y
	a	6	8	14
	b	4	0	4
	c	2	12	14
	d	0	4	4
	Sums:	12	24	36
	Means:	3	6	9
	Variances:	5	20	25

Correlation: $r_{X1Y} = .447$ $r_{X2Y} = .894$
r^2: $r_{X1Y}^2 = .20$ $r_{X2Y}^2 = .80$ ←proportion of variance accounted for

Ratio of variances:
$S_{X1}^2/S_Y^2 = 5/25 = .20 = r_{X1Y}^2$
$S_{X2}^2/S_Y^2 = 20/25 = .80 = r_{X2Y}^2$

illustrate the following: The square of the correlation in this special example is equal to the ratio of the variance of X to the variance of Y, or how much the variances overlap. Thus, since the variance of Y is 25, the variance of X_1 is 5, and the variance of X_2 is 20:

$$\frac{S^2_{X1}}{S^2_Y} = \frac{5}{25} = .20 = (.447)^2 = r^2_{X1Y}$$

$$\frac{S^2_{X2}}{S^2_Y} = \frac{20}{25} = .80 = (.894)^2 = r^2_{X2Y}$$

In real data sets (rather than contrived sets, like this one), this mathematical relationship holds only after the X variance has been converted to the same "unit of measurement" as the Y variance. Furthermore, in the real world, more than two X variables are normally needed to explain 100% of the variation in Y. Nonetheless, the idea shown here—that "proportion variance accounted for" refers to the extent that X variance overlaps with Y variance—is a generally valid way to interpret the squared correlation coefficient. And bear in mind that the more strongly two variables are associated with each other, the more overlap there is.

Comparative strength of correlations   When you are comparing two correlations to judge the relative strength of the relationships that they represent, you can get a more accurate assessment if you compare squared r's rather than unsquared r's. The reason is that r^2 is on the ratio scale of measurement,[4] whereas r is only on an ordinal scale (see the Concept Recap below). Hence, a correlation of .80 is not twice as strong as a correlation of .40; it is four times as strong, in the sense that $r = .80$ accounts for 64% of the variance in Y, whereas $r = .40$ accounts for a mere 16% of the variance. Put another way, "units of correlation" that are relatively high on the correlation scale (say, between .50 and 1.00) are "worth more" than "units" that are lower on the scale (say, up to .49). Squaring r makes all units equal.

Also, squaring r often provides a more realistic perspective on small correlations, which account for minuscule amounts of variation in Y. Gender, for example, has about a .10 correlation with scores on tests of mathematics aptitude (Westoff, 1979). Squaring this correlation accurately indicates just how trivial gender is as a predictor of quantitative ability.

Yet another approach to variance accounted for will be presented in the next chapter.

---

[4] Note that it is the r^2 index itself that is on a ratio scale, not necessarily the variables that are involved in the correlation. Those variables may be on any scale of measurement.

> **CONCEPT RECAP: Scales of Measurement**
> - A variable is measured on an *ordinal scale* if you can rank people on the variable but cannot specify how far apart two people are on the variable.
> - A variable is measured on an *interval scale* if all units on the scale are equal to one another and, therefore, you can specify distances between people on the variable.
> - A *ratio scale* of measurement is an interval scale with an absolute zero point. Therefore, you can form true ratios of scale values and say that 64% of the variance is four times as much as 16% of the variance, for example.

A Brief Note on Causality

A common mistake in the behavioral sciences is to infer that, since *X* and *Y* are strongly correlated, changes in variable *X* are causing changes in variable *Y*. By itself, *a correlation between two variables shows only that the two variables covary systematically;* they may or may not be related in any cause-and-effect way. For example, the number of people who drown tends to increase as ice cream consumption goes up. But does eating ice cream cause people to drown? Neuroticism is inversely correlated with self-esteem, but what is the direction of cause and effect here? The fact is that we don't know. Perhaps high self-esteem contributes to extinguishing one's neurotic behaviors. On the other hand, maybe an absence of neurotic habits is good for one's self-opinion. But then, again, some third set of factors (as yet unmeasured) might be responsible for "causing" the correlation.

In general, a cause-and-effect interpretation of a correlation should be avoided unless the correlation in question was produced in a well-controlled experiment designed to rule out alternative explanations of the relationship.

THE PROBLEM OF RESTRICTED RANGE

Anything that causes the range of scores on variable X or variable Y to be less than it should be also tends to make the correlation coefficient smaller than it should be. The **restriction of range** problem may occur either because the sample of people measured is too homogeneous (that is, the subjects are unusually similar to one another on one or both of the variables) or because the method of measuring the variables limits the possible range of values. It is important to be aware of this potential problem so that you can avoid automatically concluding that a low correlation means there is no relationship between *X* and *Y*.

If you have good reason to expect a substantial correlation and yet the obtained *r* is very small, always check the score ranges to see if the small coefficient is a result of too little variation on either *X* or *Y* in your particular study. Remember that *a variable must have sufficient variation in order to have good covariation with any other variable.*

The restriction of range phenomenon is illustrated in Figures 6.12 and 6.13. The figures are from an investigation of the relationship between the priority of conventional religious values (*X*) in marriages and the quality (*Y*) of those marriages, as indicated by ratings of such characteristics as intimacy, fidelity, and happiness (Splinter, 1989). The main hypothesis was that reported marriage quality would be a direct function of reported adherence to the value system. The investigation used a sample of 90 respondents, 36 of whom were currently married and 54 of whom were currently divorced. All subjects were members of the same church. The divorced subjects rated the quality of their former marriages "when they were at their best."

Figure 6.12 shows the scatter diagram for the entire sample of 90. Note that there is a substantial correlation between priority of conventional religious values (*X*) and reported quality of the marriages (*Y*), where *r* = .65. This statistically reliable finding indicates that the higher the priority respondents gave to values, the better their marriages tended to be, which was consistent with the research hypothesis. Also note that there is a substantial range of scores on both *X* (range = 17) and *Y* (range = 27).

When the 36 currently married subjects were assessed independently of their divorced counterparts, however, the correlation coefficient between priority of values and marriage quality dropped

Figure 6.12 *Quality of Marriage As a Function of Value Priorities (r = .65) (Note the large range of scores on both variable X and variable Y.)*

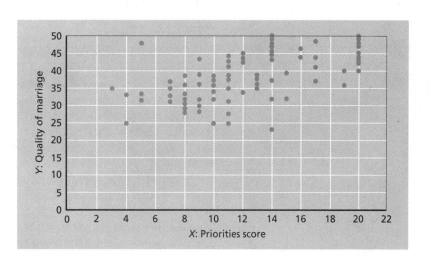

Figure 6.13 *Marriage Quality Versus Value Priorities for Marrieds Only (r = .13) (Note the restricted range of values on variables X and Y.)*

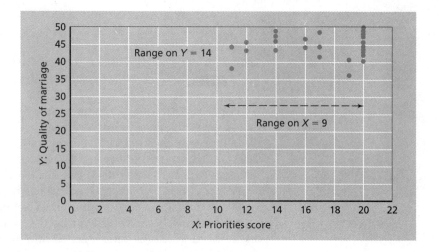

to .13, which is not statistically reliable. Figure 6.13 reveals what happened in this instance. The range of values was greatly restricted on both variable *X* (range = 9) and variable *Y* (range = 14). In other words, the married subjects varied so little on these dimensions that *X* and *Y* did not have room to covary with each other. Hence, the size of the correlation was artificially restricted.

This outcome does not necessarily mean that priority of values is unrelated to marriage quality among currently married people. It means only that the research method used in this study did not produce a range of response that was sufficient to address the research hypothesis within the context of the married subgroup. Thus, when restriction of range seems evident in a research project using the correlation coefficient, you probably should avoid drawing the conclusion that *X* and *Y* are unrelated.

THE PEARSON *r* AND DIFFERENT SCALES OF MEASUREMENT

The examples in this chapter have involved data measured on interval or ratio scales. But a real advantage of *r* is that it is "scale free"; it can be meaningfully computed on data from any scale of measurement. When *r* is calculated on pairs of ranks, it is called "Spearman's rank-order correlation coefficient" or, more simply, "*rho*" (symbolized r_S). When applied to nominal-scale data, *r* is called the "*phi* coefficient" (symbolized φ; see Chapter 15). Both *rho* and *phi* have special formulas that are not presented here. Those formulas are used for convenience only. You get the same value of *r* whether you use the special formulas or the usual *z*-score and raw-score formulas that we have been working with (see Hays, 1988; Kirk, 1990).

KEY TERMS

relationship	abscissa
correlation	line of best fit
linear	regression line
Pearson r	null relationship
independent variable	curvilinear relationship
predictor variable	spurious r
dependent variable	proportion of variance
criterion variable	accounted for
negative (inverse) correlation	coefficient of determination
positive (direct) correlation	coefficient of
scatter diagram, scatterplot	nondetermination
ordinate	restriction of range

SUMMARY

1. All empirical sciences base their theories, predictions, and applications on relationships between variables. The Pearson r, or linear correlation coefficient, is the statistical device most often used to discover and utilize these relationships.

2. The Pearson r is a numerical index of the direction and strength of the linear association between two variables. It assumes that variables X and Y have an essentially straight-line relationship and that scores on those variables exist in pairs.

3. The linear correlation coefficient has both scientific and applied uses, the latter involving the prediction of future behavior.

4. The properties of r include size (.00 to 1.00) and direction of relationship (positive or negative). The size property is interpreted independently of the direction property.

5. A graphing device called the scatter diagram provides information on the form, direction, and strength of a relationship.

6. The linear correlation coefficient, as a numerical index, is defined by the z-score formula, which conveys the idea that the size of r depends on how well people's positions on the X dimension correspond to their positions on the Y dimension. Since the z-score formula is cumbersome to use, r is most often calculated with the raw-score computational formula.

7. The linear correlation coefficient is interpreted according to its absolute magnitude, its statistical reliability (reproducibility), and the proportion of variance in variable Y that it allows us to account for on the basis of variable X alone. By itself, a correlation coefficient is not sufficient for supporting a cause-and-effect interpretation.

8. The size of r can be artificially lowered by a restriction of the range of values on either X or Y.

REVIEW QUESTIONS

1. In what sense is the linear correlation coefficient the "statistical essence of science"? Give a specific answer.

2. In your own words, define the term *relationship* as a statistician might.

3. Roy and Mary are disagreeing on the interpretation of some correlational findings. Research showed that marital satisfaction was correlated with the number of shared interests between spouses ($r = .32$), number of years of marriage ($r = .20$), and number of children living at home ($r = -.43$). Roy claims that number of shared interests is most highly correlated with satisfaction and that number of children is most weakly associated with satisfaction. Mary asserts that, to the contrary, number of children is most strongly associated with satisfaction. Who is correct in this debate, if either? Explain your answer.

4. Make up a set of pairs of data for which the Pearson r is not an appropriate index of correlation. Also, tell why r is inappropriate to use in this particular case.

5. Which of the following relationships is likely to be direct (positive) and which inverse (negative)?
   (a) Amount of television viewing per week and college grade point average
   (b) Age and visual ability
   (c) Advertising money and sales volume within a company
   (d) Height of parents and height of their children
   (e) Economic frustration and social class

6. In your own words, explain what is meant by the statement that the size of a correlation is independent of its sign.

7. In your own words, describe the relationship between z scores and correlation.

8. If there is a perfect inverse relationship between geography students' scores on exam 1 and their scores on exam 2, then what z scores would you expect the following students to have gotten on their second exam?

Student	Exam 1	Exam 2
Gary	$z = -2.2$	?
Susan	$z = 0.0$	?
Brenda	$z = -1.0$	?
Mike	$z = 2.2$	?

9. Given the following raw scores and summary statistics for variables X and Y, calculate the Pearson r using the z-score formula.

X: 10, 15, 19, 17, 21, 6, 9 $\bar{X} = 13.86$ $S_X = 5.19$

Y: 82, 72, 55, 76, 67, 91, 84 $\bar{Y} = 75.28$ $S_Y = 11.08$

10. Use the z-score formula to compute r for the data given in Table 6.3. Does your answer agree with the raw score calculation shown in Table 6.6? If there is a small discrepancy, how do you account for it?

11. Use the raw-score computational formula (6.2) to calculate r for the data in Question 9. Do the two calculation methods agree? If there is a small discrepancy, how do you account for it?

12. One theory of memory attributes the long-term retention of material to the degree of rehearsal, or the amount of time that one spends reciting information that he or she has just perceived. It has also been observed that extroverted people like to talk a lot about trivia that they have just heard or read about. Therefore, one plausible hypothesis is that subjects' knowledge of trivia is correlated directly with their scores on a test of extroversion. The data shown here are the results of a study designed to test this hypothesis. Construct a scatter diagram of these data. What does the graph suggest concerning the (a) linearity, (b) direction, and (c) strength of the relationship between the variables? Take a guess at the numerical size of the correlation.

Subject	Extroversion Score	Trivia Score
a	32	40
b	12	43
c	54	7
d	27	18
e	40	28
f	23	51
g	44	19
h	34	36
i	20	14
j	32	23

13. Use the raw-score computational formula to calculate r for the data in Question 12 to find out whether the research hypothesis is supported.

14. Interpret the result of Question 12 within the following frameworks: (a) absolute magnitude of the correlation, and (b) variance accounted for by r.

15. In terms of variance accounted for, is the difference between $r = .70$ and $r = .50$ equal to, less than, or greater than the difference between $r = .50$ and $r = .30$? Justify your answer.

16. For all freshman students at a university, the correlation between parents' income and college grade point average is .75. On this basis, is it logical to expect Mary's GPA to increase significantly since her mother was just recently promoted to a lucrative position within a large company? Why or why not? Be as specific and complete as possible.

17. Use your own words to explain the following assertion: Correlation is necessary but not sufficient for causation.

18. What is restriction of range and how might it affect the size of a correlation? Your answer should include both the concept of variation and the concept of covariation.

19. A manager trainer is interested in a possible relationship between leadership style and the verbal problem-solving strategies used by middle managers. The trainer is working with six "task-oriented" leaders and seven "social-emotional" leaders. In group problem-solving situations, the trainer notes the percent of leader verbalizations that are questions and the percent that are suggestions or assertions. The research hypothesis is that there will be a positive correlation between the social-emotional leadership style and the frequency of questions asked. For purposes of calculating the Pearson r between leadership style and question asking, the trainer identifies the social-emotional leaders with the numeral 1 and the task-oriented leaders with a 0. The results appear in the table. Compute the linear correlation coefficient between leadership style and percent of questions asked. Evaluate the hypothesis with a proportion of variance accounted for criterion.

Leadership Style	Code X	Percent of Question-Type Verbalizations Y
Task	0	20
Task	0	14
Task	0	56
Task	0	37
Task	0	29
Task	0	19
Social	1	28
Social	1	48
Social	1	35
Social	1	68
Social	1	65
Social	1	70
Social	1	20

20. For the data in Question 19, remove the three pairs with the highest percent of question-type verbalizations and recompute the correlation coefficient. What is the impact of the smaller range of scores on the correlation?

21. The following data are the ages of employees (variable *X*) paired up with the amount of job satisfaction reported by those employees on a 100-point rating scale (variable *Y*). Compute the linear correlation between *X* and *Y*. Now remove the one extreme pair ($X = 60$, $Y = 98$) and recompute the correlation. What is your conclusion about the effect of extreme observations on the correlation coefficient? What cautionary point does this exercise suggest regarding the interpretation of a linear correlation that includes one or two extreme pairs?

(*X*) Age: 20, 33, 37, 40, 41, 43, 60

(*Y*) Rated job satisfaction: 30, 32, 31, 35, 34, 32, 98

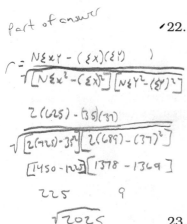

part of answer

$$r = \frac{N\Sigma xy - (\Sigma x)(\Sigma y)}{\sqrt{\left[N\Sigma x^2 - (\Sigma x)^2\right]\left[N\Sigma y^2 - (\Sigma y)^2\right]}}$$

$$\frac{2(625) - (35)(37)}{\sqrt{\left[2(725) - 35^2\right]\left[2(689) - (37)^2\right]}}$$

$$[1450 - 1225][1378 - 1369]$$

$$225 \qquad 9$$

$$\sqrt{2025}$$

$$\frac{45}{45} = 1$$

22. Calculate a Pearson *r* for the following three pairs. Now pick any two pairs and calculate the correlation for those two pairs only. Finally, discard one of the two pairs just analyzed, substitute in the remaining pair, and calculate the correlation for those two pairs only. What does this exercise suggest about using a linear correlation coefficient when there are fewer than three pairs of *X* and *Y* values?

X	x^2	Y	y^2	Cross Product
10	100	20	400	200
20		13		
25	625	17	289	425

23. The following data represent the observed relationship between class size and amount of discussion during class meetings in college courses. Compute a linear correlation coefficient for these data, and interpret the result according to both the "magnitude" criterion and the "proportion of variance accounted for" criterion.

Class Size *X*	Amount of Discussion *Y*
30	55
80	21
24	50
50	32
60	26
44	34
18	75

24. A sociologist wishes to investigate the relationship, if any, between the typical amount of education required in different

professions and the rated social prestige of those professions. The results of the study appear below. Compute and interpret a Pearson r for these data.

Profession	Years of Education	Prestige Rating
Medical doctors	20	58
Judges	18	48
Clergy	17	43
Bankers	14	37
Lawyers	18	31
School principals	19	25
Business executives	14	22
School teachers	17	18
Funeral directors	16	17
Local politicians	14	16
Advertising practitioners	15	8
Realtors	13	6

25. It has been hypothesized that there is a correlation between the number of children in a family and the average IQ of the children. Use the following data to determine whether there is such a correlation, whether the relationship is positive or negative, and the strength of the relationship, if any.

Number of Children	Average IQ of Children
1	103
2	102
3	102.5
4	101.3
5	100.6
6	102
7	98
8	98
9	95
10	93

CHAPTER 7   LINEAR REGRESSION

In Chapter 6, you learned about the Pearson *r*, or linear correlation coefficient, which is an index of the degree and direction of a straight-line relationship between two variables. I mentioned at that time that linear correlation has both scientific and applied functions. It may be used not only to discover relationships between variables (for the sake of advancing scientific knowledge) but also to *predict* future values of a *Y* variable on the basis of present values of an *X* variable. This prediction function has great value in applied behavioral science.

Using such a predictive device, for example, you can forecast students' college grade point averages from their Scholastic Aptitude Test scores. You can get reasonable estimates of a new employee's level of job performance from a knowledge of her scores on a job qualifications test. With this technique, you can even predict about how many sessions of psychotherapy will be necessary to reach a 75% rate of patient improvement (Howard, Kopta, Krause & Orlinsky, 1986).

The general procedure involved in using r to predict future outcomes is called **linear regression analysis.** Its most essential assumption is that variables *X* and *Y* have a straight-line relationship with each other. (Its other assumptions will be covered later in this chapter.) If that assumption is true for a set of pairs of scores, then *Y* values can be predicted from *X* values. The stronger the correlation between *X* and *Y*, the more accurate are the predictions.

> **CONCEPT RECAP: PAIRS OF SCORES**
> Each person in your investigation has been measured on two variables, *X* and *Y*, and each person's *X* value is aligned in a table with his or her *Y* value.

We will explore the very useful idea of linear regression in a three-step sequence: building the regression model, evaluating the regression model, and using the regression model to partition variance.

BUILDING THE REGRESSION MODEL

Our study of linear regression begins with the set of pairs of *X* and *Y* scores shown in Table 7.1. There you see fictitious data on the percent of psychology students admitted to graduate school (variable *Y*) as a function of the grade category they fell into (variable *X*) in their undergraduate statistics course. As you probably can surmise, a grade point of 1 corresponds to a D, a grade point of 2 rep-

Table 7.1

PERCENT OF PSYCHOLOGY
STUDENTS ADMITTED TO
GRADUATE SCHOOL AS A
FUNCTION OF STATISTICS
COURSE GRADE

	STATISTICS GRADE POINT, X	PERCENT ADMITTED, Y
	1	0.00
	2	33.33
	3	66.67
	4	100.00
Sum	10.00	200.00
N	4	4
Mean	2.50	50.00
Variance	1.67	1851.82
Standard deviation	1.29	43.03

resents a C grade, and so on. These data are portrayed in the scatter diagram (or scatterplot) shown in Figure 7.1. That figure clearly shows that Y has a perfect positive correlation with X: $r = +1.00$. With that kind of absolute predictability from X to Y, it is easy to construct an uncomplicated prediction model.

CONCEPT RECAP
In the context of linear correlation and linear regression, the X variable is called the *predictor* and the Y variable is called the *criterion* (see Chapter 6).

Figure 7.1 *Percent of Psychology Students Admitted to Graduate School (X) As a Function of Grade Attained in Undergraduate Statistics Course*

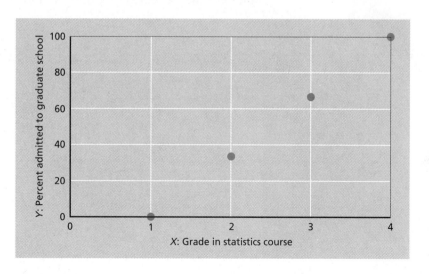

A Simple Model Based on *z* Scores

In Table 7.2 you can see the result of converting both the X and Y data to z scores. In keeping with the idea of a perfect positive correlation, the z scores on Y are equivalent to the corresponding z scores on X. Thus, in the unlikely event that two variables have a correlation of 1.00, the regression model that produces the most accurate prediction of Y from X is simply

$$z'_Y = z_X \tag{7.1}$$

where z'_Y is the predicted z score on variable Y, and z_X is the present z score on variable X. *Note that the ′ symbol on z_Y signifies that the z score is an estimated value.* The term for this combination of symbols is "z prime." *In this book, the "prime" symbol (′) always indicates an estimated value.*

Why would we want to use z scores in the prediction model? Keep in mind that the objective here is to predict values of variable Y from values of X. The problem is that X and Y are usually in different units of measurement. (Check Table 7.1 again to see what I mean.) Yet the prediction model must produce values on the Y variable scale. Using a z-score formula solves this problem because all z scores—regardless of the variables they are based on—are in the same units.

> **CONCEPT RECAP**
> All z scores are on the same scale, with a mean of 0 and a standard deviation of 1. The formula for a z score is $z_X = (X - \bar{X})/S_X$ or $z_Y = (Y - \bar{Y})/S_Y$.

The bad news is that this uncomplicated prediction model has limited utility because it works well only when two variables are

Table 7.2

z SCORES FOR PERCENT OF PSYCHOLOGY STUDENTS ADMITTED TO GRADUATE SCHOOL AS A FUNCTION OF STATISTICS COURSE GRADE		STATISTICS GRADE POINT, z_X	PERCENT ADMITTED, z_Y
		1.162	1.162
		0.387	0.387
		−0.387	−0.387
		−1.162	−1.162
	Sum	0.00	0.00
	N	4	4
	Mean	0.00	0.00
	Variance	1.00	1.00
	Standard deviation	1.00	1.00

perfectly predictable from each other. The good news is that when *X* and *Y* have a correlation of less than 1.00, we must elaborate the regression formula only slightly to have a workable model.

A Realistic *z*-Score Model

The pairs of *X* and *Y* scores shown in Table 7.3 do not represent a perfect correlation. In this case *r* = .894. The data were produced by a small class of behavioral statistics that I taught during a summer school session. The variable-*X* values are the scores that the nine students got on the first of five exams in the course; the variable-*Y* values are the corresponding final course averages achieved by those students.

The question in this example is: How well can statistics students' final grades be forecast from their performances on the very first exam in the class? This question was not posed purely for pedagogical purposes. My interest in it was stimulated by a recurring practical problem: Students who don't fare so well on the first test in my statistics classes often ask me what kind of final grade they are likely to make, considering their initial outcome—and, relatedly, whether it would be wise to withdraw from the course. Until I conducted a few regression analyses, I didn't have a very solid basis for answering this question. Now I do. Let's see how I developed it.

Figure 7.2 is a scatterplot of the statistics exams data. The graph shows that the relationship between exam 1 scores and final averages is basically linear. This feature of the data is accentuated by the "line of best fit" that I've drawn through the dots. This line, called the **regression line**, is *a kind of graphical average of the dots in the scatterplot*.[1]

Table 7.3

FINAL AVERAGES IN STATISTICS (*Y*) AS A FUNCTION OF EXAM 1 SCORES (*X*)		EXAM 1, *X*	FINAL AVERAGE, *Y*
	Student 1	31	43
	Student 2	46	65
	Student 3	54	60
	Student 4	62	62
	Student 5	68	63
	Student 6	72	67
	Student 7	78	69
	Student 8	86	87
	Student 9	88	90

---

[1] There is an exact mathematical way to determine the position of the regression line within the configuration of dots. You will learn how this is done shortly.

Figure 7.2 *Scatterplot of Final Averages As a Function of Exam 1 Grades (Note: The "line of best fit" is the regression line.)*

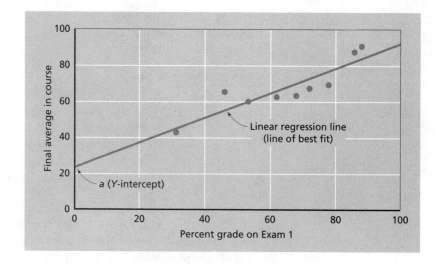

When X and Y have a linear relationship but the correlation is less than 1.00, the **definitional regression formula** for predicting values of Y is

$$z'_Y = rz_X \tag{7.2}$$

The formula says this: *To calculate an estimated z score on variable Y from a known z score on variable X, multiply the correlation coefficient between X and Y by the known z_X.* You can see that this way of estimating Y values takes into account the information provided by the correlation between X and Y.

Like most definitional formulas, expression (7.2) makes some important points about the statistical concept being considered:

- The higher the correlation between X and Y, the closer the *estimated z_Y* (that is, z'_Y) approximates the known z_X.
- When $r = 1.00$, the regression formula predicts that $z_Y = z_X$.
- When $r = .00$, the best prediction for every X value is that $z_Y = 0$—that is, that every Y value equals the mean of Y (since $z_Y = 0$ is the standard score at the mean of Y).
- When $r < 1.00$, *the predicted Y value is closer to its mean than the corresponding X value.* For example, if $r = .89$, then when $z_X = 3$ (i.e., 3 standard deviations from its mean), $z'_Y = rz_Y = .89 \cdot 3.00 = 2.67$ (i.e., 2.67 standard deviations from its mean). In a sense, the prediction model assumes that the Y value will "regress" toward its mean, in comparison to the corresponding X value. Indeed, this characteristic of the definitional formula is the basis of the terms *linear regression* and *regression analysis*.

Although the z-score formula helps you grasp some ideas underlying linear regression, you know by now that large batches of z scores are tedious to compute and work with. Therefore, it's time to move on to the computational formula.

A Raw-Score Regression Model

Some algebra transforms $z'_Y = rz_X$ into a *computational* formula for linear regression. The logic is fairly simple. Since $z'_Y = (Y' - \overline{Y})/S_Y$ and $z_X = (X - \overline{X})/S_X$, we can rewrite $z'_Y = rz_X$ as:

$$\frac{Y' - \overline{Y}}{S_Y} = r\left(\frac{X - \overline{X}}{S_X}\right)$$

Therefore,

$$Y' - \overline{Y} = r\left(\frac{S_Y}{S_X}\right)(X - \overline{X})$$

If we symbolize $r(S_Y/S_X)$ with b, then

$$Y' - \overline{Y} = b(X - \overline{X})$$
$$= bX - b\overline{X}$$

Adding $\overline{Y}$ to each side of the equation results in

$$Y' = bX + (\overline{Y} - b\overline{X})$$

If we symbolize $(\overline{Y} - b\overline{X})$ with a, we arrive at the raw-score regression equation:

$$Y' = bX + a \tag{7.3}$$

where Y' is the predicted raw-score value of Y, b is the slope of the regression line, X is a known raw score on variable X, and a is the Y-intercept (or simply "intercept"), the value of Y at which the regression line is "intercepted" by (i.e., crosses) the Y axis.

Two features of this computational model are worth noting:

1. The equation enables you to use raw scores on X to predict raw scores on Y.
2. The equation automatically converts the predicted values to raw-score units on the Y variable scale when X and Y are not on the same scale of measurement.

The raw score on X is self-evident, so let's consider the meaning and computation of the other parts of the formula.

The slope   Graphically, the **slope** refers to the steepness of the regression line. *Other things being constant*, the larger the slope index—also called the **regression coefficient**—the steeper the regression line. Numerically, *the slope refers to the number of units of change in Y' for each unit increase in variable X.*

This definition is illustrated in Figure 7.3. There you see that the slope equals +0.671. As shown in the figure, this means that the raw-score value of Y' increases by 0.671 every time the raw-score value of X goes up by 1. For instance, if $Y' = 25$ when $X = 40$, then when X increases to 41, Y' changes to 25.671.

You should also be aware that b can be either positive or negative, depending on whether X and Y are positively or negatively correlated. If the slope is a negative value, then Y' *decreases* by b units every time X increases by 1.

The slope of a regression line is calculated in the following way:

$$b = r\left(\frac{\overset{②}{\downarrow}S_Y}{\overset{①}{\downarrow}S_X}\right) \tag{7.4}$$

where r is the correlation between X and Y, S_Y is the standard deviation of the Y raw scores, and S_X is the standard deviation of the X raw scores.

FORMULA GUIDE

① Divide the standard deviation of Y by the standard deviation of X.
② Multiply r by the result of step 1.

CONCEPT RECAP

A **standard deviation** is a measure of variation in a data set. It is the square root of the variance, which is the average of the squared deviations from the mean. The more variation there is among scores in a data set, the bigger is the standard deviation. The formulas for the standard deviation of Y include:

$$S_Y = \sqrt{\frac{\Sigma(Y - \bar{Y})^2}{N}} = \sqrt{\frac{\Sigma Y^2 - \frac{(\Sigma Y)^2}{N}}{N}} = \sqrt{\frac{SS_Y}{N}}$$

The quantities needed to build a linear regression model for the statistics course data appear in Table 7.4. Note that $S_Y = 13.342$ and $S_X = 17.776$. Earlier I noted that r in this problem is .894. The calculations in Table 7.4 verify that fact. Hence, the slope for the present data is:

$$b = r\left(\frac{S_Y}{S_X}\right)$$

$$= .894\left(\frac{13.342}{17.776}\right)$$

$$= .894(0.751)$$

$$= 0.671$$

Figure 7.3 *Illustration of the Graphical Meaning of the Slope of a Regression Line*
(Note: Since the slope is 0.671, the value of Y' increases by 0.671 unit every time X increases by 1.)

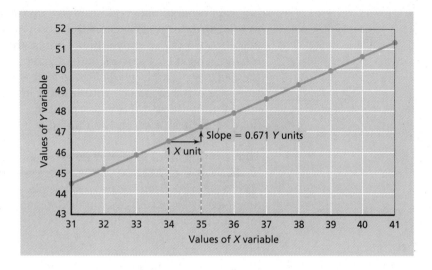

Table 7.4

FINAL AVERAGES AS A FUNCTION OF EXAM 1 SCORES: BASIC CALCULATIONS

	SCORES ON EXAM 1, X	FINAL AVERAGES, Y	X^2	Y^2	XY
Student **1**	31	43	961	1,849	1,333
Student **2**	46	65	2,116	4,225	2,990
Student **3**	54	60	2,916	3,600	3,240
Student **4**	62	62	3,844	3,844	3,844
Student **5**	68	63	4,624	3,969	4,284
Student **6**	72	67	5,184	4,489	4,824
Student **7**	78	69	6,084	4,761	5,382
Student **8**	86	87	7,396	7,569	7,482
Student **9**	88	90	7,744	8,100	7,920
Sums:	$\Sigma X = 585$	$\Sigma Y = 606$	$\Sigma X^2 = 40,869$	$\Sigma Y^2 = 42,406$	$\Sigma XY = 41,299$
Means:	$\overline{X} = 65.00$	$\overline{Y} = 67.333$			

1. Find variance of X:
$$S_X^2 = \left[\Sigma X^2 - (\Sigma X)^2/N\right]/N = \left[40,869 - (585)^2/9\right]/9 = 316.00$$

2. Find standard deviation of X:
$$S_X = \sqrt{S_X^2} = \sqrt{316.00} = 17.776$$

3. Find variance of Y:
$$S_Y^2 = \left[\Sigma Y^2 - (\Sigma Y)^2/N\right]/N = \left[42,406 - (606)^2/9\right]/9 = 178.00$$

4. Find standard deviation of Y:
$$S_Y = \sqrt{S_Y^2} = \sqrt{178.00} = 13.342$$

5. Find correlation between X and Y:
$$r = \frac{\Sigma XY - (\Sigma X \Sigma Y)/N}{\sqrt{\Sigma X^2 - (\Sigma X)^2/N}\sqrt{\Sigma Y^2 - (\Sigma Y)^2/N}} = \frac{41,299 - (585 \cdot 606)/9}{\sqrt{40,869 - (585)^2/9}\sqrt{42406 - (606)^2/9}} = .894$$

You probably noticed that I have taken the above quantities out to the third decimal place, rather than only to the second decimal place (as would be conventional when the data are integers). *It is always a good idea to work with at least the third, and perhaps even the fourth or fifth, decimal place when computing the slope.* The reason is that the slope is a multiplier in the regression equation. As such it must be as precise as possible, so that the product of bX is not distorted by a severely rounded b.

The Y-intercept   The **Y-intercept**, symbolized a, is *the value of Y at which the regression line crosses the Y axis.* Examination of either Figure 7.2 or Figure 7.4 reveals that a is the value that Y' assumes when variable X equals 0. Once you have computed b, the intercept is a cinch to calculate:

$$a = \overline{Y} - b\overline{X} \qquad\qquad (7.5)$$

For the current data set:

$$a = \overline{Y} - b\overline{X}$$
$$= 67.33 - 0.671(65.00)$$
$$= 67.33 - 43.62$$
$$= 23.71$$

As indicated in Figure 7.2, the Y axis "intercepts" the regression line at 23.71.

Using the Y-intercept to locate the regression line   Figure 7.4 shows you how to use the intercept and the means of X and Y to determine the precise location of the regression line among the points in a scatter diagram. The procedure is a 1-2-3 sequence:

Figure 7.4 *How to Locate the Regression Line from Means and the Y-Intercept*

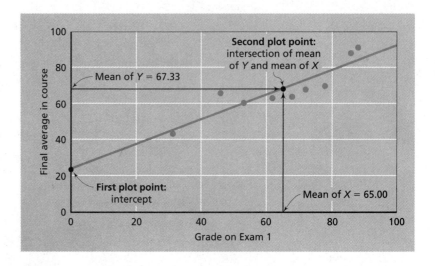

1. Place a dot where the intercept value (a) falls on the Y axis.
2. Place a dot where the mean of Y intersects with the mean of X.
3. Connect the two dots with a straight line that extends as far as the highest X value in your data set.

You may also determine the location of the regression line by using the regression equation to calculate two or more predicted Y values, plotting those Y' values against the X values used to generate them, and connecting the plot points. This will always give you an accurate plot of the regression line because *all the predicted Y values lie exactly on the regression line.* That is true because the regression equation that predicts Y values also defines the regression line.

Calculation of Predicted Y Values

Once b and a have been determined, it is simple to plug these constants into the computational formula (7.3) and generate the predicted Y values. Note that a Y' *must be calculated for each person in the sample,* as follows:

$$Y' = bX + a \qquad\qquad Y'\downarrow$$

Student 1	Y'	=	$0.671 \cdot 31$	+	23.71	=	20.80	+ 23.71	=	44.51
Student 2	Y'	=	$0.671 \cdot 46$	+	23.71	=	30.87	+ 23.71	=	54.58
Student 3	Y'	=	$0.671 \cdot 54$	+	23.71	=	36.23	+ 23.71	=	59.94
Student 4	Y'	=	$0.671 \cdot 62$	+	23.71	=	41.60	+ 23.71	=	65.31
Student 5	Y'	=	$0.671 \cdot 68$	+	23.71	=	45.63	+ 23.71	=	69.34
Student 6	Y'	=	$0.671 \cdot 72$	+	23.71	=	48.31	+ 23.71	=	72.02
Student 7	Y'	=	$0.671 \cdot 78$	+	23.71	=	52.34	+ 23.71	=	76.05
Student 8	Y'	=	$0.671 \cdot 86$	+	23.71	=	57.71	+ 23.71	=	81.42
Student 9	Y'	=	$0.671 \cdot 88$	+	23.71	=	59.05	+ 23.71	=	82.76

Now that we have used the regression formula to forecast final averages, the next step is to use these predicted values together with the actual raw scores on variable Y to check the accuracy of the model.

EVALUATING THE REGRESSION MODEL

There are three ways to assess the precision with which a regression model predicts values of Y: (1) through what is called a "least squares" analysis, (2) by computing the standard deviation of prediction errors, and (3) via cross-validation of the regression formula with a new sample of subjects.

The Least Squares Principle

Unless the correlation between X and Y is perfect (that is, $r = 1.00$ or $r = -1.00$), the regression formula you build will not work flawlessly. It will make prediction errors, where a **prediction error** is defined *as the deviation between the actual Y value that a subject has and the Y value that the formula predicts for him or her.* Symbolically, a prediction error is $(Y - Y')$.

Prediction errors are illustrated in Figure 7.5, which is based on the data from the nine statistics students. Observe that each predicted Y value lies exactly on the regression line and is symbolized by a dark dot. For every predicted Y value there is an actual Y value symbolized by a light dot. If the actual Y value happens to fall on the regression line, then $(Y - Y') = 0$. If the actual Y value is above the regression line, then $(Y - Y')$ is a positive error of prediction. If the actual Y value lies below the line, then $(Y - Y')$ is a negative error of prediction. All nine prediction errors are shown in the fourth column of Table 7.5. You can see that the sum of the $(Y - Y')$ values is 0. That will always be true because the positive errors cancel the negative errors.

Of course, you want all prediction errors to be as small as possible. Actually the linear regression equation that we have been working with was developed with the goal of minimizing the *squared* prediction errors. Specifically, the model is based on the mathematical **principle of least squares:** *Predictions produced by $Y' = bX + a$ always yield the smallest possible sum of squared prediction errors.* This characteristic of the linear regression equation can be credited entirely to the formulas for *b* and *a*. That is, $\Sigma(Y - Y')^2$ is smaller when $r\ (S_Y/S_X)$ is used to generate the slope and $(\overline{Y} - b\overline{X})$ is used to compute the intercept than when any other formulas are substituted into these calculations.

Table 7.5 shows that $\Sigma(Y - Y')^2 = 320.60$ for the present data set. Is this good or poor? The answer is: That depends on what alternative predictor variables are available and what the sum of

Figure 7.5 *Linear Regression Line Relating Final Averages to Exam 1 Score*
(Note: A prediction error is the discrepancy between a predicted value, which always lies on the regression line, and the corresponding actual Y value.)

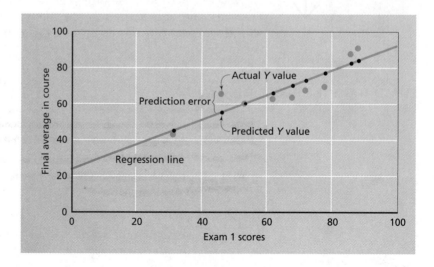

Table 7.5

PREDICTION ERRORS ($Y - Y'$) FOR THE REGRESSION OF FINAL AVERAGES ON EXAM 1 SCORES	EXAM 1 SCORES, X	FINAL AVERAGES, Y	PREDICTED FINAL AVERAGES, Y'	PREDICTION ERRORS, $Y - Y'$	SQUARED ERRORS, $(Y - Y')^2$
	31	43	44.51	−1.51	2.28
	46	65	54.58	10.42	108.66
	54	60	59.94	0.06	0.00
	62	62	65.31	−3.31	10.97
	68	63	69.34	−6.34	40.17
	72	67	72.02	−5.02	25.22
	78	69	76.05	−7.05	49.67
	86	87	81.42	5.58	31.18
	88	90	82.76	7.24	52.45

Sum of squared prediction errors: $\Sigma(Y - Y')^2 = 320.60$

Standard error of estimate:

$$S_{est.Y} = \sqrt{\frac{\Sigma(Y - Y')^2}{N - 2}} = \sqrt{\frac{320.60}{9 - 2}} = \sqrt{\frac{320.60}{7}} = \sqrt{45.80} = 6.77$$

the squared prediction errors is for each of them in comparison to the present X variable. Least squares is a comparative matter. In constructing a regression equation, you always want to use the X variable that yields the smallest $\Sigma(Y - Y')^2$. And, in a single-predictor situation, the X variable that best meets this criterion is always the one that has the highest linear correlation with Y: The higher r_{XY}, the smaller $\Sigma(Y - Y')^2$.

Standard Error of Estimate

It is also true that the higher r_{XY}, the smaller is the *standard deviation of the prediction errors*, referred to as the **standard error of estimate**. The standard error of estimate, symbolized $S_{est.Y}$, is a *numerical index of the average amount of deviation of Y from Y' along the regression line*. The more accurate the prediction equation, the more closely scatterplot points representing the actual Y values cluster near the regression line (defined by the predicted Y's), and the smaller $S_{est.Y}$ is.

Calculating the standard error of estimate The equation for $S_{est.Y}$ is:

$$S_{est.Y} = \sqrt{\frac{\Sigma(Y - Y')^2}{N - 2}} \quad \xleftarrow{} ① \qquad ③ \atop ② \tag{7.6}$$

FORMULA GUIDE

① Sum the squared prediction errors in a linear regression table.
② Divide the sum of the squared prediction errors by $N - 2$.
③ Take the square root of the result of step 2.

For the present data set:

$$S_{est.Y} = \sqrt{\frac{320.60}{9 - 2}}$$

$$= \sqrt{\frac{320.60}{7}}$$

$$= \sqrt{45.80}$$

$$= 6.77$$

A second way to calculate the standard error of estimate If you have the linear correlation coefficient and the variance of Y for a set of paired scores, then $S_{est.Y}$ may be easily computed this way:

$$S_{est.Y} = \left(\sqrt{S_Y^2(1 - r^2)}\right)\left(\sqrt{\tfrac{N}{N-2}}\right) \tag{7.7}$$

FORMULA GUIDE

① Multiply the sample variance of Y times 1 minus the squared correlation.
② Take the square root of the result of step 1.
③ Divide the number of pairs (N) by $N - 2$.
④ Take the square root of step 3.
⑤ Multiply the result of step 2 by the result of step 4.

For the data we are working with, $S_Y^2 = 178.00$ and $r = .894$ (see Table 7.4); thus, the computations are:

$$S_{est.Y} = \left(\sqrt{178.00(1 - .894^2)}\right)\left(\sqrt{\tfrac{9}{9-2}}\right)$$

$$= \left(\sqrt{178.00(1 - .800)}\right)\left(\sqrt{\tfrac{9}{7}}\right)$$

$$= \left(\sqrt{178.00(.200)}\right)\left(\sqrt{1.286}\right)$$

$$= \left(\sqrt{35.600}\right)(1.134)$$

$$= (5.967)(1.134)$$

$$= 6.77$$

Notice that the value of $S_{est.Y}$ is the same with this formula as that produced by equation (7.6), if you are willing to carry your decimal numbers to the third place. An important benefit of equation (7.7) is that it directly shows that the larger the correlation between X and Y, the less error there is in predicting Y from X. That is, the larger r^2 is, the smaller is the index of prediction errors, $S_{est.Y}$.

Basic ideas and assumptions behind $S_{est.Y}$ Although the standard error of estimate is computed from sample data, it is based on the notion that the regression equation can be applied to the population of subjects represented by the sample. The population for the present data is all students in my present and future statistics classes. The theory underlying $S_{est.Y}$ assumes the following results if we would apply the regression model to every subject in the population:

- Each X value (that is, exam 1 score) in the population would produce a particular Y' or predicted score.
- The plot point of each Y' as a function of its X would lie exactly on the regression line.
- Any given X score (and the corresponding Y') would be made by a very large number of subjects in the population.
- Because not everyone with the same exam 1 score would end up with the same final average, there would be a large distribution of actual Y scores around each Y' plot point, with Y' serving as a type of "mean" of the distribution.

The last point asserts that *each and every predicted point on the regression line would be surrounded by a distribution* of plot points that represent actual Y scores. Theoretically, each X value is associated with a population of actual Y values that vary around a mean of Y'. In effect, then, *the regression equation predicts the mean Y score for a population of persons who have a particular score on X.* This theory also rests on the following specific assumptions:

1. Variables X and Y have a straight-line relationship (the familiar **linearity assumption**).
2. *The spread, or variation, of actual Y scores around predicted Y scores is the same all along the regression line; that is, $S_{est.Y}$ has the same value for each and every Y'.* This is known as the **homoscedasticity assumption**. Figure 7.6 shows a scatterplot of data that essentially meet this assumption. Figure 7.7 illustrates data that violate it.

Figure 7.6 *Scatterplot Illustrating Data That Satisfy the Homoscedasticity Assumption*

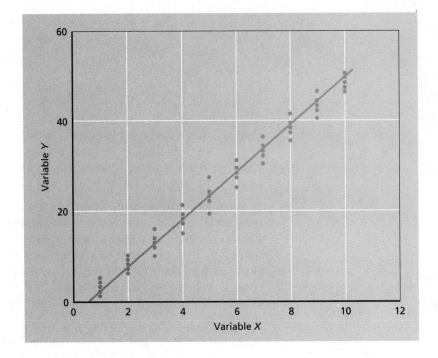

Figure 7.7 *Scatterplot Illustrating Data That Violate the Homoscedasticity Assumption*

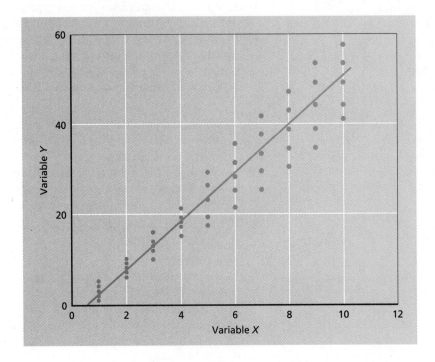

3. The dispersion of actual Y values around any given Y' has all the usual properties of the normal curve. This is the assumption of **bivariate normality**. It holds that Y *is normally distributed at every level of* X.

To the degree that these assumptions are true in a regression analysis, you can apply $S_{est.Y}$ in a most interesting way.

Applications of the standard error of estimate Consider the normal curve shown in Figure 7.8. The figure represents the distribution of plot points of actual Y scores around one predicted score. Bear in mind that the theory of $S_{est.Y}$ assumes that such a normal distribution exists for each predicted point on the regression line. As is true of any normal distribution, 68.26% of the values (in this case, actual Y's) lie between +1 standard deviation (that is, $+1S_{est.Y}$) and –1 standard deviation (that is, $-1S_{est.Y}$). Similarly 95.44% of the actual Y values are assumed to fall between $+2S_{est.Y}$ and $-2S_{est.Y}$.

This relationship between $S_{est.Y}$ and the distribution of actual Y values enables you to make the following kinds of statements:

- Approximately 95% of subjects in linear regression analysis obtain actual Y scores that do not differ from their predicted scores by more than $2S_{est.Y}$. In other words, about 95% of Y scores are contained in the interval $Y' \pm 2S_{est.Y}$.
- Since probability equals percent under the normal curve, the probability is approximately .95 that subjects' Y scores are within $2S_{est.Y}$ of their Y' values.
- About two-thirds of my statistics students should obtain final averages that are within $1S_{est.Y}$ of their predicted final averages.

Figure 7.8 *Normal Distribution of Actual Y Values Around a Particular Predicted Y at Some Point on the Regression Line (Note: Such a distribution of actual Y values is assumed to exist for each predicted Y value in the population.)*

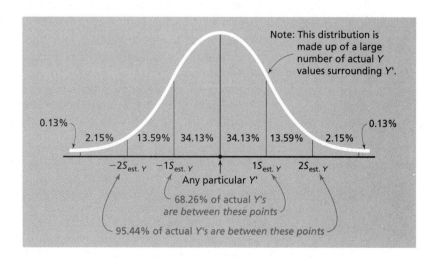

(In fact, if you look back at Table 7.5, you will find that for six of the nine students—exactly two-thirds of them—Y is within $1S_{est.Y}$ of Y'.)

In addition to gauging the accuracy of a regression model, then, the *standard error of estimate provides an objective way of stating the likely margin of error in our predictions*—that is, a way of confidently placing a limit on how much the actual Y values could reasonably deviate from predicted values. But remember that it is appropriate to use the standard error of estimate to state such limits only if the above assumptions are true. If the homoscedasticity assumption is false, then the amount of scatter among the actual Y values varies from one Y' to the next. Therefore, $S_{est.Y}$ is too large an index of error for some portions of the regression line and too small an index for other segments. If the bivariate normality assumption is false, the percent and probability statements are inaccurate because they are based on a normal distribution.

Cross-Validation of the Regression Model

A third major method of assessing the precision of a regression equation is to *examine how well it predicts future Y values with a new sample of subjects.* Such an investigation is called a **cross-validation study.**

Cross-validation is actually the second of three phases involved in the use of linear regression to forecast future behaviors and outcomes. The typical sequence is:

Phase 1: Use a "developmental" sample as an empirical foundation for constructing the regression model.

Phase 2: Cross-validate the regression model with a new sample of people to see how well the equation holds up outside the particular data set from which it was generated.[2] Normally, regression formulas lose some accuracy under cross-validation.

Phase 3: If the model shows good accuracy under cross-validation, begin applying it routinely to forecast the Y values of additional groups of people in some real-world setting; for example, an effective model might be used to predict college applicants' likely freshman grade point averages from their SAT scores.

Table 7.6 displays the results of a cross-validation study in which I used the regression equation developed from the summer

---

[2] This cross-checking with a new sample is necessary because the mathematics involved in calculating a and b "take advantage of chance variance" in the original sample in a way that maximizes the accuracy of the equation for that data set but not necessarily for other ones.

Table 7.6

EXAM 1 AND PREDICTED
VERSUS ACTUAL FINAL
AVERAGES IN A LARGE
FALL CLASS

EXAM 1, X	PREDICTED FINAL AVERAGE, Y′	PREDICTED GRADE	ACTUAL FINAL AVERAGE, Y	ACTUAL GRADE
24	39.804	F	49	F
38	49.198	F	43	F
38	49.198	F	53	D
40	50.540	D	47	F
40	50.540	D	53	D
40	50.540	D	50	D
42	51.882	D	57	D
44	53.224	D	54	D
46	54.566	D	51	D
46	54.566	D	55	D
48	55.908	D	54	D
48	55.908	D	57	D
48	55.908	D	49	F
50	57.250	D	57	D
50	57.250	D	55	D
50	57.250	D	48	F
50	57.250	D	55	D
52	58.592	D	55	D
54	59.934	D	60	C
54	59.934	D	58	D
54	59.934	D	60	C
54	59.934	D	58	D
56	61.276	C	60	C
58	62.618	C	58	D
58	62.618	C	60	C
60	63.960	C	71	C
62	65.302	C	67	C
62	65.302	C	62	C
62	65.302	C	65	C
64	66.644	C	65	C
64	66.644	C	61	C
64	66.644	C	61	C
66	67.986	C	56	D
66	67.986	C	73	C
68	69.328	C	67	C
68	69.328	C	61	C
68	69.328	C	63	C
68	69.328	C	67	C
68	69.328	C	65	C
70	70.670	C	63	C
70	70.670	C	62	C

(continued)

Table 7.6

(CONTINUED)

EXAM 1, X	PREDICTED FINAL AVERAGE, Y′	PREDICTED GRADE	ACTUAL FINAL AVERAGE, Y	ACTUAL GRADE
70	70.670	C	78	B
70	70.670	C	61	C
70	70.670	C	64	C
72	72.012	C	67	C
72	72.012	C	67	C
72	72.012	C	65	C
72	72.012	C	71	C
74	73.354	C	75	C
74	73.354	C	84	B
74	73.354	C	67	C
76	74.696	C	67	C
76	74.696	C	65	C
78	76.038	B	69	C
80	77.380	B	77	B
80	77.380	B	75	B
80	77.380	B	77	B
80	77.380	B	76	B
80	77.380	B	81	B
82	78.722	B	69	C
82	78.722	B	72	C
84	80.064	B	79	B
84	80.064	B	83	B
84	80.064	B	84	B
86	81.406	B	87	A
86	81.406	B	83	B
88	82.748	B	73	C
88	82.748	B	85	A
88	82.748	B	85	A
88	82.748	B	81	B
90	84.090	B	88	A
92	85.432	A	89	A
92	85.432	A	92	A
94	86.774	A	93	A

statistics class to predict (yes, actually predict in advance) the final averages of students in a larger fall term class. There you see not only the predicted final averages, but also the final grades that are expected on the basis of those projections. Comparison of the actual outcomes with those forecast reveals that the regression model constructed from the summer class's data enabled me to make reasonably accurate predictions of fall term final grades on the basis of fall term exam 1 scores. A little arithmetic will show that 76% of the actual grades matched the predicted grades.

Now you know how a simple linear relationship between variables can be used for behavioral estimation. But regression analysis isn't limited to just predicting behavior. It also provides a means to assess how much of the variation in Y values is associated with a particular X variable and how much is accounted for by other factors. Such information plays an important role in the development of scientific theories. It is to this second major function of linear regression that we now turn.

USING THE REGRESSION MODEL TO PARTITION VARIANCE

The results of a regression analysis allow you to break Y variance down into two parts: a proportion that is predictable from variable X and a proportion that is not "explained" or "accounted for" by X.[3] This procedure is known as **partitioning variance**. In regard to the students in my statistics course, for example, what proportion of student variation in final averages can be "known" from information provided by the first exam in the course? What proportion of final average variance is *not* associated with performance on exam 1? For another example, suppose you are told that there is a strong linear correlation between the number of psychotherapy sessions and reduction of depression in clients who suffer from major depression. What percent of the variance in depression levels can be "accounted for" by the amount of therapy per se? What percent is not predictable from the number of therapy sessions and therefore must be "explained" by other variables relevant to mood changes? We will consider this partitioning of variance at both a theoretical level and a computational level.

The Theory Behind Partitioning Variance

In a linear regression analysis, any particular Y score can be viewed as containing variance, in that the score is likely to vary from its mean. Hence, the theoretical focus of variance partitioning is on the so-called deviation score—that is, $(Y - \overline{Y})$. Each deviation score is viewed as consisting of two parts: a variance component that is predictable from variable X, represented by $(Y' - \overline{Y})$, and a variance

---

[3] In the partitioning of variance, the terms *variance explained* and *variance accounted for* do not imply the ability to ascertain cause and effect. "Explained," for example, is used only in a statistical sense that means that some Y variance is associated with X variance. Of course, we can never identify cause-and-effect processes solely on the basis of correlation.

component that is not "determinable" from X, represented by $(Y - Y')$. You will recognize the latter as a prediction error. Within this framework, then, you should think of the variance represented by a deviation score on variable Y as:

$$(Y - \overline{Y}) = (Y' - \overline{Y}) + (Y - Y') \qquad (7.8)$$

↑	↑	↑
deviation score	predicted component	error component

So there you have the general idea behind variance partitioning. To make this concept useful, however, you have to carry out variance calculations on the entire data set, rather than at the level of individual deviation scores.

Computing Variance Accounted For

In thinking about how much Y variance is explained by X, we start with the whole chunk of variance in the data set; the overall variance of Y, $S_Y^2 = \Sigma(Y - \overline{Y})^2/N$, constitutes 100% of the variation in Y values. In our summer statistics class example, $S_Y^2 = 178.00$ (from Table 7.4). How much of this total variance can be determined from information provided by variable X? We will answer that question through a reverse strategy. We first calculate the variance not accounted for by X and then use subtraction to obtain the part that is predicted by X.

Recall that we can get a sum of squared prediction errors from $\Sigma(Y - Y')^2$. Dividing this sum by N gives us a **variance of prediction errors**: $S_{\text{pred.errors}}^2 = \Sigma(Y - Y')^2/N$. Using the sum of squared prediction errors for the statistics class shown in Table 7.5, we calculate:

$$S_{\text{pred.errors}}^2 = \frac{320.60}{9} = 35.62$$

This represents the part of the Y variation not predictable from X.

Now how can we obtain the part of the Y variance that is predictable from X? Let's denote that variance with $S_{\text{explained}}^2$. It is logical to say that the total variance on variable Y is equal to the explained variance plus the variance of prediction errors: $S_Y^2 = S_{\text{explained}}^2 + S_{\text{pred.errors}}^2$. Following the same logic, we can arrive at the figure for explained variance by $S_{\text{explained}}^2 = S_Y^2 - S_{\text{pred.errors}}^2$. For statistics students,

$$S_{\text{explained}}^2 = 178.00 - 35.62 = 142.38$$

The partitioning of variance is now complete. The breakdown is:

$$S_Y^2 = S_{\text{explained}}^2 + S_{\text{pred.errors}}^2$$

$$178.00 = 142.38 + 35.62$$

This is somewhat informative, but it will be easier to interpret the results of the variance breakdown if we convert this information to a percentage scale.

Proportion of variance accounted for   *The proportion of variance accounted for,* called the **coefficient of determination**, can be found simply by forming a ratio of the explained variance to the total variance:

$$\text{Coefficient of determination} = \frac{S^2_{\text{explained}}}{S^2_Y} = \frac{142.38}{178.00} = .80$$

This means that 80% of the variance in final averages can be determined (is predictable) from exam 1 scores.

You may also be interested to know that the coefficient of determination is exactly equal to the square of the correlation coefficient. The correlation between the exam 1 scores and the final averages is .894 and

$$\text{Coefficient of determination} = r^2 = (.894)^2 = .80 = \frac{S^2_{\text{explained}}}{S^2_Y} \quad (7.9)$$

The upshot of this fact is that the quickest and most practical way to partition variance in regression is simply to square the correlation. These calculations also reinforce an important idea introduced in Chapter 6 (on linear correlation): r^2 *is the proportion of variation in Y accounted for by variable X.*

Proportion of variance not accounted for   Consistent with the preceding observations, *the proportion of variance in Y that is not determinable from variable X* is called the **coefficient of nondetermination**, symbolized as k. It can be calculated by

$$\text{Coefficient of nondetermination} = k = \frac{S^2_{\text{pred.errors}}}{S^2_Y} \quad (7.10)$$

For the data set we have been working with, $k = 35.62/178.00 = .20$. This figure is exactly the same as that produced by $(1 - r^2) = (1 - .894^2) = (1 - .80) = .20$. Hence, the more practical way to calculate k is simply to subtract the square of the linear correlation from 1:

$$\text{Coefficient of nondetermination} = k = (1 - r^2) \quad (7.11)$$

Either approach reaches the conclusion that 20% of the variation in final averages is not predictable from the exam 1 scores.

A SECOND EXAMPLE

We have covered a lot of territory in this chapter. It will be to your advantage to review the various concepts and procedures by going through another complete regression analysis. The raw material for this experience appears in Table 7.7. Seventeen pairs of scores

Table 7.7

LEVEL OF DEPRESSION AS A FUNCTION OF NUMBER OF THERAPY SESSIONS	SUBJECT	THERAPY SESSIONS, X	LEVEL OF DEPRESSION, Y
	1	0	98
	2	2	90
	3	5	92
	4	7	90
	5	8	84
	6	10	70
	7	10	76
	8	11	80
	9	12	76
	10	15	68
	11	16	70
	12	18	50
	13	21	40
	14	21	52
	15	23	30
	16	27	25
	17	30	13

represent the relationship between the number of psychotherapy sessions attended (variable X) and the current level of depression (variable Y) in 17 clients who have the diagnosis of major depression. If the therapy is effective, we expect a negative (inverse) relationship between number of therapy sessions (the predictor) and current level of depression (the criterion). That is, the more therapy sessions clients have, the less depressed they should be. Figure 7.9 verifies that, indeed, the variables are inversely correlated, and the relationship appears to be a strong one.

The regression analysis of these data appears in Boxes 7.1, 7.2, and 7.3, which take you through every step in the procedure. The objective is to construct a regression equation that will enable you to predict what levels of depression patients can be expected to exhibit after they have undergone different amounts of therapy. Box 7.1 shows calculations of the basic statistical indexes, from means and variances to the slope and intercept. Box 7.2 displays the regression equation and calculation of predicted Y values from that model. Box 7.3 presents the predicted Y scores, the prediction errors, and computation of the standard error of estimate and coefficient of determination.

Box 7.1

DEPRESSION LEVEL VERSUS NUMBER OF THERAPY SESSIONS: BASIC CALCULATIONS

SUBJECT	THERAPY SESSIONS, X	LEVEL OF DEPRESSION, Y	X^2	Y^2	XY
1	0	98	0	9,604	0
2	2	90	4	8,100	180
3	5	92	25	8,464	460
4	7	90	49	8,100	630
5	8	84	64	7,056	672
6	10	70	100	4,900	700
7	10	76	100	5,776	760
8	11	80	121	6,400	880
9	12	76	144	5,776	912
10	15	68	225	4,624	1,020
11	16	70	256	4,900	1,120
12	18	50	324	2,500	900
13	21	40	441	1,600	840
14	21	52	441	2,704	1,092
15	23	30	529	900	690
16	27	25	729	625	675
17	30	13	900	169	390

Sums: $\Sigma X = 236$ $\Sigma Y = 1104$ $\Sigma X^2 = 4452$ $\Sigma Y^2 = 82{,}198$ $\Sigma XY = 11{,}921$
Means: $\overline{X} = 13.88$ $\overline{Y} = 64.94$

1. Find variance of X:
$$S_X^2 = \left[\Sigma X^2 - (\Sigma X)^2/N\right]/N = \left[4452 - (236)^2/17\right]/17 = 69.16$$

2. Find standard deviation of X:
$$S_X = \sqrt{S_X^2} = \sqrt{69.16} = 8.32$$

3. Find variance of Y:
$$S_Y^2 = \left[\Sigma Y^2 - (\Sigma Y)^2/N\right]/N = \left[82{,}198 - (1104)^2/17\right]/17 = 617.82$$

4. Find standard deviation of Y:
$$S_Y = \sqrt{S_Y^2} = \sqrt{617.82} = 24.86$$

5. Find correlation between X and Y:
$$r = \frac{\Sigma XY - (\Sigma X \Sigma Y)/N}{\sqrt{\Sigma X^2 - (\Sigma X)^2/N}\sqrt{\Sigma Y^2 - (\Sigma Y)^2/N}} = \frac{11{,}921 - (236 \cdot 1104)/17}{\sqrt{4452 - (236)^2/17}\sqrt{82{,}198 - (1104)^2/17}} = -.969$$

6. Find slope for regression:
$$b = r(S_Y/S_X) = -.969(24.86/8.32) = -.969(2.988) = -2.895$$

7. Find Y-intercept:
$$a = \overline{Y} - b\overline{X} = 64.94 - (-2.895 \cdot 13.88) = 64.94 + 40.18 = 105.12$$

Figure 7.9 *Scatterplot of Level of Depression As a Function of Number of Psychotherapy Sessions*

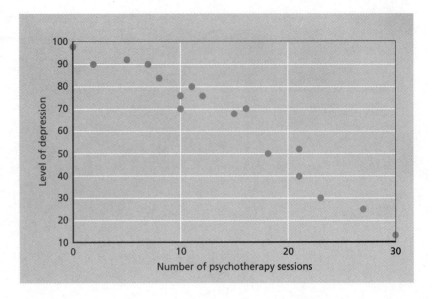

Box 7.2

CALCULATION OF PREDICTED Y VALUES (PREDICTED DEPRESSION LEVELS)

$$Y' = bX + a$$
$$Y' = -2.895X + 105.12$$

		Y'
Patient 1	$Y' = -2.895 \cdot 0 + 105.12 = 0.000 + 105.12 =$	105.120
Patient 2	$Y' = -2.895 \cdot 2 + 105.12 = -5.790 + 105.12 =$	99.330
Patient 3	$Y' = -2.895 \cdot 5 + 105.12 = -14.475 + 105.12 =$	90.645
Patient 4	$Y' = -2.895 \cdot 7 + 105.12 = -20.265 + 105.12 =$	84.855
Patient 5	$Y' = -2.895 \cdot 8 + 105.12 = -23.160 + 105.12 =$	81.960
Patient 6	$Y' = -2.895 \cdot 10 + 105.12 = -28.950 + 105.12 =$	76.170
Patient 7	$Y' = -2.895 \cdot 10 + 105.12 = -28.950 + 105.12 =$	76.170
Patient 8	$Y' = -2.895 \cdot 11 + 105.12 = -31.845 + 105.12 =$	73.275
Patient 9	$Y' = -2.895 \cdot 12 + 105.12 = -34.740 + 105.12 =$	70.380
Patient 10	$Y' = -2.895 \cdot 15 + 105.12 = -43.425 + 105.12 =$	61.695
Patient 11	$Y' = -2.895 \cdot 16 + 105.12 = -46.320 + 105.12 =$	58.800
Patient 12	$Y' = -2.895 \cdot 18 + 105.12 = -52.110 + 105.12 =$	53.010
Patient 13	$Y' = -2.895 \cdot 21 + 105.12 = -60.795 + 105.12 =$	44.325
Patient 14	$Y' = -2.895 \cdot 21 + 105.12 = -60.795 + 105.12 =$	44.325
Patient 15	$Y' = -2.895 \cdot 23 + 105.12 = -66.585 + 105.12 =$	38.535
Patient 16	$Y' = -2.895 \cdot 27 + 105.12 = -78.165 + 105.12 =$	26.955
Patient 17	$Y' = -2.895 \cdot 30 + 105.12 = -86.850 + 105.12 =$	18.270

Box 7.3

CALCULATION OF THE COEFFICIENT OF DETERMINATION, THE COEFFICIENT OF NONDETERMINATION, AND THE STANDARD ERROR OF ESTIMATE	ACTUAL LEVEL OF DEPRESSION, Y	PREDICTED LEVEL OF DEPRESSION, Y'	PREDICTION ERRORS, $Y - Y'$	SQUARED PREDICTION ERRORS, $(Y - Y')^2$
	98	105.120	–7.120	50.694
	90	99.330	–9.330	87.049
	92	90.645	1.355	1.836
	90	84.855	5.145	26.471
	84	81.960	2.040	4.162
	70	76.170	–6.170	38.069
	76	76.170	–0.170	0.029
	80	73.275	6.725	45.226
	76	70.380	5.620	31.584
	68	61.695	6.305	39.753
	70	58.800	11.200	125.440
	50	53.010	–3.010	9.060
	40	44.325	–4.325	18.706
	52	44.325	7.675	58.906
	30	38.535	–8.535	72.846
	25	26.955	–1.955	3.822
	13	18.270	–5.270	27.773

Sums:

$\Sigma(Y - Y')^2 = 641.426$

Variance of prediction errors:

$S^2_{\text{pred.errors}} = \Sigma(Y - Y')^2/N = 641.426/17 = 37.731$

Coefficient of determination:

$\left(S^2_Y - S^2_{\text{pred.errors}}\right)/S^2_Y = (617.82 - 37.731)/617.82 = .939$

Check coefficient of determination:

$\left(S^2_Y - S^2_{\text{pred.errors}}\right)/S^2_Y = r^2 = (-.969)^2 = .939$

Coefficient of nondetermination:

$k = S^2_{\text{pred.errors}}/S^2_Y = 37.731/617.82 = .061$

Check coefficient of nondetermination:

$k = S^2_{\text{pred.errors}}/S^2_Y = 1 - r^2 = 1 - .939 = .061$

Standard error of estimate:

$S_{\text{est}.Y} = \sqrt{\frac{\Sigma(Y-Y')^2}{N-2}} = \sqrt{\frac{641.426}{17-2}} = \sqrt{\frac{641.426}{15}} = 6.54$

Check standard error of estimate:

$S_{\text{est}.Y} = \left(\sqrt{S^2_Y(1 - r^2)}\right)\left(\sqrt{\frac{N}{N-2}}\right) = \left(\sqrt{617.82(1 - .939)}\right)\left(\sqrt{\frac{17}{15}}\right)$

$= \left(\sqrt{617.82 \cdot .061}\right)\left(\sqrt{1.133}\right) = \left(\sqrt{37.687}\right)(1.064) = (6.139)(1.064) = 6.53$

TAKING STOCK

You have learned how to use information provided by the linear correlation coefficient to predict values of one variable (Y) from values of another variable (X). Such a "regression analysis" is a major application of the Pearson r that is used extensively to facilitate decision making in industry, education, and various other fields.

You have also learned to use linear regression to partition variance into a portion that is accounted for by an independent variable (X) and a portion that is considered "error variance." There is an informative relationship between the prediction function and the variance-analysis function of linear regression: The stronger the correlation between X and Y, the more accurately you can predict Y from X, and the greater the percent of variation in Y values that can be explained by X. I cannot overemphasize the importance of this partitioning of variance idea. It is central to some procedures of inferential statistics that you will be using later in this course and, very likely, in your profession. For that reason, I hope you will give it a prominent place in your memory bank.

KEY TERMS

linear regression analysis

regression line

definitional regression formula

z'_Y

Y'

slope

regression coefficient

Y-intercept

prediction error

principle of least squares

standard error of estimate

linearity assumption

homoscedasticity assumption

bivariate normality assumption

cross-validation study

partitioning variance

variance of prediction errors

coefficient of determination

coefficient of nondetermination

SUMMARY

1. The general procedure involved in using r to predict future outcomes is called linear regression analysis. Linear regression can be used to predict Y values from known X values whenever data are in the form of pairs of scores and there is a correlation between variables X and Y. We explored this idea in a three-step sequence: building the regression model, evaluating the regression model, and using the regression model to partition variance.

2. The definitional formula for linear regression is $z'_Y = rz_X$, where z'_Y is the predicted z score on variable Y, and z_X is the present z score on variable X.

3. The raw-score computational formula for linear regression is $Y' = bX + a$, where Y' is the predicted raw score on variable Y, X is the known raw score on variable X, b is the slope (regression coefficient) of the regression line, and a is the Y-intercept of the regression line. The slope is the number of units of change in Y' for every unit increase in X, and the intercept is the value on the Y axis where the regression line crosses that axis.

4. There are three ways to assess the precision with which a regression model predicts values of Y: (a) through what is called "least squares" analysis, (b) by computing the standard deviation of prediction errors (known as the standard error of estimate), and (c) via cross-validation of the regression formula with a new sample of subjects.

5. The linear regression equation is based on the mathematical principle of least squares, which states that predictions produced by $Y' = bX + a$ always yield the smallest possible sum of squared prediction errors, where a prediction error is defined as $(Y - Y')$.

6. The standard deviation of prediction errors is called the standard error of estimate and is used to place a margin of error on predictions made with the regression formula. The standard error of estimate assumes that the distribution of prediction errors has a uniform range all along the regression line (homoscedasticity assumption) and that the actual Y values are normally distributed around corresponding predicted Y values (bivariate normality assumption). The regression formula also assumes that X and Y have a linear relationship.

7. Cross-validation studies are conducted to examine how well regression formulas predict future values of Y with new samples of subjects.

8. In addition to its prediction function, linear regression is used to divide Y variance into two parts: a proportion that is predictable from X (the coefficient of determination) and a proportion that is not determinable from X (the coefficient of nondetermination). This procedure is called the partitioning of variance. The coefficient of determination is r^2.

REVIEW QUESTIONS

1. Define linear regression analysis and describe its relationship to linear correlation.

2. Define or describe: z-score regression formula, regression line, slope, and intercept.

- 3. Summarize what the z-score linear regression formula tells us about the nature of linear regression.

4. If r_{XY} is 1.00, what will z'_Y equal for each z_X? If r_{XY} is .00, what will z'_Y equal for each z_X?

5. George has a mathematics aptitude score of 470. He also took a verbal aptitude test that has a mean of 530, but you don't have information on George's verbal aptitude score. You do know that the math aptitude test has a zero correlation with the verbal aptitude test. Using the z-score regression equation as the basis of your reasoning, what is the best estimate of George's verbal aptitude score? Describe the logic behind your estimate.

6. Assume that $r_{XY} = .70$. For each z_X calculate the corresponding z'_Y.

z_X	z'_Y
1.44	
0.77	
0	
−0.77	
−1.44	

7. For the data shown in Question 6, assume that $r_{XY} = -.50$. Calculate z'_Y for every z_X.

8. In a psychiatric clinic, a measure of ego strength (X) is correlated with the time required for patients to recover (in days, Y); $r = -.80$. Given the following additional information, calculate the slope and Y-intercept, and set up the raw-score (computational) linear regression equation.

	X Ego Strength	Y Recovery Time
Mean	14.2	82
Standard deviation	6	4

9. For the data in Question 8, draw a two-axis graph with Y as the vertical dimension, and draw the regression line (line of best fit) at its precise location. (*Hint:* The intersection of the X and Y means will give you one plot point for the line.)

10. Using the regression equation developed in Question 8, predict the number of recovery days needed for each of the patients in the table.

Patient	X	Y'
Nick	22	
Norman	7	
Eve	13	
Steve	16	
Carmine	10	
Ginger	20	

- 11. The pairs of values in this table represent the relationship between scores on an employee selection test (variable X) and the job performance ratings (variable Y) that those employees received in their first semiannual performance evaluations. Starting with only the raw data, compute:
  (a) The means and standard deviations of X and Y
  (b) The correlation between X and Y
  (c) The slope and intercept of the regression line

Employee	Selection Test, X	Performance Evaluation, Y	X^2	Y^2	XY
a	40	90			
b	32	74			
c	46	96			
d	50	90			
e	22	44			
f	10	50			
g	18	56			
h	30	76			
i	26	42			
j	30	64			

- 12. Use the data in Question 11.
  (a) Set up X and Y axes and precisely locate the regression line on that graph.
  (b) Construct the linear regression equation, using the values of a and b computed.
  (c) Calculate each employee's predicted performance evaluation rating.

13. For the data in Question 11, compute:
  (a) The standard error of estimate
  (b) The variance of the prediction errors
  (c) The coefficient of nondetermination, using the variance of the prediction errors to arrive at the answer
  (d) The coefficient of nondetermination, using the correlation coefficient to arrive at the answer

14. Use the data in Question 11.
  (a) Determine the percent of Y variance that is accounted for by X.
  (b) Calculate the standard error of estimate, using formula (7.7). *10.7439*
  (c) For an X score of 30, calculate the predicted Y value. *67.646 = y'*
  (d) Using the standard error of estimate, calculate two values that would be expected to bracket 95.44% of the population Y scores that cluster around the predicted Y value found above. (*Hint:* One of the values is equal to $Y' + 2S_{est.Y}$.)

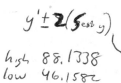

$y' \pm 2(S_{est\;y})$

high 88.1338
low 46.1582

z for 95.44 is = 2
Pred value + se of est.

15. A police officer training academy finds a correlation of .68 between a general intelligence test and student grade point average at the end of the course of study. The mean of the test for last year's class was 122 and the standard deviation was 13. The mean grade point average was 3.1 and the standard deviation was 0.4. Using $N = 30$, compute the slope and intercept, and build a linear regression model to predict the GPAs of future cadets from their intelligence test scores.

16. For the information given in Question 15, calculate:
    (a) The standard error of estimate
    (b) The coefficient of nondetermination
    (c) The percent of variance in grade point average that is determinable from the intelligence test

17. Define or describe: least squares principle, standard error of estimate, homoscedasticity assumption, bivariate normality assumption, and cross-validation study.

18. State the reasons that the homoscedasticity and bivariate normality assumptions must be true in order to correctly apply the standard error of estimate.

19. Define or describe: partitioning of variance, variance of prediction errors, coefficient of determination, and coefficient of nondetermination.

20. For each of the following sets of summary statistics, calculate the standard error of estimate, coefficient of nondetermination, and percent of variance in Y accounted for by X. Assuming $N = 20$, state which data set gives the most accurate prediction of Y values, and use the statistics you have calculated to support your answer.

	r	$\overline{X}$	$\overline{Y}$	S_X	S_Y
Set 1	.45	2.4	66	4	10
Set 2	−.77	5.8	90	3	8
Set 3	−.45	7.2	58	2	20

21. For each of the following sets of summary statistics, calculate the standard error of estimate, coefficient of nondetermination, and percent of variance in Y accounted for by X. Assuming $N = 40$, state which data set gives the most accurate prediction of Y values, and use the statistics you have calculated to support your answer.

	r	$\overline{X}$	$\overline{Y}$	S_X	S_Y
Set 1	−.50	6.7	44	5	12
Set 2	−.80	3.4	42	7	15
Set 3	.30	7.8	90	6	10

22. The pairs of values in this table represent the relationship between scores on a self-esteem test (variable X) and scores on a test of neuroticism (variable Y). Starting with only the raw data, compute:

(a) The means and standard deviations of X and Y
(b) The correlation between X and Y
(c) The slope and intercept of the regression line

Subject	Self-Esteem, X	Neuroticism, Y	X^2	Y^2	XY
a	27	25			
b	32	18			
c	40	16			
d	33	20			
e	24	24			
f	35	21			
g	20	28			
h	21	29			
i	36	17			
j	26	26			
k	30	22			
l	28	24			
m	26	22			
n	27	31			
o	29	20			

23. Use the data in Question 22.
    (a) Set up X and Y axes and precisely locate the regression line on that graph.
    (b) Construct the linear regression equation using the values of a and b computed.
    (c) Calculate each person's predicted neuroticism score.
24. For the data in Question 22, compute:
    (a) The standard error of estimate
    (b) The variance of the prediction errors
    (c) The coefficient of nondetermination, using the variance of the prediction errors to arrive at the answer
    (d) The coefficient of nondetermination, using the correlation coefficient to arrive at the answer
25. Use the data in Question 22.
    (a) Determine the percent of Y variance that is accounted for by X variance.
    (b) Calculate the standard error of estimate using formula (7.7).
    (c) For an X score of 28, calculate the predicted Y value.
    (d) Using the standard error of estimate, calculate two values that would be expected to bracket 99.7% of the population Y scores that cluster around the predicted Y value found above. (*Hint:* One of the values is equal to $Y' + 3S_{est.Y}$.)

PART 3

CONCEPTS OF INFERENTIAL STATISTICS

In Part 1 of this book you learned about the reasons for taking a systematic statistical approach to the behavioral sciences, and you saw how applied statistics dovetails with behavioral research methods. Part 2 provided you with the essential statistical tools for describing, summarizing, and predicting statistical data. You now have a conceptual and computational foundation for grasping the theory and practical implications of inferential statistics. It is to those interesting and useful topics that we now turn.

Chapter 8 presents the idea of sampling distributions and their relationship to parameter estimation—that is, the estimation of population characteristics from sample data. Chapter 9 extends the practical scope of sampling distributions by describing their role in determining the statistical significance of data. "Statistical significance testing" is also known as "hypothesis testing," in that a statistically significant outcome allows you to reject the hypothesis of chance and support the hypothesis that your data represent a reliable behavioral phenomenon.

CHAPTER 8 SAMPLING DISTRIBUTIONS

This chapter will cover the idea of sampling distributions, which is the very heart of inferential statistics. Sampling distributions (1) are theoretical distributions of random events and (2) enable you to make general statements about populations on the basis of sample data. In regard to the second point, perhaps you've seen reports of survey results like "75% of adult Americans believe in extrasensory perception" [but] "the data have a margin of error of ±3%." Reading on, you discover that the sweeping generalization about "adult Americans" was based on the responses of a sample of only 1000 adults. Haven't you wondered how behavioral scientists can confidently make general statements about a large population when they have observed only one relatively small sample from that population? And hasn't your curiosity been stimulated just a little by the "margin of error" statements attached to most properly conducted surveys? After all, where do statisticians get the "±3%" error figure?

By the time you complete this chapter, you will comprehend the rationale of the margin of error idea. You will also understand why behavioral scientists place such a high degree of confidence in their sample-to-population inferences. You'll find that the whole enterprise of sample surveys rests on the concept of sampling distributions. And if I get the message across, the results of sample surveys will be much more meaningful from now on; so will the remaining topics in this book.

> **CONCEPT RECAP**
>
> *Inferential statistics* consists of more advanced procedures that enable us to generalize from sample data to population characteristics—in other words, to *infer* something about a large set of people or situations on the basis of a small subset of people or situations.

Practical applications of sampling distributions in the behavioral sciences assume (1) that you are working with a sample of data and wish to generalize your findings to some population, and (2) that you have selected your sample according to certain rules. In fact, the conventional theory of sampling distributions does not apply very well unless you have obtained or closely approximated what is known as a "random sample." Accordingly, the earliest sections of this chapter will present an overview of the purposes of sampling and describe an appropriate sampling procedure. Next, the properties of sampling distributions will be demonstrated, and a widely used type of sampling distribution will be examined in some detail. Finally, you will learn how to use sampling distributions to estimate population characteristics from the information contained in a sample.

THE PURPOSES OF SAMPLING

In a general sense, we behavioral scientists usually draw a sample to find out something important about the larger group—the population—represented by the sample. This begs an obvious question: If we are chiefly interested in measuring certain characteristics of the population, why not just study the population directly as a whole? Why bother with sampling and, then, trying to generalize from the sample to the larger group?

> **CONCEPT RECAP**
>
> A **population** is the entire set of people, things, or events that the researcher wishes to study.
>
> A **sample** is a subset of a population—some portion of the larger group of people, things, or events targeted by the study.
>
> In inferential statistics, we use a numerical characteristic of a sample, called a **statistic**, to estimate a corresponding population characteristic, called a **parameter.**

Moore (1991) responds very aptly to the question Why sample?:

A *census* is a sample consisting of the entire population. If information is desired about a population, why not take a census? The first reason should be clear...: If the population is large, it is too expensive and time-consuming to take a census. Even the federal government, which can afford a census, uses samples to collect data on prices, employment, and many other variables. Attempting to take a census would result in this month's unemployment rate becoming available next year rather than next month.

There are also less obvious reasons.... In some cases, such as acceptance sampling of fuses or ammunition, the units in the sample are destroyed. In other cases, a relatively small sample yields more accurate data than a census [because, with a census, the task is so large that the data collectors soon become bored and careless]. (p. 7)

So in most behavioral research situations, it is absolutely necessary to work with a subset of the population. And, as you'll see next, there is a preferred way to draw samples.

RANDOM SAMPLING

Sampling procedures are usually divided into two general categories: random sampling, which normally yields good samples, and nonrandom sampling, which often results in unrepresentative samples. *Random sampling uses chance to select survey respondents,* whereas nonrandom samples are often selected on the basis of convenience—that is, on the basis of who is readily available to participate in the survey. The examples and procedures in this chapter assume that data are obtained through a particular kind of random sampling called simple random sampling.

In **simple random sampling**, *subjects are chosen from the population in such a way that every possible sample of size N has an equal chance of being drawn.* When you use simple random sampling—sometimes referred to as just "random sampling"—each person[1] in the population has an *equal* and *independent* chance of being selected: equal because each person's inclusion in the sample is determined entirely by random processes, or chance; independent because the chance selection of a particular individual has no effect on the likelihood of another person's being selected. You can conduct simple random sampling by enumerating all the members of a population and then using either a computer's random-number generator or a table of random numbers to select the desired quantity of survey respondents for your sample.

Take an example: Suppose you want to use an established test of "optimism" to assess the average level of optimism in a population of 2700 residents of a retirement community. The cost of a complete census of the residents would be prohibitive, so you decide to test a sample of 300 and then generalize your findings to the entire group of 2700. To ensure that your sample will accurately represent the population, and to feel confident about using the standard procedures of inferential statistics, you also decide that you will use simple random sampling to select your research subjects. To accomplish this, you first obtain a listing of all residents and their addresses. Next each resident is assigned a unique number within the range of 1 to 2700. Then you use a computer program to generate 300 random numbers between 1 and 2700. The residents whose numbers match those produced by the computer become your simple random sample. You test these 300 people, calculate the sample mean, and use methods of inferential statistics to draw a conclusion about the mean level of optimism in the population of 2700. Crucial to these methods of statistical inference is the idea of sampling distributions.

---

[1] For the sake of discussion, I am assuming that we are concerned with sampling people. Of course, it is also perfectly legitimate to sample objects, situations, events, or animals, depending on the purpose of your investigation.

SAMPLING DISTRIBUTIONS

Obtaining a properly drawn random sample is an important part of doing inferential statistics. But once the sample is in hand and the sample statistics have been calculated, how do you accurately relate those sample results to the population? The secret lies in a concept called the sampling distribution. Any **sampling distribution** is a *theoretical distribution of sample statistics that may be considered the result of (1) drawing all possible samples of a fixed size, N, from a population, (2) calculating a particular sample statistic (let's say the sample mean) for each of those possible samples, and (3) plotting all of the sample statistics into their own distribution.*

Some general points about sampling distributions need to be emphasized:

- Sampling distributions are different from population distributions and sample distributions. The latter two distributions consist of raw scores, whereas *sampling distributions are made up of sample statistics,* such as sample means.
- Each sampling distribution is based on all possible samples of a *fixed size* from a given population. This means that there is a large family of potential sampling distributions for any population you specify; *there is a different potential sampling distribution for each possible sample size* (this will be demonstrated momentarily), and they are all slightly different from one another.
- Samples are real and populations often are real (though some are hypothetical), but *sampling distributions are strictly conceptual* and do not exist in the real world. Nonetheless, sampling distributions work effectively to link samples with their populations so that accurate inferences can be made.

To make sampling distributions less abstract, let's construct a couple of them, using the following population of four values:

Population:[2] {0, 1, 2, 3}   Population mean = μ = 1.5

Population variance = σ^2 = 1.25

Example 1: Sampling Distribution for $N = 2$

We start by drawing all possible samples of size $N = 2$ from the population specified. Because this population is so small, we use

---

[2] Yes, this population is very small. But since we will be constructing *all* possible samples of a particular size, it is necessary to keep the number of possibilities manageable. Besides, the principles elucidated in these minimal examples apply equally well to populations of all sizes, including infinite populations.

a procedure called "sampling with replacement." That is, after selecting the first of two units, we place it back into the sampling frame so that it is again eligible to be selected as the second unit.[3]

Sample 1:	{0, 0}
Sample 2:	{0, 1}
Sample 3:	{1, 0}
Sample 4:	{1, 1}
Sample 5:	{0, 2}
Sample 6:	{2, 0}
Sample 7:	{0, 3}
Sample 8:	{3, 0}
Sample 9:	{1, 2}
Sample 10:	{2, 1}
Sample 11:	{2, 2}
Sample 12:	{1, 3}
Sample 13:	{3, 1}
Sample 14:	{2, 3}
Sample 15:	{3, 2}
Sample 16:	{3, 3}

Box 8.1 shows the mean ($\overline{X}$) for each of the 16 samples drawn. Since a sampling distribution is defined as the distribution of a statistic for all possible samples of a fixed size from a particular population, Box 8.1 actually contains the sampling distribution of $\overline{X}$. Notice that the box also displays the average of the $\overline{X}$ values across all 16 samples.

What does the information in Box 8.1 tell us about the properties of sampling distributions? Take another look at the box, and the following facts will become apparent.

μ is the mean of the $\overline{X}$ values The average (that is, mean) of the sample means is the population mean. This fact is always true of the sampling distribution of $\overline{X}$. Indeed, this statistical law is exactly what is meant by the statement "μ is the **expected value** of $\overline{X}$; in the long run, $\overline{X} = \mu$." This equivalence makes $\overline{X}$ an **unbiased estimator** of μ. A statistic is an unbiased estimator of a parameter if, in the long run, the average of that statistic is exactly equal to the parameter.

---

[3] In actual practice, behavioral researchers do not use "sampling with replacement." Since real-world populations usually are very large, it is not normally necessary to replace selected items after each draw from the population. Rather, "sampling without replacement" is used.

Box 8.1

SAMPLING
DISTRIBUTION
FOR N = 2

Population: {0, 1, 2, 3}

Mean of population: $\mu = 1.5$

Variance of population: $\sigma^2 = 1.25$

	POSSIBLE SAMPLES FOR N = 2	SAMPLE MEANS
Sample 1	{0, 0}	0.000
Sample 2	{0, 1}	0.500
Sample 3	{1, 0}	0.500
Sample 4	{1, 1}	1.000
Sample 5	{0, 2}	1.000
Sample 6	{2, 0}	1.000
Sample 7	{0, 3}	1.500
Sample 8	{3, 0}	1.500
Sample 9	{1, 2}	1.500
Sample 10	{2, 1}	1.500
Sample 11	{2, 2}	2.000
Sample 12	{1, 3}	2.000
Sample 13	{3, 1}	2.000
Sample 14	{2, 3}	2.500
Sample 15	{3, 2}	2.500
Sample 16	{3, 3}	3.000

Average 1.500

The sampling distribution of $\overline{X}$ has variability   Within the sampling distribution of the mean, also called the **sampling distribution of $\overline{X}$**, *individual sample means vary about the population mean* (in this case, $\mu = 1.5$), with a few $\overline{X}$ values being exactly equal to μ but most deviating somewhat from μ. Figure 8.1 shows how the sample means are distributed around the population mean in this example.

Example 2: Sampling Distribution for N = 3

Some interesting additional points emerge when we create sampling distributions by taking all possible samples of size $N = 3$ from the population of four values shown earlier. There are 64 such samples. Those samples are listed in Box 8.2, along with the sampling distribution of $\overline{X}$.

Again, you can see that the average sample mean is the unbiased estimator of the population mean. Additionally, the following new pieces of information are conveyed in Box 8.2:

Figure 8.1 *Sampling Distribution of $\overline{X}$ for Sample Size N = 2*

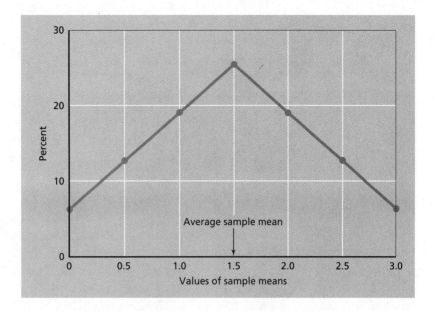

Larger N yields smaller errors of estimation Even though the average sample mean is the same when $N = 3$ as when $N = 2$, the average amount of variation of the $\overline{X}$ values around μ is smaller for $N = 3$ than for $N = 2$. Tables 8.1 and 8.2 show a comparison of the average absolute differences between the $\overline{X}$ values and μ for the larger and smaller sample sizes.[4] When $N = 2$, the average absolute difference between $\overline{X}$ and μ is 0.625, whereas when $N = 3$, the average absolute difference between $\overline{X}$ and μ is 0.531. A larger N is associated with a "skinnier" sampling distribution—one with less variation among its values. The smaller variation is desirable because $\overline{X}$ is used to estimate μ. If we define $|\overline{X} - \mu|$ as an **error of estimation**, the upshot is that *the larger the sample size, the smaller the average error of estimation.*

Larger N produces a more normal sampling distribution Still another important point is made when we plot a frequency distribution of the sample means in Box 8.2. Figure 8.2 shows the "curve" formed by that plot. Doesn't the bell shape of the distribution remind you of something you've seen quite often? Compared with the plot in Figure 8.1 (for $N = 2$), Figure 8.2 (for $N = 3$) much more closely resembles the standard normal distribution. It is a fact that *as the sample size becomes larger, the sampling distribution of $\overline{X}$ rapidly approaches a normal distribution in every respect.*

---

[4] You probably remember this, but just in case: An absolute difference between two quantities ignores any negative sign; both +0.5 and –0.5 are treated as the same positive value.

Box 8.2

SAMPLING DISTRIBUTION FOR $N = 3$

Population: {0, 1, 2, 3}

Mean of population: $\mu = 1.5$

Variance of population: $\sigma^2 = 1.25$

	POSSIBLE SAMPLES FOR $N = 3$	SAMPLE MEANS
Sample 1	{0, 0, 0}	0.000
Sample 2	{0, 0, 1}	0.333
Sample 3	{1, 0, 0}	0.333
Sample 4	{0, 1, 0}	0.333
Sample 5	{1, 1, 0}	0.667
Sample 6	{1, 0, 1}	0.667
Sample 7	{0, 1, 1}	0.667
Sample 8	{0, 0, 2}	0.667
Sample 9	{0, 2, 0}	0.667
Sample 10	{2, 0, 0}	0.667
Sample 11	{1, 1, 1}	1.000
Sample 12	{0, 1, 2}	1.000
Sample 13	{0, 2, 1}	1.000
Sample 14	{1, 0, 2}	1.000
Sample 15	{1, 2, 0}	1.000
Sample 16	{2, 0, 1}	1.000
Sample 17	{2, 1, 0}	1.000
Sample 18	{0, 0, 3}	1.000
Sample 19	{0, 3, 0}	1.000
Sample 20	{3, 0, 0}	1.000
Sample 21	{1, 1, 2}	1.333
Sample 22	{1, 2, 1}	1.333
Sample 23	{2, 1, 1}	1.333
Sample 24	{2, 2, 0}	1.333
Sample 25	{2, 0, 2}	1.333
Sample 26	{0, 2, 2}	1.333
Sample 27	{0, 1, 3}	1.333
Sample 28	{0, 3, 1}	1.333
Sample 29	{1, 0, 3}	1.333
Sample 30	{1, 3, 0}	1.333
Sample 31	{3, 0, 1}	1.333
Sample 32	{3, 1, 0}	1.333
Sample 33	{2, 2, 1}	1.667
Sample 34	{2, 1, 2}	1.667
Sample 35	{1, 2, 2}	1.667
Sample 36	{1, 1, 3}	1.667

Box 8.2

(CONTINUED)

	POSSIBLE SAMPLES FOR $N = 3$	SAMPLE MEANS
Sample 37	{1, 3, 1}	1.667
Sample 38	{3, 1, 1}	1.667
Sample 39	{0, 2, 3}	1.667
Sample 40	{0, 3, 2}	1.667
Sample 41	{2, 0, 3}	1.667
Sample 42	{2, 3, 0}	1.667
Sample 43	{3, 0, 2}	1.667
Sample 44	{3, 2, 0}	1.667
Sample 45	{2, 2, 0}	2.000
Sample 46	{1, 2, 3}	2.000
Sample 47	{1, 3, 2}	2.000
Sample 48	{2, 1, 3}	2.000
Sample 49	{2, 3, 1}	2.000
Sample 50	{3, 1, 2}	2.000
Sample 51	{3, 2, 1}	2.000
Sample 52	{3, 3, 0}	2.000
Sample 53	{3, 0, 3}	2.000
Sample 54	{0, 3, 3}	2.000
Sample 55	{2, 2, 3}	2.333
Sample 56	{2, 3, 2}	2.333
Sample 57	{3, 2, 2}	2.333
Sample 58	{3, 3, 1}	2.333
Sample 59	{3, 1, 3}	2.333
Sample 60	{1, 3, 3}	2.333
Sample 61	{3, 3, 2}	2.667
Sample 62	{3, 2, 3}	2.667
Sample 63	{2, 3, 2}	2.667
Sample 64	{3, 3, 3}	3.000
		Average 1.500

For all practical purposes, *when N is at least 30, you may assume that the sampling distribution of the mean is normal* (Hays, 1988).

Relationship Between Population Variance and the Variance of $\overline{X}$

The information in Box 8.2 allows us to discover yet one more critical fact about the sampling distribution of the mean: The variance of the sampling distribution can be derived from a knowledge of the sample size, *N*, and the population variance, σ^2.

To use the sampling distribution of $\overline{X}$ to best advantage, you need to be able to compute its variance, symbolized by $\sigma_{\overline{X}}^2$. Table

Table 8.1

AVERAGE ABSOLUTE
DIFFERENCE BETWEEN THE
SAMPLE MEAN AND THE
POPULATION MEAN: $N = 2$

SAMPLE	SAMPLE MEAN	POPULATION MEAN	ABSOLUTE DIFFERENCE
Sample 1	0.000	1.500	1.500
Sample 2	0.500	1.500	1.000
Sample 3	0.500	1.500	1.000
Sample 4	1.000	1.500	0.500
Sample 5	1.000	1.500	0.500
Sample 6	1.000	1.500	0.500
Sample 7	1.500	1.500	0.000
Sample 8	1.500	1.500	0.000
Sample 9	1.500	1.500	0.000
Sample 10	1.500	1.500	0.000
Sample 11	2.000	1.500	0.500
Sample 12	2.000	1.500	0.500
Sample 13	2.000	1.500	0.500
Sample 14	2.500	1.500	1.000
Sample 15	2.500	1.500	1.000
Sample 16	3.000	1.500	1.500

Average absolute difference between means: 0.625

Figure 8.2 *Sampling Distribution of $\overline{X}$ for Sample Size N = 3*

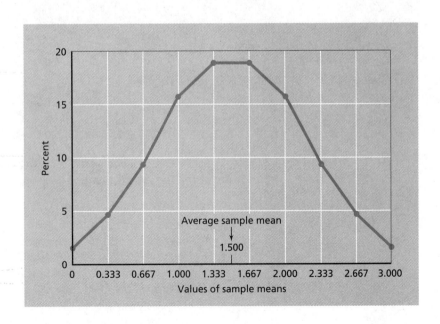

Table 8.2

AVERAGE ABSOLUTE
DIFFERENCE BETWEEN THE
SAMPLE MEAN AND THE
POPULATION MEAN: $N = 3$

SAMPLE	SAMPLE MEAN	POPULATION MEAN	ABSOLUTE DIFFERENCE
Sample 1	0.000	1.500	1.500
Sample 2	0.333	1.500	1.167
Sample 3	0.333	1.500	1.167
Sample 4	0.333	1.500	1.167
Sample 5	0.667	1.500	0.833
Sample 6	0.667	1.500	0.833
Sample 7	0.667	1.500	0.833
Sample 8	0.667	1.500	0.833
Sample 9	0.667	1.500	0.833
Sample 10	0.667	1.500	0.833
Sample 11	1.000	1.500	0.500
Sample 12	1.000	1.500	0.500
Sample 13	1.000	1.500	0.500
Sample 14	1.000	1.500	0.500
Sample 15	1.000	1.500	0.500
Sample 16	1.000	1.500	0.500
Sample 17	1.000	1.500	0.500
Sample 18	1.000	1.500	0.500
Sample 19	1.000	1.500	0.500
Sample 20	1.000	1.500	0.500
Sample 21	1.333	1.500	0.167
Sample 22	1.333	1.500	0.167
Sample 23	1.333	1.500	0.167
Sample 24	1.333	1.500	0.167
Sample 25	1.333	1.500	0.167
Sample 26	1.333	1.500	0.167
Sample 27	1.333	1.500	0.167
Sample 28	1.333	1.500	0.167
Sample 29	1.333	1.500	0.167
Sample 30	1.333	1.500	0.167
Sample 31	1.333	1.500	0.167
Sample 32	1.333	1.500	0.167
Sample 33	1.667	1.500	0.167
Sample 34	1.667	1.500	0.167
Sample 35	1.667	1.500	0.167
Sample 36	1.667	1.500	0.167
Sample 37	1.667	1.500	0.167
Sample 38	1.667	1.500	0.167
Sample 39	1.667	1.500	0.167
Sample 40	1.667	1.500	0.167
Sample 41	1.667	1.500	0.167

(continued)

Table 8.2

(CONTINUED)

SAMPLE	SAMPLE MEAN	POPULATION MEAN	ABSOLUTE DIFFERENCE
Sample 42	1.667	1.500	0.167
Sample 43	1.667	1.500	0.167
Sample 44	1.667	1.500	0.167
Sample 45	2.000	1.500	0.500
Sample 46	2.000	1.500	0.500
Sample 47	2.000	1.500	0.500
Sample 48	2.000	1.500	0.500
Sample 49	2.000	1.500	0.500
Sample 50	2.000	1.500	0.500
Sample 51	2.000	1.500	0.500
Sample 52	2.000	1.500	0.500
Sample 53	2.000	1.500	0.500
Sample 54	2.000	1.500	0.500
Sample 55	2.333	1.500	0.833
Sample 56	2.333	1.500	0.833
Sample 57	2.333	1.500	0.833
Sample 58	2.333	1.500	0.833
Sample 59	2.333	1.500	0.833
Sample 60	2.333	1.500	0.833
Sample 61	2.667	1.500	1.167
Sample 62	2.667	1.500	1.167
Sample 63	2.667	1.500	1.167
Sample 64	3.000	1.500	1.500

Average absolute difference between means: 0.531

8.3 shows how this is done for a *finite* sampling distribution of the mean—again, for the example where $N = 3$. For a finite sampling distribution, the variance is calculated in the usual way, but you are working with $\overline{X}$ values instead of raw scores. In this example, $\sigma_{\overline{X}}^2 = 0.4167$. There is a problem with this way of finding the variance of the sampling distribution, however. In everyday statistical practice, the sampling distribution of the mean is *infinite*. It is impossible to directly compute the variance of an infinite distribution from its own values because there is no divisor for the variance formula. The number of means in an infinite sampling distribution cannot be determined.

Fortunately, you can compute the variance of the sampling distribution of $\overline{X}$ another way, if you have the variance of the population from which the sample observations were drawn. The formula is

$$\sigma_{\overline{X}}^2 = \frac{\sigma^2}{N} \tag{8.1}$$

where σ^2 is the variance of the population and N is the size of the sample selected from that population. In the present example, σ^2 is 1.25 (see Box 8.2) and $N = 3$. Therefore,

$$\sigma_{\overline{X}}^2 = \frac{\sigma^2}{N} = \frac{1.25}{3} = 0.4167$$

the same result we obtained in Table 8.3. Remember that you can use this convenient formula to compute the variance of all sampling distributions of $\overline{X}$, whether infinite or finite.

Summary of the Characteristics of Sampling Distributions

The contrived small sampling distributions in Boxes 8.1 and 8.2 provided a lot of information about the nature of such distributions. Before going on to additional implications of that material, let's review what I have covered.

- A sampling distribution is a distribution of a statistic for all possible samples of a fixed size drawn from a particular population.
- In the sampling distribution of $\overline{X}$, individual sample means vary about the population mean.
- The average of the sample means is μ, the population mean.
- The larger the sample size (N), the smaller is the average difference between the sample means and their population mean. That is, larger samples yield better estimates of μ on the average. This is known as the "law of large numbers."
- As sample size becomes larger, the sampling distribution of $\overline{X}$ rapidly approaches a normal distribution in every respect.
- The variance of the sampling distribution of $\overline{X}$ can be computed directly from the variance of the population via the formula $\sigma_{\overline{X}}^2 = \sigma^2/N$.

A THEOREM FOR THE SAMPLING DISTRIBUTION OF $\overline{X}$

Earlier we manually generated two empirical sampling distributions of $\overline{X}$ by listing all possible samples of fixed sizes from an artificial population of four values.[5] When you apply the sampling distribution of the mean in statistical analysis, however, you won't have to produce your own sampling distributions for each procedure. That is because there is a mathematical theory that specifies

_____

[5] To review, *empirical* means "based on actual observations" rather than mathematical theory.

Table 8.3

CALCULATION OF THE VARIANCE OF THE SAMPLING DISTRIBUTION OF THE MEAN: $N = 3$		POSSIBLE SAMPLES FOR $N = 3$	SAMPLE MEANS	$(\overline{X} - \mu)$	$(\overline{X} - \mu)^2$
	Sample 1	{0, 0, 0}	0.000	−1.500	2.25000
	Sample 2	{0, 0, 1}	0.333	−1.167	1.36111
	Sample 3	{1, 0, 0}	0.333	−1.167	1.36111
	Sample 4	{0, 1, 0}	0.333	−1.167	1.36111
	Sample 5	{1, 1, 0}	0.667	−0.833	0.69444
	Sample 6	{1, 0, 1}	0.667	−0.833	0.69444
	Sample 7	{0, 1, 1}	0.667	−0.833	0.69444
	Sample 8	{0, 0, 2}	0.667	−0.833	0.69444
	Sample 9	{0, 2, 0}	0.667	−0.833	0.69444
	Sample 10	{2, 0, 0}	0.667	−0.833	0.69444
	Sample 11	{1, 1, 1}	1.000	−0.500	0.25000
	Sample 12	{0, 1, 2}	1.000	−0.500	0.25000
	Sample 13	{0, 2, 1}	1.000	−0.500	0.25000
	Sample 14	{1, 0, 2}	1.000	−0.500	0.25000
	Sample 15	{1, 2, 0}	1.000	−0.500	0.25000
	Sample 16	{2, 0, 1}	1.000	−0.500	0.25000
	Sample 17	{2, 1, 0}	1.000	−0.500	0.25000
	Sample 18	{0, 0, 3}	1.000	−0.500	0.25000
	Sample 19	{0, 3, 0}	1.000	−0.500	0.25000
	Sample 20	{3, 0, 0}	1.000	−0.500	0.25000
	Sample 21	{1, 1, 2}	1.333	−0.167	0.02778
	Sample 22	{1, 2, 1}	1.333	−0.167	0.02778
	Sample 23	{2, 1, 1}	1.333	−0.167	0.02778
	Sample 24	{2, 2, 0}	1.333	−0.167	0.02778
	Sample 25	{2, 0, 2}	1.333	−0.167	0.02778
	Sample 26	{0, 2, 2}	1.333	−0.167	0.02778
	Sample 27	{0, 1, 3}	1.333	−0.167	0.02778
	Sample 28	{0, 3, 1}	1.333	−0.167	0.02778
	Sample 29	{1, 0, 3}	1.333	−0.167	0.02778
	Sample 30	{1, 3, 0}	1.333	−0.167	0.02778
	Sample 31	{3, 0, 1}	1.333	−0.167	0.02778
	Sample 32	{3, 1, 0}	1.333	−0.167	0.02778
	Sample 33	{2, 2, 1}	1.667	0.167	0.02778
	Sample 34	{2, 1, 2}	1.667	0.167	0.02778
	Sample 35	{1, 2, 2}	1.667	0.167	0.02778
	Sample 36	{1, 1, 3}	1.667	0.167	0.02778
	Sample 37	{1, 3, 1}	1.667	0.167	0.02778
	Sample 38	{3, 1, 1}	1.667	0.167	0.02778
	Sample 39	{0, 2, 3}	1.667	0.167	0.02778
	Sample 40	{0, 3, 2}	1.667	0.167	0.02778

Table 8.3

(CONTINUED)

	POSSIBLE SAMPLES FOR $N = 3$	SAMPLE MEANS	$(\bar{X} - \mu)$	$(\bar{X} - \mu)^2$
Sample 41	{2, 0, 3}	1.667	0.167	0.02778
Sample 42	{2, 3, 0}	1.667	0.167	0.02778
Sample 43	{3, 0, 2}	1.667	0.167	0.02778
Sample 44	{3, 2, 0}	1.667	0.167	0.02778
Sample 45	{2, 2, 0}	2.000	0.500	0.25000
Sample 46	{1, 2, 3}	2.000	0.500	0.25000
Sample 47	{1, 3, 2}	2.000	0.500	0.25000
Sample 48	{2, 1, 3}	2.000	0.500	0.25000
Sample 49	{2, 3, 1}	2.000	0.500	0.25000
Sample 50	{3, 1, 2}	2.000	0.500	0.25000
Sample 51	{3, 2, 1}	2.000	0.500	0.25000
Sample 52	{3, 3, 0}	2.000	0.500	0.25000
Sample 53	{3, 0, 3}	2.000	0.500	0.25000
Sample 54	{0, 3, 3}	2.000	0.500	0.25000
Sample 55	{2, 2, 3}	2.333	0.833	0.69444
Sample 56	{2, 3, 2}	2.333	0.833	0.69444
Sample 57	{3, 2, 2}	2.333	0.833	0.69444
Sample 58	{3, 3, 1}	2.333	0.833	0.69444
Sample 59	{3, 1, 3}	2.333	0.833	0.69444
Sample 60	{1, 3, 3}	2.333	0.833	0.69444
Sample 61	{3, 3, 2}	2.667	1.167	1.36111
Sample 62	{3, 2, 3}	2.667	1.167	1.36111
Sample 63	{2, 3, 2}	2.667	1.167	1.36111
Sample 64	{3, 3, 3}	3.000	1.500	2.25000

$$\Sigma(\bar{X} - \mu)^2 = 26.66667$$

Variance of the sampling distribution of $\bar{X}$ *is* $\Sigma(\bar{X} - \mu)^2/64 = 26.66667/64 = 0.41667$

a generally applicable sampling distribution of $\bar{X}$ that pertains to almost all populations, from very small to infinite, and all sample sizes of at least $N = 30$.

The core principle of mathematically derived sampling distributions is the **central limit theorem**, which states that:

For any normal or nonnormal population with a mean of μ and a variance of σ^2, the distribution of means from samples of fixed size N will approach a normal distribution as N becomes larger, and that sampling distribution will have a mean of μ and a variance of σ^2/N.

Owing to examples covered in prior sections of this chapter, nearly all the information in this theorem is familiar to you. Two new items are worth noting, however:

1. The sampling distribution of $\overline{X}$ tends toward normality regardless of whether the population from which the samples are drawn is itself normal. (Recall that when N is at least 30, normality of the sampling distribution may be assumed.)
2. Since the sampling distribution specified by this theorem is truly a normal distribution, it should always be considered an infinite distribution—that is, composed of too many values to count.

Figure 8.3 depicts the type of sampling distribution of $\overline{X}$ specified by the central limit theorem. Bear in mind that that distribution has all the usual properties of the normal distribution. By design, it appears to be slightly "skinnier" than the normal distribution you are used to. That's understandable when you consider that the present distribution is made up of an infinite number of sample means and that sample means are less variable than raw scores. As the sample size becomes larger, the sampling distribution of $\overline{X}$ becomes "skinnier" still because means based on larger samples are even less variable than means based on smaller N's. *Regardless of the girth of the sampling distribution of the mean, however, it will have the properties of the normal distribution as long as N is at least 30.*

There is something else a little bit different about the distribution in Figure 8.3, isn't there? Like the usual normal distribution,

Figure 8.3 *Relationship Between $\sigma_{\overline{X}}$ and Area Under the Sampling Distribution of $\overline{X}$ (Note: The sampling distribution of the mean is normal in every respect.)*

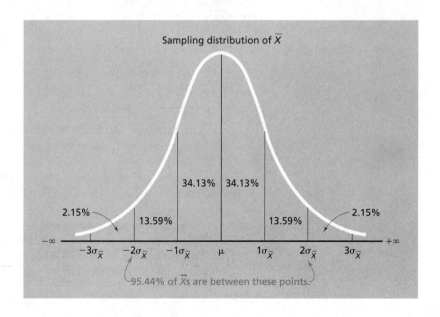

this one is marked off with a standard unit of variation (along its horizontal axis), and you can see that the percents of the area under different segments of the curve are related in a familiar way to the standard unit of variation—nothing new there. What is new is the unit of variation itself. It is not the standard deviation of the population, σ; rather, it is *the standard deviation of the sampling distribution of* $\overline{X}$, known as the **standard error of the mean**, $\sigma_{\overline{X}}$. The standard error of the mean is the square root of the variance of the mean. Thus, it is calculated as

$$\sigma_{\overline{X}} = \sqrt{\sigma_{\overline{X}}^2} \qquad\qquad (8.2)$$

or

$$\sigma_{\overline{X}} = \sqrt{\frac{\sigma^2}{N}} \qquad\qquad (8.3)$$

or

$$\sigma_{\overline{X}} = \frac{\sigma}{\sqrt{N}} \qquad\qquad (8.4)$$

CAUTION! It is very important not to confuse the variance of the sampling distribution of the mean, $\sigma_{\overline{X}}^2$, with the variance of the population, σ^2. The variance of the sampling distribution is derived from the variance of the population and will always be the smaller of the two quantities. These two parameters are *not* interchangeable.

In short, to calculate the standard error of the mean, divide the population standard deviation by the square root of the sample size. Alternatively, you can simply take the square root of the variance of the sample mean. In practice, you will want $\sigma_{\overline{X}}$ to be as small as possible: *The smaller the standard error, the more precisely you can estimate the population mean from your sample mean.* That is, $\sigma_{\overline{X}}$ is an index of the precision with which $\overline{X}$ estimates μ. Because the sample size affects the size of $\sigma_{\overline{X}}$, you have some control over the accuracy of estimation. This control comes from your ability to determine *N* in studies that you do.

Relationship of $\sigma_{\overline{X}}$ to *N*

The size of the standard error of the mean—and hence the precision of estimation—is *inversely proportional to the square root of the sample size.* This statement means that:

The size of $\sigma_{\overline{X}}$ will decrease as the sample size, *N*, is increased. The decrease in $\sigma_{\overline{X}}$ will not be in proportion to the increase in the sample size but rather in proportion to the square root of the increase in the sample size.

These points are implied by the formula $\sigma_{\overline{X}} = \sigma/\sqrt{N}$.

Next, we consider an example that demonstrates the idea. Suppose that the population standard deviation, σ, is 50 and the sample size is 100. This yields a standard error of $\sigma_{\overline{X}} = \sigma/\sqrt{N} = 50/\sqrt{100} = 50/10 = 5$.

Now let's keep σ at 50 but make the sample size 100 times larger: $N = 10,000$ (or 100×100). If the statements made above are true, then increasing N by a factor of 100 should reduce the size of $\sigma_{\overline{X}}$ by a factor of the *square root* of 100 (because the size of the standard error is an inverse function of the square root of the sample size). In other words, we would expect the first $\sigma_{\overline{X}}$ (that is, 5) to shrink by a factor of 10. The new $\sigma_{\overline{X}}$ should be $5/10 = 0.5$. Applying the formula verifies this expectation: $\sigma_{\overline{X}} = \sigma/\sqrt{N} = 50/\sqrt{10,000} = 50/100 = 0.5$.

Figure 8.4 depicts the effect of increasing the sample size from 100 to 10,000 on the spread of the sampling distribution. The figure assumes that the mean of the sampling distribution was $\mu = 200$ in the earlier examples. Notice how much more closely the sample means cluster around μ when the sample size is increased dramatically. Since the distribution is normal, 68.26% of the sample means are within 0.5 of μ when $N = 10,000$.

The *important implication* of this demonstration is that *it is usually advisable to use as large a sample as you have time and money to obtain. The larger the N, the less the average difference between $\overline{X}$ and μ and, therefore, the more precisely you can estimate μ from $\overline{X}$.*

Figure 8.4 *Relationship of Sample Size to the Standard Error of the Mean*

(Note: The size of $\sigma_{\overline{X}}$ is inversely proportional to the square root of the sample size.)

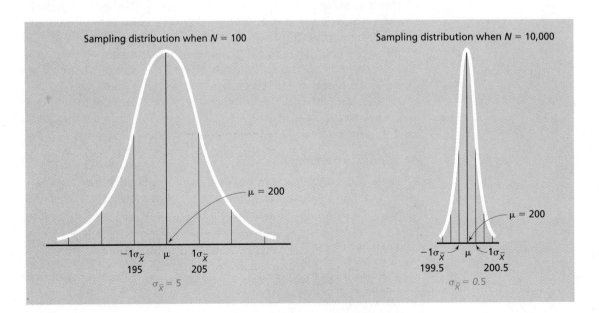

Because the sampling distribution of $\overline{X}$ is normally distributed, knowing the value of $\sigma_{\overline{x}}$ will enable you to relate the $\overline{X}$ of any one sample to μ in an exact way, using the sampling distribution as the key link. The next section will explain how.

> CAUTION: Be careful not to confuse the standard error of the mean, $\sigma_{\overline{x}}$, with the standard deviation of the population, σ. The standard error of the mean is derived from the standard deviation of the population and will always be the smaller of the two quantities. And, of course, the two measures of variation apply to different distributions.

Relating $\overline{X}$ to μ

To establish a statistical connection between a sample mean and a population mean, you need to keep three ideas in mind:

1. The sampling distribution of $\overline{X}$ is made up entirely of sample means, and any particular sample mean that you might be working with is *one* of the millions of $\overline{X}$ values that exist, theoretically, within that distribution.
2. Those sample means are normally distributed around μ.
3. Because the $\overline{X}$ values are normally distributed around μ, *the standard error of the mean, $\sigma_{\overline{x}}$, can be used to determine the probability that $\overline{X}$ deviates from μ by no more than a specific amount.*

The third idea is the crucial one: Any distribution that is normal can be represented by the standard normal distribution (a normal distribution of z scores), which you learned about in Chapter 5. Then any particular score in that distribution can be converted to a z score, and probability statements can be made about it from information provided in Table Z of Appendix A (in the back of this book). Since the sampling distribution of the mean is normal, you can convert any $\overline{X}$ in that distribution to a z score with this formula:

$$z = \frac{\overline{X} - \mu}{\sigma_{\overline{X}}} \qquad (8.5)$$

and then use the areas under the normal curve in Table Z to make probability statements about that $\overline{X}$.

The probability of $\overline{X}$: A simple example   You are working with an aptitude test for college-bound students that has a population mean of $\mu = 500$ and a standard deviation of $\sigma = 100$. You draw a simple random sample of 625 college-bound students and administer the test to them. Since you have a population and a sample, you may assume the existence of a third distribution—the sampling distribution of $\overline{X}$. Since the sample size is $N = 625$, the standard error of the sampling distribution is

$$\sigma_{\bar{X}} = \frac{\sigma}{\sqrt{N}} = \frac{100}{\sqrt{625}} = \frac{100}{25} = 4$$

The mean of the sampling distribution is the same as the mean of the population: $\mu = 500$.

Suppose your sample of students gets a mean score of $\bar{X} = 502.67$ on the aptitude test. Viewing this average in the context of the sampling distribution, you can say that the sample mean differs from the population mean by $|\bar{X} - \mu|$, or $|502.67 - 500|$, or $|2.67|$ points. (Recall that a number within vertical lines, $|\ |$, is an absolute value.) Converting this deviation to a z score, you get $z = (\bar{X} - \mu)/\sigma_{\bar{X}} = 2.67/4 = 0.67$. So the sample mean is 0.67 standard errors away from the population mean.

Once the sample mean has been converted to a z score, you may use the standard normal distribution to represent the sampling distribution of $\bar{X}$. Then it is possible to ask, What is the probability that the mean of a randomly selected sample will not differ from μ by more than $0.67\sigma_{\bar{X}}$? Equivalently, what proportion of the sample means are within the range $z = -0.67$ to $z = +0.67$ under the sampling distribution curve? You can find the answer to both of these questions by looking up a z score of 0.67 in Table Z of Appendix A. You will see the following:[6]

Column 1	Column 2	Column 3
z score	Proportion of area between z and μ	Proportion of area between z and infinity
0.67	.2486 (or about 25%)	.2514 (or about 25%)

Table Z reveals that about 25% of the sampling distribution lies between $z = +0.67$ and μ and that about 25% of the sampling distribution lies between $z = -0.67$ and μ. Therefore, the probability is .50 that the mean of a randomly selected sample will be within $0.67\sigma_{\bar{X}}$ of the population mean. To phrase the matter yet another way, we have about a 50% chance of randomly selecting a sample mean that is as close as or closer to μ than $\bar{X} = 502.67$. This fact is illustrated in Figure 8.5.

In a similar fashion, Figure 8.6 shows that about 50% of the sampling distribution falls outside $|0.67\sigma_{\bar{X}}|$. That is the same as saying that the probability is about .50 that $\{z < -0.67 \; or \; z > +0.67\}$. We have about a 50% chance of randomly selecting a sample mean that is more distant from μ than $\bar{X} = 502.67$ (or $\bar{X} = 497.33$).

---

[6] It would be a good idea to confirm these figures by looking at Table Z. Sooner than you think, you'll need to use Table Z on a regular basis.

> **CONCEPT RECAP**
> Remember that when you are working with area under the normal curve:
>
> Probability = proportion = percent/100

Making general probability statements When you get the gist of using the sampling distribution of $\overline{X}$ to relate any sample mean to its μ, it becomes apparent that there is a general formula for making probability statements about areas under the sampling distribution curve. The probability that any $\overline{X}$ will not differ from μ by more than a specified amount is given by:

Probability (that $\overline{X}$ is within $\{μ ± z_c σ_{\overline{X}}\}$) = area

where "area" is a proportion of the area in some segment of the standard normal curve (which represents the normally distributed sampling distribution), $σ_{\overline{X}}$ is the standard error of $\overline{X}$, and z_c is the **confidence factor**, defined as the *z score that corresponds to a particular normal-curve area that you're working with. Observe that the quantity $z_c σ_{\overline{X}}$ is both added to and subtracted from μ to form an interval of values around μ.* The term *confidence factor* comes from the fact that, in a theoretical sense, the size of the *z* score determines how confident you are that the interval formed by

Figure 8.5 *Sampling Distribution of $\overline{X}$ Showing That About 50% of the Sample Means Are Within 0.67 Standard Error of μ*

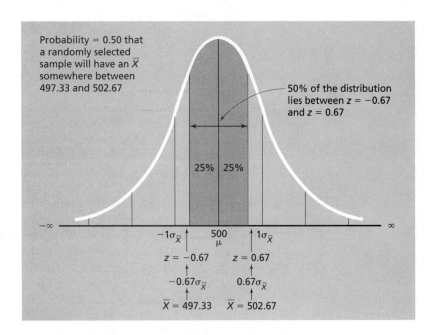

Probability = 0.50 that a randomly selected sample will have an $\overline{X}$ somewhere between 497.33 and 502.67

50% of the distribution lies between *z* = −0.67 and *z* = 0.67

Figure 8.6 *Sampling Distribution of $\overline{X}$ Showing That About 50% of the Sample Means Are at Least 0.67 Standard Error away from μ*

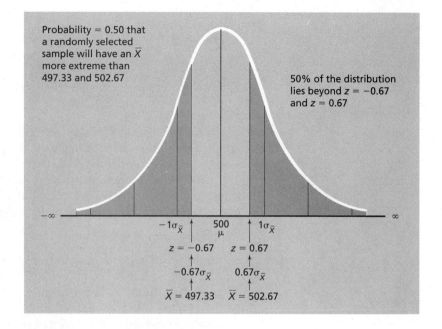

Probability = 0.50 that a randomly selected sample will have an $\overline{X}$ more extreme than 497.33 and 502.67

50% of the distribution lies beyond $z = -0.67$ and $z = 0.67$

$μ \pm z_c\sigma_{\overline{X}}$ actually contains $\overline{X}$. The larger the z score, the wider is the interval formed by $μ \pm z_c\sigma_{\overline{X}}$ and the more probable it is that the sample mean will be included within it.

In the above example, where $μ = 500$ and $\sigma_{\overline{X}} = 4$, the probability is .50 that a randomly selected $\overline{X}$ will be no smaller than

$$500 - 0.67 \cdot 4 = 500 - 2.67 = 497.33 \quad \leftarrow\text{lower boundary of interval}$$

and no larger than

$$500 + 0.67 \cdot 4 = 500 + 2.67 = 502.67. \quad \leftarrow\text{upper boundary of interval}$$

Special confidence factors   Two confidence factors are used so often in inferential statistics that they have become default standards. The 95% confidence factor is $z_c = 1.96$. As shown in Figure 8.7, the interval ranging from $μ - 1.96\sigma_{\overline{X}}$ to $μ + 1.96\sigma_{\overline{X}}$ includes the middle 95% of the sampling distribution of $\overline{X}$. In most inferential statistical procedures, behavioral scientists prefer to work with either exactly the middle 95% of the sampling distribution or exactly the most extreme 5% of the sampling distribution (which is the area *outside* $μ \pm 1.96\sigma_{\overline{X}}$; see Figure 8.7). Table Z in Appendix A indicates that 47.5% of a normal distribution lies between $z = 1.96$ and $μ$— on each side of $μ$—and that 2.5% of the distribution lies between $z = 1.96$ and infinity—again, on both the positive and the negative sides of the curve.

The other default standard in inferential statistics is the 99% confidence factor, which is $z_c = 2.58$. Figure 8.8 shows that that z value defines both the middle 99% and the most extreme 1% of the

Figure 8.7 *Sampling Distribution of $\overline{X}$ Showing That the Confidence Factor of 1.96 Includes the Middle 95% of the Distribution (Note: Probability = .95 that $\overline{X}$ will not differ from μ by more than $1.96\sigma_{\overline{X}}$.)*

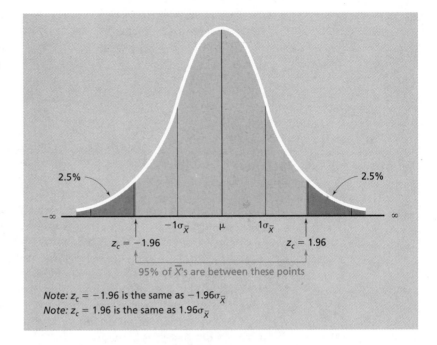

Note: $z_c = -1.96$ is the same as $-1.96\sigma_{\overline{X}}$
Note: $z_c = 1.96$ is the same as $1.96\sigma_{\overline{X}}$

sampling distribution. (Check this in Table Z.) In general, the 99% confidence factor is used instead of the 95% confidence factor when the statistician wishes to be more cautious, and more "confident," in his or her statements about the proximity of a particular $\overline{X}$ to μ.

Figure 8.8 *Sampling Distribution of $\overline{X}$ Showing That the Confidence Factor of 2.58 Includes the Middle 99% of the Distribution (Note: Probability = .99 that $\overline{X}$ will not differ from μ by more than $2.58\sigma_{\overline{X}}$.)*

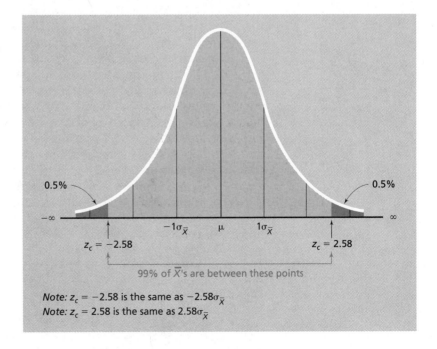

Note: $z_c = -2.58$ is the same as $-2.58\sigma_{\overline{X}}$
Note: $z_c = 2.58$ is the same as $2.58\sigma_{\overline{X}}$

Why is the statistician more confident when $z_c = 2.58$ rather than 1.96? As you'll discover next, a major function of sampling distributions is to enable us to say that a particular sample mean does not differ from μ by more than $z_c\sigma_{\bar{X}}$. When $z_c = 2.58$, the interval formed by $\mu \pm z_c\sigma_{\bar{X}}$ encompasses a greater area of the sampling distribution and, hence, a greater proportion of the sample means. This fact increases the likelihood that any randomly selected $\bar{X}$ that the statistician might be working with is within the stated interval. Therefore, an assertion that a sample mean does not differ from μ by more than $z_c\sigma_{\bar{X}}$ can be made with greater confidence.

And onward to applications   So far the discussion of the sampling distribution of $\bar{X}$ has been theoretical. This approach was necessary to give you the intellectual tools you will need to apply sampling distributions to behavioral research problems that require inferential statistics. These research problems will mark the path through most of the remainder of this book. But there is no need to wait that long to use the sampling distribution of $\bar{X}$. The next section of this chapter will show you how to use it to estimate μ from a single sample mean.

INTRODUCTION TO PARAMETER ESTIMATION

In parameter estimation, the goal is to infer something about a population from a sample that is drawn randomly from that population. The researcher commonly estimates a population mean on some psychological or social dimension from a sample mean on that dimension. For example, Stinson, DeBakey, Grant, and Dawson (1992) conducted an "AIDS knowledge" survey of random samples from several different groups of adult Americans. They found that the sample of people with less than 12 years of formal education had a mean score of $\bar{X} = 16.62$ out of a possible 30 on the AIDS knowledge test. From this sample finding, we may infer that the *population mean* for those with less than 12 years of education is "around" 16.62.

Even more commonly, the objective is to estimate the proportion (or percent) of a given population that has a certain trait or a particular opinion. For example, a May 1993 survey of 1200 randomly selected adult Americans revealed that 65% of them were satisfied with available health care services (*TIME*, July 5, 1993). This finding may be interpreted to mean that the proportion of the U.S. adult population who are satisfied with available health care options is "around" .65. The general idea behind parameter estimation is illustrated in Figure 8.9.

Figure 8.9 *The General Idea Behind Parameter Estimation: Estimating a Population Characteristic from a Sample Characteristic*

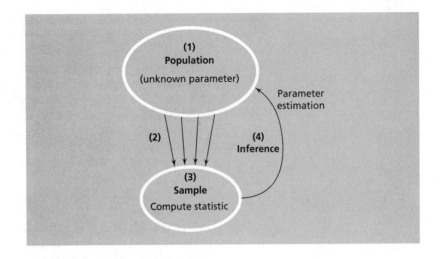

Types of Parameter Estimation

In inferring a population mean from a sample mean, you can carry out two kinds of parameter estimation: point estimation and interval estimation.

Point estimation In the first approach to parameter estimation, *a single "point," or value (namely, the sample mean), is used to estimate* μ. In the example on the AIDS knowledge scores of persons with less than a high school education, the point estimate of μ is the sample mean, $\overline{X} = 16.62$. With this approach, then, once the sample mean is computed, the parameter estimation is finished. It is an attractively simple matter. But isn't this method too simple, to the extent of being deceptive? Why?

The problem is that *sample means vary randomly from μ*, producing what statisticians call **sampling error**—*the random differences between $\overline{X}$ values and their μ*. Point estimation does not take this sampling error into account. It would be more informative to present the sample mean as an estimate of μ but then qualify the estimate by stating a *likely maximum margin of error*, telling readers of the research results how far $\overline{X}$ might reasonably be expected to deviate from the population mean. Essentially, you create a range of values to reflect the possible effect of sampling error on the accuracy of your parameter estimate. Such an "error-qualified" approach, which is favored in the behavioral sciences, is the second type of estimation.

Interval estimation When doing an interval estimate of μ, you use the sampling distribution of $\overline{X}$ to set up *a range of values*, called the **confidence interval**, *that is very likely to include the value of the* *population mean.* The sample mean is right at the center of the interval, and the upper and lower boundaries of the interval are separated from $\overline{X}$ by $z_c\sigma_{\overline{X}}$, a quantity known as the **bound on the error of estimation**. Recall that the error of estimation is the absolute difference between the sample mean and the population mean, or $|\overline{X} - \mu|$. With this approach you are, in effect, saying, "Here is the mean of my simple random sample. I think that μ is about the same as $\overline{X}$, but the two means could reasonably be expected to be as far apart as $z_c\sigma_{\overline{X}}$."

Three important ideas concerning the bound on the error of estimation should be remembered.

$z_c\sigma_{\overline{X}}$ is the *probable* limit   The bound on the error of estimation is the *likely* maximum absolute difference between $\overline{X}$ and μ. The two means could differ by even more than $z_c\sigma_{\overline{X}}$, but they are not likely to. For example, if you use the 99% confidence factor ($z_c = 2.58$) to calculate the "bound," then $\overline{X}$ will not differ from μ by more than $z_c\sigma_{\overline{X}}$ in approximately 99 of every 100 simple random samples. But in 1 of every 100 samples, on the average, the difference between the means will exceed this bound. In other words, the probability is .99 that $|\overline{X} - \mu| \leq z_c\sigma_{\overline{X}}$ when $z_c = 2.58$.

The size of the bound is determined by z_c and $\sigma_{\overline{X}}$ The size of the likely maximum error of estimation is a direct function of the confidence factor ($z_c = 1.96$ for the 95% confidence factor; $z_c = 2.58$ for the 99% confidence factor) and the size of $\sigma_{\overline{X}}$. You can make the bound smaller by using a smaller confidence factor, but there is a tradeoff: Relative to a larger confidence factor, a smaller confidence factor produces an estimation interval that is less likely to include μ.

The bigger the *N*, the better the estimate   The best way to reduce the maximum likely error of estimation is to make $\sigma_{\overline{X}}$ smaller by increasing the sample size. It is in this sense that larger samples give more precise estimates of μ than smaller samples.

Setting Up the Confidence Interval

Earlier in this chapter you were told that you can establish an interval around μ that has a known probability of containing $\overline{X}$. That formula is $\mu \pm z_c\sigma_{\overline{X}}$. "But, hey!" you say. "The problem we are looking at is not finding $\overline{X}$ from a knowledge of μ but estimating μ from $\overline{X}$." Right you are. The same logic applies whether we are relating the sample mean to the population mean or estimating the population mean from the sample mean. However, the latter situation requires us to take a different perspective on the $\overline{X}$-to-μ relationship.

Reversing perspective Let's suppose that we are working with a confidence factor of $z_c = 2.58$; the focus is on the middle 99% of the sampling distribution. It is correct to say that the probability is .99 that $\overline{X}$ does not differ from μ by more than $|2.58\sigma_{\overline{X}}|$; in other words, the probability is .99 that $|\overline{X} - \mu| \leq 2.58\sigma_{\overline{X}}$.

First, recognize that *the key fact here is that the absolute difference between $\overline{X}$ and μ will not exceed $2.58\sigma_{\overline{X}}$ in 99 of every 100 simple random samples, on the average.* Now let's reverse the perspective and look at the world from the vantage point of the sample mean. When we make $\overline{X}$ the center of the confidence interval, *it must likewise be true that the probability is .99 that μ will not differ from $\overline{X}$ by more than $2.58\sigma_{\overline{X}}$.* Thus, in about 99 of every 100 simple random samples, a confidence interval formed by $\overline{X} \pm 2.58\sigma_{\overline{X}}$ will include μ.

This idea is critical. Let's go over it one more time:

If it is true that for 99 of every 100 simple random samples, on the average, $\overline{X}$ will not differ from μ by more than $|2.58\sigma_{\overline{X}}|$, then it must also be true that for 99 of every 100 simple random samples, on the average, μ will not differ from $\overline{X}$ by more than $|2.58\sigma_{\overline{X}}|$.

This whole reversal of perspective idea is the same as saying that if the distance between the front of your classroom and the back of the room is exactly 30 feet, then it must also be true that the distance from the back to the front is exactly 30 feet. The absolute difference between the front and back is what counts in describing the relationship between the ends of the room.

The upshot of this discussion is that you can form a confidence interval around the sample mean that will include the population mean a specifiable percent of the time by using this formula:

$$
\text{Confidence interval} = \overset{\overset{④\quad①}{\downarrow\ \ \downarrow}}{\overline{X} \pm z_c\sigma_{\overline{X}}} \qquad (8.6)
$$
$$
\underset{③\ \ ②}{\uparrow\ \uparrow}
$$

FORMULA GUIDE

① Calculate the standard error of the mean.

② Multiply the confidence factor (*z* score) by the standard error to produce the *bound on the error of estimation.*

③ Subtract the "bound" from the sample mean—this is the *lower boundary of the confidence interval.*

④ Add the "bound" to the sample mean—this is the *upper boundary of the confidence interval.*

We will use formula (8.6) to set up a 99% confidence interval for the results of the Stinson et al. (1992) study on AIDS awareness that was mentioned earlier. They found that their subjects with less than a high school education had a sample mean of $\overline{X} = 16.62$ on a 30-point test of AIDS knowledge. This is a point estimate of μ, but we wish to construct an interval of values that fairly reflects the sampling error in the data. We need four quantities to do the calculations for a confidence interval: (1) the sample mean, 16.62 in this case; (2) the variance or standard deviation of the population (their data suggested that $\sigma^2 = 83.47$); (3) the sample size, $N = 6,898$ in this case; and (4) the confidence factor, which is $z_c = 2.58$ for a 99% confidence interval.

The sequence of steps in establishing a confidence interval is given here:

1. Calculate the standard error of the mean. Using formula (8.3), we get:

$$\sigma_{\overline{X}} = \sqrt{\frac{\sigma^2}{N}} = \sqrt{\frac{83.47}{6898}} = \sqrt{0.0121} = 0.11$$

2. Calculate the bound on the error of estimation:

$$z_c \sigma_{\overline{X}} = (2.58)(0.11) = 0.28$$

3. Find the lower boundary of the confidence interval:

Lower boundary $= \overline{X} - z_c \sigma_{\overline{X}} = 16.62 - 0.28 = 16.34$

4. Find the upper boundary of the confidence interval:

Upper boundary $= \overline{X} + z_c \sigma_{\overline{X}} = 16.62 + 0.28 = 16.90$

Hence, the confidence interval is 16.34 to 16.90. This means that we can be *very confident* that if we administered the AIDS knowledge test to the entire population of adults who have less than a high school education, their mean score (μ) would be at least 16.34 but not higher than 16.90.[7] Theoretically, there is only a 1% probability that confidence intervals formed in this way would not include the population mean.

This confidence interval is very small, isn't it? Why? With a sample size as large as 6898, $\sigma_{\overline{X}}$ will be tiny, and any particular $\overline{X}$ is likely to be very close to μ. This example is testimony to the assertion that large samples give very accurate estimates of μ.

A Limitation of the Method

You probably have noticed that the formula used to calculate the standard error of the mean requires that you know either the pop-

---

[7] By comparison, respondents with 12+ years of education had a mean AIDS knowledge score of 21.18. Yes, knowledge is power, and perhaps longevity!

ulation variance or the population standard deviation. More often than not in actual practice, such information is not available and, thus, you cannot use this exact approach. An alternative method for constructing a confidence interval when σ^2 is unknown will be presented in Chapter 10.

Interpreting the Confidence Interval for a Single Study

In giving you a theoretical interpretation of the 99% confidence interval, I stated that "in about 99 of every 100 simple random samples, a confidence interval formed by $\overline{X} \pm z\sigma_{\overline{x}}$ will include μ" (when $z_c = 2.58$). Does this mean that when you conduct a sample survey and set up the confidence interval with $z_c = 2.58$, you should be 99% confident that μ is within $2.58\sigma_{\overline{x}}$ of $\overline{X}$ in that specific study? Does it mean that there is a 99% chance that your particular confidence interval includes the true population mean? The answer to both questions is no. Theoretical conclusions about confidence intervals, though correct, are based on long-run averages. They refer to what we should expect if thousands of sample surveys are carried out and thousands of confidence intervals are computed. *In any particular interval estimation study, however, the confidence interval will either include or exclude the population mean—period.* If you've done the survey and the statistics correctly, it is appropriate to be "very confident" in your conclusion regarding the confidence interval, but not necessarily 99% confident.

Estimating a Population Proportion

Even though most of this chapter has focused on estimating a population mean, parameter estimation techniques are actually used more often to estimate population proportions. Estimated proportions are usually expressed as percents. Recall the survey results in *TIME* magazine showing that 65% of adult Americans are satisfied with the available health care options. The margin of error—also known as the bound on the error of estimation—reported in that study was ±3%. Hence, the confidence interval ranged from 62% to 68%. How was the 3% figure arrived at?

The formula used to set up a confidence interval for a proportion is

$$P \pm z_c\sigma_P \tag{8.7}$$

where P is the *proportion of the sample* that has a particular opinion or trait, z is the confidence factor you are using (either 1.96 or 2.58), and σ_P is the standard error of the proportion.

The sampling distribution of the proportion is made up of sample proportions rather than sample means, but the logic and procedure used to estimate a population proportion, symbolized by lowercase p, are the same as those used to estimate a population mean.

The standard error of the proportion is found with the formula

$$\sigma_P = \sqrt{\frac{pq}{N}} \qquad\qquad (8.8)$$

where p is the *population proportion* and $q = (1 - p)$. The big problem with this formula is that p is not known. It is what we are trying to estimate! The generally accepted solution to this dilemma is to make a conservative assumption about what p might be and substitute the assumed p into the formula. *By convention, p is assumed to be .50.* This assumption produces the largest possible value of σ_P and hence the largest possible confidence interval.[8] This is a cautious, yet reasonable, approach that almost guarantees that the confidence interval will include the true population proportion.

With this method, let's calculate the standard error of the proportion for the *TIME* survey. The following facts are relevant: sample size $N = 1200$, assumed $p = .5$, and assumed $q = (1 - .5) = .5$. Therefore:

$$\sigma_P = \sqrt{\frac{pq}{N}} = \sqrt{\frac{(.50)(1 - .50)}{1200}} = \sqrt{\frac{(.50)(.50)}{1200}} = \sqrt{\frac{.25}{12.00}}$$

$$= \sqrt{.00021} = 0.0145$$

Now we compute the bound on the error of estimation using a 95% confidence factor. The bound is in the form of a proportion:

$$\text{Bound} = z_c\sigma_P = 1.96(0.0145) = .0284$$

This proportion corresponds to a percent "bound" of 2.84, where percent = proportion · 100. *TIME* rounded 2.84% to 3% to produce the "margin of error" reported. Using the formula for a confidence interval of a proportion, $P \pm z_c\sigma_P$, but substituting corresponding percents for the proportions, we then subtract 3% from the sample percent (65%) to get the lower boundary of the interval (62%) and add 3% to 65% to obtain the upper boundary (68%). The conclusion is that about 65% of adult Americans are satisfied with their health care options. But the true population percent could reasonably be as low as 62% or as high as 68%. We are confident that the population percent is within these limits because in 95 of every 100 such sample surveys, the confidence interval will contain the population percent.

Using just such a simple procedure, you are now equipped to construct and interpret confidence intervals for any percent or proportion produced by a simple random sample.[9]

------

[8] Don't take my word for it. Try other values of p.

[9] *TIME*'s statisticians actually use a sampling design that is more complicated than simple random sampling; likewise, their statistical formulas are also more complicated. But the underlying logic of their approach is very similar to what is covered here.

TAKING STOCK: INFERENCE IS A TALE OF THREE DISTRIBUTIONS

You have gotten into the habit of thinking in terms of two distributions: sample and population. Now that you are using inferential statistics rather than simple descriptive statistics, you need to bear in mind that inferential procedures always involve a third distribution—the sampling distribution—which is necessary to relate the sample to the population. It is also important not to mix up these distributions and their respective measures of variation. Figure 8.10 illustrates these points. Take a good look at it, so that the image will register permanently.

KEY TERMS

simple random sampling	confidence factor
sampling distribution	point estimation
expected value	sampling error
unbiased estimator	interval estimation
sampling distribution of $\overline{X}$	confidence interval
error of estimation	bound on the error of
central limit theorem	estimation
standard error of the mean	

Figure 8.10 *A Tale of Three Distributions*

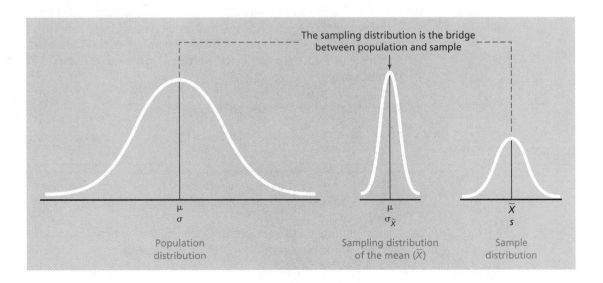

The sampling distribution is the bridge between population and sample

μ	μ	$\overline{X}$
σ	$\sigma_{\overline{X}}$	s
Population distribution	Sampling distribution of the mean ($\overline{X}$)	Sample distribution

SUMMARY

1. Sampling distributions are theoretical distributions of random events. They enable a researcher to make general statements about populations on the basis of sample data.

2. Behavioral scientists draw samples in order to find out something important about a larger group called the population. The methods of inferential statistics assume that samples have been randomly selected.

3. Sampling distributions are distributions of statistics, rather than distributions of raw scores. One of the most useful sampling distributions is the sampling distribution of the mean ($\overline{X}$), which has a mean of μ and a standard error of $\sigma_{\overline{X}}$.

4. The sample mean ($\overline{X}$) is an unbiased estimator of μ, the mean of the sampling distribution of $\overline{X}$.

5. The larger the sample, the more "normal" the sampling distribution becomes, and the smaller the variability of the sampling distribution. Samples with larger N's produce more precise estimates of μ than do samples with smaller N's.

6. The core principle of mathematically derived sampling distributions is the central limit theorem, which states: For any normal or *nonnormal* population with a mean of μ and a variance of σ^2, the distribution of means from samples of fixed size N will approach a normal distribution as N becomes larger, and that sampling distribution will have a mean of μ and a variance of σ^2/N.

7. The size of the standard error of the mean is inversely proportional to the square root of the sample size. You should usually use as large a sample as possible because smaller standard errors yield more precise estimates of μ.

8. Because $\overline{X}$ values are normally distributed around μ, the standard error of the mean can be used to determine the probability that $\overline{X}$ deviates from μ by no more than a specified amount. This specified amount, $z_c\sigma_{\overline{X}}$, is called the bound on the error of estimation.

9. The bound on the error of estimation can be used to compute an interval estimate of the population mean.

REVIEW QUESTIONS

1. Why do scientists often observe samples of people, animals, objects, or events rather than studying the whole populations of interest?

2. Define simple random sampling, and describe how you would use this type of sampling to select a representative sample of students at your college.

3. Identify each of the following methods as random or non-random sampling, and give the reason for each classification decision you make.

   (a) You use a random number table to select 50 professors from a numbered list of all faculty members at your institution.

   (b) You post yourself at the entrance to a local discount department store and arbitrarily select 200 passersby to interview.

   (c) You are interested in the average age of first-year students at your college, so you obtain the ages of all students in the three largest sections of the World Civilization course, which all students are required to take. You know that the number of students in these three sections is equal to $\frac{1}{25}$ of the total population of first-year students.

4. Consider the following population of three values: 4, 6, 8. Select all possible samples of size $N = 1$ and construct a sampling distribution of $\overline{X}$. Now construct a sampling distribution for all possible samples when $N = 2$. What is the mean of each sampling distribution? How does each of those means compare with μ? Which sampling distribution has the larger average absolute difference between $\overline{X}$ and μ? What is the practical implication of your answer to the latter question?

5. Repeat the steps in Question 4 for the following population of three values: 0, 5, 10.

6. Define or describe: simple random sampling, sampling distribution, sampling distribution of $\overline{X}$, and error of estimation.

7. Consider the following population of four values: 2, 4, 6, 8. Select all possible samples of size $N = 2$ and construct a sampling distribution of $\overline{X}$. Calculate the population variance, and then calculate the variance of the sampling distribution using the formula: $\sigma_{\overline{X}}^2 = \sigma^2/N$. Now compute the variance of the sampling distribution *directly* using the formula $\sigma_{\overline{X}}^2 = \Sigma(\overline{X} - \mu)^2/k$, where $k =$ the number of sample means in the sampling distribution. State the implications that your results have for the relationship between the population variance and the variance of the sampling distribution.

8. Repeat the steps in Question 7 for the following population of four values: 5, 6, 7, 8.

9. Referring to the data in Question 8, what is the expected value of $\overline{X}$?

10. Define or describe: expected value, unbiased estimator, central limit theorem, and standard error of the mean.

11. Assume a sampling distribution of $\overline{X}$ based on a population of IQ scores with $\mu = 100$ and $\sigma = 15$. What is the probability of randomly selecting a sample of 225 IQ scores with an $\overline{X}$ that is 103 or greater?

12. Assume a sampling distribution of $\overline{X}$ based on a population of IQ scores with $\mu = 100$ and $\sigma = 15$. What is the probability of randomly selecting a sample of 225 IQ scores with an $\overline{X}$ that is *more extreme* (in either direction from μ) than 101.47?

13. Assume a sampling distribution of $\overline{X}$ based on a population of IQ scores with $\mu = 100$ and $\sigma = 15$. What is the probability of randomly selecting a sample of 400 IQ scores with an $\overline{X}$ that is *more extreme* (in either direction from μ) than 101.47?

14. Assume a sampling distribution of $\overline{X}$ based on a population of IQ scores with $\mu = 100$ and $\sigma = 15$. What is the probability of randomly selecting a sample of 900 IQ scores with an $\overline{X}$ that does not deviate from μ by more than 0.82 IQ point?

15. Define or describe: confidence factor, point estimation, interval estimation, confidence interval, and bound on the error of estimation.

16. You have a simple random sample of 64 patients from a large psychiatric hospital. The patients have a mean score of 34 on the extraversion–introversion scale of the Myers-Briggs Type Indicator. Assume that this scale has a population variance of 25. What is the point estimate of the population mean? Set up the 95% confidence interval, and draw an appropriate conclusion regarding the value of μ. What can you do to make the confidence interval smaller in future studies of this population?

17. You are working in the office of the dean of students, and the dean gives you the assignment of determining whether the student population has enough intellectual interest to support an honors program. You search the literature and find that honors programs usually do not fare too well on campuses where the average score on the verbal scale of the Scholastic Aptitude Test is less than 550. You then draw a simple random sample of 400 student files and find that the mean SAT-Verbal score in this sample is 570. Assuming that the population standard deviation is 100, construct a 99% confidence interval to take sampling error into account. Based on the outcome, is it reasonable to conclude that an honors program probably would succeed on your campus?

18. One of my graduate students conducted a survey of 73 married midwestern women to find out how many hours per week they spent in leisure activities. Her sample mean was 22.9 hours. Assuming that the population variance was 144 and the .95 confidence factor was used, calculate the "margin of error" factor, or bound on the error of estimation, for this survey. Draw an appropriate conclusion regarding the survey outcome. Your conclusion should reflect the margin of error factor.

19. A survey of 900 randomly selected inmates of a federal prison suggested that 30% of them would repeat their crime if re-

leased today. Develop both a point estimate and an interval estimate of the percent of the prison population who would very likely repeat their offenses once released. Use the 99% confidence factor. Draw an appropriate conclusion for each estimate.

20. You are working with a simple random sample of 196 8-year-old children. The proportion of them who report having nightmares more than once a week is .72. Construct a 95% confidence interval for this proportion, and draw an appropriate conclusion about the population proportion.

21. You read in the newspaper that a high school principal has conducted a sample survey on 1000 students in her district. Her results show, among other things, that the average amount of time that students spend doing homework is 5 hours per week, with a margin of error of ±1.2 hours. Since she constructed a 95% confidence interval, the principal claims she is "95% confident that the typical student studies at least 3.8 hours per week." Is her conclusion accurate and appropriately phrased? Explain your answer.

22. In the survey described in Question 21, suppose the principal sampled only 100 students. What specific effect would the smaller N have on the margin of error? What would the new margin of error be? Is the change in the margin of error directly proportional to the reduction in the sample size? Explain your response.

23. Using a 99% confidence factor, a political scientist carries out a survey of registered voters and finds that about 77% of the sample desires a more fiscally conservative federal government, with a margin of error equal to ±4%. The researcher concludes that "there is a 99% probability that at least 73% of registered voters desire a more fiscally conservative government." Comment on the appropriateness of this conclusion. Is it correct in any sense? Explain.

24. A sample survey of 400 senior citizens reveals that 22% of the sample has sexual intercourse at least twice a week. Construct a 95% confidence interval for this result, and draw an appropriate conclusion.

25. A population has a variance of 16 and a mean of 100. Calculate the standard error of the mean for sample sizes of $N = 64$, $N = 256$, and $N = 1024$. What do the results of this exercise tell about the relationship between sample size and precision of interval estimation? Be as specific and complete as possible.

CHAPTER 9   LOGIC OF HYPOTHESIS TESTING

It was about ten years ago, on a dark, stormy night, that I encountered a most intriguing statistical phenomenon in a pile of 66 statistics exams. Grading multiple-choice tests never has been all that interesting to me, but the sight of 20 papers with a score of exactly 80% certainly grabbed my attention. Yes, 30.3% of the students got quiz scores of 80%. This outcome seemed "rare," "extreme," and "not to be expected on the basis of chance." There seemed to be something *significant* in these data. I was starting to feel uneasy—and I should say, mildly suspicious—about the unusually high incidence of a particular exam score.

From an objective statistical view, there are two possible ways to account for this seemingly improbable event. They take the form of opposing hypotheses:

Hypothesis 0: The rate of occurrence of 80% scores is merely *an instance of chance variation* to be expected when a sample of data (the set of test scores) is randomly selected from a population (all multiple-choice quiz scores in my statistics courses, both past and present).

Hypothesis 1: The rate of occurrence of 80% scores is too unusual, rare, and extreme to be reasonably attributed to chance variation. The outcome is a *systematic effect of some variable;* it is not merely a chance event.

As a behavioral scientist, I was interested in using inferential statistical analyses to evaluate the validity of these hypotheses. Specifically, I wanted to determine whether I could place more confidence in hypothesis 0, the hypothesis of chance variation, or hypothesis 1, the hypothesis asserting a nonchance outcome. As a college instructor, of course, I was interested in answering a practical question: Is it likely that some of the students cheated on the exam? Both the behavioral scientist and the professor in me wanted to determine whether the intriguing test score phenomenon was *significantly* different from chance expectations.

Many research questions fit the "competing hypothesis" model. To choose one hypothesis over the other, researchers must perform "significance tests" on data pertinent to the hypotheses. It is for this reason that "hypothesis testing," a very important aspect of inferential statistics, is also known as "significance testing." The statistical outcome of a study is "significant" if the hypothesis of chance can be rejected. A significant outcome is said to be **reliable**—that is, *likely to be repeated in its basic form if the same study is conducted again and again.* In contrast, chance outcomes vary randomly and are not consistently repeatable.

NOTE: In this textbook, the terms **statistical significance** and **statistical reliability** are used interchangeably. This is an appropriate use of the terms because a significant event is literally likely to be repeatable in replications of an investigation. A significant result is almost always a reliable result.

By reading this chapter you will learn the logic and procedures underlying hypothesis testing. Specifically you will learn (1) how to set up competing hypotheses, (2) why we test the hypothesis of chance variation rather than the hypothesis of nonchance outcomes, (3) the meaning of "levels of significance," and (4) how to properly interpret significant results; you will also learn about (5) decision errors that sometimes occur in significance tests and how to minimize the frequency of such errors. But, to avoid getting ahead of ourselves, let's see how the puzzle of unusual test scores was solved. In following the story, you will gain insight into hypothesis testing.

FOUR LOGICAL STEPS

To pursue the truth behind the statistics test phenomenon, I used a simple four-step system of logic that characterizes all hypothesis-testing procedures in behavioral statistics:

1. State the competing statistical hypotheses.
2. Compute a "test statistic" from the sample data. A test statistic, such as a z ratio, is used to compare the obtained sample data with numerical outcomes that would be expected on the basis of the hypothesis of chance.
3. Compare the test statistic with a distribution of random (i.e., chance) test statistics to determine whether the computed statistic is significant—that is, unlikely to belong to the chance distribution.
4. On the basis of the significance or nonsignificance of the test statistic, decide whether to reject or retain the hypothesis of chance; if the hypothesis of chance is rejected, the alternative (nonchance) hypothesis is supported.

Notice the reverse logic used in hypothesis testing: The researcher garners support for a nonchance explanation of her data by testing and rejecting the hypothesis of chance. By showing that random variation is not likely to be a valid account of the results, she can support the validity of some nonchance explanation of the data. The reason for using such reverse logic will be discussed later in this chapter.

Logical Step 1: Stating Competing Hypotheses

As indicated earlier, two hypotheses are at loggerheads in these unusual test scores: the *hypothesis of chance* (random variation), known as the **null hypothesis**, and the *competing hypothesis*, called the **alternative hypothesis**. Since the null hypothesis is symbolized H_0 and the alternative hypothesis is symbolized by H_1, the hypotheses are formally stated as follows:

Null hypothesis→H_0:   The rate of occurrence of 80% scores is due to chance.

Alternative hypothesis→H_1:   The rate of 80% scores differs from chance variation.

Logical Step 2: Computing a Test Statistic

A **test statistic** is *a numerical index computed from the sample data that can be readily compared with a sampling distribution of chance statistics of the same type.* Recall that 30.3% of my statistics students got a score of 80% on the exam in question. This translates to a proportion of .303. Using the idea of the sampling distribution of the proportion discussed in Chapter 8, we can compute a z score based on the sample proportion of .303. Then we can use that z value as a test statistic to evaluate the hypothesis of chance.

> **CONCEPT RECAP**
>
> A *sampling distribution* is a theoretical distribution of random events. It is made up entirely of sample statistics, the values of which are assumed to be determined entirely by chance. A sampling distribution enables you to make general statements about a population on the basis of sample data. See Chapter 8 for a complete review.

Sampling distribution of the proportion   To refresh your memory, the sampling distribution of the proportion is made up of sample proportions rather than sample means. The infinite number of sample proportions that make up the distribution are assumed to vary randomly from the population proportion, which is the average value in the distribution. Thus, *the sampling distribution of the proportion is a model of chance variation. Therefore, using a z score as the vehicle, we can compare any computed sample proportion with the sampling distribution and determine how probable the computed proportion would be if it were a chance event.*

To use the sampling distribution of the proportion to test the present null hypothesis (that the 80% scores are a chance event), we need to determine the following quantities:

N, the sample size

P, the sample proportion

p, the population proportion

σ_P, the standard error of the sampling distribution of proportions

N is the number of students in the sample: $N = 66$. P is the proportion of students who got a score of 80%. Earlier we found that $P = .303$. We'll round it to $P = .30$. So in order to proceed with the hypothesis test, all we need are the population proportion and the standard error of the sampling distribution.

Finding a population proportion   In everyday statistical practice, it is often tricky to determine the exact value of a population parameter. This time it was relatively simple. I had kept records of students' quiz grades over a span of 12 years. A brief survey of my little brown grade books revealed that, over the long run, 15% of my statistics students got scores of 80% on the multiple-choice quizzes. Therefore, the population proportion was assumed to be $p = .15$.

Calculating the standard error of the proportion   In order to determine the probability of the sample proportion in a chance (null) distribution, we need to calculate the standard error of the sampling distribution. In Chapter 8, the formula for the standard error of the proportion was given as

$$\sigma_P = \sqrt{\frac{pq}{N}}$$

where p = the population proportion, $q = (1 - p)$, and N = the size of the sample in the investigation. Since $p = .15$, $q = (1 - .15)$, or $.85$. Hence,

$$\sigma_P = \sqrt{\frac{pq}{N}} = \sqrt{\frac{(.15)(.85)}{66}} = \sqrt{\frac{.1275}{66}} = \sqrt{.001932} = 0.044$$

Calculating z as a test statistic   To find out how likely or unlikely the sample proportion would be in a sampling distribution of chance proportions that vary randomly from the population proportion, we must convert the sample proportion to a test statistic. The standard score formula for the sampling distribution of the proportion is $z = (P - p)/\sigma_p$. Doing the arithmetic, we get:

$$z = \frac{P - p}{\sigma p} = \frac{.30 - .15}{.044} = \frac{.15}{.044} = 3.41$$

Hence, the rate of occurrence of 80% test scores is *3.41 standard deviations above what would be expected on the basis of chance alone.* Is this extreme enough to be declared statistically significant? Can we reject the null hypothesis and support the validity of the alternative hypothesis? We shall find out shortly.

Logical Step 3: Comparing the Test Statistic to Chance

To decide whether the test statistic is significant, we have to compare it with a model of chance and determine whether it should be considered a probable outcome or an improbable outcome in the context of random variation.

The "substitutability" of the standard normal distribution Since (1) the sampling distribution of the proportion is a normal distribution and (2) the sample proportion of .30 has been converted to a z score within the sampling distribution, we can use the standard normal distribution as the "model of chance." *Note: Anytime your sample statistic has a sampling distribution that may be assumed to be normal, it is appropriate to substitute the standard normal distribution for the original sampling distribution.* In the present example, this conventional act of substitution means that:

- The standard normal distribution will represent the sampling distribution of the proportion.
- The computed z value of 3.41 will represent the sample proportion, .30, which is the proportion of students who got quiz scores of 80%.

So, right this minute, turn to Table Z in Appendix A, go to the last page of the table, and proceed down column 1 until you find a z score of 3.40. (There is no z = 3.41 in the table; therefore, the closest available z is used.) Now focus on the probability in column 3, which shows the proportion of random z scores at and beyond z = 3.40.

1	2	3
z	Area between z and the mean	Area beyond z
3.40	.4997	.0003
		(probability = .0003)

The proportion of the area under the normal curve that lies above z = 3.40 is .0003. Since proportion is the same as probability under the normal curve, this means that there is only a .0003 probability that 30% or more of the students would have obtained a score of 80% *by chance alone*. Should we therefore conclude that the chance probability of the outcome equals .0003? Well, not exactly. The probability is actually 2 × .0003, or .0006. Read on to see why.

The importance of wording   If you check back to the statement of hypotheses, you'll see that the alternative hypothesis (H_1) states only that "the rate of 80% scores differs from chance variation." H_1 does not specify that the rate of 80% test scores is higher or lower than what would be expected on the basis of chance alone, only that the rate *differs* (*in some direction*) from chance expectations.

This way of setting up the alternative hypothesis leads to what is called a **nondirectional test**. It is the usual way of doing tests of significance. The upshot of a nondirectional null hypothesis is that we must consider both possible ways that a z score can differ in an "extreme way" from its mean (0) within the standard normal distribution: z can be extremely high on the positive side of the distribution or extremely low on the negative side of the distribution.

As we determine the likelihood of our computed z score, the appropriate way to phrase the probability question is: What is the overall probability of a z score *as extreme as* 3.40 (in either direction from 0)? More precisely yet, What is the combined probability of z scores at or above +3.40 *or* at or below -3.40? Since the standard normal distribution is symmetrical, the probability that applies to "at or above +3.40" also applies to "at or below -3.40." Each has a likelihood of .0003 within a chance distribution. Therefore, the combined—or "two-tailed"—probability of outcomes this rare or rarer *by chance alone* equals .0003 + .0003, or .0006. The rationale of this combined probability is illustrated in Figure 9.1.

Thus, it is very unlikely that 30% or more of the students would get test scores of exactly 80% just by chance. In fact, there are only 6 chances in 10,000 that an event *as extreme as this* could have occurred as a result of random variation alone. Hmmmm ... pretty unlikely, no? But was the outcome unlikely enough within a model of chance to be considered statistically significant?

A criterion of significance A statistically significant result is *a z score (or some comparable index) that would be a "rare" event in a distribution of chance events.* But that begs the question, How do you define *rare*? By convention, the value of a test statistic is

Figure 9.1 *Location of Test Statistic in the Standard Normal Distribution (Note: The two-tailed probability of a z statistic this extreme or more extreme equals the sum of the positive and negative extremes.)*

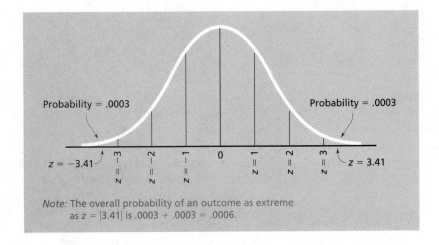

Probability = .0003   Probability = .0003

$z = -3.41$ $z = 3.41$

Note: The overall probability of an outcome as extreme as $z = |3.41|$ is .0003 + .0003 = .0006.

considered "rare" enough to be significant if its chance probability is 5% or less—in other words, when probability ≤ .05. On the basis of the law of randomness, such outcomes will occur only 5 times or fewer in every 100 chance events. *The chance probability that you select to define significant, or "rare," events* is called the **level of significance**. Thus, outcomes that are this unlikely or more unlikely in a sampling distribution are said to be "significant at the .05 level."

As shown in Figure 9.2, any z score as extreme as +1.96 or −1.96 is significant according to the .05 standard of significant events. Note that z_{crit} = |1.96| defines the most extreme—that is, "rarest"—5% of the sampling distribution. The "crit" subscript attached to z_{crit} = |1.96| stands for "critical," meaning that *the computed z must exceed* the **critical value** |1.96| in order to be significant. There are other levels of significance besides the .05 level. We will consider them and their associated z_{crit} values later in this chapter.

Critical z vs. computed z   *Important!* Don't confuse the z-score value that you compute from the data, called "computed z" or the "test statistic," with the *theoretical z value*, or **critical z**, *required for significance*. The critical z—represented by $\mathbf{z_{crit}}$—is determined by the level of significance that you use. You look it up in Table Z of Appendix A. Your test statistic is *computed* from and determined by your data. If the absolute value of the computed z is as big as or bigger than the absolute value of the critical z, then your data are statistically significant.

Figure 9.2 *Critical Values Needed for Significance at the .05 Level and Regions of Rejection of the Null Hypothesis*
(Note: Any test statistic that falls into the shaded regions of rejection is statistically significant. The null hypothesis is rejected.)

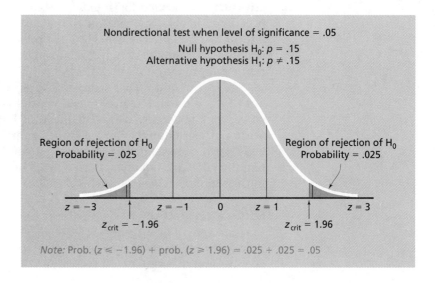

Nondirectional test when level of significance = .05
Null hypothesis H_0: p = .15
Alternative hypothesis H_1: p ≠ .15

Region of rejection of H_0
Probability = .025

Region of rejection of H_0
Probability = .025

z = −3   z = −1   0   z = 1   z = 3

z_{crit} = −1.96 z_{crit} = 1.96

Note: Prob. (z ≤ −1.96) + prob. (z ≥ 1.96) = .025 + .025 = .05

You should think of any "level of significance" as a decision criterion that is used in hypothesis testing. The result of a statistical test is considered unlikely to be a chance event, and hence is declared "significant," if its chance probability is at least as small as .05. Is our z score of 3.41 significant at the .05 level? Since its overall probability is .0006, which clearly is smaller than .05, we declare it significant. And now we are in position to make an informed decision about the null hypothesis.

Logical Step 4: Making a Decision About H_0

The fourth logical step in hypothesis testing is to make a decision about the null hypothesis. *The decision is always to either reject or not reject H_0.* There is no middle-ground alternative in conventional hypothesis testing.

Region(s) of rejection   If the absolute value of the test statistic is big enough to meet or exceed the critical value associated with the level of significance ($z_{crit} = |1.96|$ for the .05 level), then the test statistic is said to fall into the "region of rejection of H_0." The **region of rejection** is *that portion of the sampling distribution that represents significant outcomes.* The shaded areas in Figure 9.2 show that in a nondirectional test of significance, the sampling distribution has two regions of rejection. If the computed z value falls into either of these regions, you decide to reject the null hypothesis. In that case the results are deemed to be statistically reliable, and you consider the alternative hypothesis (H_1) to be valid. Usually this decision means that the data of your study were influenced in a systematic way by some variable.

If the absolute value of the test statistic is smaller than the critical value needed for significance, you decide to not reject the null hypothesis. This means that you doubt the validity of the alternative hypothesis: The chance probability of the test statistic simply is not small enough to dismiss the assertion that the data represent only random variation.

Since our computed z score of 3.41 exceeds the critical z score of $|1.96|$, it is in a region of rejection, and the high rate of 80% test scores is significant. We reject the null hypothesis that the test score pattern is merely an instance of random variation. Accordingly, we now consider the alternative hypothesis to be valid. A systematic effect of *something* determined the high rate of 80% scores in the statistics class. But what?

Some aftermath of the weird test scores   Having determined that the rate of occurrence of 80% scores was significantly higher than it should have been according to a model of random variation, I confronted the statistics class with my finding. Since the students were familiar with the logic of hypothesis testing and the meaning of statistical significance, they all understood the likely implication of the outcome: Some of them had cheated on the quiz.

Additional inspection of the 80% papers revealed that in ten of the 20 cases, the students had missed exactly the same questions with exactly the same incorrect answers! That narrowed the field of errant students to ten. The remaining ten students (15.15% of the class) had gotten their scores in a legitimate way. In view of these statistical facts, nine of the ten students with illegitimate 80% scores visited me individually. Each offered an apology for being disingenuous and agreed to take a makeup test to replace the spurious grade. The tenth disappeared from the course and was never heard from. This all took place a decade ago. Since that debacle, I have not observed any "rare" statistical phenomena in my test score distributions.

A word on the importance of replication   It is unusual for a behavioral scientist to be able to immediately confirm the correctness of a statistical decision, as I was able to do in this example. Just because a test statistic—the computed z score here—is deemed to be a rare event in the context of a model of random variation, that does not mean that the test statistic is certainly not a result of chance. Such a significant outcome simply means that the data *are very unlikely* to have been produced by chance alone. Therefore, it makes logical sense to reject the hypothesis of chance and conclude in favor of some nonchance hypothesis. However, confirmation of the correctness of a statistical significance decision normally must await successful replications of the study by other researchers. Such repeated instances of the same outcome demonstrate that the statistically significant event is, in fact, reliable, as significance implies. But sometimes a "significant" result turns out to be nonreplicable, which means that the "significance" conclusion reached in the original study was an erroneous decision. More will be said about statistical decision errors later in this chapter.

CONCEPTUAL SUMMARY

If you understood most of what was done and concluded in the preceding example, then you already know most of what you need to use the idea of statistical significance in your work. Because of the newness of the terrain, however, it will help to review the following definitions before taking a more detailed look at the components of hypothesis testing.

Statistical significance: the same as statistical reliability—namely, that the outcome of a study is likely to be repeated in its basic form if the same investigation is conducted again and again; it also means that the test statistic would be a rare event in a distribution of chance events.

Null hypothesis (H_0): hypothesis of chance; the hypothesis that you actually test.

Alternative hypothesis (H_1): the hypothesis that competes with the null hypothesis; the assertion that the sample data differ from chance expectations; the hypothesis that is supported through rejection of the null hypothesis.

Test statistic: a numerical index computed from the sample data that can be readily compared to a sampling distribution of chance statistics of the same type.

Sampling distribution: a theoretical distribution of random outcomes; the "model of chance" to which the computed test statistic can be compared; the "null distribution."

Nondirectional test of significance: a hypothesis-testing procedure that recognizes any rare value of a test statistic as significant, regardless of whether the statistic is positive or negative.

Statistically significant result: a z score (or some comparable index) that would be a "rare" event in a distribution of chance events.

Level of significance: the chance probability that you select to define significant ("rare") outcomes within the sampling distribution.

Critical value: in a sampling distribution, the absolute z value that your test statistic must equal or exceed in order to be deemed significant.

Region of rejection: that extreme portion of the sampling distribution that represents a significant outcome; this region is marked off by the critical value(s).

Now that you've had an overview of the logic and process of hypothesis testing, let's examine the null hypothesis idea in some depth.

WORKING WITH HYPOTHESES

There are numerous ways to express the null and alternative hypotheses. You will need to be familiar with these variations and recognize what each variation signifies. Every behavioral investigation requires that these hypotheses be set up in a way that specifies what the researcher's general expectations are *prior to collecting any data.* In order to acquaint you with the general conventions that govern the statement of hypotheses, I will present several research scenarios that vary in terms of how the null and alternative hypotheses must be expressed. As you consider these examples, there are some general principles of hypothesis construction that you should keep in mind.

- *Null means "chance."* By definition, the null hypothesis is the hypothesis of chance, random variation, "no difference," no reliable effect.
- *The null hypothesis is the "test" hypothesis.* The null hypothesis, not the alternative hypothesis, is what we directly test with our statistical procedure.
- *The alternative must complement the null.* For every null hypothesis that you test, you must formulate a competing alternative hypothesis that logically complements the null. For example, if the null hypothesis states that the population mean will be less than or equal to 0 ($\mu \leq 0$), then the alternative hypothesis must state that the population mean will be greater than 0 ($\mu > 0$).
- *Hypotheses usually concern parameters.* Even though you perform significance tests on sample statistics—sample means, for example—the null and alternative hypotheses are almost always symbolized with population parameter symbols; you would use $\mu = 0$ instead of $\overline{X} = 0$, for example, to state a null hypothesis about a mean. The reason for using parameter symbols is that a significant result always implies that what is true of your sample is also true of the whole population represented by your sample. So in any significance test, you are really evaluating hypotheses about a population parameter; the sample statistic you work with merely serves as a convenient vehicle for accomplishing that objective.
- *Hypotheses can be nondirectional or directional.* Alternative hypotheses can be stated in a nondirectional form, which says only that a population parameter differs *in some way* from a chance expectation (e.g., $\mu \neq 0$), or in a directional form, which specifies that a population parameter will be either greater than an expected chance value (e.g., $\mu > 0$) or less than an expected chance value (e.g., $\mu < 0$). The purpose of your study will determine whether you use a directional or a nondirectional hypothesis, although a nondirectional hypothesis is almost always appropriate (see below).

Now let's see how these general principles are put into practice.

Nondirectional Test of a Proportion

The scenario This first example comes from the story of the weird test grades that opened this chapter. You'll recall that my goal was to find out whether the results of a statistics test came from a population of test outcomes in which the proportion of students who got 80% scores is .15. The .05 level of significance was used to test the null hypothesis.

Statement of hypotheses   The null and alternative hypotheses are properly expressed as:

> Null hypothesis→H_0: $p = .15$
>
> Alternative hypothesis→H_1: $p \neq .15$

Interpretation and implications   Notice that in a nondirectional test of significance, H_0 simply states that the population parameter of interest is equal to some value that is expected on the basis of chance alone. H_1 challenges H_0 by stating that the population parameter represented by the sample data *is not equal* to the value expected on the basis of chance alone. In this case, H_1 states that "the sample data were drawn from a population in which the proportion of students with grades of 80% is *different from* .15." The direction of the postulated inequality does not matter. According to H_1, the true population proportion might be higher than .15 *or* lower than .15. Any large deviation from .15 will satisfy H_1.

It is important to recognize that in a nondirectional test, either a positive or a negative z statistic, if sufficiently large, can result in the rejection of H_0. This fact was depicted in Figure 9.2. Refer back to that figure. Notice that if the z score computed from the sample data is at least as extreme as either $z_{crit} = -1.96$ or $z_{crit} = +1.96$, then the test statistic will be in a region of rejection of the null hypothesis, and the data will be declared significant. In a nondirectional test, the standard normal distribution has two regions of rejection of H_0. This means that *significant outcomes can exist in either of the two "tails" of the distribution*. It is for this reason that nondirectional tests of significance are often called **two-tailed tests**.

Nondirectional Test of a Mean

The scenario   Newnham (1988) was interested in finding out whether the self-esteem of convicted felons *differed* from that of the general adult population. He administered the Personal Data Inventory, a test of self-esteem, to 43 inmates of a correctional institution. The population mean of the Personal Data Inventory was 30, and the population standard deviation was 6.5. Newnham used the .05 level of significance to test the null hypothesis.

Statement of hypotheses   Newnham expressed his null and alternative hypotheses as:

> H_0: $\mu = 30$
>
> H_1: $\mu \neq 30$

Interpretation and implications   The null hypothesis in this case asserts that the self-esteem scores of Newnham's convicts come from a population of self-esteem scores that has a mean of 30. The alternative hypothesis holds that the convicts' self-esteem scores belong to a population in which the mean is different from 30. Note that H_1 does not specify that the convicts' mean will be higher or lower than 30, only that it will differ from 30. Hence, the critical value of z was both -1.96 and $+1.96$. Again we see a two-tailed test.

I should point out that Newnham did expect the convicts to have somewhat lower self-esteem than adults in general. Nonetheless, he (and other researchers) would have been interested in the opposite kind of difference as well, even though he did not expect it. Therefore he performed a nondirectional test. In general, *behavioral scientists use nondirectional, or two-tailed, tests of significance unless there is a compelling reason to restrict the region of rejection to one side of the sampling distribution.* The next example illustrates a research situation that definitely justifies using a directional test.

Directional Test of a Mean: Positive Difference Expected

The scenario   I have served as an expert witness (statistician) in several age-discrimination lawsuits filed against corporations that had laid off large numbers of older employees. In one such case, the mean age of workers in a company was 40 years. The court was concerned only about the possibility that the group of laidoff employees was older than the typical employee in the company. The other possibility, that the laidoff employees were younger than average, was irrelevant to the age-discrimination charge because persons under 40 are not protected by law from being dismissed on the basis of their age alone. But if it could be shown that the furloughed group was significantly *older* than 40, then the court was prepared to impose penalties on the company and award monetary settlements to dismissed employees who were over 40. That is, only a positive age difference between the laidoff group and the companywide group was of practical interest. Accordingly, I performed a directional test of significance using the .05 level of significance.

Statement of hypotheses   The hypotheses were symbolized as:

H_0: $\mu \leq 40$

H_1: $\mu > 40$

Interpretation and implications   The null hypothesis asserts that the laidoff employees belong to a population of employees whose mean age is less than or equal to 40. Note that, *in a directional test, when the test statistic (that is, the computed z value) is*

Highlight in Class

in the direction specified by the null hypothesis, the result is always considered nonsignificant—regardless of how big the value of the test is. To be significant in a directional test, the computed z value must be big enough to exceed the critical z and have the appropriate sign. In this example, the computed z would have to be "big" and have a positive sign because the alternative hypothesis asserts that the mean age of the laidoff group is *higher than* 40.

Figure 9.3 illustrates important facts about a directional hypothesis. First notice that only one tail of the standard normal distribution is used to evaluate the test statistic. It is for this reason that a directional test is often called a **one-tailed test** of significance. In this case, only the right (positive) side of the distribution is relevant to rejecting the null hypothesis.

Second, observe that the entire "region of rejection" is located in one tail of the curve. Since the whole 5% region of rejection is restricted to the positive tail of the distribution, the portion of the tail that represents significant outcomes is larger for this one-tailed test than for two-tailed procedures.

A third fact is related to the second: The critical z value that must be met or exceeded is smaller in a directional test than it is in a nondirectional test that uses the same level of significance. You can see that, at the .05 level of significance, $z_{crit} = 1.65$ for the directional test,[1] whereas z_{crit} was $|1.96|$ for the nondirectional test. Since the test statistic z value that is computed from the sample

Figure 9.3 *Critical Value and Region of Rejection for a Directional (One-Tailed) Test When a Positive Value Is Expected by the Alternative Hypothesis*

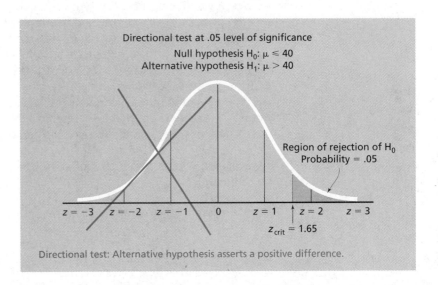

Directional test at .05 level of significance
Null hypothesis H_0: $\mu \leq 40$
Alternative hypothesis H_1: $\mu > 40$

Region of rejection of H_0
Probability = .05

$z = -3$ $z = -2$ $z = -1$ 0 $z = 1$ $z = 2$ $z = 3$

$z_{crit} = 1.65$

Directional test: Alternative hypothesis asserts a positive difference.

---

[1] Where did I get $z = 1.65$? The answer can be found in Table Z of Appendix A. Column C shows that approximately 5% of the normal curve lies above $z = 1.65$.

data is more likely to exceed 1.65 than 1.96, *it is easier to reject H₀ when you do a directional test of significance than when you conduct a nondirectional test.* Because of this, some behavioral scientists routinely use a nondirectional test, their objective being to give the null hypothesis a fairer chance of not being rejected. In effect, this practice grants H₀ the benefit of the doubt.

Directional Test of a Mean: Negative Difference Expected

The scenario A developmental psychologist thinks that a child's exposure to lead-based paint during the first five years of life results in permanent brain damage and a lowering of the IQ. The psychologist assesses the IQs of 72 six-year-old children known to have been exposed to lead-based paint. It is known that the average IQ of the population of six-year-olds in the surrounding community is 96. The psychologist expects the average IQ of the exposed children to be lower than 96. The opposite prediction—that exposure to lead enhances brain development and IQ—is not a reasonable possibility in light of the known toxicity of lead. Therefore, the psychologist will perform a directional test at the .05 level of significance.

Statement of hypotheses The psychologist's null and alternative hypotheses are:

$H_0: \mu \geq 96$

$H_1: \mu < 96$

Interpretation and implications H_0 states that the children exposed to lead belong to a population of six-year-olds in which the mean IQ is equal to or greater than 96. H_1 contends that the children come from a population in which six-year-olds have an IQ of less than 96. Figure 9.4 shows that the test statistic must be negative and at least as extreme as $z = -1.65$ in order for the null hypothesis to be rejected. The computed z will be negative if the mean IQ of the children exposed to lead is less than 96.

A List That Is Too Long

There are many other variations in the statement of hypotheses in tests of significance. They are too numerous to cover in one chapter. The examples provided here will get you started, and you will see more examples in later chapters.

Right now, let's try to answer a few of the questions that might have been raised by the preceding material.

Figure 9.4 *Critical Value and Region of Rejection for a Directional (One-Tailed) Test When a Negative Value Is Expected by the Alternative Hypothesis*

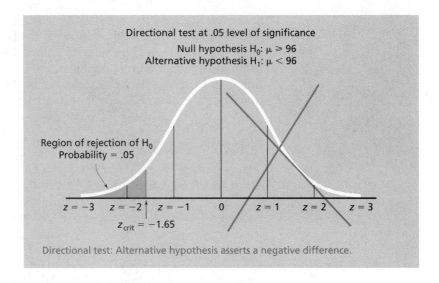

Directional test at .05 level of significance
Null hypothesis H_0: $\mu \geqslant 96$
Alternative hypothesis H_1: $\mu < 96$

Region of rejection of H_0
Probability = .05

$z = -3$ $z = -2$ $z = -1$ 0 $z = 1$ $z = 2$ $z = 3$

$z_{crit} = -1.65$

Directional test: Alternative hypothesis asserts a negative difference.

QUESTIONS OFTEN ASKED ABOUT THE NULL HYPOTHESIS

When Should I Use a Directional Test?

Given that there are two types of hypothesis-testing procedures—directional and nondirectional—students are often confused about when to use a one-tailed test rather than a two-tailed test. Unfortunately, there is no pat answer to this question. But, as you will see, there is a conventional preference, which you probably should heed.

Two viewpoints   There are two schools of thought on the question of what circumstances warrant the use of a one-tailed test of significance: "discretionary" and "conservative." According to the discretionary school, you should use a directional test any time you have enough information, from either theory or prior research, to clearly formulate a directional hypothesis (Kirk, 1990). With this model, most tests of hypotheses in the behavioral sciences would be one-tailed.

In contrast, the conservative school contends that a directional test should be used only if either of the following conditions is true: (1) just one particular direction of difference between the obtained mean and the theoretical mean has any practical value (for example, a new medicine is designed to make people better, not sicker) (Hays, 1988); or (2) a researcher's theory specifically predicts either a positive *or* a negative outcome, and the opposite outcome would be theoretically meaningless to all researchers in that field (Kimmel, 1957). With this model, most tests of hypotheses in the behavioral sciences would be two-tailed.

An additional consideration in this issue is that it is easier to get a significant result when you use a directional test of significance. As Figures 9.2–9.4 show, at the .05 level of significance, a one-tailed significance decision requires a z value of only 1.65, whereas the corresponding two-tailed decision requires a less probable z value of 1.96. Therefore, if the researcher knows that the z score is likely to be a positive value, let's say, then she can stack the deck in favor of H_1 (and against H_0) by setting up a directional hypothesis. It is for this reason that the behavioral sciences generally frown on the routine use of directional tests. If you check the professional journals in your discipline, you will see that the conservative model of significance testing is overwhelmingly favored. The upshot is that you should always use a nondirectional test of significance unless you have a compelling theoretical or practical reason for conducting a directional test. See Hays (1988) and Kirk (1990) for more discussion of these points.

Decide in advance   Related to the above recommendation, it is imperative that you decide whether you want to conduct a directional test *before* you analyze your data. And that prior decision should be based on the nature of the theoretical or practical question that your research is designed to address. It definitely is considered bad practice to switch to a one-tailed test after you discover that the test statistic isn't quite big enough to be significant under a two-tailed model. As a general principle, the ground rules of your test of significance should always be set in advance of analyzing the data. Otherwise, it is usually possible to manipulate the standards of the procedure to make most research results "come out significant." Any such illegitimate manipulation nullifies the purpose of behavioral research, which is to discover the laws of behavior (not manufacture them!).

Why Am I Testing H_0 Instead of H_1?

A behavioral scientist usually is more interested in the alternative hypothesis than in the null hypothesis because the alternative hypothesis represents his expectations regarding the outcome of the study. In most cases, if he truly expected to find only chance outcomes, he probably wouldn't carry out the investigation.

However, as mentioned earlier, a researcher tests only the null hypothesis, not the alternative hypothesis. Founded on a type of reverse logic, the rationale behind this is as follows: If you are interested in supporting research hypothesis A (represented by the alternative hypothesis), you can do so by showing that its opposite, the null hypothesis, is false. Rejection of the null hypothesis doesn't directly "prove" the research hypothesis, but it does permit you to retain the position that the research hypothesis is true

(Bakan, 1966). On the other hand, if the value of the test statistic is within the range predicted by the chance (null) hypothesis—let's say, within the middle 95% of the sampling distribution—then your research hypothesis, which asserts that nonchance processes are operating, is not likely to be true.

Why is this reverse logic used? Why not simply test the research hypothesis directly? The answer is twofold. First, it is logically easier to disprove a general assertion than to prove it. For example, suppose I assert that all nurses are extroverts. If this proposition is true, it would be impossible to directly prove it without testing *all* nurses on a measure of extroversion. But if the proposition is false, it could be disproved simply by locating one exception to it—that is, one nurse who is not extroverted. Likewise, it is much more feasible to disprove H_0, the opposite of the research hypothesis, than to try to directly prove the research hypothesis.

There is a second, and even more important, reason for testing H_0. We know the shape, properties, characteristics, and values of the null distribution, otherwise known as the sampling distribution. This information has been provided by mathematical theories of random variables. Hence, test statistics, such as z ratios, can be compared with the convenient and invariant model of chance that represents H_0. By contrast, in most research situations we have little, if any, idea about the mean, standard deviation, shape, and critical values of the "alternative hypothesis" distribution. That distribution is literally unknown and unobservable. Consequently, we usually have no appropriate, invariant model of nonchance events to compare with our data. By default, then, we must test the hypothesis of chance (H_0) instead.

Is It Ever Correct to Accept H_0?

As you saw earlier, and as you will observe later, when a test statistic does not meet or exceed the critical value for significance, it is said that we "fail to reject" the null hypothesis. Under those circumstances, why don't we say that we "accept" the null hypothesis?

Most statisticians prefer to say that one "fails to reject" H_0 rather than one "accepts" H_0 because it is just as difficult to directly prove that no real difference exists as it is to directly prove that one exists. For example, the sample of people that you test might be peculiar in some way or perhaps too small to show that there is a real difference between the sample mean and the population mean. Or you might have used an unreliable way of measuring the dependent variable you are testing in your study. Thus, although no significant difference may turn up in your sample, it is possible that a nonchance difference could be shown to exist in a more adequate sample, especially if a more reliable measure of the dependent variable is used. Considering the above points, it is almost never appropriate to draw the general conclusion that H_0 is true. However, it is perfectly acceptable to either (1) reject H_0 on the basis of evidence provided by your sample data or (2) state that there is insufficient evidence to reject H_0. Of course, if you fail to reject H_0, by implication you cannot make any statements in support of H_1.

STATISTICAL DECISIONS AND DECISION ERRORS

When you carry out a test of significance, the only statistical decision you have to make concerns the truth or falsity of the null hypothesis. The decision is an either/or one. In a particular investigation, the *reality* of the situation is also an either/or matter. Regardless of your decision, the null hypothesis is either really false or really true. The ideal in significance testing is to maximize the likelihood of matches between reality and statistical decisions. But mismatches can and do sometimes occur. In such cases, a statistical decision error has occurred.

Hits and Errors

When the two decision options (reject H_0 or not reject H_0) are combined with the two possible states of reality, it becomes clear that there are four possible outcomes of a significance test:

1. A true H_0 is not rejected: This is a correct decision.
2. A false H_0 is rejected: This is a correct decision.
3. A true H_0 is rejected: This is called a **Type I error.**
4. A false H_0 is not rejected: This is called a **Type II error.**

There are two ways to be right and two ways to be wrong in your decision about the null hypothesis. These four potential outcomes are summarized in Table 9.1. Our major concern here is with the possible decision errors.

Table 9.1

DECISION MATRIX SHOWING POSSIBLE STATISTICAL DECISION ERRORS

| | | STATISTICAL DECISION | |
		FAIL TO REJECT H_0.	REJECT H_0.
True State of Affairs	H_0 is true.	Correct decision probability = $1 - \alpha$	Type I error probability = α (level of significance)
	H_0 is false.	Type II error probability = β	Correct decision probability = $1 - \beta$ (power of the test)

Type I error   A Type I error occurs when you reject the null hypothesis even though it's true. The theoretical probability of making this error is equal to the level of significance that you use in conducting the test. The *level of significance* is commonly referred to as the **alpha level** (symbolized by α). So far, you have seen only the .05, or 5%, alpha level. But, as explained later, you can use other alpha levels, such as .01, or 1%, depending on how much risk of a Type I error you are willing to take. Table 9.2 shows three conventional levels of significance, their associated z_{crit} values, and what each level means in terms of the risk of a Type I error.

Recall that a test statistic, such as a computed z ratio, is declared significant if the value of that statistic falls into the "extreme" or "rare" portion of the sampling distribution called the region of rejection. But it is important to bear in mind that *the entire sampling distribution, including the region of rejection, is a model of random variation*; it actually represents the outcomes expected if the null hypothesis is true. Therefore:

- *A Type I error can occur only when the null hypothesis is true* and, as a result of unlikely random variation, the test statistic "lands" in the region of rejection of H_0.
- The probability of a Type I error is equal to the proportion of the sampling distribution that has been designated to represent significant events; this probability is equal to the level of significance, or α (the "alpha level").
- Since the total area of the sampling distribution equals 1, the probability of (correctly) not rejecting H_0 when it is true is equal to $1 - \alpha$. So *if H_0 is true* and $\alpha = .05$, then the probability of making a correct decision is $1 - .05$, or .95.

These facts are illustrated in Figure 9.5, which shows a graphical representation of α and $1 - \alpha$ for a one-tailed test. Note that this distribution is the null distribution. *Because H_0 is true in this case, there is no alternative distribution in the picture.*

Table 9.2

CONVENTIONAL LEVELS OF SIGNIFICANCE AND THEIR MEANING	LEVEL OF SIGNIFICANCE	TWO-TAILED z_{crit}	ONE-TAILED z_{crit}	INTERPRETATION
	.05, or 5%	1.96	1.65	Conventional risk of a Type I error
	.01, or 1%	2.58	2.33	Low risk of a Type I error
	.001, or .1%	3.30	3.08	Negligible risk of a Type I error

Figure 9.5 *The Probability of a Type I Decision Error Equals the Proportion of the Area in the Region of Rejection If the Null Hypothesis Is True*

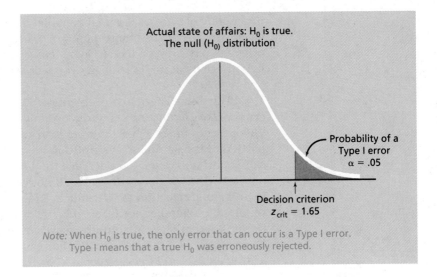

Actual state of affairs: H_0 is true.
The null (H_0) distribution

Probability of a Type I error $\alpha = .05$

Decision criterion $z_{crit} = 1.65$

Note: When H_0 is true, the only error that can occur is a Type I error. Type I means that a true H_0 was erroneously rejected.

Figure 9.6 *Regions of Rejection Shrink As the Level of Significance Decreases*

(Note: Examples pertain to a one-tailed test.)

One way to reduce the theoretical probability of a Type I error is to reduce the proportion of the sampling distribution that represents significant outcomes. For example, you could use the .01 (i.e., 1%) level of significance rather than the .05 (i.e., 5%) level. Figure 9.6 illustrates the effect of using a more conservative—that is, smaller—alpha level: The region of rejection and, hence, the

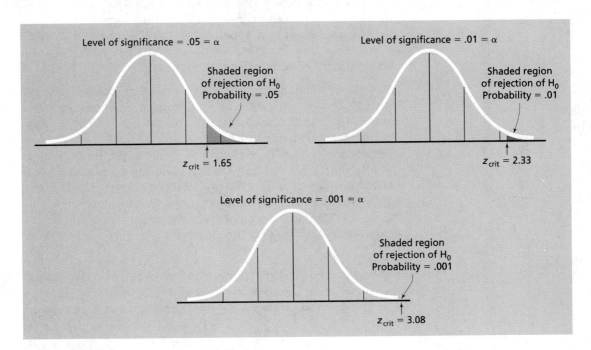

Level of significance = .05 = α

Shaded region of rejection of H_0 Probability = .05

$z_{crit} = 1.65$

Level of significance = .01 = α

Shaded region of rejection of H_0 Probability = .01

$z_{crit} = 2.33$

Level of significance = .001 = α

Shaded region of rejection of H_0 Probability = .001

$z_{crit} = 3.08$

probability of a Type I error, are smaller if you use a 1% significance level rather than the more popular 5% level. But there may be a price to pay if you use a more conservative α when the null hypothesis is actually false, as you will see next.

Type II error   There is a problem with using a more conservative criterion of significance to reduce the likelihood of a Type I error: *This practice makes Type II errors more likely.* By reducing the proportion of the sampling distribution associated with a statistically significant outcome, you decrease the likelihood of rejecting H_0 *regardless of its truth or falsity.* You'll notice, for instance, that when the region of rejection is "shrunken," a larger value of the z ratio is required for significance (see Table 9.2 or Figure 9.6). Hence, when the null hypothesis is actually false, using a more conservative α (for example, .01 rather than .05) increases the likelihood of not rejecting a false null hypothesis (a Type II error).[2]

What is the theoretical probability of a Type II error? Unfortunately, the answer cannot be as easily determined as the probability of a Type I error, which is always equal to the level of significance that you choose to use. The likelihood of a Type II error is a function of four variables: (1) sample size, (2) the level of significance that you use (.1%, 1%, or 5% level), (3) the amount of random error variability in the data, and (4) the size of the effect that the independent variable has on the dependent variable. *The probability of a Type II error becomes smaller as sample size, effect size, and the level of significance are increased (from .01 to .05, for example) and becomes larger as the error variability in the data increases.* The influence of these factors on Type II errors is discussed in the Addendum to this chapter, titled "Understanding Statistical Power." For now you need to remember the following:

- A Type II error can occur only when the null hypothesis is false. When H_0 is false, there are two overlapping distributions—the *visible null distribution* and the *unknown but real alternative distribution*—as shown in Figure 9.7
- The probability of a Type II error is equal to the proportion of the alternative distribution that lies below the critical z value needed for significance. This probability is represented by the lightly shaded area of the alternative distribution shown in Figure 9.7. The Greek symbol β, known as **beta**, is used to symbolize *the probability of a Type II error.*

---

[2] This doesn't mean that a Type II error will be avoided if you do not make α smaller. A Type II error can occur when any level of significance is used. The point being made is that the chance of a Type II error increases when a smaller α is used.

- The *proportion of the area in the alternative distribution that lies at or above the critical z value equals the probability of rejecting the false H_0*. This probability is called the **power** of the test of significance. "Power" is symbolized by $1 - \beta$ (see Figure 9.7). The higher the "power" of a statistical test, the more likely it is that the test will detect and reject a false null hypothesis.
- Any factor that inflates the probability of a Type II error decreases the power of the test.

How Type II errors come about What could cause a failure to reject H_0 when it is, in fact, false? There are four general answers to this question, and they hinge on the four factors mentioned in the previous section. For instance, using a very conservative alpha level, such as .001, can result in a failure to reject a false H_0, simply because the strict criterion of rejection (a z value of 3.30 in a nondirectional test) doesn't give the computed z value much of a chance to fall into the region of rejection. In such a case, even a sizable test statistic, reflecting a real difference between two means, could very likely fall short of significance. As a consequence, an erroneous decision to not reject H_0 would be made.

Likewise, a Type II error could result from any condition that leads to a small computed z value even though H_0 is false. Recall that a z ratio is calculated by $z = (\overline{X} - \mu)/(\sigma/\sqrt{N})$. For the moment, assume that $(\overline{X} - \mu)$ is a fixed size for a particular sample of data. In that case, the denominator of the z ratio, $\sigma/\sqrt{N}$, determines the size of the test statistic. Anything that makes the denominator larger will decrease the size of the z ratio and increase the proba-

Figure 9.7 *Type II Errors and Power of the Statistical Test (Note: When H_0 is false, there are two distributions, and Type II errors are possible. The Type II error probability equals the proportion of the alternative distribution that is below z_{crit}.)*

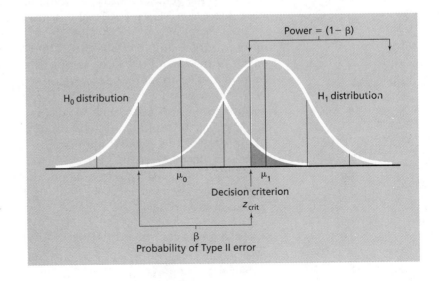

bility of a Type II error. Therefore, a very small sample size (N) could lead to a Type II error. Similarly, an inefficient research procedure could increase the amount of error variance in your data. In effect, such a problem makes σ larger than it has to be, makes the computed z value smaller than it should be, and thus increases the chance of not rejecting H_0.

Finally, consider situations in which $(\overline{X} - \mu)$ can vary. If the effect of the independent variable you are investigating is not very large—either because the variable really doesn't have a big impact on behavior or because of a weak experimental manipulation of the variable—the *numerator* of the z ratio $(\overline{X} - \mu)$ will be quite small. That is, because of a small impact by the independent variable, *the sample mean doesn't deviate very much from the hypothesized population mean*. This **small mean-difference size**, too, will cause the z ratio to be smaller than it should be, which increases the probability of a Type II error.

> NOTE: Technically, **mean-difference size** is *the difference between the true population mean, μ_1, and the population mean sponsored by a false null hypothesis, μ_0.* Thus, theoretical mean-difference size $= (\mu_1 - \mu_0)$. However, $(\overline{X} - \mu)$ estimates $(\mu_1 - \mu_0)$ and should be thought of as the empirical mean-difference size.

Minimizing both kinds of error   At this point in their consideration of hypothesis testing, some students get the impression that behavioral scientists' procedures for determining statistical significance are fraught with pitfalls, perhaps yielding erroneous decisions more often than good ones. Such is not the case. Statistical decisions are far more often correct than incorrect. I recommend the following general strategy to minimize the overall probability of decision errors: Keep the level of significance fairly low—say, not higher than 5%—to avoid most Type I errors, and use an adequately large sample of observations to avoid most Type II errors. What is an "adequately large sample size"? No answer to this question is etched in stone because unusually small mean-difference sizes must be offset with very large sample sizes. But under most circumstances, $N \geq 30$ provides adequate protection from a Type II error when the alpha level is set at .05 (see Kirk, 1995).

In view of this general approach to the dilemma posed by two kinds of decision errors, under what circumstances would you want to use the .01 level of significance, or even the .001 level, rather than the more conventional .05 level? The answers to that question are on deck.

Choosing a Level of Significance

Your choice of a particular level of significance should be governed by the amount of risk of a Type I or Type II error that you are willing to take. Although the behavioral sciences usually consider the Type I error to be more serious than the Type II error,[3] particular research circumstances will determine just how much protection against each type of error is desirable.

Considerations in basic research   In theoretical, or "basic," research, the alpha level should reflect how unlikely an outcome would have to be under H_0 to convince you, the researcher, that the data are not merely a result of chance variation (Fisher, 1966). The more doubt you have regarding the validity of the alternative hypothesis prior to conducting your study, the lower the alpha level should be. You may wish to use $\alpha = .01$ rather than $\alpha = .05$, for example, if your preinvestigation confidence in the alternative hypothesis is low. Opting for the .01 level amounts to a public statement that you require more than the usual amount of evidence before discarding the null hypothesis in favor of a dubious alternative hypothesis. In that case, you are willing to accept a higher than normal risk of a Type II error (failing to reject a false H_0), but you are not willing to allow much risk of a Type I error (rejecting a true H_0).

Considerations in applied research   In some kinds of applied research, on the other hand, the Type II error may be considered more troublesome than a Type I error. For instance, if you are testing the effectiveness of a remedial education program for disadvantaged children, then failing to reject a false null hypothesis (that is, a Type II error) would wrongly imply that the program is not effective. That implication could result in the discontinuation of a useful and sorely needed educational enterprise. In contrast, a Type I error in that situation might simply result in the continuation of a program that, though not uniquely effective, does no harm. In such a situation, then, it is sensible to use a moderate level of significance, such as $\alpha = .05$, together with a sufficiently large sample. These steps minimize the chances of a socially costly Type II error.

In any case, the level of significance is supposed to reflect your standard for rejecting H_0 *prior to data collection*. Therefore, the criterion of significance should be set in advance of data collection, should be consistent with your theoretical or practical concerns as

---

[3] The reason that a Type I error is held to be more serious than a Type II error is that Type I errors, which result in the rejection of H_0, give researchers in the field the impression that an answer to a question has been found. Thinking that the "truth" is at hand, scientists then tend to stop doing tests of that particular question (Bakan, 1966). In contrast, Type II errors tend to spur efforts toward a better way of testing the question, at least initially.

a researcher, and should not be altered after the data are in hand (Bakan, 1966). As mentioned earlier in this chapter, a weak test statistic can be "made significant" by "nudging" the level of significance after the z ratio is calculated. But that practice is a pointless kind of dishonesty that defeats the purpose of doing behavioral research.

On the other hand, there are legitimate steps that you can take to maximize the probability of rejecting a null hypothesis that is, in fact, false. These legitimate procedures fall under the general topic of statistical power. See the Addendum to this chapter for an in-depth consideration of power.

CONSOLIDATION: EIGHT PROCEDURAL STEPS IN HYPOTHESIS TESTING

To strengthen your understanding of the general process of hypothesis testing—as well as to review several specific concepts—I will present a second complete example of a test of significance. This discussion will reinforce the ideas introduced by the example that opened this chapter. You'll remember that that introductory case involving weird test scores illustrated *four logical steps* that underlie all tests of significance. In addition to having its own logic, hypothesis testing is characterized by a standard sequence of actions that you will usually carry out in your analysis of sample data. The eight procedural steps of hypothesis testing are listed here:

1. Stating the research problem and the nature of the data
2. Stating the statistical hypotheses
3. Choosing a level of significance (alpha level)
4. Specifying the type of test statistic you will use
5. Determining the critical value needed for significance
6. Stating a decision rule for rejecting H_0
7. Computing the test statistic
8. Comparing the test statistic to the decision rule and making a decision

These steps will serve as a working model of significance testing for all hypothesis-testing methods covered in this textbook. Note the following example well.

Step 1: State the Problem and Nature of the Data

A clinical social worker is investigating the effectiveness of an environmental treatment for seasonal depression. Patients who have seasonal depression (also known as the "wintertime blues") develop an inappropriately low mood mainly during the winter months, when there are fewer hours of daylight. Our social worker wants to know whether the level of depression can be lessened by exposing seasonally depressed patients to 8 hours of full-spectrum fluorescent light each day for two weeks. Accordingly, the social worker applies this "enhanced ambient lighting therapy" to 64 seasonally depressed patients on each of 14 successive days, and then measures their level of depression using the Beck Depression Inventory. The sample of 64 patients has been randomly selected from a known population of patients with the diagnosis of "seasonal depression." Assume that the population mean on the Beck Depression Inventory is 25 and that the population standard deviation is 6. Of course, the lower the score on the inventory, the less depressed the person is. The initial quantitative facts are: $N = 64$, $\mu = 25$, $\sigma = 6$.

Step 2: State the Statistical Hypotheses

Adhering to conventional practices in the behavioral sciences, the social worker formulates nondirectional statistical hypotheses:

$$H_0: \quad \mu = 25$$

$$H_1: \quad \mu \neq 25$$

The alternative hypothesis asserts that the population mean of the treated depressives will differ from the standard population mean of untreated depressives. The social worker expects that if the enhanced ambient lighting does anything, it will reduce the patients' depression. But since the opposite outcome (that is, increased depression) would also be of theoretical or practical importance, a nondirectional test is conducted. Note that the hypotheses are formulated before the data are analyzed, not as an afterthought.

Step 3: Choose a Level of Significance

Since the social worker does not have a great deal of confidence in the effectiveness of the enhanced ambient lighting therapy, a conservative level of significance is specified: $\alpha = .01$. This alpha level presents only a low risk of a Type I error. This means that the social worker requires more than the usual amount of evidence to be convinced that the therapy produces a real effect on depression. Note that the social worker decides on the level of significance before analyzing the data.

Step 4: Specify the Test Statistic

The appropriate test to use in this situation is the z ratio for a sample mean, where

$$z = \frac{\overline{X} - \mu}{\sigma_{\overline{X}}}$$

Step 5: Determine the Critical Value Needed for Significance

Consulting Table Z in Appendix A, the social worker finds that in a two-tailed test, a z_{crit} of 2.58 separates the most extreme 1% of the standard normal distribution from the middle 99%. Thus, the absolute value of the computed z ratio (the test statistic) must equal or exceed 2.58, as shown here.

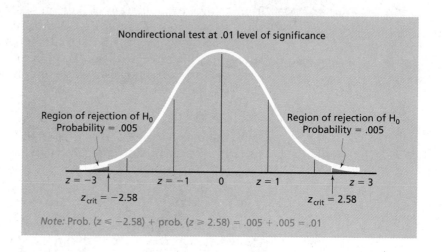

Nondirectional test at .01 level of significance

Region of rejection of H_0
Probability = .005

Region of rejection of H_0
Probability = .005

$z = -3$ $z = -1$ 0 $z = 1$ $z = 3$

$z_{crit} = -2.58$ $z_{crit} = 2.58$

Note: Prob. $(z \leqslant -2.58)$ + prob. $(z \geqslant 2.58)$ = .005 + .005 = .01

Step 6: State the Decision Rule

The decision rule is a verbal-symbolic statement of what decision will be made if the test statistic is at least as extreme as the critical value. In this case,

Decision rule: If $|z| \geq 2.58$, reject H_0.

Step 7: Compute the Test Statistic

The social worker finds that the mean Beck Depression Inventory score of the sample of 64 treated patients is 23. The test statistic, a z ratio, will be based on the difference between the $\overline{X}$ of 23 and the μ of 25.

Two computations are required here: the standard error of the mean and the z ratio itself.

1. The standard error of the sampling distribution is

$$\sigma_{\overline{X}} = \sigma/\sqrt{N} = 6/\sqrt{64} = 6/8 = 0.75.$$

2. The test statistic is given by:

$$z = \frac{\overline{X} - \mu}{\sigma_{\overline{X}}}$$

$$= \frac{23 - 25}{0.75}$$

$$= \frac{-2}{0.75}$$

$$= -2.67$$

Step 8: Consult the Decision Rule and Make a Decision

The social worker's decision rule states: If $|z| \geq 2.58$, reject H_0. Since the absolute value of the test statistic (2.67) is greater than z_{crit} (2.58), the correct decision is to reject the null hypothesis.

The outcome of the study was significant. Therefore, the social worker writes the following in the "Results" section of a research report: "Enhanced ambient lighting therapy significantly reduced the level of depression in the sample of seasonally depressed patients, $z = 2.67$, $p < .01$." Observe how the significant result is reported. The verbal description of the significant result is immediately documented with (1) the value of the test statistic followed by (2) a statement of the alpha level that was used to arrive at a statistical decision. The "p" stands for the probability that the conclusion represents a Type I error. That probability is less than .01.

This method of reporting a significant result is fairly standard in the behavioral sciences. You will see it again in later chapters. You should use it in reporting the results of statistical tests that you conduct on your own data.

THINLY VEILED SECRET REVEALED

This chapter contains some very challenging ideas. If things aren't quite clear yet, welcome to the "mere mortal" club. Virtually every student who has ever studied the theory of hypothesis testing has felt similarly intimidated upon the first exposure. Hypothesis testing isn't for the fainthearted. Go over these ideas again. Think about them, and use them to answer the review questions. The remainder of this book will teach you how to apply the logic of hypothesis testing in several specific statistical methods. It's what behavioral science is about.

KEY TERMS

reliable

statistical significance

null hypothesis, H_0

alternative hypothesis, H_1

test statistic

sampling distribution

nondirectional test of significance	Type I error
statistically significant result	Type II error
level of significance	alpha level
critical value	beta
region of rejection	power
two-tailed test	mean-difference size
one-tailed test	small mean-difference size

SUMMARY

1. The purpose of hypothesis testing is to find out whether a set of sample data is unlikely to be accounted for by random variation alone.

2. The four logical steps in all hypothesis-testing procedures are: (a) State the competing statistical hypotheses, null (chance) hypothesis versus alternative (nonchance) hypothesis; (b) compute a "test statistic" from the sample data; (c) compare the test statistic with a distribution of random test statistics to determine whether the computed statistic is significant—that is, unlikely to belong to the random distribution; and (d) on the basis of the significance or nonsignificance of the test statistic, decide whether to reject or fail to reject the hypothesis of chance; if the hypothesis of chance is rejected, the alternative (nonchance) hypothesis is supported.

3. A test of significance may be "directional," if the alternative hypothesis predicts a particular direction of difference between the sample data and chance expectations, or "nondirectional," if the alternative hypothesis predicts only that the sample data will "differ" from chance expectations. By convention, most significance tests are nondirectional, or two-tailed.

4. When a statistical decision (to reject or not reject the null hypothesis) is inconsistent with the actual truth or falsity of the null hypothesis, a statistical decision error has occurred. A Type I error occurs when a true null hypothesis is rejected. The probability of a Type I error equals the level of significance, or alpha level, that you choose to use in your hypothesis-testing procedure. Different levels of alpha reflect how much risk of a Type I error you are willing to accept in your research.

5. A Type II error occurs when a false null hypothesis is not rejected. The probability of a Type II error, known as beta, is an inverse function of the sample size, alpha level, and mean-difference size, and a direct function of data variability. A Type II error generally is considered less serious than a Type I error.

6. The power of a statistical test equals the probability of rejecting a false null hypothesis; mathematically, power = $1 - \beta$, the complement of beta. It is very desirable for tests of significance to have a high power, because false null hypotheses should be rejected. The power may be improved by increasing the alpha level or decreasing data variability. But the best way to augment power is to use larger samples.

REVIEW QUESTIONS

1. Define or describe: statistical significance, null hypothesis, alternative hypothesis, test statistic, sampling distribution, nondirectional test of significance, and statistically significant result.

2. At an urban psychiatric hospital, the average number of patients admitted on Friday evenings is 31. However, on a particular Friday evening that coincides with a full moon, the number admitted jumps to 54. A staff psychologist plans to test the statistical significance of this increase. Assuming a nondirectional test, state the relevant statistical hypotheses, both verbally and symbolically.

3. Assume that previous research has shown that 60% of depressed persons experience "spontaneous remission" (i.e., get better without formal therapy) within a 12-month period. You are investigating the effectiveness of a new therapy for depression, and you find that the 100 clients who receive the treatment have a 73% cure rate within a 12-month interval. Assuming a two-tailed test of significance, state the relevant statistical hypotheses for this study, both verbally and symbolically. Explain why you might not want to use a one-tailed test, even though you expect your therapy to improve the remission rate.

4. For the statistical test associated with the investigation described in Question 3, what critical value would you use if you tested the null hypothesis using the .05 level of significance? What critical value would be appropriate for the .03 level of significance? What does decreasing the level of significance do to the regions of rejection?

5. For the statistical test associated with the investigation described in Question 3, calculate the test statistic and assess the significance of that statistic at the .05 level of significance. Do you reject or fail to reject the null hypothesis?

6. Why is it harder to reject the null hypothesis when the level of significance is made smaller? Why is it more difficult to reject the null hypothesis in a two-tailed test than in a one-tailed test at the same α? Be specific in stating your answers.

7. Define or describe: level of significance, critical value, region of rejection, Type I error, Type II error, alpha level, beta, power, mean-difference size, and small mean-difference size.

8. You are examining the impact of passive smoking on the longevity of laboratory mice. Your dependent variable is how long mice live when they must function in a smoke-filled cubicle every day. Assume that, under normal (nonsmoking) conditions, the average laboratory mouse lives 796 days. Would you use a directional or a nondirectional test in this study? Give the rationale of your answer, and state the statistical hypotheses symbolically.

9. Produce the alternative hypotheses that correspond to each of these null hypotheses:

   (a) $H_0: p \geq .75$
   (b) $H_0: p = .42$
   (c) $H_0: \mu \leq 16.34$
   (d) $H_0: \mu = 120$
   (e) $H_0: \mu \geq 0$

10. The alpha level in a directional test of significance is set at .01. The computed z statistic is 2.44. Would you reject or fail to reject the null hypothesis? Explain your answer.

11. In a nondirectional test of significance, α is set to .001. The computed z statistic is 2.28. Would you reject or fail to reject the null hypothesis? Explain your answer.

12. A behavioral scientist wishes to conduct a two-tailed test using an unconventional level of significance, where $\alpha = .03$. What will z_{crit} be? (*Hint:* You definitely need to consult Table Z of Appendix A.)

13. Suppose that a particular state strictly enforces the physical fitness regulations that apply to its state highway patrol officers. One of the fitness standards specifies that officers must be able to run a mile in 7 minutes or less. You are the chief statistician for the state attorney general's office, and you have been asked to check on adherence to fitness regulations by testing a random sample of 40 troopers. You plan to compare the mean mile-run time of the 40 officers to the standard of 7 minutes. Would you use a directional or nondirectional test? Give the rationale for your answer, and state the null and alternative hypotheses symbolically.

14. Which of the following statements represent null hypotheses?

   (a) Dogs are smarter than cats.
   (b) Ghosts do not exist. *Null*
   (c) Henry is the baby's biological father.
   (d) The medicine works.
   N (e) Intelligence declines with age.
   (f) The AIDS test results will be negative. *Null*

15. Contrast the "conservative" and "discretionary" points of view on the question of when it is appropriate to use a directional test of significance.

16. A prominent theory of criminal behavior predicts that nine of ten violent crimes will be committed by males. Consulting U.S. Bureau of the Census records, a criminologist interested in testing the theory finds that 86% of the violent crimes in the past decade were committed by males. Is this empirical percentage consistent with theoretical expectations? Using the .01 level of significance, conduct a test of significance to find out whether the actual percentage deviates significantly from the theoretical percentage. Assume $N = 625$, and use the eight procedural steps of significance testing described on pages 256–259.

17. A sociologist has developed a theory that hypothesizes a relationship between marital happiness and the amount of "cultural commonality" within a marriage: The more partners share the same cultural values, the happier their marriage is. The sociologist knows that the population mean of a frequently used scale of marital happiness is 77 and that the standard deviation is 8. A sample of 36 culturally discordant couples is tested with the marital happiness scale. The sociologist expects the sample mean to be lower than that of the general population of marrieds. Should the sociologist conduct a directional or a nondirectional test? Defend your answer. State the appropriate null and alternative hypotheses symbolically.

18. Construct the null hypotheses that correspond to the following alternative hypotheses:
    (a) H_1: $p \neq .45$
    (b) H_1: $p < .63$
    (c) H_1: $\mu \neq 128$
    (d) H_1: $\mu > 50$

19. For $\alpha = .001$, $\alpha = .01$, and $\alpha = .05$, draw the standard normal distribution, indicate the z_{crit} values for nondirectional tests, and shade in the regions of rejection in each case.

20. For $\alpha = .001$, $\alpha = .01$, and $\alpha = .05$, respectively, draw the standard normal distribution, indicate the z_{crit} values for directional tests, and shade in the region of rejection in each case.

21. Assume that H_0 is true. If the probability of a Type I error is .10, what is the probability of failing to reject H_0?

22. If H_0 is false, what is the probability of a Type I error when $\alpha = .01$? (*Caution:* Read this question again carefully.)

23. If H_0 is true, what is the probability of a Type II error when $\alpha = .10$?

24. If $\beta = .22$, what is the power of your test?

25. Refer to the information in Question 17. Suppose the sociologist finds that the sample of culturally discordant couples has

a mean of 75 on the test of marital happiness. Using the .05 level of significance, carry out a test of the significance of that outcome. Be sure to use the eight procedural steps of significance testing described on pages 256–259.

ADDENDUM: UNDERSTANDING STATISTICAL POWER

Earlier in this chapter, I defined the **power** of a statistical test as *the probability of rejecting a false null hypothesis*. In behavioral statistics, the power of your test is very important: A good hypothesis-testing procedure has the "strength," or power, to disconfirm H_0. If the null hypothesis is false, then science progresses if H_0 is rejected. In contrast, science flounders if, because of a "weak" test, H_0 is not rejected. The following facts are relevant to the concept of power:

- The power of your test is the mathematical complement of the probability of a Type II error (β). So if $\beta = .10$, then power = $(1 - \beta)$, or $(1 - .10) = .90$. In other words, if there is a 10% chance of failing to reject a false H_0, then there is a 90% probability that it will be rejected. The important implication of this relationship is that as the probability of a Type II error increases, the power of your test decreases, and vice versa.

- Since a Type II error can occur only if H_0 is false, increasing the power of your test is of concern only when H_0 is false. You would not want to reject a true null hypothesis, so power is practically irrelevant when H_0 is true.

- When H_0 is false, there are, in theory, two sampling distributions to consider: the null, or H_0, distribution and the alternative, or H_1, distribution. In practice, only the null distribution is available to us, but in theory both distributions exist. Refer back to Figure 9.7. There you will see that the *probability of a Type II error is equal to the proportion of the alternative distribution that is to the left of the critical z value in a statistical test* (the *shaded* part of the H_1 distribution). Accordingly, the *power is equal to the proportion of the alternative distribution that is to the right of the critical z* (the unshaded portion of the H_1 distribution). These statements make sense only if you are looking at Figure 9.7.

- Any factor that reduces the probability of a Type II error enhances the power of your statistical test. These factors include (1) making alpha larger, (2) increasing the sample size, (3) decreasing data variability (that is, somehow making σ smaller), and (4) working with a large mean-difference size.

Let's examine the specific effect of each of the above factors on statistical power within the context of the overlapping H_0 and H_1

distributions. As we do, *bear in mind that the null hypothesis is false in every one of the upcoming examples.* Therefore, the principles we will derive from the examples apply only when H_0 is false.

The Research Problem

Suppose a cognitive psychologist is working with a sample of $N = 25$ gifted teenagers in a school district called Shadyvale. The psychologist is trying out an experimental training technique that is designed to increase the performance IQs (that is, "nonverbal" IQs) of intellectually precocious youngsters. A prior census of the gifted adolescents in the Shadyvale district showed that subjects in this population have an average performance IQ of 139, so the known population mean is $\mu = 139$. The population standard deviation is $\sigma = 10$.

Statement of hypotheses   Since all the training exercises serve to enrich the students' daily cognitive environment, the cognitive psychologist and all fellow researchers in the field of IQ enhancement expect either a positive effect of the training or no effect at all. Because there is no reason to expect a negative effect of the training, and a negative effect of environmental enrichment would make no sense within prevailing theory, a directional, or one-tailed, test is warranted. That is the kind of significance test the researcher decides to conduct. The pertinent hypotheses are:

$H_0: \mu \leq 139$

$H_1: \mu > 139$

The null hypothesis asserts that the sample of 25 trained adolescents belong to the general Shadyvale population of gifted adolescents, which has a mean performance IQ of 139. The alternative hypothesis states that the 25 trained adolescents belong to a different population whose mean performance IQ is greater than 139.

The chosen alpha level   The cognitive psychologist decides to use a conventional level of significance to test H_0: α is set at .05. Since a directional test is used, the critical value for significance is $z_{crit} = 1.65$ (see Table Z of Appendix A).

The outcome of the study   After exposing the students to IQ enhancement exercises for an entire semester, the cognitive psychologist has each student's performance IQ retested by an independent clinical practitioner. On the end-of-term test, the mean performance IQ is 142; hence, $\overline{X} = 142$. Is this a significant increase in nonverbal intelligence?

The actual state of affairs   As emphasized earlier, the null hypothesis is false in this particular situation. Of course, the cognitive psychologist doesn't know this yet, but we do because this is a hypothetical example that is used to make some realistic points. In a hypothetical example, we can shape the world to our liking.

Because the null hypothesis is false, *we are dealing with two sampling distributions in this situation:*

1. There is a "null" sampling distribution with a mean of $\mu_0 = 139$, the same as the mean of the general population of gifted teenagers in the Shadyvale school district.
2. Additionally, there is an "alternative" sampling distribution that, *unbeknownst to the researcher*, has a mean of $\mu_1 = 143$. This second sampling distribution exists because the special training does, in fact, increase the mean IQ from 139 (μ_0) to 143 (μ_1) *on the average*. Of course, the sample of 25 trained adolescents "belongs to" the second sampling distribution; that is, *the adolescents' sample mean of $\overline{X} = 142$ should be thought of as having been randomly selected from the sampling distribution that has a mean of 143.*

Both sampling distributions have a standard error of $\sigma_{\overline{X}} = 2$. This figure is based on the fact that the population standard deviation is $\sigma = 10$. Since $N = 25$, $\sigma_{\overline{X}} = \sigma/\sqrt{N} = 10/\sqrt{25} = 10/5 = 2$.

NOTE: In tests of significance that involve a false H_0, *it is routinely assumed that the null and the alternative distributions have equivalent standard errors of the mean.*

To summarize, then:

- Sample size = $N = 25$
- Sample mean = $\overline{X} = 142$ = the mean performance IQ of the 25 research subjects
- Mean of the null distribution = $\mu_0 = 139$ = the hypothesized mean of the population
- Standard error of the null distribution = $\sigma_{\overline{X}} = 2$
- Mean of the alternative distribution = $\mu_1 = 143$ = the *actual mean* of the population represented by the 25 specially trained students
- Standard error of the alternative distribution = $\sigma_{\overline{X}} = 2$

Now let's see what statistical decision the psychologist reaches and what the implications are.

A Bad Decision under Low Power

The statistical picture painted so far is illustrated in Figure 9.8, which shows the relevant numbers in the context of the overlapping null and alternative distributions. The only new number there is 142.3, which is how big a sample mean must be to satisfy a z_{crit}

Figure 9.8 *Probability of a Type II Error and Power in a Study of IQ Enhancement*

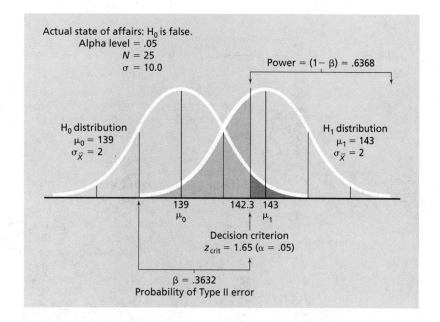

of 1.65 and thus be declared significant. This is equivalent to a value in the null distribution that is 1.65 standard errors above μ_0. Hence, $\mu_0 + z_{crit}\sigma_{\overline{X}} = 139 + 1.65 \cdot 2 = 139 + 3.3 = 142.3$.

Unfortunately, even though H_0 is false in this case, and in spite of the fact that the trained adolescents come from a population with a mean of 143, their sample mean is only 142. This isn't large enough to meet or exceed the 142.3 figure required for significance. Therefore, the cognitive psychologist decides not to reject H_0—and in doing so, unknowingly commits a Type II error.[4]

As Figure 9.8 shows, this particular combination of α, sample size, and data variability result in a sizable probability of a Type II error ($\beta = .3632$) and relatively low power ($1 - \beta = 1 - .3632 = .6368$). Ideally the probability of a Type II error should be no greater than .05, and power should be no less than .95.[5]

Again observe that β equals the proportion of the alternative distribution that is below the critical value needed for significance, and power equals $(1 - \beta)$, which is the proportion of the alternative distribution that is above the critical value. The total area of the

---

[4] Remember, the cognitive psychologist does not know that the alternative distribution exists. Only we do.

[5] The rationale behind this ideal is that the probability of a Type II error should be no higher than the probability of a Type I error when a conventional alpha level is used. Often this ideal is not met, however.

Box 9.1

HOW TO
FIGURE POWER

Wondering how the β and power were calculated in this example? It's pretty simple:

1. Find the critical value of $\overline{X}$ in the null distribution with the formula $\mu_0 + z_{crit}\sigma_{\overline{X}}$. In the present example, the formula yields 142.3. We call this $\overline{X}_{crit}$. It is the theoretical sample mean that the computed sample mean must equal or exceed to be significant.

2. Determine by how many z scores the critical value deviates from μ_1, the mean of the alternative distribution. Use the formula $z_1 = (\overline{X}_{crit} - \mu_1)/\sigma_{\overline{X}}$. In this case, we get $z_1 = (142.3 - 143)/2 = -0.7/2 = -0.35$.

3. Go to Table Z in Appendix A and find the column C proportion associated with z_1. This is the proportion of the alternative distribution that lies below $\overline{X}_{crit}$. This proportion equals the probability that a mean from the alternative distribution will fall below the critical value and, thus, will be declared nonsignificant—or β, the probability of a Type II error. In this example, the column C proportion equals .3632.

4. To calculate the power of the test, use $(1 - \beta)$. Here power = $(1 - .3632)$ = .6368. This is the proportion of the alternative distribution that is above $\overline{X}_{crit}$. This proportion equals the probability of rejecting a false H_0.

alternative distribution equals 1.00, so the probability of a Type II error and power must sum to 1.00. Therefore, power is comparatively low to the extent that β is high.

Now that we've duly considered how low power can produce a bad statistical decision, our next task will be to observe how varying α, sample size, data variability, or mean-difference size can increase the power and increase the likelihood of a correct decision.

Effect of Increasing Alpha

Figure 9.9 reveals the effect of increasing the level of significance from .05 to .10. If the cognitive psychologist uses a 10% alpha level, the effect is to lower z_{crit} to 1.28. Accordingly, the new critical value becomes $\mu_0 + z_{crit}\sigma_{\overline{X}} = 139 + 1.28 \cdot 2 = 139 + 2.56 = 141.56$. The decision standard shifts to the left, closer to the mean of the null distribution, and results in less of the alternative distribution being below the critical value needed for significance.

If you look carefully at Figure 9.9 you'll see that using the higher α reduces the probability of a Type II error from .3632 (in Figure 9.8) to .2358—a definite improvement, but not yet ideal because there still is a 23.58% chance of a Type II error. This means that the power $(1 - \beta)$ increases from .6368 (in Figure 9.8) to .7642. Now the cognitive psychologist has a better chance of rejecting a false H_0. Indeed, since the sample mean of $\overline{X} = 142$ exceeds the critical value of 141.56, the null hypothesis is rejected.

Figure 9.9 *Increasing the Level of Significance (α) from .05 to .10 Increases Power*
(Note: Making α larger to enhance power is not a recommended practice because it makes the probability of a Type I error too big in the event that H_0 is true.)

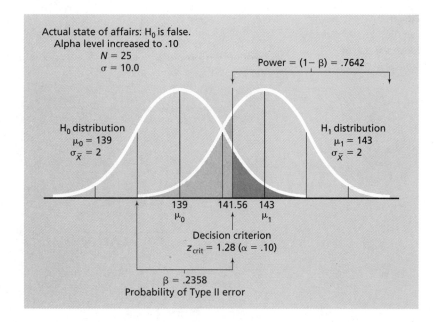

Actual state of affairs: H_0 is false.
Alpha level increased to .10
$N = 25$
$\sigma = 10.0$

Power $= (1 - \beta) = .7642$

H_0 distribution
$\mu_0 = 139$
$\sigma_{\bar{X}} = 2$

H_1 distribution
$\mu_1 = 143$
$\sigma_{\bar{X}} = 2$

139   141.56   143
μ_0 μ_1

Decision criterion
$z_{crit} = 1.28$ ($\alpha = .10$)

$\beta = .2358$
Probability of Type II error

Although using an unusually high level of significance leads to a correct decision in this example, the strategy of making α larger than .05 is not generally recommended and, in fact, is likely to be seriously questioned by professionals in your field. As stated earlier, hiking the level of significance is a risky step to take. When H_0 is true (as it frequently is, in practice), an α of .10 makes the probability of a Type I error unacceptably high. There should be a better way to increase the power of a statistical test. And there is, as will be demonstrated next.

Effect of Increasing Sample Size

Generally, the safest and most effective way to improve the power of a statistical test is to use large samples in your studies. Suppose, for example, that the cognitive psychologist is working with 100, instead of 25, gifted teenagers. This fourfold increase in N changes the standard error of the null and alternative sampling distributions from 2.0 to 1.0. Observe that $\sigma_{\bar{X}} = \sigma/\sqrt{N} = 10/\sqrt{100} = 10/10 = 1.0$.

The impact of the new larger N on the power is illustrated in Figure 9.10. The first thing you'll notice is that increasing the sample size causes the null and alternative sampling distributions to have less variation: They get "skinnier." This change is reflected in the reduced size of $\sigma_{\bar{X}}$.

The next observation of interest is that the slimmer sampling distributions overlap considerably less than they did when sample size was smaller. *Augmenting N, therefore, reduces the proportion of*

Figure 9.10 *Increasing Sample Size (N) Increases the Power of the Statistical Test*

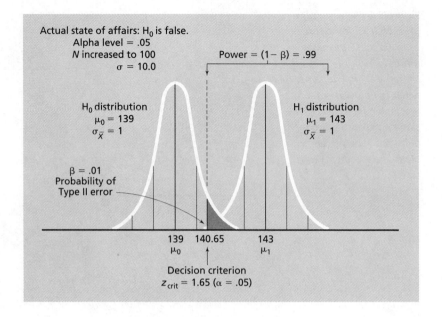

Actual state of affairs: H_0 is false.
Alpha level = .05
N increased to 100
$\sigma = 10.0$

Power = $(1 - \beta)$ = .99

H_0 distribution
$\mu_0 = 139$
$\sigma_{\overline{X}} = 1$

H_1 distribution
$\mu_1 = 143$
$\sigma_{\overline{X}} = 1$

$\beta = .01$
Probability of
Type II error

139   140.65   143
μ_0 μ_1

Decision criterion
$z_{crit} = 1.65 \ (\alpha = .05)$

the alternative distribution that lies below the critical value needed for significance. This change is reflected in the smaller critical value of $\overline{X}$ produced by the diminished standard error: $\mu_0 + z_{crit}\sigma_{\overline{X}} = 139 + 1.65 \cdot 1 = 139 + 1.65 = 140.65$. Since only 1% of the alternative distribution is below this value, $\beta = .01$. And power = $(1 - .01) = .99$. With the larger sample, the cognitive psychologist has a 99% chance of rejecting the false H_0. Since the computed sample mean is 142, H_0 is, in fact, properly rejected.

Effect of Reducing Data Variability

In another possible scenario, suppose the cognitive psychologist retains an alpha level of .05 and a sample size of $N = 25$, but uses a more efficient research procedure that reduces the amount of error variance in the IQ data. The population standard deviation (σ) now equals 5 instead of 10. This change makes the standard error smaller without increasing the sample size:

$$\sigma_{\overline{X}} = \sigma/\sqrt{N} = 5/\sqrt{25} = 5/5 = 1$$

As shown in Figure 9.11, making the standard error smaller by reducing data variability has the same effect as diminishing the standard error by using a larger N. The result is less overlap between the null and alternative distributions and, consequently, higher power.

Mean-Difference Size Considerations

A bigger mean-difference size (i.e., $\mu_1 - \mu_0$) increases the power of a statistical test not by reducing the denominator of the test statistic

Figure 9.11 *Decreasing Random Error Variability Increases the Power of the Statistical Test*

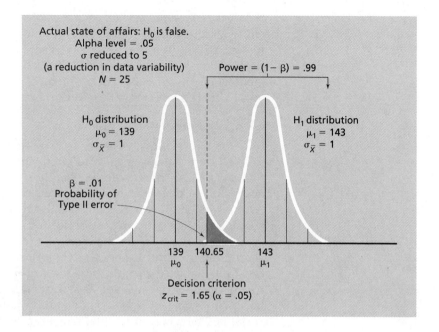

Actual state of affairs: H_0 is false.
Alpha level = .05
σ reduced to 5
(a reduction in data variability)
$N = 25$

Power = $(1 - \beta)$ = .99

H_0 distribution
$\mu_0 = 139$
$\sigma_{\bar{X}} = 1$

H_1 distribution
$\mu_1 = 143$
$\sigma_{\bar{X}} = 1$

$\beta = .01$
Probability of
Type II error

139   140.65   143
μ_0 μ_1

Decision criterion
$z_{crit} = 1.65$ ($\alpha = .05$)

but by increasing the likelihood that the numerator ($\bar{X} - \mu$) will be larger. Suppose, for example, that the researcher in the present study had used a more effective experimental training technique that brought the true population mean up to $\mu_1 = 146$, instead of 143. The impact of this enhanced mean-difference size is shown in Figure 9.12. Observe that the null and alternative sampling distributions have the same variability as they originally did, yet the

Figure 9.12 *A Larger Mean-Difference Size ($\mu_1 - \mu_0$) Increases the Power of the Statistical Test*

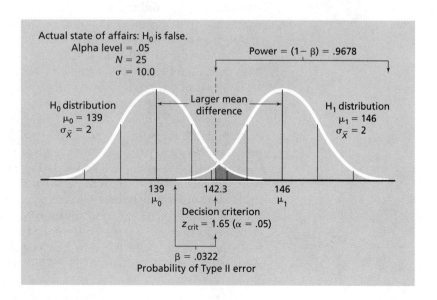

Actual state of affairs: H_0 is false.
Alpha level = .05
$N = 25$
$\sigma = 10.0$

Power = $(1 - \beta)$ = .9678

H_0 distribution
$\mu_0 = 139$
$\sigma_{\bar{X}} = 2$

Larger mean
difference

H_1 distribution
$\mu_1 = 146$
$\sigma_{\bar{X}} = 2$

139   142.3   146
μ_0 μ_1

Decision criterion
$z_{crit} = 1.65$ ($\alpha = .05$)

$\beta = .0322$
Probability of Type II error

power (i.e., $1 - \beta$) is an impressive .9678. This means that 96.78% of the $\overline{X}$ values will be above the z_{crit} value of 142.3, compared with just 63.68% in the first scenario (Figure 9.8). This outcome is entirely the result of the increased difference between μ_1 and μ_0. The greater mean-difference size "pushed the distributions farther apart," thereby reducing the overlap between them.

What's an Honest Researcher to Do?

Now you know four ways to boost the power of your statistical tests. But only one approach is feasible in most research situations—namely, using a larger sample size. Raising the level of significance above .05 is usually considered a risky, flimsy "solution" to the power problem. Reducing data variability is often considered a reasonable approach to power enhancement, but not all behavioral research situations permit the researcher to deliberately choose a research method that lowers the amount of error variance in his or her data. Likewise, mean-difference size often is dictated by the nature of the phenomenon you are studying, and hence you might not be able to change it by a stronger manipulation of the independent variable. The moral of all this: Use large N's in your investigations.

How large must N be to produce adequate power? That depends on the mean-difference size and the level of significance being used. Fortunately, devices called "power curves" enable us to plan appropriate sample sizes by taking these other factors into account.

Using power curves   Power curves, such as those presented in Kirk (1995) and Cohen (1977), are charts that show you how large your sample size must be to obtain a desirable level of statistical power (.95, for example), assuming a particular mean-difference size (i.e., $\mu_1 - \mu_0$) and alpha level. The advantage of power curves is that they enable you to make a conservative assumption about mean-difference size (that is, assume it will be very small) and then look up an N that will ensure good power. To avoid wasting your time with significance tests that are incapable of rejecting a false H_0, it is always wise to consult these charts while your investigations are still in the planning stages.

Is it possible for samples to be too large?   Although behavioral researchers are more often concerned about using a sample that is too small, it is possible, in a sense, to have an excessive sample size in a test of significance. With a huge sample, such as $N = 400$, nearly any difference between $\overline{X}$ and μ_0 will be significant. Although this might not be a problem in and of itself, it can lead us to ascribe importance to a trivial difference that has no practical value. Suppose, for example, that the cognitive psychologist had a sample of $N = 400$ instead of $N = 25$. This would have made the standard error of the mean $\sigma_{\bar{x}} = 10 / \sqrt{400} = 10 / 20 = 0.5$. Then the critical value of the mean would have been $\mu_0 + 1.65\sigma_{\bar{x}} = 139 + 1.65 \cdot 0.5$, or $139 + .0.825 = 139.825$. Therefore, the effect of the experimental training program would have been "statistically significant" if the average IQ increase had been a mere 0.83 of a point relative to the population. It is questionable whether such a small impact, even though statistically reliable, should be accorded much attention, let alone funding.

PART 4

METHODS OF INFERENTIAL STATISTICS

Having studied the theory underlying statistical inference (Chapters 8 and 9), you are now prepared to understand how that theory is applied in the analysis of behavioral data. Chapter 10 presents hypothesis-testing procedures that you should use when your investigation involves just one sample or treatment condition. In that chapter, you'll also learn more about setting up confidence intervals. Chapter 11 discusses and illustrates hypothesis-testing techniques for two-sample studies, and Chapters 12–14 cover procedures for experiments that have more than two conditions. Sometimes your data won't satisfy the assumptions of conventional tests of significance. Under such circumstances, "nonparametric tests" must be used. Chapter 15 tells you when and how to apply those special kinds of tests. Finally, Chapter 16 "brings it all together" by summarizing how to select an appropriate statistical procedure for analyzing just about any set of behavioral data you're likely to encounter.

CHAPTER 10 ONE-SAMPLE *t* STATISTIC

You are now ready to study a very handy statistic that substitutes for the *z* statistic when it is not appropriate to use the standard normal distribution to test hypotheses and estimate parameters. The inferential tool that I'm referring to is called Student's *t* statistic, or simply the *t* statistic.[1] *The t ratio was designed to be used instead of z whenever the population variance (or population standard deviation) is unknown and, therefore, must be estimated.* And it is used in the same general way as the *z* statistic. In fact, formulas that require *t* are practically identical to those that use *z*. So you already know how to work with *t*-statistic formulas. You'll find that the *t*-statistic procedures differ from the *z*-statistic procedures in only two ways:

1. *t*-statistic procedures involve estimating the standard error of the sampling distribution of $\overline{X}$ rather than calculating it directly.
2. *t*-statistic procedures require you to use a random distribution of *t* statistics to represent the sampling distribution of $\overline{X}$, instead of using the standard normal distribution to represent the sampling distribution of $\overline{X}$.

In this chapter you will learn (1) why the *z* ratio is not always a feasible way to do inferential statistics, (2) what to use as an estimate of the population variance (σ^2) when it is unavailable, (3) why the *t* statistic must be substituted for *z* when σ^2 is unknown, (4) how the *t* distribution differs from the standard normal distribution, (5) how to use *t* to test hypotheses, and (6) how to use *t* to set up a confidence interval.

WHEN A *z* RATIO IS NOT PRACTICAL

Consider this example: A biological psychologist wishes to find out whether learning can be physically transferred from trained rats to naive rats via brain chemicals. Accordingly, the researcher trains 25 rats to avoid *either* the left *or* the right arm of a T-maze by exposing each rat to a bright flashing light for turning a particular way at the choice point. Nocturnal creatures by nature, rats prefer dark areas over bright areas and are annoyed by bright flashing lights. Twelve of the rats are consistently "flashed" for turning right, and the remaining 13 rats are "flashed" for turning left. For each animal, then, one arm of the T-maze is "annoying" and the

---

[1] The *t* statistic is named after the developer of the *t*-distribution idea, William Gossett, who was an Irish chemist. Gossett wrote under the pen name "Student." Accordingly, the *t* statistic is often called "Student's *t*."

opposite side is "safe." Once all of the subjects are choosing the safe arm at least 90% of the time, the training phase of the experiment is finished.

Next, a small amount of ribonucleic acid (RNA) is carefully extracted from the brain of each trained rat. At an appointed time, the RNA from each trained animal is injected into a naive (that is, untrained) rat. The rats injected with this RNA are the "recipient" rats, and the rats that contributed the RNA are referred to as the "donors."

The testing phase of the experiment now starts. Each of the $N = 25$ recipient rats—none of which has ever been "flashed" in a maze—is placed in a T-maze 120 times in succession. The question is: Will the recipient rats behave like naive rats or like trained rats? If the brain chemicals have no effect, the recipient rats will go to each arm of the T-maze an equal number of times, on the average. That is, they will go to the left an average of 60 times and to the right an average of 60 times. On the other hand, if the brain chemicals did result in a physical transfer of learning to the recipient rats, the recipient rats will show a statistical preference for the arm of the maze considered "safe" by the corresponding donor rats. Therefore, the statistical hypotheses are:

Null hypothesis→H_0: $\mu = 60$

Alternative hypothesis→H_1: $\mu \neq 60$

CONCEPT RECAP

Remember that the null hypothesis is the hypothesis of no difference or no effect. The null hypothesis states that the mean of the population equals some value that is predicted by a model of random variation. The alternative hypothesis opposes the null by asserting that the mean of the population differs from the value expected on the basis of chance alone. Even though the researcher almost always believes in the alternative hypothesis more than in the null hypothesis, the null hypothesis is actually tested. In the logic of hypothesis testing, you can support the alternative hypothesis by rejecting the null hypothesis. See Chapter 9 for a complete review of these ideas.

The null hypothesis in this study contends that the recipient rats belong to a population of rats that are completely naive and respond to the T-maze in a completely random fashion overall. This is exactly what would be expected if the injected brain chemicals have no effect on the behavior of the recipient rats. The alternative hypothesis counters by asserting that the recipient rats come from a population in which the average response to the T-maze differs from that produced by random choices. This deviation from chance responding is what would be expected if H_0 is false—that is, if the injected brain chemicals did have a systematic effect on the rats' choices. Since H_1 doesn't specify a particular direction of deviation from a mean of 60 turns in the safe direction, the researcher is conducting a nondirectional, or two-tailed, test of the null hypothesis (see Chapter 9). This means that H_0 can be rejected by the finding of an average number of "safe-arm" choices that is either significantly higher or significantly lower than the mean of 60 predicted by the chance model. Either type of deviation from 60 is consistent with the conclusion that the brain chemicals affected the recipients' behavior.

The researcher's findings appear in Table 10.1, which displays the number of times that each of the 25 recipient rats elected

Table 10.1

NUMBER OF TIMES (IN 120 TRIALS) THAT EACH OF 25 RECIPIENT RATS CHOSE THE ARM OF A T-MAZE CONSIDERED SAFE BY DONOR RATS	X
	76
	61
	74
	45
	68
	83
	54
	52
	77
	56
	77
	64
	79
	53
	60
	61
	75
	80
	63
	82
	57
	66
	44
	66
	78

Sum of raw scores: 1651
Mean of raw scores ($\overline{X}$): 66.04

to enter the arm that would have been considered safe by the corresponding donor rat. As the table shows, the mean number of safe-arm turns is $\overline{X} = 66$, which is higher than the population mean of 60 predicted by the null hypothesis. The recipients of trained-brain chemicals *seem* to prefer the same side of the T-maze as their donor counterparts. But is this discrepancy large enough to be significant, or is it within the range of random variation? Let's see if the z ratio can help us answer this question.

Woe Is z

The z-statistic formula for this type of hypothesis-testing situation is $z = (\overline{X} - \mu)/\sigma_{\overline{X}}$ [formula (8.5)]. Here $\overline{X} - \mu$ is easy to find and equals 66 – 60, or 6. But consider the denominator of the z ratio: $\sigma_{\overline{X}} = \sqrt{\sigma^2/N} = \sigma/\sqrt{N}$ [formulas (8.3) and (8.4)]. We must have either the population variance or the population standard deviation in order to use the z ratio in inferential statistics. Since we have neither here, we are at an impasse unless σ^2 can be estimated in an unbiased fashion.

An Unbiased Estimator of σ^2

Since the sample mean is an unbiased estimator of the population mean, perhaps we could use the sample variance (S^2) to estimate the population variance (σ^2). But, unfortunately...

S^2 is not an unbiased estimator of σ^2   The sample variance, S^2, is calculated by $S^2 = \Sigma(X - \overline{X})^2/N$ [formula (4.7)]. It can be shown algebraically that S^2 underestimates σ^2 on the average (Hays, 1988). Although the algebra is beyond the scope of this book, I will demonstrate that S^2 is not an unbiased estimator of σ^2. Consider a population of four values: {0, 1, 2, 3}. This population has a mean of $\mu = 1.5$ and a variance of $\sigma^2 = 1.25$. Suppose we take all possible samples of size $N = 2$ from this population (as we did in Chapter 8) and calculate S^2 for each of the possible samples. If S^2 is a good (i.e., unbiased) estimator of σ^2, then the average of all possible sample variances should be equal to σ^2. The results of such an exercise are shown in Box 10.1. Note that the average S^2 for the 16 possible samples is 0.625, whereas $\sigma^2 = 1.25$. *On the average, the sample variance, S^2, underestimates σ^2.* So we can't use the sample variance to estimate the population variance.

> **CONCEPT RECAP**
> Recall that an **unbiased estimator** is *a sample statistic that has a long-run average exactly equal to the corresponding population parameter.* For example, since the average sample mean equals μ, $\overline{X}$ is an unbiased estimator of μ.

Box 10.1

LIST OF ALL POSSIBLE SAMPLE VARIANCES WHEN *N* = 2

Population: {0, 1, 2, 3}

Mean of population: $\mu = 1.5$

Variance of population: $\sigma^2 = 1.25$

	POSSIBLE SAMPLES FOR *N* = 2	SAMPLE VARIANCES, S^2
Sample 1	{0, 0}	0.000
Sample 2	{0, 1}	0.250
Sample 3	{1, 0}	0.250
Sample 4	{1, 1}	0.000
Sample 5	{0, 2}	1.000
Sample 6	{2, 0}	1.000
Sample 7	{1, 2}	0.250
Sample 8	{2, 1}	0.250
Sample 9	{0, 3}	2.250
Sample 10	{3, 0}	2.250
Sample 11	{2, 2}	0.000
Sample 12	{1, 3}	1.000
Sample 13	{3, 1}	1.000
Sample 14	{2, 3}	0.250
Sample 15	{3, 2}	0.250
Sample 16	{3, 3}	0.000
	Average	0.625

It's inconvenient that S^2 cannot be used as a substitute for σ^2. But the good news is that . . .

s^2 **is an unbiased estimator of** σ^2 A modified kind of sample variance, symbolized by a lowercase s^2 rather than a capital S^2, does accurately estimate σ^2 on the average. s^2 is called the **unbiased estimator of the population variance** or, more simply, the **unbiased variance estimator**. The definitional formula for s^2 is:

$$\text{Unbiased estimator of } \sigma^2 = s^2 = \frac{\overset{\text{③}\;\text{①}\;\text{②}}{\Sigma\left(X - \overline{X}\right)^2}}{\underset{\text{④}}{N - 1}} \tag{10.1}$$

> **FORMULA GUIDE**
> ① Subtract the sample mean from each of the N raw scores to get N deviation scores.
> ② Square each deviation score to get N squared deviations.
> ③ Sum the N squared deviations.
> ④ Divide the sum of the squared deviations by $(N - 1)$.

Consider the population of four values: {0, 1, 2, 3}. This population has a mean of $\mu = 1.5$ and a variance of $\sigma^2 = 1.25$. Again, let's take all possible samples of size $N = 2$ from this population and calculate s^2 for each of those samples. If s^2 is an unbiased estimator of σ^2, then the average of all possible s^2 values should be equal to σ^2. The results of such an exercise are shown in Box 10.2. Note that

Box 10.2

LIST OF ALL POSSIBLE UNBIASED VARIANCE ESTIMATES WHEN $N = 2$

Population: {0, 1, 2, 3}

Mean of population: $\mu = 1.5$

Variance of population: $\sigma^2 = 1.25$

	POSSIBLE SAMPLES FOR $N = 2$	UNBIASED VARIANCE ESTIMATES, S^2
Sample 1	{0, 0}	0.000
Sample 2	{0, 1}	0.500
Sample 3	{1, 0}	0.500
Sample 4	{1, 1}	0.000
Sample 5	{0, 2}	2.000
Sample 6	{2, 0}	2.000
Sample 7	{1, 2}	0.500
Sample 8	{2, 1}	0.500
Sample 9	{0, 3}	4.500
Sample 10	{3, 0}	4.500
Sample 11	{2, 2}	0.000
Sample 12	{1, 3}	2.000
Sample 13	{3, 1}	2.000
Sample 14	{2, 3}	0.500
Sample 15	{3, 2}	0.500
Sample 16	{3, 3}	0.000
	Average	1.250

the average s^2 for the 16 possible samples is 1.25, exactly the same as σ^2. *On the average, the unbiased variance estimator, s^2, is exactly equal to σ^2.* In statistical jargon, it is correct to say that σ^2 is the long-run "expected value" of s^2. So we can use s^2 to estimate the population variance whenever the latter is unknown.

> NOTE: You will always know whether the sample variance or the unbiased variance estimator is being referred to by noticing the "case" of the letter used to symbolize the statistic. Capital S^2 always represents the sample variance, and lowercase s^2 always represents the unbiased estimator of the population variance. Also, S^2 is always calculated with a divisor of N, whereas s^2 is always calculated with a divisor of $N - 1$.

The degrees of freedom concept   The unbiased variance estimator (s^2) is computed in much the same way as the sample variance (S^2), except that the unbiased estimator formula requires you to divide the sum of the squared deviations by $N - 1$ instead of just N. $N - 1$ is called the **degrees of freedom** of the sample, defined as *the number of independent observations in the sample minus the number of parameters estimated in the formula.* When you use s^2 to estimate σ^2, you are implicitly using $\overline{X}$ to estimate μ. In other words, $\Sigma(X - \overline{X})^2$ is actually an estimate of $\Sigma(X - \mu)^2$. So 1 is subtracted from N to "mathematically compensate" for the fact that you are estimating μ in the formula for s^2. (See Hays, 1988, for the mathematical proof that underlies the degrees of freedom concept.)

Degrees of freedom can also be conceived of in the following way: Any sample of N observations has a certain value of $\overline{X}$ that is known. And relative to that known $\overline{X}$, only $(N - 1)$ raw scores are "free to vary." The *one* remaining observation is fixed, in the sense that it can have only a particular value. For example, suppose a sample of $N = 4$ raw scores has a mean of $\overline{X} = 7$. Now, we can arbitrarily specify the values of the first $(N - 1)$ values because they are "free to vary." Let's make these first three raw scores $X_1 = 9$, $X_2 = 8$, and $X_3 = 5$. What is the fourth raw score? Since the sample mean is 7, X_4 must equal 6. (The X values must sum to 28 to have a mean of 7.) To repeat: Given the known $\overline{X}$ and the first $(N - 1)$ raw scores, the last raw score must be one particular value. It is "fixed." Only $(N - 1)$ of the observations are free to vary relative to $\overline{X}$.

There is still another interesting perspective on the degrees of freedom idea. Recall that on the average, S^2 is too small an estimate of σ^2; that is, when $\Sigma(X - \overline{X})^2$ is divided by N, the resulting quotient is too small to be a good estimate of σ^2. But dividing $\Sigma(X - \overline{X})^2$ by $N - 1$ makes the quotient a little larger because $N - 1$ is a smaller divisor than N; in fact, the quotient is just enough larger to make the average s^2 perfectly equal to σ^2.

Computing s^2   Since the unbiased variance estimator is a viable substitute for σ^2, it plays a star role in many hypothesis-testing scenarios. You definitely need to know how to calculate s^2. You have seen and, I hope, understand formula (10.1). But that is a definitional formula. It reveals the mathematical meaning of s^2, but it is somewhat cumbersome to use because it requires you to work with deviation scores (that is, $X - \bar{X}$ values). In contrast, a computational formula enables you to find the value of s^2 using only raw scores and squared raw scores. To set up a computational formula, we need to first review the meaning of a quantity called the *sum of squares*. From Chapter 4, formula (4.10) shows that

$$\text{Sum of squares} = SS = \Sigma\left(X - \bar{X}\right)^2 = \Sigma X^2 - \frac{\left(\Sigma X\right)^2}{N}$$

After you get SS (the sum of squares) using only raw scores and squared raw scores, it's a cinch to compute the unbiased variance estimate through

$$s^2 = \frac{SS}{N - 1} \tag{10.2}$$

Let's apply formula (10.2) to the present research problem. Table 10.2 shows the raw scores, squared raw scores, sum of raw scores, and sum of squared raw scores for the 25 recipient rats. To obtain SS, we use

$$SS = \Sigma X^2 - \frac{\left(\Sigma X\right)^2}{N}$$

$$= 112,291 - \frac{(1651)^2}{25}$$

$$= 112,291 - \frac{2,725,801}{25}$$

$$= 112,291 - 109,032$$

$$= 3259$$

With SS in hand, s^2 is computed simply by:

$$s^2 = \frac{SS}{N - 1} = \frac{3259}{25 - 1} = \frac{3259}{24} = 135.79$$

Now that we have an unbiased estimate of the population variance, can we proceed with the test of significance using the z ratio? Well, if things were that easy, I wouldn't need a whole chapter. The z statistic still does not quite handle the kind of situation we are dealing with this time. The shortcoming of the z statistic stems from the assumptions that underlie the standard normal distribution and its relationship to the sampling distribution of $\bar{X}$.

Table 10.2

RAW SCORES AND
SQUARED RAW SCORES
FOR 25 RECIPIENT RATS

NUMBER OF SAFE TURNS, X	X^2
76	5,776
61	3,721
74	5,476
45	2,025
68	4,624
83	6,889
54	2,916
52	2,704
77	5,929
56	3,136
77	5,929
64	4,096
79	6,241
53	2,809
60	3,600
61	3,721
75	5,625
80	6,400
63	3,969
82	6,724
57	3,249
66	4,356
44	1,936
66	4,356
78	6,084
$\Sigma X = 1651$	$\Sigma X^2 = 112{,}291$

Limiting Assumptions of the z Statistic

To review a bit, we can use the z statistic to test a hypothesis about a mean because the sampling distribution of $\overline{X}$ is often normal (when the population distribution is normal or when the central limit theorem holds). Hence, our hypothesis-testing procedures involve substituting the standard normal distribution (made up of z values) for the sampling distribution of the mean (made up of $\overline{X}$ values). The substitution occurs when we convert a sample mean from the sampling distribution to a z statistic in the standard normal distribution through formula (8.5): $z = (\overline{X} - \mu)/\sigma_{\overline{X}}$. Once this conversion is made, we can obtain the probability associated with the sample mean by looking up certain areas in the standard normal distribution (in Table Z). If the z score has a low probability of occurrence (e.g., less than .05) in the standard normal distribution, then we conclude that the sample mean has the same low probability of occurrence in the sampling distribution of $\overline{X}$. Based on that low probability, we say that the sample mean is a "statistically significant" outcome.

With this rationale in mind, take a look at a slightly expanded version of formula (8.5):

the only random variable
↓

$$\text{Test statistic} = z = \frac{\overline{X} - \mu}{\sqrt{\sigma^2/N}}$$

This ratio *is based on the assumptions that* (1) $\overline{X}$ *is the only random variable in the sampling distribution*, and that, since $\overline{X}$ is normally distributed around μ, (2) z is normally distributed around its mean of 0. Thus, the standard normal distribution, with its mean of 0 and standard deviation of 1, is held to be an accurate substitute for the sampling distribution of $\overline{X}$. However, when we don't know what the population variance is and must estimate it with s^2, the test-statistic ratio changes to:

first of two random variables
↓

$$\text{Test statistic} = \frac{\overline{X} - \mu}{\sqrt{s^2/N}}$$

↑ second of two random variables

Now there is not just one but two random variables that contribute to variation in the test statistic; that is, when σ^2 must be estimated by s^2, both $\overline{X}$ and s^2 contribute to random variability in the test statistic. This is true because both s^2 and $\overline{X}$ will vary from one sample to the next as a function of the sampling error. Consequently, the distribution of the test statistic is no longer the standard normal distribution but is *a more variable distribution that is not normal.* As Figure 10.1 shows, this distribution is flatter and more spread out than the standard normal distribution.

The implication of all this is that when σ^2 must be estimated, we can't use the standard normal distribution to test hypotheses. We must use a new distribution instead. But just what is this new distribution, and how can we use it to test hypotheses?

NOTE: **Sampling error** as it is used here, refers to the fact that, *as a result of the random selection of observations from a population, almost all samples have a mean and variance estimate that deviate from the mean and variance of the population.* The discrepancy between each statistic and the corresponding parameter is an "error" inherent in the process of random sampling and, therefore, is a "sampling error."

Figure 10.1 *Comparison of a New Test Statistic Distribution with the Standard Normal Distribution (Note: In the standard normal distribution only the sample mean contributes to variability, but in the new distribution two random variables contribute to variability. The result is a flatter, broader distribution.)*

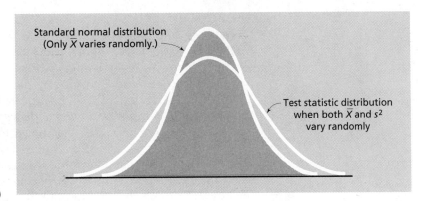

Standard normal distribution
(Only $\bar{X}$ varies randomly.)

Test statistic distribution
when both $\bar{X}$ and s^2
vary randomly

THE *t* DISTRIBUTION

The test statistic distribution that results when we use s^2 to estimate σ^2 is the *t* distribution. As Figure 10.2 suggests, the *t* distribution is characteristically more variable than the standard normal distribution. Nonetheless, the *t* distribution can easily be used to do hypothesis testing and parameter estimation under circumstances that make the standard normal distribution unusable.

Properties of the *t* Distribution

Even though the *t* distribution is not a normal distribution, it does have known properties that we can count on.

The average *t* equals 0 As was true of the standard normal distribution, the mean of the *t* distribution is always 0. Thus, the long-run "expected value" of *t* is 0.

Figure 10.2 *Comparison of a t Distribution with the Standard Normal Distribution (Note: A t distribution has more variability.)*

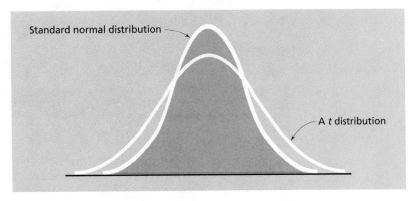

Standard normal distribution

A *t* distribution

The variance of *t* is a function of degrees of freedom   Although there is one and only one standard normal distribution, there actually is an entire family of *t* distributions. *A different t distribution exists for every particular sample size and corresponding degrees of freedom.* Keep in mind that for any particular *t* statistic, the degrees of freedom is usually $N - 1$.[2] And, as illustrated in Figure 10.3, *the amount of variability in the t distribution decreases as the degrees of freedom increases.* As shown in Figure 10.4, when the degrees of freedom equals 120 (that is, when $N = 121$), the *t* distribution virtually overlaps the standard normal distribution. In fact, when the theoretical sample size becomes infinitely large ($N = \infty$), the *t* distribution is exactly equal to the standard normal distribution (see Figure 10.3).

In the same vein, it is interesting to note that the variance of the *t* distribution is equal to

$$\frac{\text{degrees of freedom}}{\text{degrees of freedom} - 2} \qquad (10.3)$$

If we use *df* to stand for degrees of freedom, the variance of the *t* distribution when degrees of freedom equals 3 is $df/(df - 2) = 3/(3 - 2) = 3/1 = 3$. In other words, when the sample size is as small as $N = 4$, the *t* distribution is three times as variable as the standard normal distribution. When *N* is increased to 10 and *df* becomes 9, the variance of the *t* distribution is $9/(9 - 2) = 9/7 = 1.29$. Hence, by increasing the sample size by a few observations, we can make the *t* distribution much less variable.

Figure 10.3 *Three t Distributions with Different Degrees of Freedom (Note: The t distributions become more similar to the standard normal distribution as degrees of freedom increases.)*

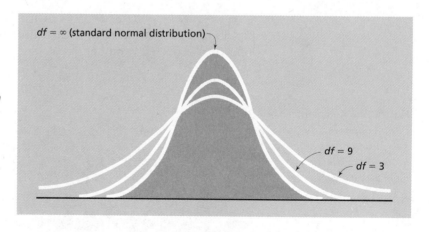

[2] Even though the word *degrees* is plural, the singular form of the verb ("is") is appropriate when referring to degrees of freedom because degrees of freedom is a single value.

Figure 10.4 *When the Sample Size Is Large, a t Distribution Is Virtually Identical to the Standard Normal Distribution*

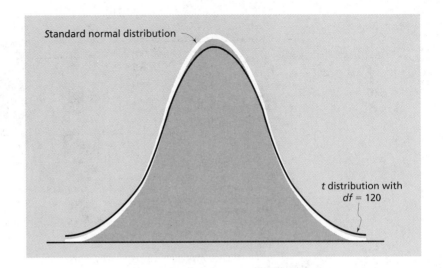

When the sample size becomes quite large, the variance of t approaches 1—the variance of z—and the t distribution closely approximates the standard normal distribution. For example, when $N = 121$, $df = 120$ and the variance of t becomes $120/118 = 1.02$. This fact reinforces the point made by Figure 10.4—namely, that when $df = 120$, the t distribution and the standard normal distribution are practically identical.

> NOTE: df is a standard way of abbreviating "degrees of freedom." In this book, df always refers to the degrees of freedom associated with whatever test statistic we are considering.

Finding a Critical *t* Value

To conduct a test of significance with the t statistic, you need to look up a "critical t statistic"—just as you had to look up z_{crit} when you were working with the standard normal distribution. You may remember that the "critical value" is the absolute value in the sampling distribution that the test statistic must equal or exceed in order to be statistically significant.

Critical t values are listed in Table T of Appendix A. A portion of Table T appears in Box 10.3. Look at that table and notice that the critical value of t—called t_{crit}—is determined by two factors: the degrees of freedom of the sample and the level of significance, or alpha level, that you have chosen. Various alpha levels are given at the top of the table. Note that the levels of significance for one-tailed tests are shown in the top row of the table header, and the levels of significance for two-tailed tests are given in the second row. The t_{crit} value you are looking for lies precisely at the point in the table where the df of your sample intersects with the alpha level that you selected.

Box 10.3

A PORTION OF TABLE T FROM APPENDIX A—CRITICAL VALUES IN THE t DISTRIBUTION

DEGREES OF FREEDOM	LEVEL OF SIGNIFICANCE: ONE-TAILED (DIRECTIONAL) TEST					
	.10	.05	.025	.01	.005	.0005
	LEVEL OF SIGNIFICANCE: TWO-TAILED (NONDIRECTIONAL) TEST					
	.20	.10	.05	.02	.01	.001
1	3.078	6.314	12.706	31.821	63.657	636.619
2	1.886	2.920	4.303	6.965	9.925	31.599
3	1.638	2.353	3.182	4.541	5.841	12.924
4	1.533	2.132	2.776	3.747	4.604	8.610
5	1.476	2.015	2.571	3.365	4.032	6.869
6	1.440	1.943	2.447	3.143	3.707	5.959
7	1.415	1.895	2.365	2.998	3.499	5.408
8	1.397	1.860	2.306	2.896	3.355	5.041
9	1.383	1.833	2.262	2.821	3.250	4.781
10	1.372	1.812	2.228	2.764	3.169	4.587
•						
•						
•						
23	1.319	1.714	2.069	2.500	2.807	3.768
24	1.318	1.711	2.064	2.492	2.797	3.745
25	1.316	1.708	2.060	2.485	2.787	3.725
•						
•						
•						
40	1.303	1.684	2.021	2.423	2.704	3.551
60	1.296	1.671	2.000	2.390	2.660	3.460
120	1.289	1.658	1.980	2.358	2.617	3.373
∞	1.282	1.645	1.960	2.327	2.576	3.291

For instance, recall the research example on the T-maze choices of rats that were injected with brain chemicals from maze-trained donor rats. The sample size is $N = 25$ and, therefore, $df = N - 1 = 25 - 1 = 24$. Assume that we wish to test the null hypothesis at the .05 level of significance using a two-tailed (nondirectional) model.

You can find the critical value by:

1. Locating the .05 column under "Level of significance for a two-tailed test"
2. Locating 24 degrees of freedom at the left of the table
3. Tracing down the .05 column and over from 24 degrees of freedom until you find the place in the table where these two values intersect

There you find the t_{crit} value of 2.064. Since we are doing a non-directional test, this means that a computed t at least as extreme as either −2.064 or +2.064 will be declared significant. Hence, the decision rule is: If $|t| \geq 2.064$, reject H_0. This decision rule is depicted in Figure 10.5. Observe that the critical value in that distribution is larger than the absolute critical value of $|1.96|$ that we would need if we were able to use a z statistic to test H_0. Because the t distribution has a larger variance than the standard normal distribution, t_{crit} *is always larger than the corresponding* z_{crit}.

A note on missing numbers   It is obvious that Table T can't display critical values for all possible degrees of freedom. For example, if $N = 75$, you will find no t_{crit} for $df = 74$ in Table T. But don't despair. When a particular df is not listed in a table of t values, it is acceptable to simply "drop back" to the first available *smaller df*. Thus, if $df = 74$, then you would use the t_{crit} value for $df = 60$ because that is the first available smaller df.

When the exact degrees of freedom for a test is not listed in the table of critical values, why is the preferred practice to use the first available *smaller df* rather than the next larger df? The answer is related to the the tendency of statisticians to give the null hypothesis the benefit of the doubt whenever a compromise must be made. Note that the smaller the degrees of freedom in Table T, the

Figure 10.5 *Two-Tailed Critical Values of t with 24 Degrees of Freedom*

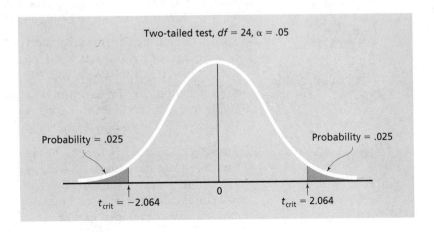

larger the associated t_{crit}. Therefore, when you use a smaller rather than a larger df, the computed t must be somewhat bigger to be significant. In effect, dropping back to the next smaller df makes it more difficult to reject H_0. This reduces the probability of a Type I error (i.e., rejecting a true H_0) slightly below the stated level of significance. In contrast, if you used the next larger df, it would be easier than it should be to reject H_0. That means that the probability of a Type I error would be greater than the stated level of significance, which is considered bad practice.

Computing *t*

Computing a t statistic from sample data is very similar to computing a z statistic. You use this formula:

$$t = \frac{\overline{X} - \mu}{s_{\overline{X}}} \tag{10.4}$$

Formula (10.4) is very similar to formula (8.5) for computing the z statistic: $z = (\overline{X} - \mu)/\sigma_{\overline{X}}$. The only difference is that instead of having the standard error of the mean in the denominator, the t ratio uses $s_{\overline{X}}$ known as the **estimated standard error of the mean** or, simply, the **estimated standard error**. It is computed this way:

$$s_{\overline{X}} = \sqrt{\frac{s^2}{N}} \tag{10.5}$$

where s^2 is the unbiased estimator of the population variance. Alternatively, the estimated standard error may be calculated by

$$s_{\overline{X}} = \frac{s}{\sqrt{N}} \tag{10.6}$$

As you use the t statistic, you'll discover that calculating it is normally a three-step process. I will illustrate this sequence of computations with the results of the study on the biochemical transfer of learning that we've been following.

Step 1: Calculate the unbiased variance estimator. We did this calculation earlier, but it is worth seeing again: $s^2 = SS/(N-1) = 3259/(25-1) = 3259/24 = 135.79$

Step 2: Calculate the estimated standard error of the mean. Using formula (10.5) for this step, we get

$s_{\overline{X}} = \sqrt{s^2/N} = \sqrt{135.79/25} = \sqrt{5.432} = 2.33$.

Step 3: Calculate the t ratio. Finally, we use formula (10.4) to compute the t statistic: $t = (\overline{X} - \mu)/s_{\overline{X}} = (66 - 60)/2.33 = 6/2.33 = 2.58$. Is this t ratio statistically significant? To answer that, we will compare the computed t statistic to t_{crit}, as specified in the decision rule developed earlier.

Was There a Biochemical Transfer of Learning?

Recall the decision rule developed earlier: If $|t| \geq t_{crit}$, reject H_0. Since the absolute value of the computed *t* ratio is 2.58, and t_{crit} is only 2.064, we reject the null hypothesis. The sample mean of 66 maze turns in the safe direction is significantly different from the μ of 60 expected under H_0. Therefore, it is likely that the rats that were injected with brain chemicals belong to a population in which the mean number of safe turns differs reliably from 60.

In the subsequent research report on the study, the researcher would write "The recipient rats' preference for the safe area of the T-maze was significant, $t(24) = 2.58$, $p < .05$. Apparently learning can be physically transferred from one animal to another via the transfer of easily extracted brain chemicals."

Note how the results of the *t* test are reported. *It is conventional to indicate the degrees of freedom in parentheses immediately after the t symbol.* As usual, "$p < .05$" communicates what level of significance was used.

Now that you have a general overview of the *why* and *how* of the *t* distribution, I will present several additional examples to demonstrate the adaptability of the *t* statistic.

FURTHER APPLICATIONS OF *t*

This section will show how the *t* statistic and the *t* distribution are used to (1) conduct a directional test of a hypothesis, (2) test the statistical significance of a correlation coefficient, and (3) construct a confidence interval in a parameter estimation study. In each example, be prepared to use Table T in Appendix A.

Doing a One-Tailed *t* Test

As an action guide for each of the following tests of significance, we will use the eight procedural steps of hypothesis testing introduced in Chapter 9. This example concerns the effect of a procedure designed to help AIDS patients gain weight. The inability to regain lost weight is a major health problem for persons who are in the later stages of AIDS. Let's assume that a medical researcher wishes to find out whether administering human growth hormone to AIDS patients will cause them to gain weight over a three-month period.

Step 1: State the problem and nature of the data   Fourteen AIDS patients being treated at a major university hospital are randomly chosen to participate in the study. They all agree to take part, which gives a sample size of $N = 14$. Every ten days each patient receives an injection of human growth hormone. The researcher records each patient's weight gain (or loss) over a three-month span. Since growth hormone is a major anabolic (i.e., tissue-building) agent, there are only two medically possible effects of the drug: either it will result in a weight gain or it will have no effect at all. It would make no sense, either practically or theoretically, to entertain the possibility that an anabolic agent might tear down tissue and produce a weight loss.

Step 2: State the statistical hypotheses Because only one direction of difference is meaningful in this context (that is, that the average weight gain of the patients will be greater than 0), a directional test is undertaken. The pertinent statistical hypotheses are

H_0: $\mu \leq 0$

H_1: $\mu > 0$

Step 3: Choose a level of significance The alpha level is set at .05. Therefore, the test statistic will be declared significant if it is so large that it would occur by chance alone only 5% of the time or less.

Step 4: Specify the test statistic A one-tailed t test will be conducted, with degrees of freedom = $df = (N - 1) = (14 - 1) = 13$. Why are we using a t test instead of a z test? Because we don't know the population variance and must estimate it. Since s^2 will be in the denominator of the test statistic, we have to use t instead of z. If we knew σ^2, we could carry out a simple z test.

Step 5: Determine the critical value needed for significance Consulting Table T in Appendix A, we find that with $\alpha = .05$ and $df = 13$, the t_{crit} for a one-tailed test is 1.771.

Step 6: State the decision rule If $t \geq 1.771$, reject H_0. Notice that because this is a directional test, the computed t *statistic must be a positive value* and at least as large as 1.771 in order for the data to be significant.

Step 7: Compute the test statistic The weights (X) and squared weights (X^2) of the $N = 14$ patients are given in Table 10.3. Observe that the mean weight in the sample is $\overline{X} = 1.36$. Is that mean far enough above 0 to be significant? A computed t ratio will provide the answer.

First we calculate the sum of squares (SS) so that it can be used to calculate the unbiased variance estimator:

$$SS = \Sigma X^2 - \frac{(\Sigma X)^2}{N} = 191 - \frac{(19)^2}{14} = 191 - \frac{361}{14}$$

$$= 191 - 25.79 = 165.21$$

Next we compute the unbiased estimator of the population variance:

$$s^2 = \frac{SS}{N-1} = \frac{165.21}{14-1} = \frac{165.21}{13} = 12.71$$

Now we get the estimated standard error of the mean with formula (10.5):

$$s_{\overline{X}} = \sqrt{\frac{s^2}{N}} = \sqrt{\frac{12.71}{14}} = \sqrt{0.9079} = 0.95$$

And, now that we have the "error term" for our ratio, we can calculate the *t* statistic:

$$t = \frac{\overline{X} - \mu}{s_{\overline{X}}} = \frac{1.36 - 0}{0.95} = \frac{1.36}{0.95} = 1.43$$

Step 8: Consult the decision rule and make a decision
Although the computed *t* is in the predicted (positive) direction, it is smaller than $t_{crit} = 1.771$. The computed *t* does not fall into the region of rejection of the *t* distribution. Therefore, we cannot reject H_0. The proper way to report the outcome is: "The mean weight gain of the patients was not significantly greater than 0, $t(13) = 1.43$, $p > .05$." Notice that because the test statistic was not significant,

Table 10.3

WEIGHT GAINS AND SQUARED WEIGHT GAINS FOR 14 AIDS PATIENTS

PATIENT	WEIGHT GAIN (POUNDS), X	X^2
1	0	0
2	−2	4
3	−1	1
4	5	25
5	3	9
6	6	36
7	9	81
8	−3	9
9	1	1
10	2	4
11	2	4
12	1	1
13	−4	16
14	0	0
	$\Sigma X = 19$	$\Sigma X^2 = 191$

$$\overline{X} = \frac{19}{14} = 1.36$$

its probability of occurring by chance alone is reported as "> .05," or "greater than .05." To be significant, a test statistic must have a chance probability that is at or below the chosen alpha level.

Using *t* to Test the Significance of a Correlation Coefficient

In Chapter 6 you learned about an index of the linear relationship between two variables. That index is called the Pearson *r*, or the linear correlation coefficient. You probably remember that the absolute value of *r* can vary between .00 and 1.00, with values closer to 1.00 reflecting stronger covariation between two variables. In fact, a popular method of interpreting *r* bases conclusions on the magnitude of the correlation coefficient. But, strictly speaking, you should conclude nothing about the strength of the relationship between two variables until you verify that the correlation coefficient is statistically significant. With small samples, even a fairly big *r* can be merely a result of chance. Hence, "chance" should be ruled out before more elaborate statements are made about a correlation. Fortunately, you can use the *t* statistic to easily test the significance of any Pearson correlation. Read on to find out how.

State the problem and nature of the data Haas and Sweeney (1992) were interested in variables that predict (i.e., are correlated with) the age of onset of symptoms in schizophrenic patients. That is, they wanted to find out what background factors were associated with the relatively early or relatively late occurrence of psychotic symptoms. One of the background variables they examined was "degree of economic and environmental independence from one's family of origin," as measured with an instrument called the Premorbid Adjustment Scale. This was variable *X* in the correlation. Variable *Y*, the dependent variable, was "age of onset of first psychotic symptoms" (expressed in years). There were $N = 71$ schizophrenic patients in this investigation.

State the statistical hypotheses Since null and alternative hypotheses are almost always expressed in terms of population parameters, the symbol for a population correlation is used to state the statistical hypotheses. *r* is a sample statistic. The corresponding population parameter is the Greek "rho," or ρ. ρ symbolizes the correlation between variables *X* and Y for the entire population that is represented by the sample of patients. When the significance of a single correlation coefficient is tested, the null hypothesis usually asserts that the correlation equals 0. The alternative hypothesis contends that the correlation differs from 0. Normally, then, the statistical hypotheses are nondirectional, so we have:

H_0: $\rho = 0$

H_1: $\rho \neq 0$

Choose a level of significance We'll assume that Haas and Sweeney set α = .01.[3]

Specify the test statistic We will assess the statistical reliability of this correlation with a two-tailed t test. *Important:* A correlation coefficient has $N - 2$, rather than $N - 1$, degrees of freedom. Since there were 71 subjects in Haas and Sweeney's study, $df = N - 2 = 71 - 2 = 69$.

Determine the critical value needed for significance If you now look at Table T in Appendix A, you'll find that 69 degrees of freedom is not listed. Consistent with convention in such cases, we "drop back" to the next lower degrees of freedom listed, which is 60. Combining $df = 60$ with the .01 level of significance in a two-tailed test, we get a t_{crit} of 2.66.

State the decision rule Based on the above, the decision rule in this example is: If $|t| \geq 2.66$, reject H_0. If the computed t ratio is at least as extreme as $|2.66|$ in either a positive or negative direction, then the linear correlation coefficient will be significant.

Compute the test statistic Haas and Sweeney found a correlation of $r = .54$ between the patients' degree of independence and the age of onset of psychotic symptoms. If significant, this correlation would suggest that the greater the patients' independence, the older they were before psychotic symptoms appeared. To test the significance of a correlation coefficient with a t statistic, we use:

$$t = \frac{\overset{\textcircled{1}}{\overset{\downarrow}{r\sqrt{N-2}}}}{\underset{\underset{\textcircled{3}\quad\textcircled{2}}{\uparrow\quad\uparrow}}{\sqrt{1-r^2}}} \leftarrow \textcircled{4} \tag{10.7}$$

FORMULA GUIDE

① Multiply the correlation coefficient by the square root of the degrees of freedom.
② Subtract the squared correlation from 1.00.
③ Take the square root of step 2.
④ Divide the result of step 3 into the result of step 1.

---

[3] Haas and Sweeney actually used a very conservative alpha level of .0001 to test this correlation.

Applying formula (10.7) to the present problem, we find:

$$t = \frac{r\sqrt{N-2}}{\sqrt{1-r^2}} = \frac{.54\sqrt{71-2}}{\sqrt{1-(.54)^2}} = \frac{.54\sqrt{69}}{\sqrt{1-.2916}} = \frac{.54(8.31)}{\sqrt{.7084}}$$

$$= \frac{4.49}{.842} = 5.33$$

Consult the decision rule and make a decision   Since the decision rule required only that the absolute value of the test statistic equal or exceed 2.66, the computed t of 5.33 is clearly significant. The report would read: "There was a significant positive correlation between the degree of premorbid independence and the age of first onset of psychotic symptoms, $r(69) = .54$, $p < .01$." This correlation means that the more economic and environmental independence the patients had, the longer they were able to avoid developing psychotic symptoms.

Using t to Construct a Confidence Interval

In Chapter 8 we explored the topic of parameter estimation, defined as "estimating population characteristics from sample statistics." In "interval estimation" you establish a range of values around $\overline{X}$ that is likely to include the population mean. The smaller the "confidence interval," the more precisely you can estimate μ from $\overline{X}$. The formula for the confidence interval is $\overline{X} \pm z_c\sigma_{\overline{X}}$ [formula (8.6)]. In Chapter 8 I mentioned that formula (8.6) has a definite limitation: It can be used only when we know the value of the population variance or standard deviation because the formula for $\sigma_{\overline{X}}$ requires one or the other of these two parameters. Specifically, $\sigma_{\overline{X}} = \sqrt{\sigma^2/N} = \sigma/\sqrt{N}$.

Unfortunately, in many efforts to estimate μ, neither the population variance nor the population standard deviation is available. Therefore, s^2 must be used to estimate σ^2. But by making this substitution, you invalidate formula (8.6), because z scores from the standard normal distribution are accurate only when $\sigma_{\overline{X}}$ can be directly calculated from σ or σ^2.

Even when the population variance is unknown, you can construct a confidence interval through the following device:

$$\text{Confidence interval} = \overline{X} \pm t_{\text{crit}}s_{\overline{X}} \tag{10.8}$$

The only differences between formulas (8.6) and (10.8) are that the estimated standard error, $s_{\overline{X}}$, must take the place of $\sigma_{\overline{X}}$ and, because of that, t must stand in for z. The following example shows how simple it is to construct a confidence interval with the t statistic.

The research problem   Rocco (1987) was interested in estimating how many hours per week midwestern women spend in planned leisure activities. Additionally, she wanted to construct a 95% confidence interval around that estimate. She obtained a simple random sample of 73 women from Missouri and Illinois and had each respondent keep a diary of her planned leisure time activities for a period of two weeks. Her sample recorded a weekly mean of 22.6 hours of planned leisure events. The unbiased estimate of the population variance was $s^2 = 144$.

Selection of t_{crit}   In parameter estimation, the critical value of t is determined in the same way it is in hypothesis testing: You find the t value in Table T that is at the intersection of degrees of freedom and alpha. It is important to note that:

- For a 99% confidence interval, you use the two-tailed ".01" column of Table T.
- For a 95% confidence interval, you use the two-tailed ".05" column of Table T.

Since Rocco intended to construct a 95% confidence interval, she worked with the .05 alpha level. Her degrees of freedom was $df = (N - 1) = (73 - 1) = 72$. Since $df = 72$ is not listed in Table T, Rocco dropped back to 60 degrees of freedom. That gave her a critical t value of 2.00.

Computation of $s_{\bar{X}}$ The estimated standard error of $\bar{X}$ was calculated in the usual way: $s_{\bar{X}} = \sqrt{s^2/N} = \sqrt{144/73} = \sqrt{1.97} = 1.4$.

Computation of the bound on the error of estimation   From Chapter 8, the "bound on the error of estimation" is the maximum likely deviation of $\bar{X}$ from μ. In the framework of formula (10.8), the bound = $t_{crit}s_{\bar{X}} = 2.00 \cdot 1.4 = 2.8$.

Construction of the confidence interval   The lower boundary of the confidence interval is $\bar{X} - t_{crit}s_{\bar{X}} = 22.6 - 2.8 = 19.8$. Similarly, the upper boundary of the interval is given by $\bar{X} + t_{crit}s_{\bar{X}} = 22.6 + 2.8 = 25.4$. The confidence interval is 19.8 hours/week to 25.4 hours/week. In other words, we can be reasonably confident that the typical respondent in the population studied by Rocco spends between 19.8 hours and 25.4 hours per week in planned leisure activities.

TAKING STOCK

If you have gotten what was intended from this chapter, then you know:

- When a t statistic must be substituted for a z statistic in inferential procedures
- Why a t statistic must often be used instead of z
- The characteristics of the t distribution and how it differs from the standard normal distribution
- How to find the degrees of freedom in a sample
- How to locate t_{crit} in Table T
- How to calculate the estimated standard error of the mean
- How to compute the t ratio as a test statistic
- How to use t_{crit} and the estimated standard error to set up a confidence interval around a sample mean

This chapter has described when and how to use the t statistic in single-sample research projects. The next chapter will show you how to apply the *t-distribution* idea to studies in which you are testing the significance of the difference between two sample means.

KEY TERMS

unbiased variance estimator
degrees of freedom
estimated standard error of the mean

SUMMARY

1. The t statistic is designed to be used instead of the z statistic whenever the population variance (or the population standard deviation) is unknown. The t statistic is used in both hypothesis testing and parameter estimation.

2. The z statistic and standard normal distribution cannot be used in inferential statistical procedures when σ^2 is unknown. Such situations require that you calculate an unbiased variance estimate (s^2) from sample data. As a result, two random variables are entered into the test statistic formula: $\overline{X}$ and s^2. This renders the standard normal distribution unusable because that distribution, and the z statistics based on it, assume that $\overline{X}$ is the only random variable in the test statistic formula. However, the t statistic and its sampling distribution can be appropriately used under such circumstances because they take both random variables ($\overline{X}$ and s^2) into account.

3. Like the standard normal distribution, a t distribution is symmetrical and has an average value of 0. However, it is more variable than the standard normal distribution, with the variance of the t distribution being equal to: degrees of freedom/(degrees of freedom – 2). The larger the sample size (N), the more closely

the t distribution approximates the standard normal distribution. So there is actually a family of t distributions, with a different t distribution existing for every possible sample size.

4. The degrees of freedom associated with a t statistic usually equals $N - 1$, with degrees of freedom defined as the number of independent observations in a sample minus the number of estimated parameters used in a statistic's formula. The t test of a linear correlation has $N - 2$ degrees of freedom.

5. Critical t statistics can be found in Table T and are a function of the level of significance and the number of degrees of freedom.

6. A t statistic is calculated by dividing $\overline{X} - \mu$ by the estimated standard error of the mean: $s_{\overline{X}} = \sqrt{s^2/N}$. s^2 is the unbiased estimator of σ^2, calculated via $s^2 = SS/(N - 1)$.

7. In hypothesis testing, the t ratio is calculated through $t = (\overline{X} - \mu)/s_{\overline{X}}$.

8. t_{crit} and the estimated standard error can be used to set up a confidence interval through the following device: $\overline{X} \pm t_{crit}s_{\overline{X}}$.

REVIEW QUESTIONS

1. Define or describe: unbiased variance estimator, degrees of freedom, and estimated standard error.

2. Under what circumstances must the t statistic and the t distribution be used in place of the z statistic and the standard normal distribution?

3. In your own words, explain why it is often necessary to base inferential statistics on the t distribution rather than on the standard normal distribution. Be specific and refer to the concepts of a standard error, random variables, and test statistic ratios.

4. How is a t distribution both similar to and different from the standard normal distribution? How many t distributions are there? What determines the degree of similarity between a t distribution and the standard normal distribution?

5. Describe how you would determine the degrees of freedom in a t test of a mean. Also, state how the variance of a t distribution is related to the degrees of freedom.

6. A sample size is $N = 18$, and the researcher is using an alpha level of .01. What is t_{crit} in a two-tailed t test of a mean?

7. In a t test of a mean, what are the respective t_{crit} values for:
   (a) $\alpha = .05$, $N = 10$, two-tailed test
   (b) $\alpha = .01$, $N = 31$, one-tailed test
   (c) $\alpha = .025$, $N = 40$, one-tailed test
   (d) $\alpha = .001$, $N = 125$, two-tailed test

8. Consider the following population of four values: 2, 4, 6, 8. Select all possible samples of size $N = 2$ and compute the regular sample variance of each of the 16 samples using

$$S^2 = \frac{\Sigma X^2 - \frac{(\Sigma X)^2}{N}}{N}$$

How does the average of the sample variances compare with the population variance? What implication does your finding have for estimating the population variance from the sample variance?

9. Consider the following population of four values: 2, 4, 6, 8. Select all possible samples of size $N = 2$ and calculate the unbiased variance estimate for each of the 16 samples using

$$s^2 = \frac{\Sigma X^2 - \frac{(\Sigma X)^2}{N}}{N - 1}$$

How does the average of the unbiased variance estimates compare with the population variance? What implication does your finding have for estimating the population variance from the unbiased variance estimate?

10. In a t test of a mean, what are the respective t_{crit} values for:
    (a) $\alpha = .001$, $N = 10$, two-tailed test
    (b) $\alpha = .01$, $N = 55$, one-tailed test
    (c) $\alpha = .025$, $N = 40$, one-tailed test
    (d) $\alpha = .05$, $N = 130$, two-tailed test

11. Compute $\overline{X}$ and the unbiased variance estimate for the following sample of IQ test scores: 90, 110, 130, 60, 95, 113, 140, 37, 126, 72, 98, 107, 92.

12. Eleven infants classified as "ambivalent-attached" in their interpersonal orientation were observed for several months. Their ages (in months) at the time they began to walk were: 15, 18, 16, 17, 14, 13, 20, 17, 16, 17, 14. Compute the mean and unbiased variance estimate of this set of observations.

13. Suppose that a particular state strictly enforces the physical fitness regulations that apply to its state highway patrol officers. One of the fitness standards specifies that officers must be able to run a mile in 7 minutes or less. You are the chief statistician for the state attorney general's office, and you have been asked to check on adherence to fitness regulations by testing a random sample of ten troopers. You plan to compare the mean mile-run time of the ten officers to the standard of 7 minutes. The run times (in minutes) are: 5.2, 5.0, 6.8, 9.3, 11.1, 7.0, 8.4, 8.0, 9.9, 8.4. Calculate the mean and unbiased variance estimate for these data.

14. Using the eight procedural steps from this chapter, conduct a directional test of the data in Question 13, using $\alpha = .05$. Draw an appropriate conclusion.

15. It is known that the mean number of fatal motorcycle accidents per state per year is 334. On a random basis, 20 of the 50 states are selected to participate in a motorcycle safety-enhancement program sponsored by the federal government. Participating states receive large federal grants in return for implementing special regulations and procedures designed to reduce motorcycling fatalities. At the end of the first year of this program, the participating states have the following numbers of fatalities on record: 290, 350, 321, 304, 308, 282, 295, 170, 399, 402, 300, 193, 245, 276, 267, 202, 130, 312, 298, 248. Compute the mean and the unbiased variance estimate for these data.

16. Carry out an appropriate one-tailed test of the effectiveness of the government-sponsored safety-enhancement program described in Question 15. Use an alpha level of .05, and draw an appropriate conclusion. Be sure to use the eight procedural steps from this chapter.

17. A psychobiologist hypothesizes that the diastolic blood pressure of Type A persons differs from average. The psychobiologist takes the blood pressure of 22 Type A men whose ages range between 21 and 29. The sample mean diastolic pressure is 88, with an unbiased variance estimate of 16. Under H_0, $\mu = 80$. Using the .001 level of significance, conduct a t test of the null hypothesis. Be sure to use the eight procedural steps explained in this chapter.

2 - tailed

18. A social worker wants to find out whether the average IQ of a group of teenage prison inmates is below 70. The social worker assesses the IQ of a simple random sample of 38 inmates who are between the ages of 17 and 19. The mean IQ in this sample is 67, and the unbiased variance estimate is 144. Conduct a two-tailed test of $H_0 : \mu = 70$. Use $\alpha = .05$. Be sure to use the eight procedural steps.

No just. for predicting one way or other
↑
2-tailed

Directional test if expect outcome in only one direction.

19. Suppose the linear correlation between two different tests of nonverbal intelligence is .40 for a sample of 21 students. Is the correlation significant at the .05 level?

20. A behavioral scientist finds a correlation of .40 between frequency of sexual intercourse in adolescence and frequency of sexual intercourse during the retirement years. There are $N = 42$ people in the sample. With $\alpha = .05$, is the correlation statistically significant?

21. A personality psychologist reports that $r = .60$ for the linear relationship between the level of achievement motivation and scores on a valid test of Type A behavior. There are $N = 10$ subjects in the sample. The psychologist reports that "there was no significant relationship between achievement motivation and Type A tendencies in this sample." Assuming that alpha = .05, is this conclusion statistically correct? Explain.

22. Assume that in a sample of 135 high school students, the correlation between grade point average and scores on a test of word fluency is .18. Is this outcome statistically reliable at the .05 level? At the .01 level?

23. In a simple random sample of 500 elementary school children from a large urban district, the mean score on a test of abstract reasoning is 28. The unbiased variance estimate is 4.7. Set up a 99% confidence interval around the sample mean, and write an interpretation of the confidence interval.

24. A social worker selects a random sample of 57 homeless persons in a large metropolitan area. The mean age of the sample is 31.2, and the unbiased estimate of the population variance is 8.67. Construct and interpret the 95% confidence interval for these data.

25. You are investigating the MMPI psychopathic deviate scores of a simple random sample of 64 professional actors. The mean raw score in the sample is 28, and the unbiased variance estimate is 9.6. Construct a 95% confidence interval around the sample mean, and interpret the interval.

CHAPTER 11 TWO-SAMPLE *t* TESTS

In Chapter 10 you applied the *t* statistic to situations in which a sample mean was compared to a hypothesized population mean. Much more frequently, however, you will encounter studies in which there are two samples, and the first sample mean is compared to the second sample mean. The universal prototype of this kind of study is the two-sample experiment, in which the mean of the "experimental group"—a sample of subjects treated in some special way—is tested against the mean of the "control group"—a sample of subjects treated in some conventional or standard way. In such a case, the familiar single-sample *t* test is not usable because it assumes that you are working with just one $\overline{X}$. Instead, you must apply a somewhat more involved procedure called the two-sample *t* test. It enables you to test the significance of the difference between the experimental group mean, $\overline{X}_1$, and the control group mean, $\overline{X}_2$.

Two types of research design call for a two-sample *t* test:

1. Independent-samples designs, in which the data in the experimental condition are completely independent of the data in the control condition
2. Correlated-samples designs, in which the data in the experimental condition are literally correlated with the data in the control condition

For each category of research design, you will learn about the following topics: (1) the defining characteristics of each kind of research design, (2) the nature and parameters of the sampling distribution associated with each design, (3) how to use the *t* statistic to test two-group hypotheses, (4) how to interpret the results of the two-sample *t* test, and (5) the assumptions of two-sample tests.

COMPLETELY RANDOMIZED TWO-GROUP DESIGN

One approach to behavioral research that requires an independent-samples *t* test is called the **completely randomized two-group design**. In this research method, *each member of a "pool" of research subjects is randomly assigned to one of two treatment conditions.* Then an independent variable is systematically manipulated by treating the two groups differently. Afterward, some target behavior, the dependent variable, is observed and measured. The researcher is interested in finding out whether the dependent-variable mean of the first group is different from the dependent-variable mean of the second group. If those means have a numerical difference, then the role of the independent-samples *t* test is to determine whether the difference is reliable or whether it is merely a result of "random sampling error."[1]

Note two very important characteristics of the completely randomized two-group design:

1. Because subjects are randomly assigned to treatments, the two groups are assumed to be statistically equivalent at the start of the study. Any significant difference between the groups on the dependent variable can usually be attributed, *in a cause-and-effect sense*, to the influence of the independent variable. In short, the completely randomized design is a true experimental design. (See Chapter 2 for a review of these ideas.)

2. Because of the entirely random way in which the two samples are formed, the data in the first group of the experiment are truly *independent* of the data in the second group. There is no correlation whatsoever between the two samples. This fact justifies using an independent-samples *t* statistic to test the significance of any difference between the sample means.

An Example from the Research Literature

Sheehan, Green, and Truesdale (1992) were interested in studying the effect of rapport between hypnotists and their subjects on the incidence of hypnotically induced "pseudomemories." In order to understand what was investigated, you need to be familiar with these definitions: *Hypnosis* is a state of selective attention brought about by repeated verbal suggestions. Hypnosis typically produces feelings of relaxation and makes the hypnotized person more likely to comply with suggestions and commands issued by the hypnotist. *Rapport* "expresses the positive interaction of hypnotist and subject" (Sheehan et al., 1992, p. 691). In general, rapport refers to the good feelings and mutual trust that ideally exist between the hypnotist and the hypnotized person. *Pseudomemories* are subjects' reports of distorted or false memories brought about by suggestions made during hypnosis. The pseudomemory phenomenon "illustrates major evidence of memory distortion and raises important questions about the interface between memory and hypnosis" (Sheehan et al., 1992, p. 691).

Several studies have revealed that hypnotic suggestion can be used to make subjects think that they saw or heard something

---

[1] Even though most experiments do not require subjects to be randomly *selected*, the process of randomly assigning subjects to treatment groups is assumed to produce the kind of chance variation between sample means that would occur as a result of random sampling.

that, in fact, did not take place. Sheehan and colleagues hypothe-sized that, since hypnotic phenomena tend to be affected by the quality of the hypnotist–subject relationship, poor rapport between the hypnotist and the subject would weaken the ability of hypnotic suggestion to induce false memories. To test this research hypoth-esis, 22 of 44 hypnosis-susceptible subjects were randomly as-signed to each of two treatment conditions: (1) rapport present or (2) rapport reduced.[2] Note that since different persons served in the two treatment groups, and since assignment to treatments was en-tirely random, this research approach is a completely randomized two-group design. The independent variable was hypnotist–subject rapport, and it had two levels (i.e., the two treatment conditions), as noted above. The dependent variable was the number of post-hypnotic pseudomemories reported by subjects in each condition. Since three false memories were hypnotically "implanted," each subject had a potential pseudomemory score of 0 to 3.

Before I examine and analyze the findings of Sheehan and col-leagues, a brief lesson is in order regarding the theory behind the independent-samples *t* test.

COMPARING ONE-SAMPLE AND TWO-SAMPLE TESTS

An independent-samples *t* test differs from the one-sample *t* test in six ways: (1) the structure of the null and alternative hypotheses, (2) the kinds of means that are compared, (3) the kind of sampling distribution that is assumed, (4) estimation of the standard error of the sampling distribution, (5) computation of the *t* ratio, and (6) how the degrees of freedom is determined.

Statement of Hypotheses

To review, in the one-sample investigation the opposing hypothe-ses are:

Null hypothesis→H_0: μ = some theoretical value (normally 0)

Alternative hypothesis→H_1: $\mu \neq$ some theoretical value (nor-mally 0)

In contrast, the two-sample study has these hypotheses:

H_0: $\mu_1 = \mu_2$

H_1: $\mu_1 \neq \mu_2$

---

[2] The Sheehan et al. (1992) investigation was actually more complicated than this. You are encouraged to examine their article for all the details.

In the two-sample case, the null hypothesis asserts that the data of the two samples come from populations that have equivalent means. Since (as will be discussed later) the null hypothesis also assumes that the populations have equal variances, H_0 is, in *effect, stating that there is only one population and, furthermore, that both samples in the study were drawn from that particular population,* which has a mean of μ. Hence, $\mu_1 = \mu_2 = \mu$.

In contrast, the two-sample alternative hypothesis contends that the two groups of observations represent nonequivalent population means. This implies that there are two different populations, not just one. According to H_1, each of these supposedly separate populations corresponds to a level of the independent variable.

Important: The theory underlying the independent-samples *t* test assumes that the null hypothesis is true.[3] As we proceed through this discussion, all points will be based on the assumption that the null hypothesis is true.

Types of Means Compared

The one-sample comparison In a one-sample investigation, a sample mean, $\overline{X}$, is computed from a sample of observations that are randomly selected from a population. Then, when you compute the test statistic, $\overline{X}$ is compared with a hypothesized population mean:

$$\text{Test statistic} = \frac{\overline{X} - \mu}{\text{standard error}}$$

The fact that the one-sample test is based, in part, on the deviation of $\overline{X}$ from μ has an important implication for the amount of variation that can exist in the sampling distribution of $\overline{X}$: The maximum conceivable difference between $\overline{X}$ and μ will limit the size of the standard error of the sampling distribution. Let's assume, for example, that we are taking a simple random sample of $N = 9$ observations from a population with a mean of $\mu = 100$ and a standard deviation of $\sigma = 15$. Let's also assume that, for all practical purposes, the maximum conceivable deviation of a sample mean from the population mean, *under the most extreme kind of sampling error*, is three standard deviations in either a positive or a negative direction. In this hypothetical one-sample case, then, no *single sample mean* based on observations from this population is likely

---

[3] This notion suggests an interesting paradox inherent in the meaning of statistical significance, doesn't it? If the theory behind a statistical test is founded on the assumption that the null hypothesis is valid, then a significant outcome represents a violation of the theory of chance that justified the use of the test in the first place!

to deviate from μ by more than $3 \times \sigma = 3 \times 15 = 45$—and virtually all of the random $\bar{X}$ values from the population will be closer than 45 units to μ. This one-sample situation is illustrated in Figure 11.1.

The two-sample comparison   In a two-sample study, of course, two groups of observations are drawn. Instead of comparing $\bar{X}$ to μ, the test statistic is actually based on the deviation between two sample means, $\bar{X}_1$ and $\bar{X}_2$, relative to the difference between the corresponding population means:

$$\text{Test statistic} = \frac{\left(\bar{X}_1 - \mu_1\right) - \left(\bar{X}_2 - \mu_2\right)}{\text{standard error}} = \frac{\left(\bar{X}_1 - \bar{X}_2\right) - \left(\mu_1 - \mu_2\right)}{\text{standard error}}$$

Technically, the difference between sample means is compared with the hypothesized difference between population means. But bear in mind that we are assuming that H_0 is true—that the two samples come from the same population with a mean of μ; that is, we assume that $\mu_1 = \mu_2 = \mu$. Therefore, $(\mu_1 - \mu_2) = 0$, and the formula can be simplified to:

$$\text{Test statistic} = \frac{\left(\bar{X}_1 - \bar{X}_2\right)}{\text{standard error}}$$

For the sake of contrasting the one- and two-sample situations, let's further assume that we have drawn two simple random samples from the same hypothetical population that we used earlier. The population mean, μ, equals 100, and the population standard deviation equals 15.

Figure 11.1 *Population Distribution from Which a Single Sample Is Randomly Selected*

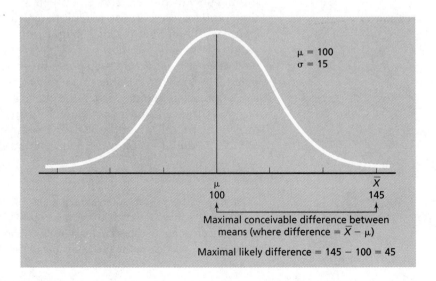

$\mu = 100$
$\sigma = 15$

μ
100

$\bar{X}$
145

Maximal conceivable difference between means (where difference $= \bar{X} - \mu$)

Maximal likely difference $= 145 - 100 = 45$

In this two-group example, what is the maximum conceivable deviation between $\overline{X}_1$ and $\overline{X}_2$? In this kind of research situation, two samples are independently drawn from the population, so *either of the sample means might be above or below μ*. In the most extreme situation, therefore, one $\overline{X}$ could conceivably be as much as three standard deviations *above* μ when the other $\overline{X}$ is as much as three standard deviations *below* μ. Since we have set σ = 15, the maximum conceivable difference between the two sample means in this hypothetical example is $(3 \times \sigma) + (3 \times \sigma) = 6 \times \sigma = 6 \times 15 = 90$—twice as large as the maximum conceivable difference between any one $\overline{X}$ and μ. This idea is depicted in Figure 11.2.

Although the difference between $\overline{X}_1$ and $\overline{X}_2$ is almost never as extreme as the possibility considered here, these general concepts should be grasped:

- In a two-sample test of significance the critical statistical comparison is the difference between $\overline{X}_1$ and $\overline{X}_2$, not between $\overline{X}$ and μ.

- The $\overline{X}_1 - \overline{X}_2$ deviation has a *greater range of variation* than the $\overline{X} - \mu$ deviation.

- On the average, the absolute difference between $\overline{X}_1$ and $\overline{X}_2$ is greater than the absolute difference between $\overline{X}$ and μ; in other words, the typical absolute mean difference in a two-sample test is expected to be larger than the typical mean difference in a one-sample test.

Figure 11.2 *Population Distribution from Which Two Independent Samples Are Selected*

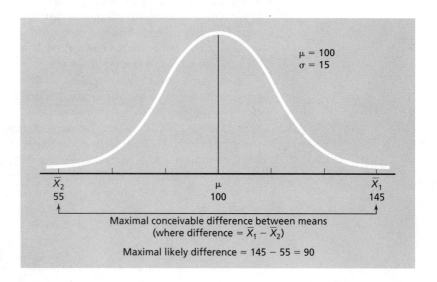

$\mu = 100$
$\sigma = 15$

$\overline{X}_2$ μ $\overline{X}_1$
55   100   145

Maximal conceivable difference between means
(where difference = $\overline{X}_1 - \overline{X}_2$)

Maximal likely difference = 145 − 55 = 90

> **CONCEPT RECAP**
> Just a reminder: **Absolute differences** between means ignore any negative signs. Thus, all absolute mean differences are treated as positive deviations.

The points made in this section will help you understand the differences between the respective sampling distributions associated with one-sample and two-sample tests.

Sampling Distribution Differences

From reading previous chapters you know that sampling distributions are theoretical distributions made up of random statistics (e.g., means or mean differences). Sampling distributions enable us to link sample statistics with corresponding population parameters. They also serve as models of random variation in hypothesis testing. The sampling distributions associated with independent-samples tests differ from those of one-sample tests in several ways.

Distribution contents   In single-sample studies, the sampling distribution is made up entirely of $\overline{X}$ values—that is, sample means. In studies that use two independent samples, *the relevant sampling distribution is comprised of* $(\overline{X}_1 - \overline{X}_2)$ values—that is, *mean differences*. In fact, the latter distribution is called the **sampling distribution of the mean difference**.

Distribution average   As you have now read many times, the average value in the sampling distribution of $\overline{X}$ is μ, the population mean. But the average value of the sampling distribution is different in independent-samples studies: In the *sampling distribution of* $\overline{X}_1 - \overline{X}_2$, the average value is 0—yes, *zero*. Why is that? Remember that the sampling distribution of the mean difference is made up of an infinite number of random mean differences. If only random processes determine $\overline{X}_1 - \overline{X}_2$, then about half of these mean differences will be positive and about half will be negative,[4] since $\overline{X}_1$ will sometimes be larger than $\overline{X}_2$ and other times smaller than $\overline{X}_2$. Although the sizes of the positive and negative differences will fluctuate randomly, on the average the positive and negative values will cancel one another, resulting in a distribution mean of 0.

---

[4] Some of the mean differences will exactly equal 0.

Distribution variation As established earlier, the range of variation is greater for $\overline{X}_1 - \overline{X}_2$ than for $\overline{X} - \mu$. Since the sampling distribution of the mean difference is made up of $(\overline{X}_1 - \overline{X}_2)$ values, we would expect its values to be more variable than those of the sampling distribution of $\overline{X}$. And that is the case. Figure 11.3 illustrates the sampling distribution of the mean when $N = 9$ and $\sigma = 15$. Compare that illustration with Figure 11.4, which shows the sampling distribution of the mean difference, where the size of the first sample is $N_1 = 9$, the size of the second sample is $N_2 = 9$, and $\sigma = 15$. The greater variability associated with the second kind of sampling distribution is also reflected in the formula used to compute its standard error.

Standard errors As shown in Figure 11.3, the standard error of the sampling distribution of $\overline{X}$ is computed with the now familiar formula (8.3): $\sigma_{\overline{X}} = \sqrt{\sigma^2/N}$. For the present hypothetical example (in which $\sigma = 15$ and $N = 9$), $\sigma_{\overline{X}} = \sqrt{\sigma^2/N} = \sqrt{(15)^2/9} = \sqrt{225/9} = \sqrt{25} = 5$.

Reflecting the relatively greater variation in the sampling distribution of $\overline{X}_1 - \overline{X}_2$, the **standard error of the mean difference** is calculated via:

$$\sigma_{\overline{X}_1-\overline{X}_2} = \sqrt{\frac{\sigma^2}{N_1} + \frac{\sigma^2}{N_2}}$$

(11.1)

or, equivalently,

$$\sigma_{\overline{X}_1-\overline{X}_2} = \sqrt{\sigma^2\left(\frac{1}{N_1} + \frac{1}{N_2}\right)}$$

(11.2)

Figure 11.3 *Sampling Distribution of the Mean and Standard Error of the Mean When N = 9 and σ = 15*

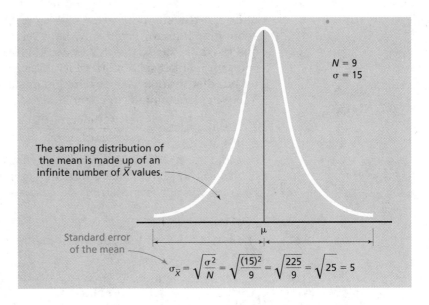

The sampling distribution of the mean is made up of an infinite number of X values.

$N = 9$
$\sigma = 15$

Standard error of the mean

$$\sigma_{\overline{X}} = \sqrt{\frac{\sigma^2}{N}} = \sqrt{\frac{(15)^2}{9}} = \sqrt{\frac{225}{9}} = \sqrt{25} = 5$$

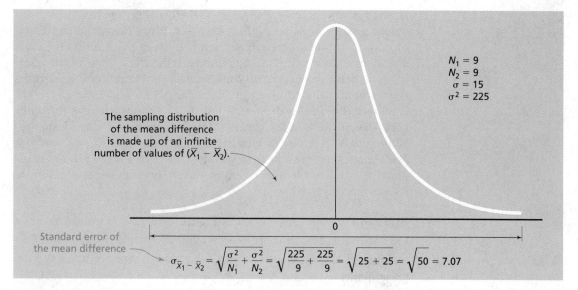

The sampling distribution of the mean difference is made up of an infinite number of values of $(\bar{X}_1 - \bar{X}_2)$.

$N_1 = 9$
$N_2 = 9$
$\sigma = 15$
$\sigma^2 = 225$

0

Standard error of the mean difference

$$\sigma_{\bar{X}_1 - \bar{X}_2} = \sqrt{\frac{\sigma^2}{N_1} + \frac{\sigma^2}{N_2}} = \sqrt{\frac{225}{9} + \frac{225}{9}} = \sqrt{25 + 25} = \sqrt{50} = 7.07$$

Figure 11.4 *Sampling Distribution of the Mean Difference When Both N's = 9 and σ = 15 (Note: The two-independent-samples sampling distribution is more variable than the one-sample sampling distribution.)*

Just by visually comparing formulas (11.2) and (8.3), you would expect the standard error of the mean difference to be larger than the standard error of the mean, when both are based on the same population variance and the same sample size. That expectation is correct, as shown by:

Class
$$\sigma_{\bar{X}_1 - \bar{X}_2} = \sqrt{\sigma^2\left(\frac{1}{N_1} + \frac{1}{N_2}\right)} = \sqrt{(15)^2\left(\frac{1}{9} + \frac{1}{9}\right)} = \sqrt{225\left(\frac{2}{9}\right)} = \sqrt{\frac{450}{9}}$$
$$= \sqrt{50} = 7.07$$

Although the standard error of $\bar{X}_1 - \bar{X}_2$ is larger than the corresponding standard error of $\bar{X}$, both "error terms" are interpreted and used in exactly the same way. Each is the standard deviation of its sampling distribution, which indexes the amount of variation in that distribution. Also, each kind of standard error is supposed to be the denominator in a test statistic. Recall that

$$\text{Test statistic} = \frac{\text{mean difference}}{\text{standard error}}$$

But like $\sigma_{\bar{X}}$, $\sigma_{\bar{X}_1 - \bar{X}_2}$ usually cannot be computed, because in actual behavioral research the population variance is rarely known. Fortunately, however, there is an easy way to estimate $\sigma_{\bar{X}_1 - \bar{X}_2}$.

Estimating the Standard Error of the Mean Difference

The formula for estimating the standard error of the mean difference is:

$$\text{Estimated standard error} = s_{\bar{X}_1 - \bar{X}_2} = \sqrt{s_p^2\left(\frac{1}{N_1} + \frac{1}{N_2}\right)} \qquad (11.3)$$

The only new expression in this formula is s_p^2, which is the **pooled estimator of the population variance**. When you are working with two independent samples, you can compute a separate unbiased estimate of σ^2 from each sample; that is, using formula (10.1), you can calculate s_1^2 and s_2^2. The best overall estimate of the population variance, however, comes from a *weighted average* of the two independent variance estimates, through this formula:

Pooled estimate of the population variance =

$$
s_p^2 = \frac{(N_1 - 1)s_1^2 + (N_2 - 1)s_2^2}{(N_1 - 1) + (N_2 - 1)} \leftarrow ⑤ \tag{11.4}
$$

with markers ① ③ ② pointing to the numerator terms and ④ pointing to the denominator.

FORMULA GUIDE

① Multiply the degrees of freedom for sample 1 by the unbiased variance estimate for sample 1.
② Multiply the degrees of freedom for sample 2 by the unbiased variance estimate for sample 2.
③ Add the result of step 2 to the result of step 1.
④ Add the degrees of freedom for sample 1 to the degrees of freedom for sample 2.
⑤ Divide the result of step 4 into the result of step 3.

CONCEPT RECAP

From Chapter 10, you'll remember that the **unbiased estimator of the population variance**, symbolized by a lowercase s^2, is calculated with $SS/(N-1)$, where SS is the sum of squares for a sample of observations, and $N-1$ is the degrees of freedom in the sample.

Also from Chapter 10, you'll remember that **degrees of freedom** *equals the number of independent observations in a sample minus the number of parameters estimated in the formula used to calculate a statistic.* Since the formula used to compute each sample's unbiased variance estimate utilizes the sample mean to estimate μ, each sample has $N-1$ degrees of freedom.

Once you have computed the pooled variance estimate, calculation of the estimated standard error of the mean difference is a simple, mechanical exercise with formula (11.3). However, since you are using an *estimated standard error* rather than the actual standard error of the sampling distribution, it is necessary to use a t test—instead of the somewhat simpler z test—to assess the significance of the sample data. (See Chapter 10 for a review of the rationale behind the latter statement.)

Computing the Independent-Samples t Statistic

By now you are seeing a familiar pattern in all t ratios:

$$t = \frac{\text{difference between two means}}{\text{estimated standard error}}$$

In the one-sample t test, the specific incarnation of this model is given by formula (10.4):

$$t = \frac{\overline{X} - \mu}{s_{\overline{X}}}$$

Similarly, the two-independent-samples t ratio is obtained from

$$t = \frac{\overline{X}_1 - \overline{X}_2}{s_{\overline{X}_1 - \overline{X}_2}} \tag{11.5}$$

or

$$t = \frac{\overline{X}_1 - \overline{X}_2}{\sqrt{s_p^2\left(\frac{1}{N_1} + \frac{1}{N_2}\right)}} \tag{11.6}$$

Of course, to test the significance of a computed t, we must compare it with a critical t. And the value of t_{crit} depends on the alpha level used and the degrees of freedom in the data.

Degrees of Freedom for Two Independent Samples

Since, in an independent-samples study, each of the two samples has $N - 1$ degrees of freedom, the *overall degrees of freedom* for the independent-samples t test is a "pooled," or "combined," value:

$$\text{Degrees of freedom} = df = (N_1 - 1) + (N_2 - 1) \tag{11.7}$$

You might recognize the degrees of freedom expression as the denominator in formula (11.4), which calculates the pooled estimate of the population variance.

Now you have the theory behind the independent-samples t test. Let's apply your new learning to the research problem that opened this chapter.

DOES RAPPORT AFFECT PSEUDOMEMORIES?

In carrying out the test of significance on the results reported by Sheehan and colleagues (1992), I will follow the eight procedural steps in hypothesis testing that were introduced in Chapter 9.

Step 1: State the Problem and Nature of the Data

Sheehan and colleagues hypothesized that reducing the rapport between hypnotists and their subjects would diminish the hypnotists' ability to implant false or distorted memories during the hypnotic state. Twenty-two hypnosis-susceptible subjects were randomly assigned to each of the following conditions: rapport present and rapport reduced. Each hypnotist attempted to create three pseudomemories during hypnosis. Upon awakening, each subject was given a test to assess the incidence of pseudomemories. Each subject had a pseudomemory raw score that was 0, 1, 2, or 3. The pseudomemory means and unbiased variance estimates were computed for each sample.

Step 2: State the Statistical Hypotheses

The null hypothesis was that the rapport-present and rapport-reduced subjects come from populations with equal means—in other words, that the rapport variable would have no effect on the incidence of pseudomemories, or $H_0: \mu_1 = \mu_2$. Another way to express this is $H_0: \mu_1 - \mu_2 = 0$.

The alternative hypothesis was that the rapport-present and rapport-reduced subjects represent populations with different means—in other words, that the rapport variable would influence the incidence of pseudomemories, or $H_1: \mu_1 \neq \mu_2$, or $H_1: \mu_1 - \mu_2 \neq 0$. From the way these hypotheses are set up, we can infer that a nondirectional test was used.

Step 3: Choose a Level of Significance

The researchers used the conventional .05 level of significance. Hence, they were willing to assume a 5% risk of a Type I error.

> **CONCEPT RECAP**
> A Type I error occurs when you reject a null hypothesis that is true.

Step 4: Specify the Test Statistic

This research method is a completely randomized two-group design. Therefore, an independent-samples *t* test is used. Degrees of freedom equals $(N_1 - 1) + (N_2 - 1)$. Since there were 22 observations in each sample, $df = (22 - 1) + (22 - 1) = 21 + 21 = 42$.

Step 5: Determine the Critical Value Needed for Significance

The value of t_{crit} can be found in Table T of Appendix A. Since Table T does not list $df = 42$, we drop back to $df = 40$, which is listed. Going across to alpha = .05 for a two-tailed test, we find the critical value of 2.021.

−1.3

−5.98

Step 6: State the Decision Rule

If $|t| \geq 2.021$, reject H_0. If the computed t ratio is at least as "extreme" as -2.021 or $+2.021$, we will declare the sample means to be significantly different from each other.

Step 7: Compute the Test Statistic

Computing the independent-samples t test is a three-step process. We need to use the means, variance estimates, and sample sizes shown in Table 11.1.

Find the pooled estimate of the population variance Using formula (11.4), we get

$$s_p^2 = \frac{(N_1 - 1)s_1^2 + (N_2 - 1)s_2^2}{(N_1 - 1) + (N_2 - 1)} = \frac{(22 - 1)1.37 + (22 - 1)1.08}{(22 - 1) + (22 - 1)}$$

$$= \frac{28.77 + 22.68}{42} = \frac{51.45}{42} = 1.225$$

Find the estimated standard error of the mean difference Using formula (11.3), we find

$$s_{\overline{X}_1 - \overline{X}_2} = \sqrt{s_p^2\left(\frac{1}{N_1} + \frac{1}{N_2}\right)} = \sqrt{1.225\left(\frac{1}{22} + \frac{1}{22}\right)} = \sqrt{1.225\left(\frac{2}{22}\right)}$$

$$= \sqrt{\frac{2.45}{22}} = \sqrt{0.1114} = 0.3338$$

Compute the t ratio Formula (11.5) yields the computed t statistic:

$$t = \frac{\overline{X}_1 - \overline{X}_2}{s_{\overline{X}_1 - \overline{X}_2}} = \frac{1.68 - 0.86}{0.3338} = \frac{0.82}{0.3338} = 2.46$$

Step 8: Consult the Decision Rule and Make a Decision

Since the computed t value of 2.46 is more extreme than $t_{crit} = 2.021$, the null hypothesis is rejected. "Specifically, results showed that strength of pseudomemory was significantly related to rapport manipulation" (Sheehan et al., 1992, p. 694).

Table 11.1

MEANS AND UNBIASED VARIANCE ESTIMATES ON A PSEUDOMEMORY TEST FOR RAPPORT-PRESENT AND RAPPORT-REDUCED SAMPLES (SHEEHAN ET AL., 1992)	RAPPORT	MEAN	VARIANCE ESTIMATE
	Present*	1.68	1.37
	Reduced†	0.86	1.08
	*$N_1 = 22$ †$N_2 = 22$		

ASSUMPTIONS OF THE INDEPENDENT-SAMPLES *t* TEST

The independent-samples *t* test rests on four assumptions:

1. The dependent variable's scale of measurement is at the *interval or ratio level.* (Interval scales have equal units of measurement all along the measurement dimension. Ratio scales are interval scales with true zero points. See Chapter 4 for a review of these ideas.)
2. The *observations in the sample must be independent* of one another; this means that no subject's score on the dependent variable should influence any other subject's score.
3. The dependent variable is *normally distributed* in the populations represented by the samples.
4. The *variances* of the two populations are *equal.* This is called the **homogeneity of variance assumption**.

Assumption 1 is important; if you cannot assume that your measure of behavior at least approximates an interval scale, then the *t* test might not give you valid information. Assumption 2 is critical, but it is usually satisfied in an independent-groups study by randomly assigning subjects to conditions and using good research methodology. Assumption 3 does not have much effect on the accuracy of a *t* test (Hays, 1988). Furthermore, it can be presumed to be true whenever there are at least 30 observations in each sample. Assumption 4 may be violated to some degree as long as the two samples have equal *N* values (Hays, 1988). What's more, it is reasonable to assume that the population variances are homogeneous if the larger of the variance estimates (s_1^2 and s_2^2) is less than twice the size of the smaller one.

OTHER KINDS OF INDEPENDENT-SAMPLES RESEARCH

The completely randomized two-group research design is only one kind of data collection method that calls for an independent-samples *t* test. The same test is appropriate for testing the significance of a mean difference in a **static-group comparison design**, in which subjects are not randomly assigned to treatment conditions. Rather, conditions are defined by preexisting characteristics of the subjects: *An intact group of subjects who have a particular trait is compared on some dependent measure to a second group of subjects who do not possess the defining trait.*

For example, a researcher might hypothesize that obese college students have greater taste sensitivity than normal-weight students. The researcher might then obtain a measure of taste sensitivity from each of 50 obese and 50 normal-weight subjects. It would be appropriate to use an independent-samples *t* test to evaluate the statistical reliability (i.e., significance) of the mean difference in taste sensitivity between these two independent groups. Because subjects were not randomly assigned to conditions, however, a significant difference between the groups could not be given a cause-and-effect interpretation. All this type of study provides is evidence of an association between the level of a trait and scores on some other variable. It is possible, for instance, that some physical or psychological cause of obesity also causes overweight persons to be more taste sensitive. On the other hand, perhaps overweight individuals are simply more likely to be dieting and hence more reactive to any stimulus variation connected with food. In this type of investigation, a significant t ratio tells us only that there is a reliable difference between the sample means. Causal interpretations must await more incisive research methods.

CORRELATED-SAMPLES t TESTS

In addition to the independent-samples test, a second type of t test is widely used in behavioral research. The **correlated-samples t test**[5] is used when the data in the "experimental" condition are literally correlated with the data in the "control" condition. One variety of research method that yields correlated columns of data is called the **repeated-measures design**. In repeated-measures studies, *each subject is measured in more than one condition, and in a real sense serves as both an "experimental" subject and a "control" subject.* Since each person's performance in the control condition is likely to be similar to his or her performance in the experimental condition, the raw scores in one sample are statistically correlated with those in the other. In fact, when you analyze repeated-measures data, it is appropriate and necessary to list the data in pairs of scores, just as you would to calculate a linear correlation coefficient (see Chapter 6). Such a systematic ordering of data from one sample to the other necessitates the use of a special kind of two-sample t ratio to assess the significance of the difference between the sample means. An example will clarify these ideas.

> **CONCEPT RECAP**
> Data are arranged in **pairs** when (1) each subject has two scores in a study, (2) the scores are arranged in two columns, and (3) *each subject's score in the first column is lined up, horizontally, with his or her score in the second column.*

---

[5] Other terms for this test are *dependent-samples, correlated-groups,* and *related-samples.*

Devoe (1990) wanted to find out whether a "pro-life" tape, titled "The Silent Scream," would significantly affect attitudes toward the abortion issue. He conducted a repeated-measures study in which each of $N = 20$ college students was (1) pretested on an abortion-attitude scale, (2) shown "The Silent Scream" tape, and then (3) posttested on an equivalent abortion-attitude scale.[6] True to the form of a repeated-measures design, *each subject was measured twice and, thereby, provided data for each of two samples.* The control-group data were generated by the pretest (absence of exposure to the tape being the control treatment), and the experimental-group data came from the posttest (exposure to the tape being the experimental treatment). Notice that each subject served as his or her own control. We would expect subjects' posttest attitude scores to be correlated with their respective pretest scores. Nonetheless, if the tape had an effect, we would also expect the mean of the posttest attitude scores to be significantly higher or lower than the pretest mean.

The raw scores from Devoe's abortion-attitude study are exhibited in Table 11.2, together with the means and unbiased variance estimates for each group. *Important:* Note that, in addition to the pretest raw scores and posttest raw scores, there is a third column of **difference scores**. Each difference score—symbolized *D*—is *the result of subtracting a subject's posttest score from his or her pretest score.* Thus, $D = \overline{X}_1 - \overline{X}_2$. There are three things to remember about the *D* scores:

1. The *D* scores—*not the raw scores*—are the actual data to be analyzed in a correlated-samples *t* test.
2. As shown in Table 11.2, the mean *D* score, symbolized $\overline{D}$, is exactly equal to the difference between the two raw-score means: $\overline{D} = \overline{X}_1 - \overline{X}_2 = 56.25 - 63.85 = -7.6$.
3. The variance estimate computed from the *D* scores is considerably smaller than the variance estimates calculated from the respective columns of raw scores. Keep this point in mind, and I will establish its importance soon.

There is a pleasant surprise in all of this: *A correlated-samples t test is conducted in exactly the same fashion, and according to exactly the same logic, as a one-sample t test*—something you already know about. Only the symbols change; everything else is identical, right down to determining the degrees of freedom. Let's see why and how.

---

[6] This example pertains to only a portion of Devoe's investigation, which featured a relatively complex experimental design overall. You are encouraged to read the original research for complete details.

Table 11.2

		PRETEST SCORES, X_1	POSTTEST SCORES, X_2	DIFFERENCE SCORES, $X_1 - X_2$,
PRETEST RAW SCORES, POSTTEST RAW SCORES, AND DIFFERENCE SCORES FOR 20 SUBJECTS IN A STUDY OF ABORTION ATTITUDES (DEVOE, 1990)	SUBJECT			D
	1	74	85	−11
	2	68	84	−16
	3	41	42	−1
	4	54	56	−2
	5	50	66	−16
	6	90	99	−9
	7	79	103	−24
	8	98	98	0
	9	37	34	3
	10	44	37	7
	11	44	45	−1
	12	24	35	−11
	13	53	58	−5
	14	38	46	−8
	15	41	63	−22
	16	52	61	−9
	17	86	86	0
	18	48	39	9
	19	48	73	−25
	20	56	67	−11
	Sums:	1125	1277	−152
	Means:	56.25	63.85	−7.6
	Unbiased variance estimates:	392.408	510.239	94.779

Parallels Between the One-Sample and Correlated-Samples Tests

The sampling distribution of $\overline{D}$ is like the sampling distribution of $\overline{X}$ *In a one-sample study,* the data are a single column of raw scores that yield a sample average, $\overline{X}$. That sample mean is evaluated by comparing it to the average value of the sampling distribution of $\overline{X}$. That sampling distribution has the following characteristics:

- It is composed of $\overline{X}$ values, with each sample mean based on a sample of a fixed size, N.
- It is a normal distribution in every respect, so that the exact probability of a given range of sample means can be determined from the standard normal distribution.
- It has a mean of μ, which is precisely equal to the mean of the population under H_0.

- It has a standard error of $\sigma_{\bar{X}} = \sqrt{\sigma^2/N}$, where σ^2 is the variance of the population of raw scores from which the single sample of raw scores was selected.

Now we'll compare the features of the two-correlated-samples type of investigation (which includes repeated-measures studies) with the traits of the one-sample study discussed above.

In a correlated-samples study, the data are a single column of difference (*D*) scores that yield a sample average, $\bar{D}$, the **sample mean of the difference scores**. That sample mean is evaluated by comparing it with the average value of the sampling distribution of $\bar{D}$. That sampling distribution has the following characteristics:

- It is composed of $\bar{D}$ values, with each sample mean based on a sample of a fixed size, *N*.
- It is a normal distribution in every respect, so that the exact probability of a given range of $\bar{D}$ values can be determined from the standard normal distribution.
- It has a mean of μ_D, which is precisely equal to the mean of the population under H_0. What's more, *the null hypothesis holds that the average difference score equals 0*; that is, H_0: $\mu_D = 0$.
- It has a standard error of $\sigma_{\bar{D}} = \sqrt{\sigma_D^2/N}$, where σ_D^2 is the variance of the population of difference scores from which the sample *D* values were drawn.

Figure 11.5 summarizes the characteristics of the sampling distribution of $\bar{D}$. Since you already know how to work with the sampling distribution of $\bar{X}$, you also know how to use the sampling distribution of $\bar{D}$ to carry out tests of statistical significance. It's a simple matter of substituting symbols.

Figure 11.5 *Sampling Distribution of $\bar{D}$ (Note: Under H_0, the average $\bar{D}$ equals 0.)*

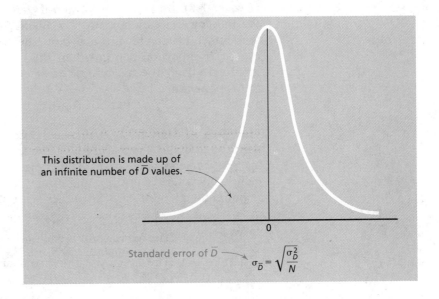

This distribution is made up of an infinite number of $\bar{D}$ values.

0

Standard error of $\bar{D}$ → $\sigma_{\bar{D}} = \sqrt{\dfrac{\sigma_D^2}{N}}$

The Estimated Standard Error of $\overline{D}$ Is Like $s_{\overline{X}}$ Since the variance of the population of difference scores usually is unknown, the standard error of the sampling distribution of $\overline{D}$ generally cannot be calculated. Rather it must be estimated through:

$$\text{Estimated standard error of } \overline{D} = s_{\overline{D}} = \sqrt{\frac{s^2}{N}} \qquad (11.8)$$

where s^2 is the unbiased variance estimator of σ_D^2, the variance of the population of difference scores, and N is the number of D scores in the sample—that is, the sample size. The unbiased variance estimate, s^2, is computed from the sample data in the usual way: $s^2 = SS/(N-1)$, as will be demonstrated later.

The t ratio for $\overline{D}$ is like the t ratio for $\overline{X}$ Functionally identical to the t ratio in a one-sample t test, the correlated-samples t statistic is computed through:

$$t = \frac{\overline{D} - \mu_D}{s_{\overline{D}}}$$

But since $\mu_D = 0$ under the null hypothesis, the formula simplifies to

$$t = \frac{\overline{D}}{s_{\overline{D}}} \qquad (11.9)$$

with degrees of freedom equal to $N - 1$.

Armed with this information about correlated-samples t tests, we apply this knowledge to Devoe's repeated-measures experiment.

Testing the Effect of "The Silent Scream"

State the problem and nature of the data As described above, Devoe (1990) conducted a repeated-measures experiment to assess the influence of exposure to "The Silent Scream" tape on college students' attitudes toward abortion. The tape was produced to increase antiabortion sentiment in the viewer. If the tape had its intended effect, the average posttest score (X_2) would be higher than the average pretest score (X_1). Thus, the average difference score ($D = X_1 - X_2$) would be negative.

State the statistical hypotheses Even though Devoe expected the tape to make the average posttest score higher than the average pretest score, the opposite outcome would have been theoretically (and practically) meaningful as well. Therefore, Devoe used nondirectional statistical hypotheses. A two-tailed test was in order:

H_0: $\mu_D = 0$ (The average difference score in the population is 0.)

H_1: $\mu_D \neq 0$ (The average difference score in the population is not 0.)

Choose a level of significance   Devoe used a conventional level of significance, assuming a moderate risk of a Type I error: $\alpha = .05$.

Specify the test statistic   Like Devoe, we will use the correlated-samples t test, with degrees of freedom = $N - 1$. Since there are 20 D scores in the sample, $df = (20 - 1) = 19$.

Determine the critical value needed for significance Table T shows that when $df = 19$ and $\alpha = .05$, $t_{crit} = 2.093$ for a two-tailed test.

State the decision rule   If $|t| \geq 2.093$, reject H_0.

Compute the test statistic   There are four steps involved in computing a correlated samples t ratio. To carry out these computations, we use the data shown in Table 11.3.

Table 11.3

		PRETEST SCORES, X_1	POSTTEST SCORES, X_2	DIFFERENCE SCORES, $X_1 - X_2$, D	SQUARED DIFFERENCE SCORES, D^2
PRETEST RAW SCORES, POSTTEST RAW SCORES, DIFFERENCE SCORES, AND SQUARED DIFFERENCE SCORES FOR 20 SUBJECTS IN A STUDY OF ABORTION ATTITUDES (DEVOE, 1990)	SUBJECT				
	1	74	85	−11	121
	2	68	84	−16	256
	3	41	42	−1	1
	4	54	56	−2	4
	5	50	66	−16	256
	6	90	99	−9	81
	7	79	103	−24	576
	8	98	98	0	0
	9	37	34	3	9
	10	44	37	7	49
	11	44	45	−1	1
	12	24	35	−11	121
	13	53	58	−5	25
	14	38	46	−8	64
	15	41	63	−22	484
	16	52	61	−9	81
	17	86	86	0	0
	18	48	39	9	81
	19	48	73	−25	625
	20	56	67	−11	121
	Sums:	1125	1277	−152	2956
	Means:	56.25	63.85	−7.6	
	Unbiased variance estimates:	392.408	510.239	94.779	

First, we must compute the sum of squares of the difference scores using a version of formula (4.8):

$$SS = \Sigma D^2 - \frac{(\Sigma D)^2}{N} = 2956 - \frac{(-152)^2}{20} = 2956 - \frac{23,104}{20}$$

$$= 2956 - 1155.2 = 1800.8$$

Second, we compute the unbiased estimate of the population variance σ_D^2:

$$s^2 = \frac{SS}{N-1} = \frac{1800.8}{20-1} = \frac{1800.8}{19} = 94.78$$

Third, we find the estimated standard error of $\overline{D}$ from formula (11.8):

$$s_{\overline{D}} = \sqrt{\frac{s^2}{N}} = \sqrt{\frac{94.78}{20}} = \sqrt{4.74} = 2.18$$

 Finally, we calculate t via formula (11.9).

$$t = \frac{\overline{D}}{s_{\overline{D}}} = \frac{-7.6}{2.18} = -3.49$$

Consult the decision rule and make a decision Since the absolute value of the computed t (i.e., $|t| = 3.49$) exceeds the $t_{crit} = 2.093$ required by the decision rule, we reject H_0. "Viewing 'The Silent Scream' brought about a significant increase in antiabortion sentiment in immediate posttest, $t(19) = -3.49$, $p < .05$." Again, note the appropriate way to express the outcome of the significance test.

A SECOND CORRELATED-SAMPLES RESEARCH DESIGN

The two-condition repeated-measures design is one of two kinds of research method that produce data that are best analyzed with a correlated-samples t test. That type of t test also is appropriate for testing hypotheses with a **randomized matched-groups design**. In the latter method, *pairs of subjects are formed by matching two persons at a time on some variable that is strongly correlated with the dependent measure in a study; then one member of each matched pair is randomly assigned to the experimental condition, while the other goes into the control group.*

For example, you may be interested in comparing two different methods of teaching concepts to preschool children. Since the rate of concept learning is strongly correlated with general intelligence, you could form matched pairs of children on the basis of their IQ scores. Then you would randomly assign one member of each matched pair to the first method of teaching concepts, while assigning the other member to the second method. This would result in two samples that are literally correlated on the matching variable and two columns of dependent-measure raw scores that are similarly correlated. Thus, it is proper to analyze the difference scores using a correlated-samples *t* test.

The main purpose of the randomized matched-groups design is to ensure nearly identical treatment groups in an experiment. But, like the repeated-measures design, it has a very important additional benefit: It enables you to use a *t* test that is usually more "powerful" than the test used in an independent-samples research design.

CORRELATED-SAMPLES TESTS AND STATISTICAL POWER

Chapter 9 defined the power of a statistical test as the probability of rejecting a false null hypothesis. It is usually desirable to make the power of your hypothesis-testing procedures as high as you can. Chapter 9 also presented four ways to enhance power:

1. Increase alpha, the level of significance (not recommended because this strategy also increases the probability of a Type I error—rejection of a true H_0).
2. Use a larger sample size—that is, increase N (almost always strongly recommended).
3. Decrease error variability in your data.
4. Increase the effect size by using a more powerful method of measuring or manipulating your independent variable.

The third approach to power enhancement is not implemented very frequently because it depends on your ability to use a particular class of research design that is not practical for many research problems. Fortunately, however, correlated-samples research designs are among those methods that are capable of improving power by reducing error variance. Let's see how.

The General Concept Behind Test Statistics

The following model applies to the various "ratio-type" test statistics in two-sample studies:

$$\text{Test statistic} = \frac{\text{difference between sample means}}{\text{standard error}}$$

For purposes of both understanding the information I am about to dispense and preparing your mind for what is to come in the next chapter, I would like you to think of the standard error in t ratios as the **error term** in tests of significance. It represents *an index of random error, which is always the denominator in a test statistic.* The general logic behind the test statistic ratio is that:

- A numerator representing between-sample mean differences is compared with a denominator representing random error.
- If the sample means differ by an amount that greatly exceeds the index of the random error, then the difference between sample means probably represents something more than random variation.
- Therefore, the sample means should be declared reliably different; the results are statistically significant.

Since each sample mean represents the typical score for that sample, the difference between the sample means is referred to as "between-samples variation." Now, recall that the *estimated* standard error used in a t ratio is based on the unbiased variance estimator (s^2). Since s^2 is calculated from the squared deviations of raw scores *within* a sample $\left[\text{i.e., } s^2 = \sqrt{\Sigma(X - \overline{X})^2/(N-1)}\right]$, the standard error in a t test is called "within-samples variation." In light of this new terminology, then, a test statistic can be thought of as:

$$\text{Test statistic} = \frac{\text{difference between means}}{\text{standard error}}$$

$$= \frac{\text{between-samples variation}}{\text{within-samples variation}}$$

So the "error term" in a t ratio represents random variation within samples. In turn, this within-samples variation is made up of two components:

1. "Random sampling error," represented by naturally occurring *individual differences between subjects* within a sample
2. "Experimental error," caused by confusing instructions, unreliable measurement, distractions, inattention, equipment malfunction, and so on.

What would happen to the t ratio if you could use a research design that eliminated one of these two sources of random error from the denominator of the t ratio? Right! The denominator would be smaller, while the numerator remains the same. Consequently, *the t ratio would become larger and, therefore, more likely to exceed the t_{crit} value*—resulting in rejection of the null hypothesis. As you will see next, this is precisely the advantage of using a correlated-samples t test rather than an independent-samples t test.

Enhancing Power by Reducing Error Variance

Since reducing the denominator of the *t* ratio makes rejection of H_0 more likely, such a change effectively increases the power of the statistical test. A correlated-samples *t* test diminishes the *t*-ratio denominator, or the "within-samples variation" factor, by allowing you to work with a column of difference (*D*) scores rather than two columns of raw scores. As you compute each *D* by subtracting X_2 from X_1, *you remove the part of random variation caused by naturally occurring individual differences between subjects within a sample.* Since this procedure does not affect the difference between the sample means at all, the ultimate impact of working with *D* scores instead of raw scores is to make the *t* ratio larger. And, to repeat, statistical power is enhanced because a larger computed *t* has a better chance of discrediting the null hypothesis.

To illustrate the variance-reduction effect of computing $D = (X_1 - X_2)$, I refer you back to Table 11.2, which shows the unbiased variance estimates for pretest scores, posttest scores, and *D* scores in Devoe's study of abortion attitudes. As you can see in the table:

Variance estimate based on pretest raw scores: $s_1^2 = 392.41$

Variance estimate based on posttest raw scores: $s_2^2 = 510.24$

Variance estimate based on the *D* scores: $s^2 = 94.78$

In this example, any "error term" computed from the variance estimate of either the pretest or the posttest raw scores—or their combination—will be considerably larger than an error term based on the *D* scores. This statement will be true, in general, to the extent that the two samples of raw scores are correlated.

The key is in the research design   The increase in power that often is associated with the correlated-samples *t* test is made possible by certain kinds of research designs that involve legitimate pairs of observations—repeated-measures and matched-groups designs, for example. And the correlated-samples *t* test should be used only in connection with such correlated-samples designs. It is not appropriate to use this kind of *t* test with independent-samples designs, such as a completely randomized design, even if the columns of data happen to be correlated just by chance. Nor is it permissible to force a correlation between independent groups of data (by rearranging the observations) just so that you can apply the correlated-groups *t* test. Doing this will give you erroneous results.

And the plan doesn't always work   Although repeated-measures and matched-groups research designs most often yield correlated samples, they don't always. When the research design fails to produce substantially correlated columns of data, the correlated-samples t test is actually less powerful than the independent-samples t test. The reason is that the correlated-samples t test has 50% fewer degrees of freedom $(N-1)$ than a corresponding independent-samples t test $[(N_1 - 1) + (N_2 - 1)]$ and, hence, has a larger t_{crit} value (see Table T in Appendix A). This means that a larger computed t value is required for significance in the correlated-samples procedure—a feature that tends to reduce the likelihood of rejecting H_0. So *the possible power advantage of this test is realized only if error reduction is big enough to more than offset the effect of working with 50% fewer degrees of freedom.*

ASSUMPTIONS OF THE CORRELATED-SAMPLES t TEST

The correlated-samples t test has four assumptions:

1. The two columns of raw scores are significantly correlated with each other.
2. The D scores are on an interval or ratio scale of measurement.
3. The D scores in the sample must be independent of one another; this means that no subject's D score should influence any other subject's D score.
4. The D scores are *normally distributed* in the population represented by the sample.

All of the qualifying statements made in reference to the assumptions of an independent-samples test also apply to the assumptions of the correlated-samples test.

KEY TERMS

completely randomized two-group design

sampling distribution of the mean difference

standard error of the mean difference

pooled estimator of the population variance

homogeneity of variance assumption

static-group comparison design

correlated-samples t test

repeated-measures design

difference scores

sample mean of the difference scores

randomized matched-groups design

error term

SUMMARY

1. This chapter focuses on the two-sample *t* test, in which the mean of an experimental group is tested against the mean of a control group. Two types of research design call for a two-sample *t* test: independent-samples designs, in which the data in the experimental condition are completely independent of the data in the control condition, and correlated-samples designs, in which the data in the experimental condition are literally correlated with the data in the control condition.

2. One research method that requires a two-sample *t* test is called the completely randomized two-group design, in which each member of a pool of research subjects is randomly assigned to one of two treatment conditions.

3. An independent-samples *t* test differs from a one-sample *t* test in several ways: The null and alternative hypotheses address the values of two population means, not just one. In the test itself, two sample means are compared with each other, rather than a sample mean being compared with a hypothesized population mean. The relevant sampling distribution is the sampling distribution of the mean difference (with an average value of 0), not the sampling distribution of the mean (with an average value of μ). The sampling distribution of the mean difference is more variable than the sampling distribution of the mean. The estimated standard error of the mean difference is found through $s_{\overline{X}_1 - \overline{X}_2} = \sqrt{s_p^2(1/N_1 + 1/N_2)}$, where s_p^2 is the pooled estimator of the population variance. The independent-samples *t* ratio is calculated via $t = (\overline{X}_1 - \overline{X}_2)/s_{\overline{X}_1 - \overline{X}_2}$, with degrees of freedom equal to $(N_1 - 1) + (N_2 - 1)$.

4. The independent-samples *t* test assumes an interval or ratio scale of measurement, independence of observations, normality of the population, and equality of population variances.

5. The correlated-samples *t* test is used when the data in the experimental condition are literally correlated with the data in the control condition. Such a *t* test is necessary in a repeated-measures experiment, in which each subject is measured in more than one condition. It is also necessary in studies that use matched groups of subjects.

6. The correlated-samples *t* test is carried out in exactly the same fashion as a one-sample *t* test, except that the data analyzed are difference (*D*) scores rather than raw scores. The correlated-samples *t* test has $N - 1$ degrees of freedom, where *N* is the number of difference scores.

7. A correlated-samples *t* test is often more powerful than a corresponding independent-samples test because the correlated-samples test usually has a smaller "error term" in the denominator of the *t* ratio. The result of this smaller denominator is a

larger test statistic with an improved chance of rejecting the null hypothesis. The error term is reduced through the use of D scores in the analysis. The computation of D scores removes a large amount of individual difference variation from the unbiased estimate of the population variance and, hence, from the estimated standard error.

REVIEW QUESTIONS

1. Define or describe: completely randomized two-group design, sampling distribution of the mean difference, standard error of the mean difference, pooled estimator of the population variance, and static-group comparison design.

2. Explain the role of random assignment of subjects to treatment conditions in (a) enabling researchers to draw cause-and-effect conclusions within the context of two-sample experiments and (b) making the experimental and control conditions "independent" samples.

[handwritten margin note: a) random ass - assume 2 groups same prior to ind varie. So after add ind. ver see diff or figure thats what condl ↳]
[handwritten note: corr =0 are ind. if cor. is 0 thats how you know E&C cond. are ind.]

3. A physiological psychologist randomly assigns three-week-old rats to two conditions: intense daily handling by lab assistants versus once weekly handling. At the end of 12 months of this differential treatment, the rats in both samples are subjected to the same stressful conditions for a six-week period. Finally, a sample of each rat's blood is assessed for "level of stress chemicals," the dependent variable. The researcher feels that early differences in handling will create different levels of stress tolerance in the two groups of rats. State the null and alternative hypotheses for this investigation, *both verbally and symbolically.*

4. In the investigation described in Question 3, what sort of comparison of means will be involved in the test of the null hypothesis? How will this comparison differ from the type conducted in one-sample studies? Be specific and complete in your response.

5. In your own words, describe the contents and average value of the sampling distribution of $\overline{X}_1 - \overline{X}_2$. How do these items differ from corresponding items in the sampling distribution of $\overline{X}$? Also, present the conceptual rationale underlying the expected average value of the sampling distribution of $\overline{X}_1 - \overline{X}_2$.

6. The textbook asserts that the sampling distribution of the mean difference is more variable than the sampling distribution of the mean. State in your own words the conceptual logic behind this expected difference.

7. Make up an original example of a kind of behavioral investigation in which an independent-samples t test is appropriate. This must be your own idea. Make your example specific and complete.

8. Make up an original example of a kind of behavioral investigation in which a correlated-samples t test is appropriate. Make your example specific and complete.

9. Assume that $\sigma^2 = 490$ and that the size of a single sample randomly drawn from the population is $N = 10$. Calculate the standard error of the mean. Now assume that a second random sample is drawn, so that $N_1 = 10$ and $N_2 = 10$. Compute the standard error of the mean difference for the two-sample situation. Is the difference between $\sigma_{\bar{X}}$ and $\sigma_{\bar{X}_1 - \bar{X}_2}$ consistent with expectations created by the text's comparison of the two standard errors? Explain your response.

10. Assume that $\sigma^2 = 900$ and that the size of a single sample randomly drawn from the population is $N = 25$. Calculate the standard error of the mean. Now assume that a second random sample is drawn, so that $N_1 = 25$ and $N_2 = 25$. Compute the standard error of the mean difference for the two-sample situation. Is the difference between $\sigma_{\bar{X}}$ and $\sigma_{\bar{X}_1 - \bar{X}_2}$ consistent with expectations created by the text's comparison of the two standard errors? Explain your response.

11. Rapee and Lim (1992) asked 28 persons with social phobias and 33 nonclinical subjects to rate themselves on a public speaking performance that they gave. Observers of both groups rated them equivalently on public speaking competence. Nonetheless, the sample of phobics gave themselves a mean rating of 12.5, whereas the nonclinical sample had a mean self-rating of 9.4, with higher ratings indicating *worse* performance. The unbiased variance estimates were $s_1^2 = 9.61$ for the phobics and $s_2^2 = 10.24$ for the nonclinical subjects. State the null and alternative hypotheses for this study, both verbally and symbolically. Also, using the .05 level of significance, locate t_{crit} for a two-tailed test. Finally, state the decision rule for the significance test.

12. For the study described in Question 11, calculate the pooled estimate of the population variance and the estimated standard error of the mean.

13. For the study described in Question 11, compute the t ratio and draw an appropriate conclusion.

14. In Chapters 1 and 2 of this book, you read about a study of the effect of personal feedback on exam performance in a large college class. To review, on a random basis one-half of the students received encouraging comments on their returned tests, whereas the remaining students received only numerical grades on their exams. The professor expected the personal feedback to motivate students so that they would try harder on subsequent tests and, he hoped, get better test grades. There were 46 students in each of the two samples. The results of this

experiment are given in the table. Use both symbols and words to state nondirectional hypotheses for this study. Set $\alpha = .01$, and state the decision rule for rejecting H_0. How many degrees of freedom are there in the total data set (both samples combined)?

Feedback Condition	Exam 1 Mean	s^2	Exam 2 Mean	s^2	Exam 3 Mean	s^2
Personal feedback	35.39	25.63	32.93	38.53	34.33	22.91
Control group	33.67	40.41	32.46	29.77	31.30	40.00

15. For the results reported in Question 14, use a t test to verify that the two samples were statistically equivalent on exam performance at the start of the course. Use the 5% level of significance.

16. For the problem presented in Question 14, use the eight procedural steps to test the hypothesis that the exam 3 means are significantly different. Set the alpha level equal to .05, and conduct a nondirectional test. Be sure to draw an appropriate conclusion.

17. Define or describe: homogeneity of variance assumption, correlated-samples t test, repeated-measures design, difference scores, sample mean of the difference scores, randomized matched-group design, and error term.

18. List the four assumptions of an independent-samples t test. Which of these assumptions are critical to the validity of the test? What can you do in setting up and conducting your behavioral research to maximize the chances that each of the assumptions will be met?

19. In what fundamental way does the data configuration of correlated-samples research designs differ from the data configuration of independent-samples designs?

20. Describe the parallels between correlated-samples t tests and one-sample t tests, with specific reference to sampling distributions, standard errors, estimated standard error, and the t ratio.

21. A sociologist has reason to believe that U.S. women are generally better drivers than their male partners. The sociologist randomly selects 12 married heterosexual couples and has a state highway officer administer a "road test" to each of the 24 subjects. The scores of the couples are given in the table. Is this study set up more like an independent-samples investigation or a correlated-groups investigation? Justify your answer. State appropriate null and alternative hypotheses, assuming a nondirectional test. With alpha = .05, state a decision rule for rejecting the null hypothesis.

Couple	Female Partner	Male Partner	D	D²
1	70	72	2	4
2	93	90	-3	9
3	84	84	0	0
4	67	57	-10	100
5	98	96	-2	4
6	47	53	6	36
7	59	61	2	4
8	64	60	-4	16
9	70	62	-8	64
10	80	68	-12	144
11	76	84	8	64
12	74	70	-4	16
	882	857		461

22. Conduct an appropriate test of significance on the results of the study described in Question 21.

23. A pediatric nurse wants to find out whether the age at which children begin walking can be influenced by special training. Eight pairs of identical twin girls under the age of four months serve as subjects. One member of each twin pair is randomly assigned to a special training condition and her co-twin goes to the control condition. The ages (in months) at which the children began to walk are shown in the table. Using the eight procedural steps, carry out an appropriate test of these data, and develop an appropriate conclusion in regard to the nurse's question. Use $\alpha = .05$.

Twin Pair	Special Training	Control Condition
1	12	12
2	14	16
3	15	14
4	12	11
5	16	14
6	15	18
7	13	16
8	10	11

24. Treat Devoe's pretest and posttest raw scores in Table 11.2 as though they belong to a completely randomized two-group research design. Reanalyze the data with an independent-samples *t* test, rather than a correlated-samples test. What effect does the independent-samples approach have on the size of the denominator of the *t* ratio? Are the data still statistically significant? What implications does the outcome of this exercise have for the relative power of correlated-samples and independent-samples tests?

25. In your own words, explain why correlated-samples *t* tests have relatively high power.

CHAPTER 12 ANALYSIS OF VARIANCE

In Chapter 11 you learned how to determine whether two sample means are significantly different from each other—that is, how to do two-sample hypothesis testing. But what kind of hypothesis-testing procedure can you use in studies that involve more than two samples? Consider the following research situation.

There are three general theories to explain the psychological processes behind the behaviors of hypnotized people. The "deep relaxation" school of thought holds that the repetitious suggestions made by the hypnotist merely place the subject in a state of pleasant relaxation, much like that experienced when one meditates. The tendency of the hypnotized subject to readily obey the hypnotist's suggestions is merely the happy compliance of a relaxed person who does not wish to disrupt a pleasant situation by disobeying. A major implication of this first theory is that meditating subjects will be just as responsive to suggestions as hypnotized subjects. The second theory, labeled the "enhanced motivation" model, contends that hypnotized persons are simply in a state of high motivation brought about by the exhortations of an authority figure (Barber, 1976). This model predicts that highly motivated subjects will show the same level of suggestibility as hypnotized persons but greater suggestibility than someone who is merely relaxed. The third view is that hypnosis is a unique altered state of consciousness that is qualitatively different from both meditative states and states of high motivation. This "trance theory" predicts that hypnotized subjects will be more responsive to suggestions than either meditating subjects or highly motivated subjects.

A THREE-SAMPLE STUDY

Suppose a researcher tests these three theories by setting up a three-sample experiment. Let's assume that the researcher randomly selects 15 subjects from a population of undergraduate students at a large private university; these persons are the "subject pool" in this study.[1] The researcher randomly assigns 5 of the 15 subjects to each of three experimental conditions:

- In the first sample, or "treatment group," the subjects are given brief training in meditation. Then, with the assistance of the researcher, each subject enters a deep state of meditation. During

---

[1] The sample size used in this example and the next would be entirely too small in real behavioral investigations. Samples are kept small in these textbook examples merely to conserve space and simplify the illustrations of calculation procedures. They are not intended to serve as ideal models of research methodology.

meditation, the researcher issues ten standard suggestions to the subject and notes whether the subject responds positively to each suggestion. *The ten suggestions are considered to be of equal difficulty; therefore, the subjects' scores are on an interval scale.*

- In the second sample ("treatment group"), each subject receives strong and compelling motivational instructions about the importance of being receptive to the suggestions he or she is about to receive. Then the researcher issues the ten standard suggestions and records each subject's compliance.

- In the third treatment group, the researcher actually hypnotizes each subject using a conventional hypnotic induction procedure, and then administers the same suggestibility test that was used in the other two samples.

The type of experiment described here is called a **completely randomized independent-samples design**. You might recall from Chapter 11 that an independent-samples research design is being used whenever *subjects are randomly assigned to different treatments and each subject serves in only one treatment condition.*

Statistical Hypotheses

In a multiple-sample study, the null and alternative hypotheses must be set up differently from those used in a two-sample investigation. In the three-sample experiment described here, the hypotheses are:

Null hypothesis→H_0: $\mu_1 = \mu_2 = \mu_3$

Alternative hypothesis→H_1: Not all the μ's are equal.

The null hypothesis asserts that all of the treatment groups come from populations with equivalent means, the implication being that all of the treatment groups belong to the *same* population. The alternative hypothesis in multiple-sample studies simply states that not all of the population means are the same, implying that more than one population is represented by the treatment groups. It is normal for alternative hypotheses to be vague like this in multiple-sample investigations. Because there are so many ways in which differences may exist among three or more means, it is usually not practical for H_1 to specify all the possible mean differences. For example, review the predictions made by the three competing theories of hypnosis effects (described earlier), and you will discover that several different patterns of mean differences are possible. The null hypothesis will be rejected if any one of the many possible mean differences is found to be significant. Before we can test the null hypothesis, however, we will have to decide what statistical procedure to use.

Will a *t* Test Do?

The results of the experiment appear in Table 12.1, which shows the raw scores, sample means, and unbiased variance estimates for each of the three treatment groups. You can see that the three means are numerically different, with the highest average suggestibility score belonging to the hypnosis group and the lowest average to the meditation group. Are these mean differences significant? There are two ways to answer that question. Since you just finished studying two-sample *t* tests, you know that you could do *three t tests* to compare $\bar{X}_{hypnosis}$ to $\bar{X}_{motivation}$, $\bar{X}_{hypnosis}$ to $\bar{X}_{meditation}$, and $\bar{X}_{motivation}$ to $\bar{X}_{meditation}$. Or, you could carry out *one analysis of variance* (abbreviated ANOVA). **Analysis of variance is** *a statistical procedure that simultaneously tests the significance of all possible mean differences within a given set of sample means.* Thus, an analysis of variance procedure makes all three mean comparisons referred to above in a single test. And, importantly, *the analysis of variance (ANOVA) test will come out significant if any sample mean in the set differs significantly from any of the other sample means.*

The last point is worth emphasizing. Analysis of variance is a very powerful procedure. An ANOVA test will be significant if any two means in a set of means differ reliably from each other, regardless of which particular means they are. The ANOVA will also be significant if several means differ significantly from one another. In short, ANOVA is sensitive to any and all reliable differences between and among several sample means. Because ANOVA *indiscriminately tests for all possible differences in a set of means*, it is called an **omnibus test**.

Table 12.1

SUGGESTIBILITY RAW SCORES, MEANS, AND VARIANCE ESTIMATES FOR THREE TREATMENT GROUPS

	TREATMENT GROUPS		
	HYPNOSIS	MOTIVATION	MEDITATION
	8	7	2
	9	4	3
	10	6	0
	6	5	0
	7	8	0
Sums:	40	30	5
Means:	8	6	1
Variance estimates:	2.5	2.5	2
N's:	5	5	5

Why Use ANOVA Instead of *t*?

In analyzing data from studies that involve more than two samples, researchers usually use analysis of variance rather than multiple *t* tests, even though either approach is possible. There is a very good reason for choosing ANOVA in such situations: avoiding excessive statistical decision errors.

Consider how many *t* tests you would have to carry out in a three-sample experiment. If *k* stands for the number of treatments (i.e., samples) in an experiment, then the number of *t* tests required to analyze all of the mean differences in such an experiment equals $C = k(k - 1)/2$. Thus, for a three-treatment study, there are $C = 3 (3 - 1)/2$, or 3, *t* tests to be done.

Earlier in this book you read that the alpha level (i.e., level of significance) used in a hypothesis-testing procedure specifies the probability of a Type I decision error, defined as rejecting a true null hypothesis. *The Type I error probability for any one t test* is called the **testwise error rate**. In general, the testwise error rate is the same as the stated alpha level.

If you carry out more than one significance test on the same set of data—as you would in multiple-treatment experiments—a second kind of error rate must be considered. The **familywise error rate** refers to *the overall probability of committing at least one Type I error when you conduct several significance tests on the same set of data.*

When you carry out multiple *t* tests on a set of, say, three means, *each one* of the individual *t* tests has a testwise error rate equal to your level of significance. So if $\alpha = .05$, then you have a 5% probability of making a Type I error in each *t* test. However, the *familywise error rate for the "family" of all t tests performed on the data set* is equal to $1 - (1 - \alpha)^C$ (Kirk, 1990), where *C* is the number of *t* tests being done and α is the level of significance. Hence, if $\alpha = .05$, then the probability of at least one Type I error in a collection of three *t* tests is $1 - (1 - \alpha)^C = 1 - (1 - .05)^3 = 1 - (.95)^3 = 1 - .857 = .143$. You can see that the familywise error rate (14.3%) is much higher than the stated alpha level (5%). This situation is unacceptable in scientific research. Furthermore, the familywise error rate becomes progressively higher as the number of *t* tests increases in larger investigations. In a five-sample design, for example, there are ten possible *t* tests. If all of these tests are conducted, there would be approximately a 40% likelihood of at least one Type I error.

Fortunately, *the mathematics of analysis of variance ensures that the familywise error rate will be equal to the stated alpha level,* regardless of how many mean comparisons the ANOVA must execute. Even if there are five treatment means being compared in the ANOVA, for instance, the overall error rate never exceeds the chosen level of significance.

Beyond the necessity of using it in certain data-analysis situations, ANOVA has an inherently pleasing logic that makes it interesting to work with. To make the best use of the procedure, you will need to grasp that underlying logic.

LOGICAL CONNECTIONS

The test statistic that you compute in an analysis of variance is called the F ratio. The structure of this F statistic follows directly from the elegant theory underlying ANOVA, which, in turn, is intimately interlocked with the rationale of the experimental method itself. To make both the method and the results of ANOVA more meaningful to you, we will now examine this logical sequence. When you arrive at the end of this section, glance into the rearview mirror for a moment and notice how the F ratio quantifies the answer to the fundamental question of any true experiment: Does the effect of the independent variable make the treatment-group means differ from one another by more than we would expect on the basis of random variation alone?

Much of what we will be going over next is a recapitulation of material on the rationale of experimentation that was introduced earlier in this text. A review will help you grasp ANOVA.

Logic of the Experiment

Language of the experiment Let's first review some terminology introduced in Chapter 2. Here are conventional terms and concepts used in the experimental method:

Treatment groups: The samples of people who are treated differently in an experiment. Usually one sample receives a standard treatment and is called the "control group." The remaining samples, the "experimental groups," receive some nonstandard treatments.

Subject pool: The original collection of persons who agree to be subjects, or participants, in an experiment. In the best of situations, the subject pool is randomly selected from some large population, but more typically the subject pool consists of research volunteers. Regardless of how the subject pool is obtained, it is considered to represent *some population*—a general population when subjects are randomly selected, and a specific population when the subjects are volunteers.

Random assignment: Using a chance process or a random number table to assign members of the subject pool to the various treatment groups. Random assignment of subjects to treatments usually makes the different groups of subjects statistically equivalent at the beginning of the experiment. This initial equating of treatment groups is what allows researchers to draw cause-and-effect conclusions from experiments.

Independent variable: The conditions that are systematically manipulated in an experiment. In the "hypnosis effects" study described earlier, the independent variable is the type of instructions subjects receive: meditation instructions, motivation instructions, or hypnosis instructions. The independent variable is the "factor" in a **one-factor ANOVA**, which is what this chapter is all about.

Treatment: One level, or condition, of the independent variable—meditation instructions, for example.

Dependent variable: The specific behavior of subjects that is measured in an experiment. In the "hypnosis effects" experiment, the dependent variable is the "suggestibility" (i.e., responsiveness to hypnotic suggestion) that is measured with a ten-point test. For purposes of this example, *we assume that the suggestibility scores are on an interval scale of measurement and are normally distributed in the population.* (See the "Assumptions of ANOVA" section later in this chapter.)

Now let's see how the vocabulary of the experiment helps reveal its underlying logic.

Rationale of the experiment   Figure 12.1 summarizes the logic of the behavioral experiment:

- Phase 1: Members of a subject pool are randomly assigned to treatment groups that represent the different levels of the independent variable. At this point, *the treatment groups would have approximately equal means* if measured on the dependent variable; random assignment practically ensures that. So between-groups mean differences are very small to nil. At the beginning of the experiment, all treatment groups are said to represent one and the same population. Note, however, that as a result of the random assignment procedure, *there is individual variation on the dependent variable within each treatment group.* That is, even at the start of the study, people within each treatment group tend to show essentially random differences from one another on the dependent variable. *This type of individual-difference variation is considered random error.*
- Phase 2: Next the experimenter manipulates the independent variable by treating the statistically equivalent treatment groups differently. In the experiment described earlier, the independent variable is manipulated by giving some subjects meditation instructions, giving others motivational instructions, while giving still others hypnotic instructions.

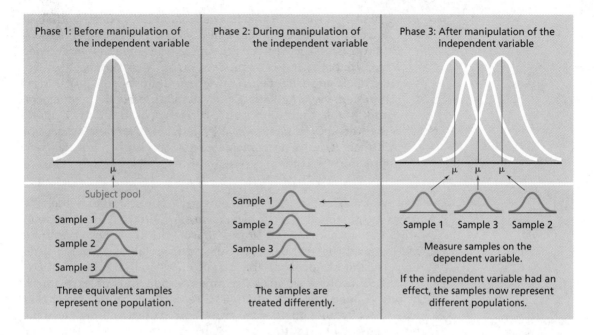

Phase 1: Before manipulation of the independent variable	Phase 2: During manipulation of the independent variable	Phase 3: After manipulation of the independent variable

Subject pool

Sample 1

Sample 2

Sample 3

Three equivalent samples represent one population.

Sample 1

Sample 2

Sample 3

The samples are treated differently.

μ μ μ

Sample 1   Sample 3   Sample 2

Measure samples on the dependent variable.

If the independent variable had an effect, the samples now represent different populations.

Figure 12.1 *Logic of the Experiment*

- Phase 3: Finally, the experimenter measures the dependent variable—"suggestibility" in the example we've been using. If the independent variable had a significant effect on the dependent variable, the treatment groups will now have different means on the dependent variable. Between-treatments differences will now be much larger than they were at the start of the study, whereas the individual-difference variation within each treatment will still be the same as it was initially. Note that *when the independent variable has an effect on the dependent variable, the treatment groups are considered to represent different populations when the experiment is finished.* In a sense, manipulation of the independent variable has created several populations where once there was only one.

Logic of ANOVA

The theory of analysis of variance rests primarily on one huge idea: that *you can tell whether variation between treatment-group means is significant by comparing that variation to an index of random variation.* A corollary to this idea is that *variation among raw scores within treatment groups can serve as an index of random variation.* Given these assumptions, it follows that:

- If the between-treatments variance is about the same size as the index of random variation (based on within-treatments variance), then both between-treatments variation and within-treatments variation reflect nothing but random error variance.

In that case the independent variable had no effect, and all of the treatment groups represent one and the same population.

- By contrast, if the between-treatments variance is several times larger than the index of random error, then the mean differences between treatments reflect the influence of two things: (1) random variation *plus* (2) a systematic effect of the independent variable that pushes the treatment means farther apart. Since the independent variable had an effect, the respective treatment groups are considered to represent more than one population.

Picture no effect   Figure 12.2 illustrates the situation that exists when the independent variable has no effect on the dependent variable. The top panel of the figure—that is, the area above the white line—shows the true situation in a particular experiment. This time, the independent variable had no effect on the dependent variable, and all three samples (treatment groups) in the experiment belong to just one population. The researcher cannot know this truth directly, though. She can only observe the data from her samples, represented by the three distributions shown in the lower panel of Figure 12.2. Observe that each of the sample distributions has *within-treatments variation* among its raw scores, *represented by the shaded areas* within the sample distributions, and *between-treatments variation, reflected in the distances among the sample means*. When the researcher statistically analyzes her data, what

Figure 12.2 *The Independent Variable Has No Effect: Average Variation Between Treatments Is About the Same As Average Variation Within Treatments*

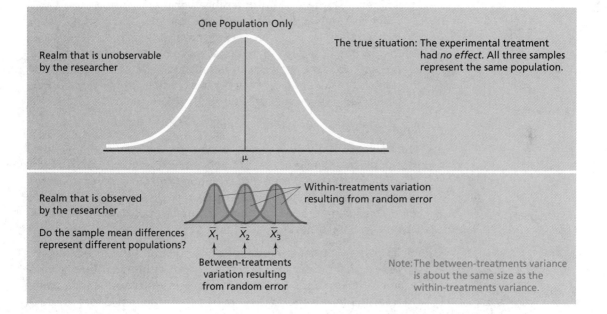

does she find? In this instance, the average between-samples variation is about the same size as the average within-samples variation. Since within-samples variation is considered an index of purely random error, the researcher concludes that the between-samples mean differences reflect only random error variation; the independent variable had no significant effect.

The preceding paragraph makes an important point about the results of experiments: Even when the independent variable has absolutely no effect on the dependent variable, there will nonetheless be some small differences among the samples means. Why? Because the very same random error processes that cause unsystematic variation among persons within a treatment group also produce some variation between treatment-group means. Thus, the main question in an analysis of variance is not whether there is any difference at all between treatment means; rather, the question is whether between-treatments variation is about the same as within-treatments variation (a nonsignificant outcome) or several times larger than within-treatments variation (a significant outcome).

Picture a real effect   Now take a look at the hypothetical research scenario shown in Figure 12.3, which illustrates the expected outcome when the independent variable does have an effect on the dependent variable. How does this figure differ from Figure 12.2? The one conspicuous difference is that there are three populations in the realm of the unobservable. That is, because the independent variable had a reliable impact on behavior, the separate treatment groups belong to different populations. But bear in mind that the researcher cannot observe the populations. She must infer them from her sample data. What does she see in her sample data? Put more specifically, how does the depiction of the treatment-group data in this instance differ from that in Figure 12.2? Clearly, the amount of variation in scores *within* the treatment groups is the same in both figures. The major difference at the level of observable data is that the *sample means are farther apart* in Figure 12.3. Random variation has caused a small amount of spread among the means. But, in addition, the reliable effect of the independent variable has "pushed" the means even farther from one another. Now the between-treatments variation is considerably larger than within-treatments variation. Therefore, a *mathematical ratio of between-treatments variation to within-treatments variation*—known as the **F ratio**—is several times larger than 1.0.

Logic of the F Ratio

The culmination of an ANOVA is a test statistic called the *F* ratio. The *F* ratio is obtained through dividing one variance that is based

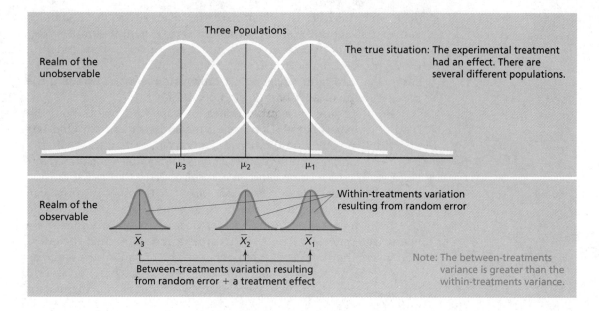

Figure 12.3 *The Independent Variable Has an Effect on the Dependent Variable: Between-Treatments Variation Is Greater Than the Average Variation Within Treatments*

on mean differences between treatments by a second variance that is based on raw-score differences within the treatment groups. Since you have read about the logic of the experiment and the related theory of ANOVA, the general rationale for the F ratio should be almost self-evident. The objective of this section will be to link the general theory of ANOVA and the calculation of the F ratio that you will learn about in a later section.

Language of the F ratio   I keep talking about within-treatments and between-treatments variation, and I've shown you figures portraying these two kinds of variation. But what do within-treatments and between-treatments variance mean in terms of the numerical data that you work with in a multiple-treatments experiment? Consider the raw scores from the hypnosis treatment group mentioned earlier:

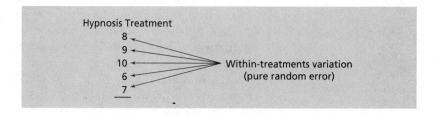

This sample has a mean, and the raw scores vary randomly around that mean *within the sample*. The same is true of the raw scores in the motivation treatment and the meditation treatment. Each sample has its own variance based on this kind of within-samples variation among the raw scores. Keep in mind that this is considered random error variation. Indeed, as you'll see later, the numerical index of random error in ANOVA is calculated from the squared deviations between the raw scores and their respective means [i.e., $(X - \overline{X})^2$]. This correctly implies that the more the raw scores tend to vary from their means, the larger the index of random error.

In contrast, between-treatments variation refers to numerical differences among sample means, as shown here:

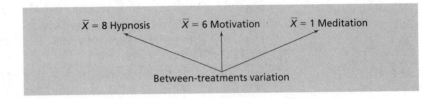

$\overline{X} = 8$ Hypnosis $\overline{X} = 6$ Motivation $\overline{X} = 1$ Meditation

Between-treatments variation

The average of the treatment means is called the **grand mean** of the data set. In this example, the grand mean of the three treatment means is $\overline{X}_G = (8 + 6 + 1)/3 = 5$. Observe that the grand mean is symbolized as $\overline{X}_G$, whereas a treatment mean is symbolized simply as $\overline{X}$. Later in the chapter I will show you that the index of between-treatments variation is computed from the squared deviations between the treatment means and the grand mean [i.e., $(\overline{X} - \overline{X}_G)^2$]. Since the grand mean is the average of the treatment means, the more the treatment means vary from one another, the larger are the squared differences between them and the grand mean; and, thus, the greater is the index of between-treatments variance.

Mean squares and the F ratio   A verbal expression of the F ratio is:

$$F = \frac{\text{between-treatments variance}}{\text{within-treatments variance}}$$

The numerator of the F statistic is always a variance based on between-samples mean differences, and the denominator of the F statistic is usually a variance based on within-samples differences among raw scores.

In ANOVA, a variance is called a **mean square**. Therefore, you should also think of the F ratio as:

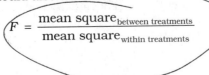

$$F = \frac{\text{mean square}_{\text{between treatments}}}{\text{mean square}_{\text{within treatments}}}$$

Since a mean square is often abbreviated *MS*, the *F* ratio is appropriately symbolized as:

$$F = \frac{MS_B}{MS_W},$$ (12.1)

where MS_B = mean square$_{\text{between treatments}}$ and MS_W = mean square $_{\text{within treatments}}$.

Expected *F* ratios   As mentioned earlier, the MS_W (or mean square$_{\text{within treatments}}$) is assumed to represent nothing but random error variation: MS_W = error variance. However, the MS_B (or mean square$_{\text{between treatments}}$) represents error variance + ?, where ? is a possible effect of the independent variable. ? equals 0 when the independent variable has no reliable effect on the dependent variable, and it is greater than 0 when the independent variable does affect the dependent variable.

In light of the above facts, when there is no "effect" in an experiment, the *F* ratio is expected to be:

$$F = \frac{\text{error variance} + 0}{\text{error variance}} = \frac{\text{error variance}}{\text{error variance}} = 1$$

Therefore, if your computed *F* statistic is close to 1.0, it is safe to conclude that your results are not significant.

On the other hand, when the independent variable does reliably influence the dependent variable, then the expected *F* ratio is:

$$F = \frac{\text{error variance} + \text{effect of the independent variable}}{\text{error variance}} > 1$$

Hence, if your computed *F* statistic is greater than 1.0, particularly if it is several times larger than 1.0, it is likely that your results are significant—that is, that they reflect more than just random error.

Just how large does your computed *F* have to be in order to be declared significant? The answer to that question is next on our agenda.

F_{crit} AND THE *F* DISTRIBUTION

As you know from experience with other test statistics, to demonstrate significance, the computed *F* ratio must equal or exceed some critical value of *F* that delineates the rarest 5% of some sampling distribution—an "*F* distribution" in this case. Having worked with the *t* statistic, you might expect that the value of F_{crit} to be a function of the level of significance (α) and the "degrees of freedom" in the sample data. If so, your expectation is correct. However, determining F_{crit} is a little more complicated than finding t_{crit}. The reason is that *an F statistic has three kinds of degrees of freedom.*

Degrees of Freedom in ANOVA

To comprehend these multiple types of degrees of freedom in analysis of variance, you need to become familiar with the following system of symbols:

- N represents *the number of independent observations in each treatment group* of an ANOVA. In the investigation of hypnosis effects that we're using as an example, $N = 5$ for each of the three groups. Also, note that *each of the treatment groups has $N - 1$ degrees of freedom within its N observations.* In our example, then, each sample of five observations has four degrees of freedom.
- N_{tot} represents *the total number of independent observations in the experiment as a whole.* Thus, $N_{tot} = \Sigma N$. In our example, $N_{tot} = N + N + N = 5 + 5 + 5 = 15$.
- k represents *the number of treatment groups in the experiment.* In the hypnosis effects study, $k = 3$.

From the above definitions, the following are three kinds of degrees of freedom in ANOVA.

Total degrees of freedom The overall degrees of freedom in an ANOVA is $N_{tot} - 1$. If there are N_{tot} independent observations in an experiment, then only $N_{tot} - 1$ of the observations are free to vary—given that the N_{tot} observations must sum to a particular total. In the hypnosis effects investigation, for instance, there are 15 total observations; therefore, the total degrees of freedom $= df_{tot} = 15 - 1 = 14$.

Between-treatments degrees of freedom The second type of degrees of freedom in an ANOVA is called between-treatments degrees of freedom and is abbreviated df_B. The between-treatments degrees of freedom equals $k - 1$, the number of treatment groups minus 1. The thinking behind $df_B = k - 1$ is that since the average of the k treatment group means ($\overline{X}$ values) must equal the grand mean ($\overline{X}_G$), only $k - 1$ of the group means are free to vary. Given the values of any $k - 1$ means, the remaining one must have a particular fixed value, so that $\Sigma \overline{X}/k = \overline{X}_G$. (*Note:* This formula is valid only when all treatment groups have equal N's.) In the hypnosis effects experiment, there are $k = 3$ group means; therefore, $df_B = 3 - 1 = 2$.

Within-treatments degrees of freedom The within-treatments degrees of freedom, known as df_W, is a "pooled" value; $df_W = \Sigma(N - 1)$. Because each treatment group has $N - 1$ degrees of freedom in it, you can get a combined within-treatments degrees of freedom by summing the degrees of freedom within the various samples. An algebraically equivalent expression for the degrees of freedom is $df_W = N_{tot} - k.$ The hypnosis effects study has $df_W = N_{tot} - k = 15 - 3 = 12.$ Equivalently, $df_W = \Sigma(N - 1) = (5 - 1) + (5 - 1) + (5 - 1) = 4 + 4 + 4 = 12.$

> POINT WORTH REMEMBERING: The total degrees of freedom always equals the sum of the within-treatments and between-treatments degrees of freedom; that is, $df_{tot} = df_W + df_B$. In our present example, $14 = 12 + 2$. The implication of this fact is that once you have determined df_{tot} and df_B, df_W can be found through simple subtraction: $df_W = df_{tot} - df_B$ (that is, $12 = 14 - 2$). You'll find that the one very handy characteristic of ANOVA is that everything adds up. More on this later.

Looking Up F_{crit}

The total degrees of freedom is *not* necessary for determining the critical value of F, although it is useful in other ANOVA operations. Rather, F_{crit} is a function of the following three criteria: the level of significance (alpha level), df_B, and df_W. Turn to Table F in Appendix A. The F_{crit} values for the .05 level of significance are in lightface (i.e., regular) type, and the F_{crit} values for the .01 alpha level are in boldface type. As you work with Table F, remember that:

- "Degrees of Freedom: Numerator" refers to df_B (between-treatments degrees of freedom).
- "Degrees of Freedom: Denominator" refers to df_W (within-treatments degrees of freedom).

Let's assume that $\alpha = .05$. To find F_{crit}:

1. Go to the leftmost column of Table F and find df_W ("Degrees of Freedom: Denominator") for the observations in the experiment. If your df_W is not listed in the table, "drop back" to the next *lower* df_W listed.
2. Go to the top of Table F and find df_B ("Degrees of Freedom: Numerator") for the experiment.
3. Find the intersection of df_W and df_B in the table.

The value in lightface type at that point in the table is the F_{crit} that must be satisfied for significance to be declared at the 5% alpha level.

Applying this procedure to the hypnosis effects experiment, where $df_W = 12$ and $df_B = 2$, we find that $F_{crit} = 3.89$ at the .05 level of significance.

The F Distribution

As suggested by Table F, there is a different F distribution for every possible combination of df_W and df_B, even though only a finite number of the possible F distributions are represented by the critical values in the table. The F distribution that we will be working with in connection with the hypnosis effects experiment appears in Figure 12.4. This is a typical *F* distribution, in the sense that most of the *F* distributions that behavioral scientists use are conspicuously skewed. These are other noteworthy characteristics of *F* distributions:

- Every *F* distribution is a sampling distribution, a theoretical random variable distribution that represents the expected distribution of *F* values if the hypothesis of chance (i.e., the null hypothesis) is true.
- All *F* values must range between 0 and positive infinity (+∞). *F* statistics cannot have negative values because they are ratios of variances, which are calculated from squared scores.
- The most likely *F* value under the hypothesis of chance is 1.0; that is, the highest point in the distribution is exactly over 1.0.
- Since only the right side of an *F* distribution can have extreme values, there is only one "region of rejection" of H_0.

Now that you understand the why and wherefore of ANOVA, you're ready for the mechanics.

DOING ANOVA

As we proceed through the various preliminary and computational steps in analysis of variance, it will be immensely helpful if *you constantly remind yourself that the objective of this sequence of operations is to develop a ratio of a between-treatments variance to a within-treatments variance.* You won't lose your way if you simply keep this objective in mind.

Figure 12.4 *A Typical F Distribution*

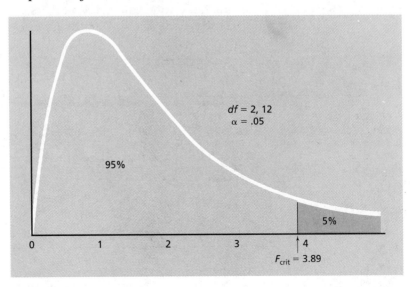

df = 2, 12
α = .05

95%

5%

0   1   2   3   4

F_{crit} = 3.89

The ANOVA Summary Table

The first step in conducting an analysis of variance is to construct an **analysis of variance summary table**, which is *a device for both displaying the results of an ANOVA and guiding the sequence of operations that yields those results.* Table 12.2 shows a typical ANOVA summary table. It contains all of the components necessary to carry out the test of significance. I've numbered the columns of Table 12.2 to indicate the order in which different kinds of information are to be inserted. We will gradually replace each label in the table with the corresponding numerical results of our calculations.

As we proceed, *note well how the various entries in the ANOVA summary table are mathematically and logically related to one another.* You will be richly repaid for doing so, for to understand the summary table is to know a lot about analysis of variance.

Find the Sums of Squares

A previous chapter of this book showed you how to "partition variance" in the context of linear regression. You might remember that the total variation on some dependent variable, Y, was divided into a component that was predictable from X, the independent variable, and a second component that represented "error," the portion of Y variance that was not "explained" by the independent variable. ANOVA involves the same kind of variance partitioning. Total variation on the dependent variable is represented by a quantity called the **total sum of squares**, symbolized by SS_{tot}. The total sum of squares is partitioned into a **within-treatments sum of squares**, or SS_W, and a **between-treatments sum of squares**, or SS_B. As you might have anticipated, SS_B and SS_W eventually produce the two variances used to compute the F ratio.

Table 12.2

SCHEMATIC ANALYSIS OF VARIANCE SUMMARY TABLE

SOURCES OF VARIANCE	(1) SUMS OF SQUARES	(2) DEGREES OF FREEDOM	(3) MEAN SQUARES	(4) F RATIO
Between treatments	SS_B	$df_B = k - 1$	$MS_B = SS_B/df_B$	$F = MS_B/MS_W$
Within treatments	SS_W	$df_W = N_{tot} - k$	$MS_W = SS_W/df_W$	
TOTAL	SS_{tot}	$df_{tot} = N_{tot} - 1$		

At the level of an *individual deviation* of a raw score from the grand mean, the partitioning of variance model is:

**deviation of
raw score from
grand mean**
$\downarrow$

$$(X - \overline{X}_G) = (\overline{X} - \overline{X}_G) + (X - \overline{X})$$
$\uparrow$ $\uparrow$

**deviation of   deviation of
treatment mean   raw score from
from grand mean   treatment mean**

The above expression states that the variation of a raw score from the overall mean in an experiment is partly determined by a between-treatments component $(\overline{X} - \overline{X}_G)$, which is "error variance + a possible effect of the independent variable," and partly by a within-treatments component $(X - \overline{X})$, which is purely "error variance."

At the level of sample data sets, the partitioning of variance model is:

$$SS_{tot} = SS_B + SS_W$$

The ANOVA procedure involves finding SS_{tot}, and then partitioning it into SS_B and SS_W. The first step in the partitioning procedure is to find the grand mean.

Compute the grand mean   The mean of all data summed across all treatments in an experiment is called the *grand mean*. The grand mean is necessary for the calculation of all the sums of squares in ANOVA. The formula for the grand mean is:

$$\overline{X}_G = \frac{\Sigma X}{N_{tot}} \tag{12.2}$$

This formula says: To get the grand mean, sum all of the raw scores in the experiment and divide that sum by the total number of scores. If you look at Table 12.3, you find that the sum of all the raw scores is 75 and that there are 15 raw scores altogether. Therefore, the grand mean is $\overline{X}_G = \Sigma X / N_{tot} = 75/15 = 5$.

When all the sample sizes in an experiment are the same, you may use the following equation to obtain the grand mean from the sample means:

$$\overline{X}_G = \frac{\Sigma \overline{X}}{k} \tag{12.3}$$

This expression instructs you to add up the sample means and divide by k, the number of means. In our example, $\overline{X}_G = \Sigma \overline{X}/k = (8 + 6 + 1)/3 = 15/3 = 5$. As you'll see next, having the grand mean is the key to finding the sums of squares in ANOVA.

Table 12.3

CALCULATION OF THE GRAND MEAN IN AN EXPERIMENT ON HYPNOSIS EFFECTS	TREATMENT GROUPS	RAW SCORES, X
	Group 1: Hypnosis	8
		9
		10
		6
		7
	Group 2: Motivation	7
		4
		6
		5
		8
	Group 3: Meditation	2
		3
		0
		0
		0
	Sum:	75

$$\text{Grand mean} = \overline{X}_G = \frac{\Sigma X}{N_{tot}} = \frac{75}{15} = 5$$

Compute the total sum of squares   The total sum of squares represents the variation of all the raw scores in an experiment around the grand mean of all the scores. It is the "total mass of variation" in an experiment that you partition into a between-samples component of variance and a within-samples component of variance. You calculate the total sum of squares via:

$$\overset{\text{③}}{\underset{\downarrow}{}} \quad \overset{\text{①}}{\underset{\downarrow}{}} \quad \overset{\text{②}}{\underset{\downarrow}{}}$$

$$SS_{tot} = \Sigma(X - \overline{X}_G)^2 \tag{12.4}$$

FORMULA GUIDE

① Subtract the grand mean from each raw score in the experiment to get N_{tot} deviation scores.

② Square each of the deviation scores.

③ Sum all of the squared deviations.

The easiest way to conduct an analysis of variance by hand is to set up a **worksheet** for any lengthy computational procedure. A worksheet *uses a sequence of columns in a table to represent the successive steps in a formula.* Then you carry out the computational procedure by simply filling in the numbers for each column. The worksheet in Table 12.4 shows that the sum of the squared deviations from the grand mean is 158; hence, $SS_{tot} = 158$. Observe that SS_{tot} literally is a "sum of squares." (Some statistical labels actually meet with common sense!) Now let's partition this sum of squares into its components.

Compute the between-treatments sum of squares   One source that contributes to the total variation in the data of an experiment is the variation among sample means. This between-treatments variation is reflected in the amount of deviation between sample means and the grand mean. It follows that the formula for the between-treatments sum of squares (SS_B) should include the squared deviations between treatment means and the grand mean:

$$\overset{\text{④③} \quad \text{①} \quad \text{②}}{\downarrow\downarrow \quad \downarrow \quad \downarrow}$$

$$\text{Sum of squares between treatments} = SS_B = \Sigma[N(\overline{X} - \overline{X}_G)^2] \quad (12.5)$$

> **FORMULA GUIDE**
> ① Subtract the grand mean from each treatment mean to get *k* mean deviations.
> ② Square each mean deviation.
> ③ Multiply each squared mean deviation by *N, the number of observations in that treatment.*
> ④ Sum the result of step 3.

The following facts are relevant to formula (12.5):

Hypnosis treatment: $\overline{X} = 8$

Motivation treatment: $\overline{X} = 6$

Meditation treatment: $\overline{X} = 1$

Since the grand mean, $\overline{X}_G$, is 5, we have

$$SS_B = \Sigma[N(\overline{X} - \overline{X}_G)^2] = 5(8 - 5)^2 + 5(6 - 5)^2 + 5(1 - 5)^2$$

$$= 5(3)^2 + 5(1)^2 + 5(-4)^2$$

$$= 5(9) + 5(1) + 5(16)$$

$$= 45 + 5 + 80$$

$$= 130$$

Table 12.4

CALCULATION OF THE TOTAL SUM OF SQUARES IN AN EXPERIMENT ON HYPNOSIS EFFECTS	TREATMENT GROUPS	RAW SCORES, X	DEVIATION SCORES, $(X - \bar{X}_G)$	SQUARED DEVIATIONS, $(X - \bar{X}_G)^2$
	Group 1: Hypnosis	8	3	9
		9	4	16
		10	5	25
		6	1	1
		7	2	4
	Group 2: Motivation	7	2	4
		4	−1	1
		6	1	1
		5	0	0
		8	3	9
	Group 3: Meditation	2	−3	9
		3	−2	4
		0	−5	25
		0	−5	25
		0	−5	25
	Sums:	75	0	158

$$SS_{tot} = \Sigma(X - \bar{X}_G)^2 = 158$$

As you can see, of the 158 units of the total sum of squares, 130 units of variance are due to differences between treatment-group means. Therefore, according to the partitioning of variance model, the within-treatments sum of squares should equal:

$$SS_W = SS_{tot} - SS_B$$

$$= 158 - 130$$

$$= 28$$

Next we'll see if this expectation holds true.

Compute the within-treatments sum of squares The formula for that portion of the total sum of squares that represents purely random error is:

Within-treatment sum of squares $= SS_W = \Sigma[\Sigma(X - \bar{X})^2]$ (12.6)

FORMULA GUIDE

① *Within each treatment group*, subtract the treatment-group mean from each raw score to get N deviation scores *within each group.*
② *Within each group*, square each deviation score.
③ *Within each group*, sum the squared deviations.
④ Sum the sums of squared deviations across the k treatment groups

Table 12.5 is a worksheet demonstrating how to find the within-treatments sum of squares. First we find the sum of squares within each of the samples (referring to the worksheet):

Hypnosis: $\Sigma(X - \overline{X})^2 = 10$

Motivation: $\Sigma(X - \overline{X})^2 = 10$

Meditation: $\Sigma(X - \overline{X})^2 = 8$

Now, formula (12.6) yields $SS_W = \Sigma[\Sigma(X - \overline{X})^2] = 10 + 10 + 8 = 28$, just as we projected earlier from the partitioning of variance model. It's always a good idea to check your calculation of SS_W against the figure you get by subtracting SS_B from SS_{tot}.

Updating the ANOVA summary table   It is good practice to insert information into the appropriate parts of the ANOVA summary table as you go along. Table 12.6 is an updated summary table that includes the three sums of squares we've calculated. The next step is to figure out the degrees of freedom for each "source of variation."

Table 12.5

CALCULATION OF THE WITHIN-TREATMENTS SUM OF SQUARES IN THE EXPERIMENT ON HYPNOSIS EFFECTS

	HYPNOSIS GROUP			MOTIVATION GROUP			MEDITATION GROUP	
X	$(X - \overline{X})$	$(X - \overline{X})^2$	X	$(X - \overline{X})$	$(X - \overline{X})^2$	X	$(X - \overline{X})$	$(X - \overline{X})^2$
8	0	0	7	1	1	2	1	1
9	1	1	4	−2	4	3	2	4
10	2	4	6	0	0	0	−1	1
6	−2	4	5	−1	1	0	−1	1
7	−1	1	8	2	4	0	−1	1
$\overline{X} = 8$		10	$\overline{X} = 6$		10	$\overline{X} = 1$		8

$$SS_W = \Sigma[\Sigma(X - \overline{X})^2] = 10 + 10 + 8 = 28$$

Table 12.6

ANOVA SUMMARY TABLE SHOWING SUMS OF SQUARES FOR HYPNOSIS EFFECTS EXPERIMENT	SOURCES OF VARIANCE	SUMS OF SQUARES	DEGREES OF FREEDOM	MEAN SQUARES	F RATIO
	Between treatments	130	$df_B = k - 1$	$MS_B = SS_B/df_B$	$F = MS_B/MS_W$
	Within treatments	28	$df_W = N_{tot} - k$	$MS_W = SS_W/df_W$	
	TOTAL	158	$df_{tot} = N_{tot} - 1$		

Find the Degrees of Freedom

In a previous section of this chapter, we determined the following degrees of freedom in the present ANOVA problem:

Total degrees of freedom: $df_{tot} = N_{tot} - 1 = 15 - 1 = 14$

Between-treatments degrees of freedom: $df_B = k - 1 = 3 - 1 = 2$

Within-treatments degrees of freedom: $df_W = N_{tot} - k = 15 - 3 = 12$

Table 12.7 shows the ANOVA summary table with this information added.

Find the Mean Squares

I have stated that the F ratio is the test statistic produced by an analysis of variance. Furthermore, F is a ratio of two variances, or "mean squares," MS_B/MS_W. To find each of those mean squares, you simply divide each sum of squares by its corresponding degrees of freedom.

Table 12.7

ANOVA SUMMARY TABLE SHOWING DEGREES OF FREEDOM FOR HYPNOSIS EFFECTS EXPERIMENT	SOURCES OF VARIANCE	SUMS OF SQUARES	DEGREES OF FREEDOM	MEAN SQUARES	F RATIO
	Between treatments	130	2	$MS_B = SS_B/df_B$	$F = MS_B/MS_W$
	Within treatments	28	12	$MS_W = SS_W/df_W$	
	TOTAL	158	14		

Calculate the between-treatments mean square   The mean square in the numerator of the F ratio is:

$$\text{Between-treatments mean square} = MS_B = \frac{SS_B}{df_B} \qquad (12.7)$$

In the present experiment, $MS_B = SS_B/df_B = 130/2 = 65$. This is our measure of "random error variation + a possible effect of the independent variable." Now to find the F ratio, all we need is the mean square for the denominator.

Calculate the within-treatments mean square   The mean square in the denominator of the F ratio is found through:

$$\text{Within-treatments mean square} = MS_W = \frac{SS_W}{df_W} \qquad (12.8)$$

In the hypnosis effects experiment, $MS_W = SS_W/df_W = 28/12 = 2.33$.
    You may be interested in knowing that, *when all treatment group N values are equal, the within-treatments mean square is simply the average of the "variance estimates"* of the treatment groups:

$$\text{Within-treatments mean square} = MS_W = \frac{\Sigma s^2}{k} \qquad (12.9)$$

From Table 12.1, we know that:

Hypnosis group: $s^2 = 2.5$

Motivation group: $s^2 = 2.5$

Meditation group: $s^2 = 2.0$

From formula (12.9), $MS_W = \Sigma s^2/k = (2.5 + 2.5 + 2.0)/3 = 7/3 = 2.33$, exactly the same result that was found from formula (12.8). The obvious implication is that, when all sample N values are the same, you can compute the denominator of the F ratio directly from the variance estimates.[2]

> **CONCEPT RECAP**
> The *variance estimate* refers to the unbiased estimate of the population variance, given by $s^2 = \Sigma(X - \overline{X})^2/(N - 1)$. One such unbiased estimate is computed for each sample.

---

[2] Important: This method is not accurate when the sample sizes are unequal. Then the following "weighted average" formula must be used: $MS_W = \Sigma[(N - 1)s^2]/\Sigma(N - 1)$, where s^2 is the unbiased variance estimate of any given sample, and $(N - 1)$ is the degrees of freedom for that sample.

Table 12.8 updates the ANOVA summary table by including the mean squares. All that is left to do is compute F and make a decision about the null hypothesis.

Compute the F Statistic

From formula (12.1), the computed F statistic in the hypnosis effects study is:

$$F = \frac{MS_B}{MS_W} = \frac{65}{2.33} = 27.90$$

The F_{crit} value that we looked up previously is 3.89. Since the computed F of 27.90 (shown in Table 12.9) exceeds F_{crit}, the results of the experiment are significant, and we reject H_0. It would be appropriate to conclude that "The treatments produced reliably different levels of responsiveness to suggestions, $F(2,12) = 27.90$, $p < .05$." Since there are three means in the investigation, the significant F ratio gives only part of the information we need to interpret the findings completely. The significant "omnibus F test" tells us only that *some effect occurred in the study.* We still need to find out which specific mean differences caused the F statistic to be significant so that we can draw more specific conclusions about the relative impacts of the different treatments.

Note the correct way to report F →

TRACKING DOWN THE SOURCE OF A SIGNIFICANT F RATIO

The occurrence of a significant F statistic normally prompts several "pairwise comparison" tests designed to identify particular mean differences that are reliable. A **pairwise comparison** is *a test that assesses the significance of the difference between two means at a time.* If there are k treatment means in a study, then there are $k(k - 1)/2$ possible pairwise comparisons that can be made as a

Table 12.8

ANOVA SUMMARY TABLE SHOWING MEAN SQUARES FOR HYPNOSIS EFFECTS EXPERIMENT	SOURCES OF VARIANCE	SUMS OF SQUARES	DEGREES OF FREEDOM	MEAN SQUARES	F RATIO
	Between treatments	130	2	65	$F = MS_B/MS_W$
	Within treatments	28	12	2.33	
	TOTAL	158	14		

Box 12.1

THE NINE STEPS OF ANOVA

In performing a one-way (i.e., one-factor) analysis of variance, you will usually carry out the following nine operations in the order given.

1. Compute the total sum of squares: $SS_{tot} = \Sigma(X - \bar{X}_G)^2$

2. Compute the between-treatments sum of squares: $SS_B = \Sigma[N(\bar{X} - \bar{X}_G)^2]$

3. Compute the within-treatments sum of squares: $SS_W = \Sigma[\Sigma(X - \bar{X})^2]$

4. Determine the total degrees of freedom: $df_{tot} = N_{tot} - 1$

5. Determine the between-treatments degrees of freedom: $df_B = k - 1$

6. Determine the within-treatments degrees of freedom: $df_W = N_{tot} - k$

7. Compute the between-treatments mean square: $MS_B = SS_B/df_B$

8. Compute the within-treatments mean square: $MS_W = SS_W/df_W$

9. Compute the F ratio: $F = MS_B/MS_W$

Table 12.9

COMPLETE ANOVA SUMMARY TABLE FOR HYPNOSIS EFFECTS EXPERIMENT

SOURCES OF VARIANCE	SUMS OF SQUARES	DEGREES OF FREEDOM	MEAN SQUARES	F RATIO
Between treatments	130	2	65	27.90
Within treatments	28	12	2.33	
TOTAL	158	14		

followup to a significant ANOVA. In the study that serves as our main example, we have the following mean differences to test for significance:

$$\bar{X}_{hypnosis} - \bar{X}_{motivation} = 8 - 6 = 2$$

$$\bar{X}_{hypnosis} - \bar{X}_{meditation} = 8 - 1 = 7$$

$$\bar{X}_{motivation} - \bar{X}_{meditation} = 6 - 1 = 5$$

So the largest mean difference here is 7, and the smallest one is 2. The critical question is: How large a mean difference is required for significance?

One way to answer the question is to compute a statistic referred to as **Tukey's honestly significant difference** (*HSD*). The *HSD is the minimum significant mean difference in a set of k means, conditional upon a constant familywise error rate.* Tukey's *HSD* can be used to make all possible pairwise comparisons while holding the probability of a Type I error at the stated alpha level—something that can't be done with the *t* test. The formula for Tukey's honestly significant difference is:

$$HSD = \underset{③}{q}\underset{②}{\sqrt{\frac{MS_W}{N}}} \leftarrow ①$$

(12.10)

FORMULA GUIDE

① Divide the within-treatments mean square by the number of observations in each treatment group.
② Take the square root of the result of step 1.
③ Multiply the result of step 2 by Tukey's *q* statistic.

In the hypnosis effects study, $MS_W = 2.33$ and the group size is $N = 5$. The only unknown quantity, then, is Tukey's *q* statistic, which can be obtained from Table Q in Appendix A. (*Note:* Tukey's *q* statistic is also known as the studentized range statistic.) To find the appropriate *q*, go to the top of Table Q and locate the column that corresponds to *k*, the number of means in the study. Now follow down that column until you get to the row of the table that corresponds to the number of degrees of freedom for within-treatments variation (df_W). At the intersection of *k* and df_W, you will see two critical values of *q*. The top value is *q* when the alpha level is .05, and the bottom value is *q* when alpha is .01. Since we are working with $\alpha = .05$ in the present example, the critical value of *q* is 3.77. Therefore, the *minimum significant difference* between two means in the hypnosis effects study is:

$$HSD = q\sqrt{\frac{MS_W}{N}} = 3.77\sqrt{\frac{2.33}{5}} = 3.77\sqrt{0.466}$$

$$= 3.77 \cdot 0.6826 = 2.57$$

Hence, *any two means differ significantly only if they are at least 2.57 units apart.*

In our example, both the hypnosis mean and the motivation mean are reliably higher than the meditation mean, but the hypnosis and motivation groups do not differ significantly from each other. Motivational instructions were as effective as hypnotic induction. This outcome is most supportive of the "enhanced motivation" theory of suggestibility that was described at the outset of this chapter: The effects of hypnosis are a result of the motivational impact of hypnotic instructions, not relaxation per se or some special trance state.

Generally speaking, *you should use Tukey's test only if your omnibus F test is significant.* If F is not significant, it is very unlikely that Tukey's test will show a "significant" difference between any two means in the study. Furthermore, if a "significant" comparison is found in the absence of a significant F ratio, the probability that it represents a Type I error is greater than the stated alpha level (see Hays, 1988).

Tukey's Test Versus the *t* Test

If we had chosen to disregard the problem of a larger-than-acceptable familywise Type I error rate, we could have established a "least significant difference" using the t ratio and then employed that critical difference to determine which particular pairs of means differ reliably. The t-test formula in this situation is:

$$\text{Least significant difference} = LSD = t_{\text{crit}}\sqrt{\frac{2MS_{\text{w}}}{N}} \qquad (12.11)$$

where t_{crit} is the two-tailed t value necessary for significance at the desired alpha level when degrees of freedom is equal to df_{w}. In the hypnosis study, $df_{\text{w}} = 12$. At $\alpha = .05$, Table T of Appendix A shows that $t_{\text{crit}} = 2.179$, so

$$LSD = 2.179\sqrt{\frac{2 \cdot 2.33}{5}} = 2.179\sqrt{0.932}$$

$$= 2.179 \cdot 0.965 = 2.103$$

You can see that the critical difference necessary for declaring any two means reliably different is considerably smaller when we use the t statistic rather than Tukey's q statistic to set up the range of values that represent a significant deviation between means. Recall that Tukey's honestly significant difference was 2.57. This larger critical difference reduces the likelihood of rejecting true null hypotheses, which provides more protection against a Type I error when you conduct many pairwise comparisons.

The drawback is that Tukey's *HSD* procedure is also less likely than the t test to reject *false* null hypotheses. In short, it has less power than the t test. But bear in mind that the main purpose of using the *HSD* is to protect against an unacceptably high familywise rate of Type I errors. The tradeoff is a somewhat larger chance of Type II errors.

In light of this discussion, you might wonder whether it is ever acceptable to conduct a *t* test in an investigation that involves many samples.

Planned Comparisons

In fact, it is sometimes both permissible and desirable to use the *t* test to compare two means in a multiple-treatments study—if the test is a planned comparison. A planned, or **a priori**, comparison occurs when *you test one sample mean against another to assess a specific hypothesis that existed prior to data collection* (Kirk, 1995). For example, if we had explicitly hypothesized that the motivation group would have higher responsiveness than the meditation group, then it would have been okay for us to use the *t* statistic to compare just those two conditions, either before or after carrying out the omnibus *F* test. Indeed, if a researcher is working with several means but has only two or three very specific expectations and no interest in other comparisons, he or she might legitimately use a couple of *t* tests *instead of* an overall *F* test.

The *t* test is just one type of *a priori* mean-comparison method. Other approaches are described in Kirk (1995) and Hays (1988). Those sources also note a number of restrictions on the use of a priori tests that exceed the scope of this textbook. Generally speaking, the number of planned comparisons you can perform is quite limited.

For reasons covered earlier, it not acceptable to conduct multiple *t* tests to "fish" for unanticipated significant mean differences as a sequel to an ANOVA. Such *unplanned tests* are called **post hoc** (i.e., after the fact) **comparisons**, and they require methods that are more conservative than the *t* test. Tukey's *HSD* is just one of a variety of post hoc procedures.

Post Hoc Mean-Comparison Procedures

You can use post hoc procedures to make a large number of mean comparisons, both planned and unplanned. As their label suggests, you usually should apply post hoc tests only after you obtain a significant *F* ratio (Scheffé's test is an exception; see below). Post hoc tests usually are less powerful than planned tests, which means that they are more likely to result in Type II errors.

I decided to illustrate this class of mean-comparison technique with Tukey's *HSD* because it is a popular, middle-of-the-road post hoc test. That is, it provides adequate protection from excessive Type I errors while not being overly apt to commit Type II errors. Two other post hoc methods are described next.

Newman-Keuls test   An older pairwise comparison procedure, the Newman-Keuls test is very similar to Tukey's test, both procedurally and in terms of "power." Like the *HSD*, this test uses the q statistic. A chief distinction is that the Newman-Keuls method requires that the sample means be ordered by magnitude. Relatedly, the critical difference in each pairwise comparison is not constant but increases as a function of the number of other means that lie between the two that are compared.

Scheffé's test   Under most circumstances, this test is the most conservative post hoc method. You can use Scheffé's test to make all possible comparisons among a set of k means while keeping the overall (i.e., cumulative) Type I error rate at or below the stated alpha level. Since this technique affords so much protection against Type I errors, you can apply it even if the omnibus F test is not significant. The disadvantage of this test is that it can lead to an excessive number of Type II errors—failing to reject a false H_0—particularly if you are working with a large number of means. Scheffé's critical difference, S, is given by:

$$S = \sqrt{(df_B)(F_{crit})(MS_W)\left(\frac{1}{N_i} + \frac{1}{N_j}\right)} \qquad (12.12)$$

where N_i and N_j are the sample sizes of the two conditions being compared, and F_{crit} is the critical F value used in the omnibus (i.e., overall) F test.

For our example,

$$S = \sqrt{(2)(3.89)(2.33)(\tfrac{1}{5} + \tfrac{1}{5})} = \sqrt{(18.13)(\tfrac{2}{5})} = \sqrt{(18.13)(0.40)}$$
$$= \sqrt{7.252} = 2.69$$

Any mean difference with an *absolute value* of at least 2.69 is considered significant. You can see that the critical difference required by this technique is larger than that required by *HSD* (which was 2.57). This reflects the greater conservatism and lower power of Scheffé's method. So you should use Scheffé's test for post hoc comparisons whenever you are especially concerned about avoiding Type I errors. Otherwise, use Tukey's *HSD*, the Newman-Keuls test, or some similarly moderate post hoc procedure.

Whew! We need to pause and reexamine the big picture for a moment.

TAKING STOCK

Here is a review of the purpose, rationale, and procedure of analysis of variance:

1. Use ANOVA to test the significance of all possible mean differences in an experiment that involves more than two treatments or conditions.

2. The rationale of ANOVA is to form a ratio of a between-treatments variance to a within-treatments variance, which is assumed to represent random-error variation. If the F ratio is several times larger than 1.0, then you conclude that the between-treatments variance reflects more than merely random error—that is, that there are reliable differences among the treatment-group means.

3. To do an ANOVA, find a crude index of total variation among all the raw scores: the total sum of squares. Then partition the total sum of squares into a between-treatments component (SS_B) and a within-treatments component (SS_W).

4. Convert SS_B and SS_W to actual variances, called "mean squares," by dividing them by their respective degrees of freedom: $MS_B = SS_B / df_B$; $MS_W = SS_W / df_W$.

5. Compute the F ratio by dividing MS_B by MS_W. Then determine whether the computed F statistic is significant by comparing it with F_{crit}, which is the F ratio expected on the basis of chance alone.

6. If the F is significant, you carry out pairwise-comparison tests to find out which particular mean differences caused the F to be significant.

A SECOND EXAMPLE

An additional example will consolidate the main ideas we have covered.

State the Problem and Nature of the Data

A psychologist designs an experiment to investigate the effect of "social reinforcers and punishers" on verbal behavior. All subjects are asked to describe their personal philosophies of life for ten minutes. The dependent variable is the number of optimistic statements subjects make in the ten-minute session. The independent variable is the social reaction that the experimenter makes in response to each optimistic statement. There are four "social reaction" treatment groups, each one corresponding to a different level of the independent variable:

1. *Control group*: The experimenter shows no reaction to any of the subjects' statements.

2. *"Um-huh" group*: The experimenter makes an encouraging grunt, "um-huh," in response to each optimistic statement the subject emits.

3. *"Huh-uh" group*: The experimenter makes a discouraging grunt, "huh-uh," in response to each optimistic statement the subject emits.

4. *Smile group*: The experimenter smiles each time the subject makes an optimistic statement.

Relative to the control condition, the encouraging-grunt condition and the smile condition are expected to increase the number of optimistic statements, and the discouraging-grunt condition is expected to decrease the number of optimistic statements.
Other relevant facts to note are:

- k symbolizes the number of treatments in the study; $k = 4$.
- Each treatment group has eight subjects; that is, each $N = 8$.
- The total number of observations in the study equals kN; that is, $N_{tot} = kN = 4 \cdot 8 = 32$.

State the Statistical Hypotheses

H_0: $\mu_1 = \mu_2 = \mu_3 = \mu_4$

H_1: Not all of the μ's are equal.

Choose a Level of Significance

Assume a conventional level of significance; alpha = .05.

Specify the Test Statistic

We will conduct a one-factor analysis of variance with 3 and 28 degrees of freedom. The between-treatments degrees of freedom = $df_B = k - 1 = 4 - 1 = 3$. The within-treatments degrees of freedom = $df_W = N_{tot} - k = 32 - 4 = 28$.

Determine the Critical Value Needed for Significance

Since the degrees of freedom for the numerator = $df_B = 3$, the degrees of freedom for the denominator = $df_W = 28$, and $\alpha = .05$, Table F tells us that F_{crit} is 2.95.

State the Decision Rule

If $F \geq 2.95$, reject H_0.

Compute the Test Statistic

The results of the experiment are shown in Table 12.10. There are obvious numerical differences among the treatment means. Our omnibus ANOVA will tell us whether any of those differences are significant.

Compute the grand mean   The data in Table 12.11 show that the grand mean is:

$$\overline{X}_G = \frac{\Sigma X}{N_{tot}} = \frac{416}{32} = 13$$

Table 12.10

RAW SCORES, MEANS, AND
VARIANCE ESTIMATES FROM
AN EXPERIMENT ON THE
EFFECT OF SOCIAL REACTION
ON THE INCIDENCE OF
OPTIMISTIC STATEMENTS

	CONTROL	UM-HUH	HUH-UH	SMILE
	15	15	9	16
	12	19	4	19
	14	13	7	11
	11	22	11	18
	14	21	3	16
	9	20	4	13
	14	20	4	14
	7	14	6	21
Sums:	96	144	48	128
N:	8	8	8	8
$\overline{X}$:	12	18	6	16
s^2:	8	12	8	10.86

Compute the sums of squares Table 12.12 provides the calculation of the total sum of squares:

$$SS_{tot} = \Sigma(X - \overline{X}_G)^2 = 944$$

The between-treatments sum of squares is found through:

$$SS_B = \Sigma[N(\overline{X} - \overline{X}_G)^2]$$
$$= 8(12 - 13)^2 + 8(18 - 13)^2 + 8(6 - 13)^2 + 8(16 - 13)^2$$
$$= 8(-1)^2 + 8(5)^2 + 8(-7)^2 + 8(3)^2$$
$$= 8(1) + 8(25) + 8(49) + 8(9)$$
$$= 8 + 200 + 392 + 72$$
$$= 672$$

We'll use the shortcut (i.e., subtraction) method to get the within-treatments sum of squares:

$$SS_W = SS_{tot} - SS_B = 944 - 672 = 272$$

Compute the mean squares The two variances that will be used in the F ratio are:

$$\text{Between-treatments mean square} = MS_B = \frac{SS_B}{df_B} = \frac{672}{3} = 224$$

$$\text{Within-treatments mean square} = MS_W = \frac{SS_W}{df_W} = \frac{272}{28} = 9.71$$

Table 12.11

CALCULATION OF THE GRAND MEAN IN AN EXPERIMENT ON SOCIAL REACTION EFFECTS	TREATMENT GROUPS	X
	Control	15
		12
		14
		11
		14
		9
		14
		7
	Um-huh	15
		19
		13
		22
		21
		20
		20
		14
	Huh-uh	9
		4
		7
		11
		3
		4
		4
		6
	Smile	16
		19
		11
		18
		16
		13
		14
		21
	Sum:	416
	Mean:	13

Compute the F ratio $F = MS_B/MS_W = 224/9.71 = 23.07$. This ANOVA is summarized in Table 12.13.

Consult the Decision Rule and Make a Decision

Since the computed F of 23.07 is larger than $F_{crit} = 2.95$, we reject H_0. An appropriate conclusion is: "Subtle social reinforcers and punishers significantly affect the frequency of optimistic statements uttered by subjects in free operant situations, $F(3, 28) = 23.07$, $p < .05$." The only remaining question is: Which specific mean differences are significant?

Table 12.12

CALCULATION OF TOTAL SUM OF SQUARES IN AN EXPERIMENT ON SOCIAL REACTION EFFECTS

	X	$(X - \bar{X}_G)$	$(X - \bar{X}_G)^2$
	15	2	4
	12	−1	1
	14	1	1
Control	11	−2	4
	14	1	1
	9	−4	16
	14	1	1
	7	−6	36
	15	2	4
	19	6	36
	13	0	0
Um-huh	22	9	81
	21	8	64
	20	7	49
	20	7	49
	14	1	1
	9	−4	16
	4	−9	81
	7	−6	36
Huh-uh	11	−2	4
	3	−10	100
	4	−9	81
	4	−9	81
	6	−7	49
	16	3	9
	19	6	36
	11	−2	4
Smile	18	5	25
	16	3	9
	13	0	0
	14	1	1
	21	8	64
Sum:		0	944

Table 12.13

ANOVA SUMMARY TABLE FOR
EXPERIMENT ON SOCIAL
REACTION EFFECTS

SOURCES OF VARIANCE	SUMS OF SQUARES	DEGREES OF FREEDOM	MEAN SQUARES	F RATIO
Between treatments	672	3	224	23.07
Within treatments	272	28	9.71	
TOTAL	944	31		

Pairwise Comparisons

Tukey's honestly significant difference in the present example[3] is:

$$HSD = q\sqrt{\frac{MS_W}{N}} = 3.90\sqrt{\frac{9.71}{8}} = 3.90\sqrt{1.214}$$

$$= 3.90(1.10) = 4.29$$

Thus, if any two treatment means in Table 12.10 differ *by at least 4.29*, then they are significantly different. The mean of the "Huh-uh" group was significantly lower than all the others. Moreover, a discouraging reaction ("Huh-uh") reliably decreased optimistic statements relative to no response at all (i.e., the control condition). Although the encouraging "Um-huh" response reliably increased optimistic statements in comparison with the control condition (18 versus 12), simply smiling did not (16 versus 12).

EFFECT SIZE

A significant *F* ratio tells us that the independent variable had a significant effect on the dependent variable. But given that outcome, it is sometimes useful to ask: How big was the effect? This is the "effect size" question, and it can be answered in terms of the proportion (or percent) of variance in the dependent variable that is explained by the independent variable. The greater the proportion of variance accounted for, the larger the effect the independent variable had on the dependent variable.

When your ANOVA yields a significant outcome, you can easily

---

[3] Note that $df_W = 28$ for this problem. Table Q does not list a degrees of freedom of 28. In such situations we simply "drop back" to the next lower degrees of freedom listed—24 in this case. This "drop back" rule was introduced in Chapter 11.

gauge the effect size through a statistic called **eta-squared**, abbreviated η^2. Eta-squared is *the proportion of variance in the dependent variable that is explained by the independent variable for the sample of data used*. The formula for this index is:

$$\eta^2 = \frac{SS_B}{SS_{tot}} \qquad (12.13)$$

For the results shown in Table 12.13, $\eta^2 = SS_B/SS_{tot} = 672/944 = .71$ Hence, a whopping 71% of the variation in the frequency of optimistic statements was explained by the experimenter's social reactions to that type of statement.

η^2 has the same general interpretation as r^2 (see Chapters 6 and 7) . However, r^2 is applicable only when X and Y have a linear relationship, whereas η^2 reflects the strength of both linear and nonlinear covariation between the independent and dependent variables.

NOTE: Eta-squared expresses the proportion of variance accounted for *in the sample only*. It does not necessarily represent the proportion of variance explained in the population from which the sample was drawn. A different statistic, called "omega- squared," is an unbiased estimator of the proportion of variance accounted for in the population (see Hays, 1988).

ASSUMPTIONS OF ANOVA

The assumptions of a simple one-factor analysis of variance procedure, such as you have been studying, are the same as those underlying an independent-samples t test (see Chapter 11). They are:

1. The dependent variable's scale of measurement is at the *interval or ratio level*. (Interval scales have equal units of measurement all along the measurement dimension. Ratio scales are interval scales with true zero points. See Chapter 4 for a review of these ideas.)
2. The *observations in each sample must be independent* of one another; this means that no subject's score on the dependent variable should influence any other subject's score.
3. The dependent variable is *normally distributed* in the populations represented by the samples.
4. The *variances* of the sampled populations are *equal*. This is the "homogeneity of variance" assumption that was introduced in Chapter 11.

Assumption 1 is important. If you cannot assume that your measure of behavior at least approximates an interval scale, then the F test might not give you valid information. Assumption 2 is critical, but it is usually satisfied in an independent-groups study by randomly assigning subjects to conditions and using good research methodology. Assumption 3 does not have much effect on the accuracy of an F test (Hays, 1988). Furthermore, it can be presumed to be true whenever there are at least 30 observations in each sample. Assumption 4 may be violated to some degree as long as the samples have equal N values (Hays, 1988). What's more, it is reasonable to assume that the population variances are equal if the largest of the sample variance estimates is less than twice the size of the smallest one—that is, if largest $s^2 <$ smallest $s^2 \cdot 2$.

OTHER KINDS OF ANOVA

This chapter has introduced you to one-factor, or "one-way," analysis of variance, so named because it tests the effect of just one independent variable. More complex research designs, called "factorial" designs, examine the effects of two or more independent variables in a single experiment, with special attention given to the joint effect—or interaction—among the independent variables. Factorial experiments require a slightly more complex "two-factor" ANOVA, which will be featured in the next chapter.

Chapter 14 will cover "repeated-measures" ANOVA, which is designed to analyze data from multiple-treatment experiments in which each subject serves in all of the treatments. That type of experimental design is an extension of the correlated-samples design that you learned about in Chapter 11 on two-sample t tests. You will discover that the kind of ANOVA used to analyze repeated-measures experiments is logically similar to the ANOVA procedure that applies to factorial experiments.

KEY TERMS

completely randomized
   independent-samples design
analysis of variance
omnibus test
testwise error rate
familywise error rate
one-factor ANOVA
F ratio
grand mean
mean square
N

N_{tot}
k
analysis of variance summary
   table
total sum of squares
within-treatments sum of
   squares
between-treatments sum of
   squares
worksheet
pairwise comparison

Tukey's honestly significant
   difference

a priori test

post hoc comparison

eta-squared

SUMMARY

1. You use ANOVA to test the significance of all possible mean differences in an experiment that involves more than two treatments or conditions.

2. ANOVA is preferable to doing multiple t tests because it provides protection against an excessive number of Type I decision errors.

3. The logic of ANOVA reflects the logic of the experimental method. An experiment starts with equivalent treatment groups, then manipulates an independent variable, and finally measures subjects on a dependent variable. If the treatment-group means are reliably different, then that difference is ascribed to the effect of the independent variable.

4. The rationale of ANOVA is to form a ratio of a between-treatments variance to a within-treatments variance, which is assumed to represent random error variation. If the F ratio is several times larger than 1.0, then you conclude that the between-treatments variance reflects more than merely random error—that is, that there are reliable differences among the treatment-group means.

5. To do an ANOVA, you find a crude index of the total variation among raw scores: the total sum of squares. Then you partition the total sum of squares into a between-treatments component (SS_B) and a within-treatments component (SS_W).

6. You convert SS_B and SS_W to actual variances, called "mean squares," by dividing them by their respective degrees of freedom. $MS_B = SS_B / df_B$; $MS_W = SS_W / df_W$.

7. You compute the F ratio by dividing MS_B by MS_W. Then you determine whether the computed F statistic is significant by comparing it to F_{crit}, which is the F ratio expected on the basis of chance alone.

8. If the F is significant, you carry out Tukey's HSD test to ascertain which particular mean differences caused the F to be significant: $HSD = q\sqrt{MS_W/N}$. *Pairwise comparison* is the term applied to two-mean tests that you conduct as a followup to a significant F test.

9. You may perform planned pairwise comparisons with a t test. You can do unplanned comparisons with post hoc methods, such as Tukey's HSD and Scheffé's test.

10. When you wish to ascertain the magnitude of the effect that an independent variable had on a dependent variable, you use eta-squared: $\eta^2 = SS_B/SS_{tot}$. η^2 is the proportion of variance in the dependent variable that is accounted for by the independent variable *in the sample data*. Proportion of variance accounted for is a conventional measure of "effect size."

REVIEW QUESTIONS

1. Define or describe: completely randomized independent-samples design, analysis of variance, testwise error rate, familywise error rate, and one-factor ANOVA.
2. Under what circumstances would it be desirable to use an analysis of variance rather than *t* tests to assess the statistical significance of mean differences? What consideration makes ANOVA preferable?
3. Distinguish between the testwise error rate and the familywise error rate. Which of these two error rates is vulnerable to inflation by multiple *t* tests performed on a single set of means, and how is the error-rate inflation figured?
4. Describe the logic of the experimental method in your own words. Relate that logic to the rationale underlying ANOVA.
5. Define or describe: *F* ratio, grand mean, mean square, *N*, N_{tot}, and *k*.
6. Describe the theory of ANOVA and relate it to the rationale underlying the *F* ratio.
7. Define or describe: analysis of variance summary table, total sum of squares, within-treatments sum of squares, and between-treatments sum of squares.
8. In what sense does ANOVA involve a "partitioning of variance"? Be specific and thorough in formulating your answer.
9. Define or describe: worksheet, pairwise comparison, Tukey's honestly significant difference, a priori comparisons, post hoc comparisons, and eta-squared.
10. Why is the *F* test called an "omnibus test," and why is it necessary to do pairwise comparisons after obtaining a significant *F* statistic?
11. A pool of subjects was randomly divided into five treatment groups. The groups were administered daily doses of vitamin C over a 12-month period. The data in the table represent the number of cold and flu viruses reported by the subjects as a function of their vitamin C dosage. Using the .05 level of significance, carry out a complete ANOVA on these data, and display the results in an ANOVA summary table. Draw an appropriate conclusion.

0 mg	250 mg	500 mg	1000 mg	2000 mg
6	3	3	4	1
5	4	3	1	0
3	5	4	0	2
2	4	2	3	1

12. Use Tukey's test to perform pairwise comparisons on the means of the data in Question 11. Do the same comparisons with Scheffé's test, and note any different decisions that result. Calculate and interpret eta-squared.

13. A comparative psychologist has attempted to selectively breed rats for high and low levels of open-field emotionality. The psychologist randomly selects five specimens each from the high- and low-emotionality populations. The psychologist then tests the samples for emotionality in an open-field situation and compares their behavior to that of control animals. The results appear in the table. Using the .05 level of significance, carry out a complete ANOVA on these data, and display the results in an ANOVA summary table. Draw an appropriate conclusion.

High Emotionality	Control	Low Emotionality
13	12	10
17	15	15
15	13	14
14	16	20
16	14	16

14. A psychiatrist randomly assigns several Alzheimer's disease sufferers to four treatment groups and administers different brain chemical supplements to each group. The psychiatrist then gives each group a standard test of immediate memory. The results of the memory test are given in the table, with higher scores signifying better memory. Using the .05 level of significance, carry out a complete ANOVA on these data, and display the results in an ANOVA summary table. Draw an appropriate conclusion.

Brain Chemical Administered

Acetylcholine	L-Dopa	Norepinephrine	Tryptophan
30	36	24	26
25	28	28	22
40	20	32	22
45	24	36	17
35	32	25	25
35	28	35	18

15. A parapsychologist hypothesizes that a person's extrasensory perception (ESP) ability varies as a function of his or her zodiac

sign. From a database that's been under development for years, the parapsychologist randomly selects five college students from each of six signs and administers an ESP test to each of them. The students' ESP scores are listed in the table, with higher scores representing greater ESP ability. Using the .05 level of significance, carry out a complete ANOVA on these data, and display the results in an ANOVA summary table. Draw an appropriate conclusion.

Libra	Pisces	Gemini	Cancer	Capricorn	Leo
30	20	21	23	14	30
22	22	25	18	23	24
14	20	26	13	20	18
19	20	23	19	26	24
25	18	20	17	17	24

16. In a forensic psychology experiment, college students see a videotape of a sports car sideswiping a limousine. Then they are asked to estimate how fast the sports car was traveling at the instant of impact. The independent variable is the specific wording used by the person asking for the speed estimate. On a random basis, one third of the subjects are asked how fast the sports car was traveling when it "brushed" the limousine, one third are asked how fast the sports car was traveling when it "hit" the limousine, and one third are asked how fast the sports car was traveling when it "smashed" the limousine. The subjects' estimates of the speed of the vehicle at the instant of impact appear in the table. Using the .05 level of significance, carry out a complete ANOVA on these data, and display the results in an ANOVA summary table. Draw an appropriate conclusion.

"Brushed"	"Hit"	"Smashed"
37	42	51
39	45	48
41	48	45

17. Use Table F to determine F_{crit} for each of the following situations:
    (a) $\alpha = .01$, $df_B = 7$, $df_W = 60$
    (b) $\alpha = .01$, $df_B = 4$, $df_W = 30$
    (c) $\alpha = .05$, $df_B = 5$, $df_W = 120$
    (d) $\alpha = .05$, $df_B = 3$, $df_W = 25$

18. The following summary statistics are from an investigation of human memory. Free recall was measured in three treatment groups that used different methods of studying a list of names that they were instructed to memorize. Given only the information in the table, do an analysis of variance. Use the .01 level

of significance, and display the results of your work in an ANOVA summary table. [*Hint*: You will need to use formulas (12.3) and (12.9), among others.]

Memorizing Strategy

	Rote Rehearsal	Imagery	Control
$\overline{X}$	32	40	36
s^2	4	6	5
N	20	20	20

19. A clinical psychologist expects self-efficacy scores to differ among psychodiagnostic categories. The psychologist randomly selects four patients from each of four diagnostic groups in a university medical center. The self-efficacy scores of the groups are listed in the table. Using the .05 level of significance, carry out a complete ANOVA on these data, and display the results in an ANOVA summary table. If the F statistic is significant, conduct the necessary pairwise comparisons with Scheffé's test, and draw appropriate conclusions.

Depression	Mania	Schizophrenia	Paranoia
37	90	60	76
42	84	58	86
38	87	56	84
43	87	54	78

20. Ten of 30 seasonally depressed patients were randomly assigned to each of three types of artificial lighting. The patients worked eight hours per day in their respective lighting environments over a nine-week period. It was hypothesized that "general happiness" scores would vary as a function of the type of lighting used. The summary statistics on "happiness scores" from that investigation appear in the table. Given only this information, do an analysis of variance. Use the .01 level of significance, and display the results of your work in an ANOVA summary table. [*Hint:* You will need to use formulas (12.3) and (12.9), among others.]

Type of Ambient Lighting

	Incandescent	Standard Fluorescent	Full Spectrum
$\overline{X}$	4	6	10
s^2	3	2	3
N	10	10	10

21. $k = 7$, $N = 10$ for all treatment groups. What are df_{tot}, df_W, and df_B?

22. $k = 5$, $N = 5$ for all treatment groups, $SS_B = 90$, $SS_W = 30$. What are df_{tot} and SS_{tot}?

23. Complete the following ANOVA summary table.

Sources of Variation	Sums of Squares	Degrees of Freedom	Mean Squares	F Ratio
Between treatments	80		40	
Within treatments		12		
TOTAL	100	14		

24. Complete the following ANOVA summary table.

Sources of Variation	Sums of Squares	Degrees of Freedom	Mean Squares	F Ratio
Between treatments	100	5		
Within treatments	160			
TOTAL		47		

25. With alpha = .01, $F = 17.19$. $MS_W = 4$, $df_W = 30$, $\overline{X}_1 = 22.5$, $\overline{X}_2 = 25$, and $\overline{X}_3 = 20$. Calculate Tukey's *HSD* and determine which particular mean differences caused the significant *F*.

CHAPTER 13 TWO-WAY ANALYSIS OF VARIANCE

Many interesting questions in the behavioral sciences don't have simple answers. The reason is that the effect of one independent variable often changes as a function of other variables. For example, what impact do stimulant drugs have on behavior? The answer is: It depends. Most people who take a stimulant drug become more energetic and active. But if such a drug is taken by someone who has an attention deficit disorder (with hyperactivity), that person actually becomes calmer and more sedate (Safer & Krager, 1988). Consider a second example: How does frustration affect a person's tendency to become aggressive (Dollard et al., 1939)? Again the most accurate answer is: It depends. A frustrated person might not exhibit any increase in aggression at all if the cause of the frustration is "understandable"—that is, viewed as resulting from normal, unavoidable circumstances. On the other hand, even mild frustration will likely produce an aggressive response in the presence of aggressive cues, such as weapons (see Myers, 1990). To repeat, the effect of a particular variable is often conditional on a second variable: The second variable is said to "interact" with the first to produce a unique outcome.

The topic you are about to study—"two-way" analysis of variance—is the behavioral scientist's tool for statistically identifying the "It depends" answers to behavioral questions. Chapter 12 introduced you to one-way analysis of variance (ANOVA), which tests the effect that one independent variable has on behavior. In this chapter, the basic logic of ANOVA will be extended to assess the joint influence—or "interaction effect"—of two independent variables. In addition to testing the combined effect of two variables acting together, two-way ANOVA assesses the significance of each independent variable alone. In short, two-way ANOVA uses one global statistical analysis to give you information on:

- The impact of independent variable *A* (factor *A*)
- The influence of independent variable *B* (factor *B*)
- The effect of the interaction of factor *A* and factor *B*

An example will help illustrate what is meant by these ideas.

A FACTORIAL EXPERIMENT

Formal Characteristics of Factorial Studies

You will use two-way analysis of variance (i.e., two-way ANOVA) in "factorial" research designs. In a two-way **factorial design**, you

- Set up *a* levels, or values, of one independent variable, called factor *A*.
- Set up *b* levels, or values, of a second independent variable, called factor *B*.

- Combine each level of factor *A* with every level of factor *B*, to produce $a \cdot b$ "treatment groups" in your study.

As you can see, *the defining characteristic of a factorial design is that all designated levels of one independent variable are combined with all designated levels of a second independent variable.* It follows that the number of treatment combinations, or groups, in a two-way factorial study is equal to the number of levels of the first independent variable multiplied by the number of levels of the second independent variable. If independent variable *A* has two levels, *A1* and *A2*, and independent variable *B* has two levels, *B1* and *B2*, then the study has 2 × 2, or 4, treatment combinations: group *A1B1*, group *A1B2*, group *A2B1*, and group *A2B2*. This two-by-two arrangement is shown in Figure 13.1.

A factorial design is called a **factorial experiment** when *subjects are randomly assigned to the various levels of at least one of the independent variables in the study.* As in normal experimental methodology, that variable is then manipulated while all other influences are held constant. In a **completely randomized factorial experiment**, *subjects are randomly assigned to all levels of both independent variables, and each subject serves in only one of the possible treatment combinations (for example, a subject would serve in condition A1B1 but not in A1B2, A2B1, or A2B2).* The example we will consider here illustrates such a design.

Example 1: Drugs and Productivity

Anxiolytic drugs are designed to reduce anxiety. This class of drugs includes such well-known medications as Librium, Valium, and Xanax. Anxiolytics are widely prescribed as treatment for a variety of anxiety conditions, including work-related stress disorders. Unfortunately, some of the side effects of antianxiety medications are lethargy and deterioration in thinking, decision making, reading, and carrying out skilled movements (Lickey & Gordon, 1991). This list of side effects prompts us to ask whether the daily use of,

Figure 13.1 *Structure of a 2 x 2 Factorial Research Design (Note: All levels of variable A are combined with all levels of variable B.)*

		Factor *B*	
		B1	*B2*
Factor *A*	*A1*	*A1B1* Treatment group 1	*A1B2* Treatment group 2
	A2	*A2B1* Treatment group 3	*A2B2* Treatment group 4

say, Valium, might make an employee less productive (if calmer) than he or she normally would be. If the general answer to this question is yes, we might also ask whether there are ever situations in which the use of Valium might actually make someone more productive than he or she would be without the drug—very stressful or aversive working conditions, for example. More precisely, *does the effect of Valium on productivity change with different levels of work-environment aversiveness?* This kind of slightly complicated question requires a factorial experimental design and a two-way ANOVA, where one of the two independent variables is "dose of Valium" (some versus none) and the other independent variable is "aversiveness of the work situation" (normal versus aversive).

To study this question, we'll move to the animal laboratory, where we can consider a comparative study of productivity in trained rats as a function of medication and environmental aversiveness. Suppose that a psychopharmacologist trains 20 rats to press a bar at a high rate in order to receive an occasional sip of milk. Thus, the number of bar presses per minute is the measure of the rats' "productivity," which is the dependent variable in this study. Under normal "work" conditions, we can expect rats to make about 50 bar presses per minute.

Bear in mind that the researcher intends to test the effect on productivity (bar pressing) of (1) the presence versus the absence of Valium in the brain (factor *A*) and (2) the aversiveness of the work situation (normal versus aversive, factor *B*). The researcher also wants to look at the joint effect of different combinations of factor *A* and factor *B*—that is, the interaction effect of these variables. Since factor *A* (the drug condition) has $a = 2$ levels and factor *B* (aversiveness of the work situation) has $b = 2$ levels, combining all levels of both independent variables results in the following $ab = 4$ treatment combinations:

A1B1: placebo/normal situation

A1B2: placebo/aversive situation

A2B1: Valium/normal situation

A2B2: Valium/aversive situation

The structure of this 2×2 (called "two-by-two") factorial design is shown in Figure 13.2. Note that each "cell" of the figure corresponds to one of the four treatment groups.

This study used a completely randomized factorial design because five of the 20 rats were randomly assigned to each treatment combination. That is, five rats were assigned to *A1B1*, five to *A1B2*, five to *A2B1*, and five to *A2B2*, entirely on a random basis. Hence,

A1B1: Five rats received no drug and worked in a nonaversive situation.

Figure 13.2 *Schematic 2 × 2 Factorial Structure of Experiment on Valium Effects*

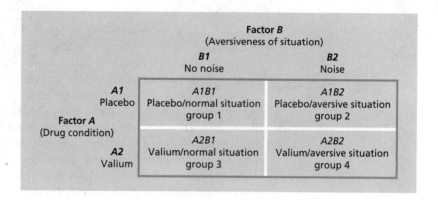

A1B2: Five received no drug and worked in an aversive situation.

A2B1: Five were administered Valium and worked in a non-aversive situation.

A2B2: Five were administered Valium and worked in an aversive situation.

Valium was administered via injection. The "placebo" (no drug) rats received injections of an inert solution. Rats in the nonaversive work situation pressed the bar at will and received a sip of milk periodically for their efforts. Rats in the aversive work situation received a moderately loud blast of foghorn noise whenever they pressed the bar; they also received a sip of milk periodically for their efforts. Essentially, then, the rats that were blasted with a noxious noise for pressing the bar found it painful to do their job—a condition to which many humans would have little trouble relating.

The results of this investigation appear in Table 13.1, which shows the raw scores in each treatment combination, the means of levels of factor *A* (drug condition), the means of levels of factor *B* (situational aversiveness), and the grand mean for the entire experiment.

"EFFECTS" IN A TWO-WAY FACTORIAL DESIGN

As I mentioned earlier, you will examine three kinds of effects in a two-way factorial study. Two are referred to as *main effects*, and the third is, of course, the *effect of the interaction* of the two independent variables. Let's define these terms more completely.

Table 13.1

NUMBER OF BAR PRESSES PER
MINUTE AS A FUNCTION OF
DRUG CONDITION AND
SITUATIONAL AVERSIVENESS
(NO NOISE VERSUS NOISE)

		FACTOR B		
		NO NOISE B1	NOISE B2	FACTOR A MEANS
	PLACEBO A1	A1B1 50 46 42 58 54	A1B2 3 4 12 13 8	29
FACTOR A	VALIUM A2	A2B1 38 46 37 47 42	A2B2 16 23 9 11 21	29
	FACTOR B MEANS	46	12	
			Grand mean = 29	

Main Effect

Each of the two independent variables, or factors, can have a main effect on the dependent variable. A **main effect** has all of these meanings:

- The effect of one factor averaged across the other factor
- The effect of one factor when the other independent variable is held constant statistically
- The overall effect of one independent variable irrespective of its interaction with the other independent variable

As you will see, a main effect is usually expressed as a difference between the means of the levels of an independent variable. In a two-way factorial design, you will check for the presence of a main effect of factor A (e.g., drug condition) and a main effect of factor B (e.g., level of aversiveness). The overall means of factor A are found by averaging across all levels of factor B, and the overall means of factor B are calculated by averaging across all levels of factor A.

Example: Factor B will have a main effect on productivity if the overall productivity mean in the nonaversive work condition (B1) is significantly higher (or significantly lower) than the overall productivity mean in the aversive work condition (B2).

Interaction Effect

In a two-way factorial design, the **interaction effect** has these meanings:

- The unique joint effect of two (or more) independent variables
- The combined effect of two factors acting together
- The tendency of the effect of one independent variable to change at different levels of another independent variable

An interaction is usually identified by "different mean differences." If factors A and B interact, for example, then $(\overline{X}_{A1} - \overline{X}_{A2})$ will be different at $B2$ than at $B1$. Interactions can take two forms:

1. A mean difference associated with one factor changes in *size* from one level to the next of the second factor
2. A mean difference associated with one factor changes in *direction*—from a positive difference to a negative one, for instance— across levels of the second factor.

If the difference changes in *size*, then $(\overline{X}_{A1} - \overline{X}_{A2})$ is relatively larger (or smaller) at $B1$ than it is at $B2$. *Example*: This kind of interaction exists if the productivity of the placebo subjects is much better than that of the Valium subjects under normal work conditions but only slightly better under aversive conditions.

If the difference changes in *direction*, then $(\overline{X}_{A1} - \overline{X}_{A2})$ is, say, positive at $B1$ but negative at $B2$. *Example*: This kind of interaction exists if the productivity of the placebo subjects is higher than that of the Valium subjects under normal work conditions but lower than that of the Valium subjects under aversive conditions.

You should note that *interactions are always bidirectional* in the following sense: If factor A has different mean differences at the respective levels of B, then factor B will have different mean differences at the respective levels of A. In short, if A interacts with B, then B must also interact with A.

A Note on the Independence of Effects in Two-Way ANOVA

It is important to realize that the three possible effects in a two-way ANOVA are both additive and independent. The model of effects is:

The Interaction
↓

Raw score = grand mean + effect of A + effect of B + effect of $A \times B$ + error

As this expression suggests, all the possible effects add together to determine each score in the data set of a 2×2 factorial study. But an effect might not be significant, meaning that it does not influence the score in any systematic way.

If a main effect is not significant, does that mean that the $A \times B$ interaction can't be significant either? Not at all. All three possible effects—A, B, and $A \times B$—are *completely independent of one another*. If one of the three effects is not significant, it simply becomes a 0 in the ANOVA model shown above, but the remaining effects can still contribute to the score. So it is possible, and indeed fairly common, to have a significant $A \times B$ interaction when neither factor A nor factor B has a significant effect by itself. In that case, we have:

Raw score = grand mean + 0 + 0 + effect of $A \times B$ + error

What's more, either factor A or factor B, or both, can have a reliable impact on behavior even when the interaction of the two variables is nonsignificant, as represented by:

Raw score = grand mean + effect of A + effect of B + 0 + error

Setting up null and alternative hypotheses in factorial studies will reinforce the concepts of main effect and interaction.

STATEMENT OF STATISTICAL HYPOTHESES IN THE VALIUM EXPERIMENT

In a factorial experiment, you should establish a set of null and alternative hypotheses for each of the main effects and for the interaction.

Hypotheses for a Main Effect of Factor A

The null hypothesis regarding the drug condition states that the rats that received Valium belong to the same population as the rats that received the placebo injections. The two population means for bar presses per minute are expected to be equal. This expectation is the same as contending that Valium will not affect the rats' productivity:

Null hypothesis→H_0: $\mu_{A1} - \mu_{A2} = 0$

Notice that factor B (situational aversiveness) is ignored in any hypothesis that concerns the main effect of factor A. Also observe that the main effect of factor A centers on the difference between the overall mean of $A1$ and the overall mean of $A2$, where each of those means is obtained by averaging across both levels of factor B. Averaging across all levels of an independent variable is the statistical equivalent of ignoring that variable. Ignoring B is exactly what you want to do when you are considering a main effect of A.

The alternative hypothesis that is pitted against the null hypothesis looks like this:

Alternative hypothesis→H_1: $\mu_{A1} - \mu_{A2} \neq 0$

It is worth noting that F tests are always nondirectional. Statistical significance can result from either a positive or a negative deviation between levels of factor A.

> **CONCEPT RECAP**
> You will recall from reading Chapter 12 that the F test is the final product of an analysis of variance. The F statistic is the ratio of a between-treatments variance to a within-treatments variance, or mean square$_{\text{between treatments}}$ divided by mean square$_{\text{within treatments}}$.

Hypotheses for a Main Effect of Factor B

The statistical hypotheses that pertain to factor B (situational aversiveness) parallel those for factor A:

Null hypothesis→H_0: $\mu_{B1} - \mu_{B2} = 0$

Alternative hypothesis→H_1: $\mu_{B1} - \mu_{B2} \neq 0$

Note that it is the mean difference between levels of situational aversiveness that defines the main effect of factor B, and that the two factor B means are obtained by averaging across both levels of factor A.

Hypotheses for the Effect of the Interaction of Factors A and B

The null hypothesis for the $A \times B$ interaction holds that the effect of factor A will be the same at level 1 of factor B (i.e., the nonaversive situation) as it is at level 2 of factor B (i.e., the aversive situation). In other words, H_0 contends that the mean difference between $A1$ and $A2$ will *not* differ at the respective levels of B:

Null hypothesis→H_0: $(\mu_{A1B1} - \mu_{A2B1}) - (\mu_{A1B2} - \mu_{A2B2}) = 0$

In plain English, this hypothesis asserts that the performance difference between drugged and sober rats will be basically the same whether the rats are in an aversive work situation or in a nonaversive work situation.

If there actually is an interaction between the drug condition and the level of aversiveness, however, we would expect the drugged versus sober performance difference to vary depending on whether the rats were punished or not punished for doing their job. This alternative expectation is expressed as:

Alternative hypothesis→H_1: $(\mu_{A1B1} - \mu_{A2B1}) - (\mu_{A1B2} - \mu_{A2B2}) \neq 0$

It should now start to become clear that an interaction effect literally reflects "different mean differences." If there is an interaction, for example, then the mean performance difference between drugged and sober rats might be larger under nonaversive work conditions than under aversive conditions. Or the drugged rats might perform worse than sober rats in the nonaversive situation but better than the sober rats in the aversive situation; that is, the mean difference might go from a positive one to a negative one. (Entertain the latter possibility for a while. You just might encounter it again later on.)

Now that you have a general idea of what is meant by main effects and interactions, we are ready to look at ways of identifying these effects in the statistical data. Learning how to think about, recognize, and verify "effects" in factorial studies will also make these concepts more meaningful to you.

There are three ways of examining main effects and interactions in a behavioral investigation: (1) inspection of numerical mean differences, (2) inspection of graphical trends, and (3) statistical analysis (i.e., the actual two-way ANOVA). The first two approaches—examination of means and graphs—provide preliminary evidence of effects in factorial studies. In short, they merely suggest that a main effect or interaction exists in the data. However, *only the outcome of the ANOVA itself can verify that reliable effects have, in fact, occurred in the investigation.*

NUMERICAL INDICATORS OF EFFECTS

Main Effect of Factor A

The drug-condition means listed in Table 13.1 show that the mean number of bar presses by rats in the placebo condition was 29. The rats on Valium also had a mean of 29. It is worth repeating that these factor A means were obtained by averaging across both levels of factor B. Since $\overline{X}_{A1} - \overline{X}_{A2} = 29 - 29 = 0$, there clearly is no main effect of the drug condition in this study. When averaged across situational aversiveness, the drug condition had no overall effect on the rats' productivity.

When the overall mean difference between levels of a variable is 0, you can be certain that that variable had no main effect on behavior. When the overall mean difference is small, it is also likely that the variable had no significant effect. In contrast, a large mean difference often is indicative of a main effect. However, even such a large difference must be tested for significance in the two-way ANOVA procedure. When sample sizes are small, as they are in this example, even a sizable mean difference might be nothing more than a chance outcome.

Main Effect of Factor *B*

The situational aversiveness variable is associated with a large overall mean difference for factor *B*. Referring to Table 13.1, we see that the mean number of bar presses per minute in the normal work situation was 46, whereas the average rat in the aversive condition produced only 12 bar presses per minute. Numerically, this main effect of factor *B* is represented by $\overline{X}_{B1} - \overline{X}_{B2} = 46 - 12 = 34$. This large mean difference clearly suggests a main effect of situational aversiveness in this study. When averaged across drug conditions, situational aversiveness appears to have had an overall effect on the rats' productivity. A little later, our ANOVA computations will verify that this is a reliable effect.

A × *B* Interaction Effect

To get a tentative appraisal of an interaction effect, we need to turn our attention to Table 13.2, which shows the mean number of bar presses for each of the four treatment groups in the Valium study. It is important that the table also displays the *A1* (placebo) versus *A2* (Valium) mean difference for each level of factor *B*. Notice that the placebo rats averaged 8 more bar presses per minute in the normal work environment, but 8 *fewer* bar presses per minute in the aversive situation. Thus, the *direction of the mean differences changed* from positive to negative as the work environment was altered from normal to aversive. This outcome means that the aversiveness variable modified the effect of the drug variable. In a word, the variables *interacted*. Formally, the "different mean differences" indicative of an interaction effect are given by:

$$(\overline{X}_{A1B1} - \overline{X}_{A2B1}) - (\overline{X}_{A1B2} - \overline{X}_{A2B2}) =$$

$$(50 - 42) - (8 - 16) = (8) - (-8) = 16$$

Table 13.2

GROUP MEANS FOR ALL DRUG-CONDITION/ SITUATIONAL-AVERSIVENESS COMBINATIONS			FACTOR *B*: SITUATIONAL AVERSIVENESS	
			NO NOISE *B1*	NOISE *B2*
	PLACEBO	*A1*	*A1B1* $\overline{X}_{A1B1} = 50$	*A1B2* $\overline{X}_{A1B2} = 8$
FACTOR *A*: DRUG CONDITION	VALIUM	*A2*	*A2B1* $\overline{X}_{A2B1} = 42$	*A2B2* $\overline{X}_{A2B2} = 16$
	MEAN DIFFERENCE:		8	−8

The apparent interaction effect is shown by a change in the direction of the factor *A* mean difference from positive to negative. I wish to remind you, however, that some interactions take the form of a change in the *size* (rather than the direction) of a mean difference. In the latter case, it is said that factor *A* has a bigger (or smaller) effect at *B1* than at *B2*. By contrast, a change in the direction of a mean difference, such as we have here, says that factor *A* has a *different kind* of effect at *B1* than at *B2*. In the context of our example, Valium appears to harm productivity in a normal work environment but facilitate it in an aversive work situation.

If the "difference between mean differences" is very small or 0, then there is no interaction between the independent variables. In contrast, if the "difference between mean differences" is a large number, as in this example, then it is likely that there is an interaction effect in the data (which must be verified by an appropriate *F* test).

GRAPHICAL INDICATORS OF EFFECTS

Line graphs are another device that you can use to get preliminary indications of main and interaction effects. Such graphs should also strengthen your understanding of those effects.

Main Effect of Factor *A*

Figure 13.3 shows a line graph in which the mean number of bar presses per minute is plotted against the levels of factor *A*. Since the placebo and Valium groups had exactly the same means ($\overline{X}_{A1} = \overline{X}_{A2} = 29$), the line function is parallel to the horizontal axis of the graph. As a general rule, *to the degree that a plot of a variable's*

Figure 13.3 *Line Graph of Factor A (Drug Condition) Effects: Example of No Main Effect*

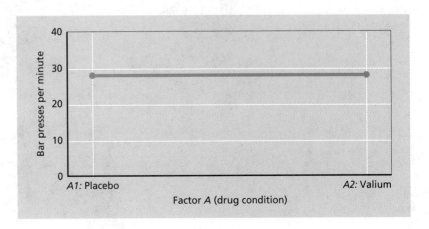

means produces a function that tends to be parallel to the X axis of the graph, there is no main effect of that variable. Thus, Figure 13.3 gives the same indication that we derived from examining the mean difference between levels of factor A: There is no main effect of the drug condition.

Main Effect of Factor B

In contrast to a plot of factor A, a graphical depiction of the factor B means clearly suggests a main effect of situational aversiveness. Figure 13.4 shows that, overall, the normal work environment was associated with dramatically higher productivity than the aversive work situation. In general, *when the plot of overall factor means results in a line that is clearly not parallel to the horizontal axis, it is likely that that factor has a main effect on behavior.*

A × B Interaction Effect

It is a little trickier to show an interaction in a graph. One independent variable has to be represented on the horizontal axis: We'll place factor B there. The other independent variable must be represented inside the plot area itself; that is, each level of the second independent variable must have its own line function. Therefore, there must be as many functions in the plot area as there are levels of the second independent variable. In our example, factor A will be represented in this way. In effect, we will plot a point for each of the four treatment-combination means in Table 13.2. The result appears in Figure 13.5. Observe that there is a separate line function for A1 (placebo) and A2 (Valium).

Figure 13.4 *Line Graph of Factor B (Situational Aversiveness) Effects: Example of a Main Effect*

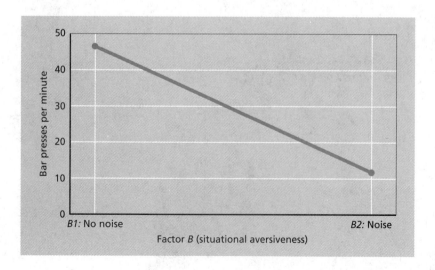

The most obvious characteristic of Figure 13.5 is that the *A1* and *A2* functions cross as they run from *B1* to *B2*. This reflects the fact that Valium has an opposite effect on productivity in an aversive situation than in a normal work environment. As a general rule, *any time a plot of treatment-combination means results in steeply crossed functions, it is likely that there is an interaction effect.* Sometimes an interaction effect is represented by lines that converge but do not cross. Converging functions would occur if the interaction involves a change in the size of the factor *A* effect at different levels of factor *B*, rather than a change in the direction of the factor *A* effect.

It should also prove instructive to see what the plot of the four treatment-combination means would look like if there were no graphical interaction effect. That hypothetical outcome is depicted in Figure 13.6. Note that the two functions neither converge nor diverge; they are parallel.

If you would like to improve your ability to recognize various main effects and interactions in 2×2 factorial investigations, I recommend that you spend a few minutes studying Figure 13.7, which shows you graphical indicators of all possible outcomes in the 2×2 factorial study.

Examining mean differences and graphs helps you to better comprehend the meaning of main effects and interactions. It also provides you with some preliminary indications of possible effects in a data set. Ultimately, however, any effects suggested by these preliminary exercises must be verified by tests of significance. It is to that crucial verification process that we will now turn.

Figure 13.5 *Line Graph of Factor B Means at Each Level of Factor A: Example of an Interaction Effect*

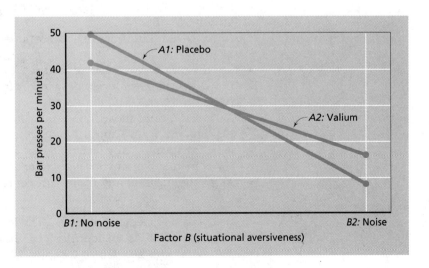

Figure 13.6 *How Figure 13.5 Would Look If There Were No A × B Interaction*

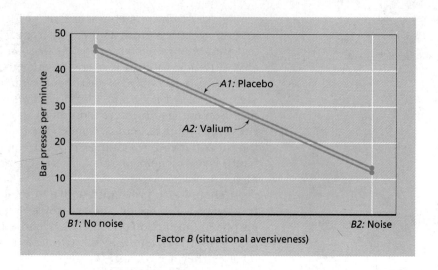

TESTING THE SIGNIFICANCE OF EFFECTS

Some Preliminaries

Before computing a two-way ANOVA, we need to address three matters: (1) understanding critical terms and symbols in two-way ANOVA, (2) computing degrees of freedom, and (3) determining the critical value of *F* required for significance.

Terminology and notation   To conduct an analysis of variance on the data from the Valium study, you should familiarize yourself with the following variables:

- N = the number of observations in each treatment combination (group): $N = 5$
- N_{tot} = the total number of observations in the experiment: $N_{tot} = \Sigma N = 20$
- a = the number of levels of factor A (drug condition): $a = 2$
- b = the number of levels of factor B (situational aversiveness): $b = 2$
- $\overline{X}_G$ = the "grand mean," or mean of all the observations in the experiment; from Table 13.1, we find that $\overline{X}_G = 29$
- $\overline{X}_A$ = the mean of any level of factor A
- $\overline{X}_B$ = the mean of any level of factor B
- $\overline{X}_{AB}$ = the mean of any treatment combination (group)

Computing degrees of freedom   The logic of degrees of freedom in two-way ANOVAs is identical to that used in the one-way ANOVAs that you studied in Chapter 12. But, as you might have

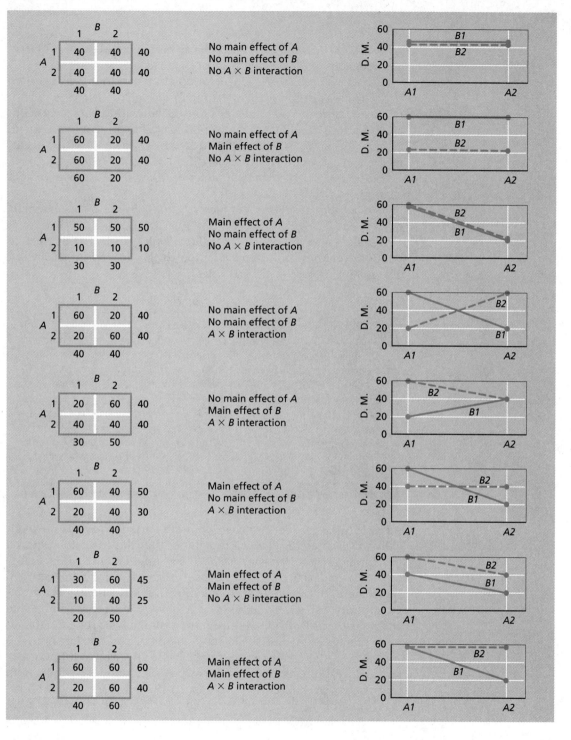

Figure 13.7 *Examples of All Possible Types of Main Effects and Interactions in a 2 × 2 Factorial Design. (Source: Figure adapted from* Invitation to psychological research *by James D. Evans, copyright © 1985 by Holt, Rinehart and Winston, Inc., reproduced by permission of the publisher.)*

guessed, in two-way ANOVA there are more types of degrees of free-dom because there are more sources of variance.

Because there are N_{tot} observations in a factorial investigation, $df_{tot} = N_{tot} - 1$. In the present example, $df_{tot} = 20 - 1 = 19$.

Since there are $N - 1$ degrees of freedom within each treatment combination, and since there are $a \times b$ treatment combinations in a two-way factorial design, the degrees of freedom within treat-ments = $df_{within} = ab(N - 1)$. Here $df_{within} = 2 \cdot 2(5 - 1) = 4(4) = 16$.

Since there are a levels of factor A, $df_A = a - 1$. In the present example, $df_A = 2 - 1 = 1$.

Since there are b levels of factor B, $df_B = b - 1$. In this example, $df_B = 2 - 1 = 1$.

Finally, *something new. Note the following well*: The **degrees of freedom for the interaction effect** is $df_{AxB} = (a - 1)(b - 1)$. In our example, $df_{AxB} = (2 - 1)(2 - 1) = 1 \cdot 1 = 1$.

CONCEPT RECAP: The Logic of ANOVA

Recall that in analysis of variance, our objective is to use the sample data to develop two indexes of variance: (1) one based on between-treatments mean differences, thought to reflect random variation plus a possible effect of some independent variable, and (2) one based on random variation of raw scores within treatments, thought to represent pure "error variation." The *F* statistic we compute is a ratio of between-treatments variance to within-treatments variance. To the extent that the computed *F* exceeds 1.0, the between-treatments variance is larger than the index of random error and probably reflects a real effect of the independent variable.

Determining F_{crit} If you close your eyes and think back nostalgi-cally to Chapter 12, you will remember that the critical *F* required for significance in one-way ANOVA is a function of the level of sig-nificance (α level) that you choose, degrees of freedom between treatments, and degrees of freedom within treatments. Those de-terminants also govern in two-way ANOVA, except that you need to consider three F_{crit} values in the two-way analysis: one for the ef-fect of factor A, one for the effect of factor B, and one for the $A \times B$ interaction effect. When these three effects involve different de-grees of freedom, as is often the case, you actually have to look up and use three different F_{crit} values. In the present problem, how-ever, each of the effects to be tested has 1 and 16 degrees of free-dom. So one critical *F* will suffice for all three of our *F* tests this time.

Refer to Table F in Appendix A. Using $\alpha = .05$, degrees of free-dom for numerator = 1, and degrees of freedom for denominator =

16, we find that F_{crit} = 4.49. Therefore, any computed F ratio that equals or exceeds 4.49 will be significant.

We are now poised to carry out the ANOVA calculations and make some decisions about the effect of Valium on productivity. As we proceed, we will be working toward the construction of an ANOVA summary table. Table 13.3 is a preview version of that table. It displays the degrees of freedom we've already come up with plus the symbolic representations of the quantities that we will need to compute. When we finish with our calculations, we will replace the symbols with the quantities we find.

Computing Sums of Squares

 In the initial stages of a two-way ANOVA, computations proceed according to the same logical sequence that was followed in doing a one-way ANOVA. First, we will find the "total mass of variation," known as the total sum of squares. Then we will partition SS_{tot} into the between-treatments sum of squares $(SS_{between})$ and the within-treatments sum of squares (SS_{within}).

Computing the total sum of squares   We calculate SS_{tot} with formula (12.4):

$$SS_{tot} = \Sigma(X - \overline{X}_G)^2$$

This formula tells us to sum the squared differences between all raw scores and the grand mean. The actual calculations appear in Table 13.4, which shows that SS_{tot} = 6572.

Computing the Between-Treatments Sum of Squares   We obtain $SS_{between\ treatments}$ with a version of formula (12.5):

$$SS_{between\ treatments} = \Sigma[N(\overline{X}_{AB} - \overline{X}_G)^2]$$

Table 13.3

SCHEMATIC ANOVA SUMMARY TABLE FOR VALIUM STUDY	SOURCES OF VARIATION	SUMS OF SQUARES	DEGREES OF FREEDOM	MEAN SQUARES	F RATIO
	A: Drug condition	SS_A	1	MS_A	F_A
	B: Situational aversiveness	SS_B	1	MS_B	F_B
	$A \times B$ interaction	$SS_{A \times B}$	1	$MS_{A \times B}$	$F_{A \times B}$
	Within treatments (error term)	$SS_{within\ treatments}$	16	$MS_{within\ treatments}$	
	TOTAL	SS_{tot}	19		

Table 13.4

CALCULATION OF THE
TOTAL SUM OF SQUARES
IN THE VALIUM STUDY

	X	$(X - \overline{X}_G)$	$(X - \overline{X}_G)^2$
	50	21	441
A1B1	46	17	289
Placebo/no noise	42	13	169
	58	29	841
	54	25	625
	3	−26	676
A1B2	4	−25	625
Placebo/noise	12	−17	289
	13	−16	256
	8	−21	441
	38	9	81
A2B1	46	17	289
Valium/no noise	37	8	64
	47	18	324
	42	13	169
	16	−13	169
A2B2	23	− 6	36
Valium/noise	9	−20	400
	11	−18	324
	21	− 8	64
			Sum: 6572 = SS_{tot}

Grand mean = $\overline{X}_G = 29$

The central operation here is to subtract the grand mean from each treatment-combination mean, square the difference, and multiply the squared deviation times N, the number of observations in the treatment group. The treatment combination means are given in Table 13.2. Proceeding with this sequence, we get:

$$SS_{between\ treatments} = \Sigma[N(\overline{X}_{AB} - \overline{X}_G)^2]$$

$$= 5(50 - 29)^2 + 5(8 - 29)^2 + 5(42 - 29)^2$$

$$+ 5(16 - 29)^2$$

$$= 5(21)^2 + 5(-21)^2 + 5(13)^2 + 5(-13)^2$$

$$= 5(441) + 5(441) + 5(169) + 5(169)$$

$$= 2205 + 2205 + 845 + 845$$

$$= 6100$$

Computing the within-treatments sum of squares   Since $SS_{tot} = SS_{between\ treatments} + SS_{within\ treatments}$, we ought to be able to find the within-treatments sum of squares via:

$$SS_{within\ treatments} = SS_{tot} - SS_{between\ treatments}$$

$$= 6572 - 6100$$

$$= 472$$

We can check the accuracy of this figure by using formula (12.6) to directly compute the within-treatments sum of squares:

$$SS_{within\ treatments} = \Sigma[\Sigma(X - \overline{X}_{AB})^2]$$

These calculations are shown in Table 13.5, and the result is as expected.

Partitioning the Between-Treatments Sum of Squares

In two-way ANOVA, there is a *second level of variance partitioning*. The $SS_{between\ treatments}$ actually consists of three sources of variance that we must break out: SS_A, SS_B, and $SS_{A \times B}$. Variation among treatment-combination means actually reflects three influences: a main effect of factor A, a main effect of factor B, and an effect of the interaction of factors A and B. This notion of a second level of partitioning is illustrated in Figure 13.8, which schematizes the overall structure of two-way ANOVA.

Calculating the SS for the main effect of factor A   The sum of squares for the main effect of factor A is computed through:

$$SS_A = \Sigma[bN(\overline{X}_A - \overline{X}_G)^2] \tag{13.1}$$

Table 13.5

CALCULATION OF THE
WITHIN-TREATMENTS
SUM OF SQUARES IN
THE VALIUM STUDY

$\overline{X} = 50$ A1B1 PLACEBO/NO NOISE			$\overline{X} = 42$ A2B1 VALIUM/NO NOISE			$\overline{X} = 8$ A1B2 PLACEBO/NOISE			$\overline{X} = 16$ A2B2 VALIUM/NOISE		
X	$(X - \overline{X})$	$(X - \overline{X})^2$	X	$(X - \overline{X})$	$(X - \overline{X})^2$	X	$(X - \overline{X})$	$(X - \overline{X})^2$	X	$(X - \overline{X})$	$(X - \overline{X})^2$
50	0	0	38	−4	16	3	−5	25	16	0	0
46	−4	16	46	4	16	4	−4	16	23	7	49
42	−8	64	37	−5	25	12	4	16	9	−7	49
58	8	64	47	5	25	13	5	25	11	−5	25
54	4	16	42	0	0	8	0	0	21	5	25
SS		160			82			82			148

$$SS_{within\ treatments} = 160 + 82 + 82 + 148 = 472$$

Figure 13.8 *Conceptual Depiction of Two Levels of Variance Partitioning in a Two-Way ANOVA*

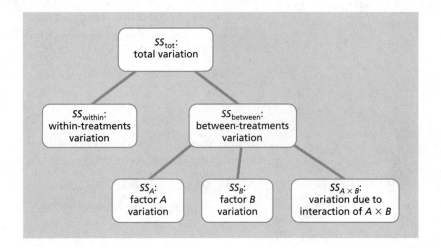

The critical operation in formula (13.1) is to square the difference between each factor *A* mean and the grand mean. Since the factor *A* means are calculated by summing across subjects within treatment groups *and* levels of the factor *b*, each of the squared differences must be multiplied by *bN*. Table 13.1 shows that the mean of *A1* is 29 and the mean of *A2* is 29. Therefore, we get:

$$SS_A = \Sigma[bN(\overline{X}_A - \overline{X}_G)^2]$$

$$= 2 \cdot 5(29 - 29)^2 + 2 \cdot 5(29 - 29)^2$$

$$= 10(0) + 10(0)$$

$$= 0$$

Since the factor *A* means did not differ from each other, it is not surprising that the factor *A* sum of squares is 0.

Calculating the SS for the main effect of factor *B* We can calculate the factor *B* sum of squares in a similar fashion, using

$$SS_B = \Sigma[aN(\overline{X}_B - \overline{X}_G)^2] \tag{13.2}$$

Here, we have summed across *a* levels of factor *A* to compute each of the factor *B* means. Table 13.1 shows that $\overline{X}_{B1} = 46$ and $\overline{X}_{B2} = 12$. Hence, we obtain:

$$SS_B = 2 \cdot 5(46 - 29)^2 + 2 \cdot 5(12 - 29)^2$$

$$= 10(17)^2 + 10(-17)^2$$

$$= 10(289) + 10(289)$$

$$= 2890 + 2890$$

$$= 5780$$

Calculating the SS for the $A \times B$ interaction effect Since $SS_{\text{between treatments}} = SS_A + SS_B + SS_{A \times B}$, we are able to find the $A \times B$ interaction sum of squares very simply through:

$$SS_{A \times B} = SS_{\text{between treatments}} - SS_A - SS_B \qquad (13.3)$$

Thus, $SS_{A \times B} = 6100 - 0 - 5780 = 6100 - 5780 = 320$.

Computing Mean Squares

In two-way ANOVA we compute mean squares in the same way that we did in one-way analyses. The basic logic applies: A mean square is a variance calculated by dividing a sum of squares by its degrees of freedom. Compared to the one-way ANOVA, however, we have more mean squares to figure. The mean squares in the present problem are:

Factor A mean square = $MS_A = SS_A / df_A = 0/1 = 0$

Factor B mean square = $MS_B = SS_B / df_B = 5780/1 = 5780$

$A \times B$ interaction mean square = $MS_{A \times B} = SS_{A \times B} / df_{A \times B} = 320/1 = 320$

Within-treatments mean square = $MS_{\text{within treatments}}$
= $SS_{\text{within treatments}} / df_{\text{within treatments}} = 472/16 = 29.5$.

Computing F and Making Decisions

You know that an F statistic is the ratio of two variances, or mean squares. In a two-way analysis, there are three F tests to conduct. In each case, $MS_{\text{within treatments}}$ is the denominator, or "error term" in the ratio.

Main effect of factor A: $F_A = MS_A / MS_{\text{within treatments}} = 0/29.5 = 0$

Main effect of factor B: $F_B = MS_B / MS_{\text{within treatments}} = 5780/29.5 = 195.93$

$A \times B$ interaction effect: $F_{A \times B} = MS_{A \times B} / MS_{\text{within treatments}} = 320/29.5 = 10.85$

REMINDER: Bear in mind that the main effects and interaction in a two-way ANOVA are independent of one another. As you can see, the interaction of factors A and B is significant even though the main effect of factor A is 0.

Table 13.6 is the ANOVA summary table for this problem. Since the F_{crit} value is 4.49, both the main effect of factor B (situational aversiveness) and the interaction of the drug condition with situational aversiveness are significant. To no one's surprise, rats are less productive when performing their job is painful. Of greater interest is the answer to the question, How does taking Valium affect

Table 13.6

ANOVA SUMMARY TABLE FOR THE VALIUM STUDY					

SOURCES OF VARIATION	SUMS OF SQUARES	DEGREES OF FREEDOM	MEAN SQUARES	F RATIO
A: Drug condition	0	1	0	0
B: Situational aversiveness	5780	1	5780	195.93*
A × B interaction	320	1	320	10.85*
Within treatments (error term)	472	16	29.5	
TOTAL	6572	19		

*Significant F ratios: $p < .05$.

productivity? The answer is: It depends. The significant $A \times B$ interaction verifies that, at least for our four-footed furry friends, Valium depresses performance under normal work conditions but can improve performance when it hurts to work.

Reporting Results in Two-Way ANOVA

In the scientific journals of your field, there are just a few fairly standard ways of reporting the F statistics in a two-way analysis of variance.

Stating main effects   In the present study, for instance, the main effects should be described in this way: "There was a main effect of situational aversiveness, $F(1, 16) = 195.93$, $p < .05$. Productivity was lower when performing the job was associated with an unpleasant consequence. However, there was no main effect of the drug condition, $F(1,16) = 0$, $p > .05$. Overall, then, Valium use, by itself, neither harmed nor boosted productivity."

There are three features to note in the preceding paragraph. First, the degrees of freedom associated with each F test are given in parentheses (in df_{effect}, $df_{\text{within treatments}}$ order) immediately after the letter "F." Second, the level of significance is presented with each value of the F ratio in the form "$p < \alpha$" when the F is significant, and in the form of "$p > \alpha$" when F is nonsignificant. Third, a simple verbal interpretation of the test of significance follows the report of the test statistic.

Stating the interaction effect   In this example, the interaction effect is reported as: "Drug condition interacted significantly with situational aversiveness, $F(1, 16) = 10.85$, $p < .05$. The use of Valium was associated with relatively poorer performance under normal work conditions but relatively better performance under unpleasant work conditions."

The interaction qualifies the main effects As you reread the report of the interaction effect, note how the existence of an interaction qualifies the meaning of the main effects. It is true, for example, that the rats were less productive when the situation was aversive than when the situation was nonaversive. But the drop in productivity was less in the rats that had been given Valium than in the sober animals. This modifying influence of the Valium requires us to qualify any statement made about *how much* an aversive consequence reduces productivity.

The interaction also requires that we not dismiss drug use as a factor in productivity. Even though there was no main effect of drug condition, can we say that Valium use on the job is unimportant? Clearly not. The interaction tells us that the impact of Valium in the bloodstream is a function of what other factors are operating. Sometimes the drug helps; sometimes it impairs. *It depends.*

Effect Size

In Chapter 12 you learned that behavioral investigators sometimes want to know more than that an *F* ratio is significant. They might also wish to make some statement about the "size" of the significant effect, in terms of the proportion of variance accounted for by a particular independent variable or interaction. As you discovered in Chapter 12, the proportion of variance accounted for in an ANOVA data set is given by eta-squared, symbolized η^2. In two-way ANOVA, the formula for η^2 is:

$$\eta^2 = \frac{SS_{\text{effect}}}{SS_{\text{tot}}} \tag{13.4}$$

where "effect" refers to any main effect or interaction that is significant.

For example, we find that the interaction of the drug condition and situational aversiveness accounts for $\eta^2 = 320/6572 = .049$, or 4.9%, of the variation in productivity. You can see that although the interaction was significant, it was not an overwhelming determinant of productivity. In contrast, the impact of situational aversiveness was a powerful determinant of behavior, accounting for an impressive $\eta^2 = 5780/6572 = .88$, or 88%, of the variance.

TAKING STOCK

It's time to put things into perspective:

1. Two-way ANOVA is a procedure for analyzing data from studies that involve two independent variables called factor *A* and factor *B*. The procedure enables you to test the main effect of each of the variables plus the effect of their interaction.

2. Two-way ANOVA is used in factorial studies in which every designated level of factor *A* is combined with all designated levels of factor *B*, producing *ab* treatment combinations.

3. A main effect is the effect that one independent variable has on the dependent variable irrespective of the influence of the second independent variable; it is the effect of one factor literally averaged across the other factor.

4. An interaction effect means that the influence of one independent variable is modified when it is combined with a second independent variable; the effect of factor *A* changes at different levels of factor *B*. An *A* × *B* interaction is indicated when the mean differences on factor *A* change in size or direction (i.e., positive versus negative) across different levels of factor *B*. An *A* × *B* interaction is also indicated by steeply converging or crossing lines in a graph of the treatment-combination means.

5. Like a one-way ANOVA, a two-way ANOVA involves partitioning the total data variance into a between-treatments component and a within-treatments component. In two-way ANOVA, however, there is a secondary partitioning procedure, in which the between-treatments sum of squares is broken down into sums of squares for a factor *A* effect, a factor *B* effect, and an *A* × *B* interaction effect.

6. A significant interaction means that the effect of variable *A* *depends* on which level of variable *B* is present.

A SECOND EXAMPLE: CAN SOCIOPATHS LEARN TO AVOID PUNISHMENT?

This next example is intended to firm up your recently acquired understanding of two-way analysis of variance. It extends the procedures to a slightly more complicated research design, in which one of the independent variables has three levels instead of just two. To make this next experience as clear as possible for you, I'll use the eight procedural steps in significance testing that you've seen frequently in this text.

State the Problem and Nature of the Data

Sociopaths are persons who have not experienced normal socialization. As a result, they often seem to lack conscience, are notorious for their tendencies to lie and manipulate, and frequently run afoul of the law. One proposed explanation of the sociopath's deficient socialization is that such a person is less sensitive to aversive stimuli and, therefore, is less responsive to emotional conditioning than normal individuals are. Hare (1970), for example, theorized

Box 13.1

SIXTEEN STEPS OF TWO-WAY ANOVA

1. Determine the total degrees of freedom: $df_{tot} = N_{tot} - 1$

2. Determine the within-treatments degrees of freedom: $df_{within} = ab(N - 1)$

3. Determine the factor A degrees of freedom: $df_A = a - 1$

4. Determine the factor B degrees of freedom: $df_B = b - 1$

5. Determine the degrees of freedom for the $A \times B$ interaction: $df_{A \times B} = (a - 1)(b - 1)$

6. Compute the total sum of squares: $SS_{tot} = \Sigma(X - \overline{X}_G)^2$

7. Compute the between-treatments sum of squares: $SS_{between\ treatments} = \Sigma[N(\overline{X}_{AB} - \overline{X}_G)^2]$

8. Compute the within-treatments sum of squares: $SS_{within\ treatments} = SS_{tot} - SS_{between\ treatments}$

9. Partition the between-treatments sum of squares:

   a. Find the sum of squares for factor A: $SS_A = \Sigma[bN(\overline{X}_A - \overline{X}_G)^2]$

   b. Find the sum of squares for factor B: $SS_B = \Sigma[aN(\overline{X}_B - \overline{X}_G)^2]$

   c. Find the sum of squares for the $A \times B$ interaction: $SS_{A \times B} = SS_{between\ treatments} - SS_A - SS_B$

10. Compute the mean square for factor A: $MS_A = SS_A/df_A$

11. Compute the mean square for factor B: $MS_B = SS_B/df_B$

12. Compute the mean square for the $A \times B$ interaction: $MS_{A \times B} = SS_{A \times B}/df_{A \times B}$

13. Compute the within-treatments mean square: $MS_{within\ treatments} = SS_{within\ treatments}/df_{within\ treatments}$

14. Compute the F ratio for the effect of factor A: $F_A = MS_A/MS_{within\ treatments}$

15. Compute the F ratio for the effect of factor B: $F_B = MS_B/MS_{within\ treatments}$

16. Compute the F ratio for the $A \times B$ interaction effect: $F_{A \times B} = MS_{A \times B}/MS_{within\ treatments}$

that, compared to "normals," sociopaths do not show normal anxiety reactions and have trouble learning to avoid physical and social punishments. It is interesting, however, that sociopaths seem to be better than average at learning to avoid the loss of money.

The present investigation tests Hare's conceptualization of sociopathy in a 2 × 3 factorial design. Factor A was "sociopathy," which had two levels: One group of subjects consisted of 12 incarcerated individuals diagnosed as sociopathic, whereas the second group consisted of 12 "normals." Thus, there were $N_{tot} = 24$ subjects in all. The two groups of subjects were equated on age, intelligence, and socioeconomic class. Factor B was "type of punishment" used in a learning task. This variable had three levels: electric shock, social disapproval, and loss of money.

Within each diagnostic group (i.e., sociopathic versus normal), 4 of the 12 subjects were randomly assigned to each punishment condition. The particular experimental design used is called a **mixed factorial,** *in that subjects were randomly assigned to one independent variable* (type of punishment) *but not the other* (sociopathy). Because sociopathy is an intrinsic, or "organismic," variable to which persons cannot be randomly assigned, the study could not be a completely randomized factorial. Since factor A has $a = 2$ levels and factor B has $b = 3$ levels, there are $ab = 6$ treatment combinations in this investigation. So with $N_{tot} = 24$ subjects, altogether, divided among 6 treatment combinations, there were $N = 4$ subjects per treatment cell.

All subjects were tested individually in a concept identification task. A subject received a punishment each time he or she made an error in the task. The dependent variable was the number of errors made prior to correctly identifying the concept. Thus, lower error scores mean faster learning, and higher scores mean slower learning. According to Hare's theory, the normals ought to have fewer errors when electric shock or social disapproval was used to discourage incorrect responses, but the sociopaths should make fewer mistakes when loss of money was the punisher. *Hare's theory specifically predicts an interaction effect.*

State the Statistical Hypotheses

Three sets of null and alternative hypotheses correspond to the main effect of factor A (sociopathy), the main effect of factor B (type of punishment), and the effect of the interaction of factors A and B.

1. Sociopathy effect

    H_0: $\mu_{A1} = \mu_{A2}$

    H_1: $\mu_{A1} \neq \mu_{A2}$

2. Type of punishment effect

    H_0: $\mu_{B1} = \mu_{B2} = \mu_{B3}$

    H_1: Not all the type-of-punishment means are the same.

3. Sociopathy × type-of-punishment interaction

    H_0: $(\mu_{A1B1} - \mu_{A2B1}) = (\mu_{A1B2} - \mu_{A2B2}) = (\mu_{A1B3} - \mu_{A2B3})$

    H_1: Not all the mean differences are the same.

Choose a Level of Significance

Let's play the role of skeptic and require a higher than usual criterion of significance for this problem: $\alpha = .01$.

Specify the Test Statistic

Since this is a factorial experiment, we will use a two-way ANOVA. We will compute F statistics to evaluate each of the following effects:

1. Main effect of A (sociopathy): Degrees of freedom = $df_A = (a - 1) = (2 - 1) = 1$.
2. Main effect of B (type of punishment): Degrees of freedom = $df_B = (b - 1) = (3 - 1) = 2$.
3. $A \times B$ interaction effect: Degrees of freedom = $df_{A \times B} = (a - 1)(b - 1) = (2 - 1)(3 - 1) = (1)(2) = 2$.

Determine the Critical Value Needed for Significance

Since three F tests will be required in this analysis, we need three F_{crit} values. These critical values are determined by the alpha level (.01), the degrees of freedom for the effect itself (listed above), and the within-treatments degrees of freedom, where $df_{\text{within}} = ab(N - 1) = 2 \cdot 3(4 - 1) = 6(3) = 18$.

Using the .01 critical values in Table F of Appendix A, we find:

With 1 and 18 degrees of freedom, $F_{\text{crit}} = 8.29$ for the main effect of sociopathy.

With 2 and 18 degrees of freedom, $F_{\text{crit}} = 6.01$ for the main effect of type of punishment.

With 2 and 18 degrees of freedom, $F_{\text{crit}} = 6.01$ for the interaction of sociopathy and type of punishment.

State the Decision Rules

1. Main effect of sociopathy: If $F_A \geq 8.29$, reject H_0.
2. Main effect of type of punishment: If $F_B \geq 6.01$, reject H_0.
3. Sociopathy × punishment type interaction: If $F_{A \times B} \geq 6.01$, reject H_0.

Compute the Test Statistics

The results of the experiment are shown in Table 13.7, which displays the raw scores in each of the six treatment conditions, the overall means for factor A and factor B, and the grand mean. Figure 13.9 displays a line graph of the treatment-combination means, which suggests that there was an interaction between sociopathy and type of punishment, as predicted by Hare (1970).

Calculate the total sum of squares Table 13.8 presents the calculation of SS_{tot}. Formula (12.4) yields $SS_{\text{tot}} = \Sigma(X - \overline{X}_G)^2 = 212$.

Table 13.7

NUMBER OF ERRORS MADE PRIOR TO CONCEPT IDENTIFICATION PROBLEM AS A FUNCTION OF SOCIOPATHY AND TYPE OF PUNISHMENT		FACTOR B			
		ELECTRIC SHOCK *B1*	SOCIAL PUNISHMENT *B2*	LOSS OF MONEY *B3*	FACTOR A MEANS
		A1B1	*A1B2*	*A1B3*	
		10	8	2	
	SOCIOPATHS *A1*	7	11	1	7
		8	10	5	
		11	7	4	
FACTOR *A*		*A2B1*	*A2B2*	*A2B3*	
		4	3	8	
	CONTROLS *A2*	2	4	5	5
		1	7	6	
		5	6	9	
FACTOR *B* MEANS		6	7	5	
					Grand mean = 6

Calculate the between-treatments sum of squares The treatment-combination means and mean differences appear in Table 13.9. Note that the mean difference between *A1* and *A2* changes at different levels of factor *B*, which suggests an interaction effect—more on this later.

We will use the group means to compute the between-treatments sum of squares. Formula (12.5) gives us:

Figure 13.9 *Mean Number of Errors in a Concept Identification Task As a Function of Sociopathy and Type of Punishment*

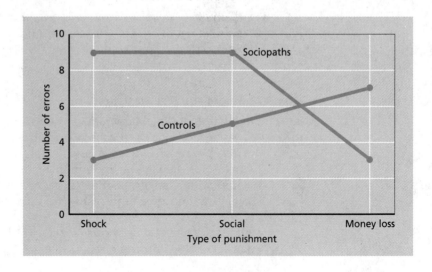

Table 13.8

CALCULATION OF THE TOTAL
SUM OF SQUARES FOR THE
EXPERIMENT ON
SOCIOPATHY AND
PUNISHMENT EFFECTIVENESS

		X	$(X - \bar{X}_G)$	$(X - \bar{X}_G)^2$
Sociopath/electric shock	(A1B1)	10	4	16
		7	1	1
		8	2	4
		11	5	25
Sociopath/social punishment	(A1B2)	8	2	4
		11	5	25
		10	4	16
		7	1	1
Sociopath/money loss	(A1B3)	2	−4	16
		1	−5	25
		5	−1	1
		4	−2	4
Control/shock	(A2B1)	4	−2	4
		2	−4	16
		1	−5	25
		5	−1	1
Control/social punishment	(A2B2)	3	−3	9
		4	−2	4
		7	1	1
		6	0	0
Control/money loss	(A2B3)	8	2	4
		5	−1	1
		6	0	0
		9	3	9

Total sum of squares: 212

$$SS_{\text{between treatments}} = \Sigma[N(\bar{X}_{AB} - \bar{X}_G)^2]$$

$$= 4(9 - 6)^2 + 4(9 - 6)^2 + 4(3 - 6)^2 + 4(3 - 6)^2 + 4(5 - 6)^2 + 4(7 - 6)^2$$

$$= 4(3)^2 + 4(3)^2 + 4(-3)^2 + 4(-3)^2 + 4(-1)^2 + 4(1)^2$$

$$= 4(9) + 4(9) + 4(9) + 4(9) + 4(1) + 4(1)$$

$$= 36 + 36 + 36 + 36 + 4 + 4$$

$$= 152$$

Calculate the within-treatments sum of squares Through the shortcut method,

$$SS_{\text{within treatments}} = SS_{\text{tot}} - SS_{\text{between treatments}} = 212 - 152 = 60$$

Table 13.9

LEARNING-ERROR MEANS
AND MEAN DIFFERENCES AS
A FUNCTION OF SOCIOPATHY
AND TYPE OF PUNISHMENT

		FACTOR *B*: TYPE OF PUNISHMENT		
		SHOCK *B1*	SOCIAL *B2*	MONEY LOSS *B3*
FACTOR *A*: SOCIOPATHY	SOCIOPATHS *A1*	*A1B1* $\overline{X} = 9$	*A1B2* $\overline{X} = 9$	*A1B3* $\overline{X} = 3$
	CONTROLS *A2*	*A2B1* $\overline{X} = 3$	*A2B2* $\overline{X} = 5$	*A2B3* $\overline{X} = 7$
	MEAN DIFFERENCES	6	4	−4

Partition the between-treatments sum of squares Next we will divide the between-treatments sum of squares into its three components.

$$SS_A = \Sigma[bN(\overline{X}_A - \overline{X}_G)^2]$$

$$= [3 \cdot 4(7 - 6)^2] + [3 \cdot 4(5 - 6)^2]$$

$$= [12(1)] + [12(1)]$$

$$= 24$$

$$SS_B = \Sigma[aN(\overline{X}_B - \overline{X}_G)^2]$$

$$= [2 \cdot 4(6 - 6)^2] + [2 \cdot 4(7 - 6)^2] + [2 \cdot 4(5 - 6)^2]$$

$$= [8(0)] + [8(1)] + [8(1)]$$

$$= 16$$

$$SS_{A \times B} = SS_{\text{between treatments}} - SS_A - SS_B$$

$$= 152 - 24 - 16$$

$$= 112$$

Compute mean squares and *F* ratios By now you are thoroughly aware that each mean square is computed by dividing a sum of squares by its degrees of freedom, and that each *F* statistic is a ratio of an "effect" mean square to $MS_{\text{within treatments}}$. Therefore, I simply refer you to Table 13.10, which summarizes the mean squares and *F* ratios in this problem.

Consult the Decision Rules and Make a Decision

General decisions regarding the null hypotheses If you compare the computed *F* statistics with the decision rules established

Table 13.10

ANOVA SUMMARY TABLE
FOR EXPERIMENT ON
SOCIOPATHY AND
PUNISHMENT EFFECTIVENESS

SOURCES OF VARIATION	SUMS OF SQUARES	DEGREES OF FREEDOM	MEAN SQUARES	F RATIO
A: Sociopathy	24	1	24	7.2^{NS}
B: Type of punishment	16	2	8	2.4^{NS}
A × B interaction	112	2	56	16.8*
Within treatments (error term)	60	18	3.33	
TOTAL	212	23		

NS Not significant.
*Significant F ratio: $p < .01$.

earlier, you will see that neither sociopathy nor type of punishment had a significant impact on the number of errors made in the concept identification task. Therefore, we fail to reject both null hypotheses concerning main effects. However, the sociopathy × type-of-punishment interaction was significant, and we reject the corresponding H_0. This outcome suggests that the effect that type of punishment has on learning efficiency depends on whether the learner is "normal" or sociopathic.

Just what is the specific nature of this interaction? That is, which particular mean differences caused the interaction to be significant? You can find the answer to this question through pairwise comparison tests.

Tukey's test and specific conclusions As you discovered in Chapter 12, when a significant effect in ANOVA involves more than two means, it is often necessary to use pairwise comparisons to track down the particular mean differences responsible for the significant F. To review, a pairwise comparison is a test that assesses the significance of the difference between two means at a time. The purpose is to develop a more precise interpretation of the significant outcome.

In Chapter 12 you learned how to carry out Tukey's honestly significant difference (HSD) test. Recall that the HSD is the minimum significant mean difference in a set of k means. In the context of two-way interactions, $k = ab$, the number of treatment combinations. You can use Tukey's HSD to make all possible pairwise comparisons while holding the probability of a Type I error constant at the stated alpha level. Formula (12.10) specifies that $HSD = q\sqrt{MS_W/N}$, where MS_W is the within-treatments mean square and N is the number of observations included in the calculation of any one mean in the comparison. In the sociopathy × type-of-punishment interaction, $MS_{\text{within treatments}}$ is 3.33, and $N = 4$ within each treatment combination. The q in the formula is the "studentized range statistic," which you can locate in Table Q of Appendix A. There you will find that for $k = 6$ means, $\alpha = .01$, and $df_{\text{within treatments}} = 18$, the critical value of q is 5.6.

For the current example, then, $HSD = 5.6\sqrt{3.33/4} = 5.6\sqrt{0.8325} = 5.6 \cdot 0.91 = 5.11$. Therefore, any difference between treatment-combination means that equals or exceeds $HSD = 5.11$ is a significant difference at the .01 level and has contributed to the significant interaction effect. If you refer back to Table 13.9, you'll see that the only significant differences are those between the group means of 3 and 9. All other differences in the table are not big enough to be considered reliable at an alpha level of .01. One way to interpret the interaction in light of this new information is to conclude that using loss of money as a punisher significantly improved learning in sociopaths but not in "normals." What, then, is the effect of using loss of money rather than, say, social disapproval as a punisher in learning situations? As you might anticipate, the answer is "it depends" on the diagnostic category of the learner.

It is also noteworthy that the sociopaths' mean number of errors was significantly higher than the normal subjects' mean when shock was used. In agreement with Hare's (1970) theory, physically aversive stimulation was not an effective conditioning device for the sociopaths.

ASSUMPTIONS OF TWO-WAY ANOVA

The assumptions behind two-way ANOVA are essentially the same as those underlying simple one-way ANOVA, as described in Chapter 12. Specifically, in using two-way ANOVA, you are assuming that:

- The dependent variable's scale of measurement is at the interval or ratio level.
- The observations within each treatment combination are independent of one another.
- The dependent variable is normally distributed in the populations represented by the treatment combinations.
- The variances of the populations are equal.

The conditions under which some of these assumptions may be eased are the same as those discussed in Chapter 12.

KEY TERMS

factorial design

factorial experiment

completely randomized factorial experiment

main effect

interaction effect

degrees of freedom for the interaction effect

mixed factorial design

SUMMARY

1. Two-way ANOVA is a procedure for analyzing data from studies that involve two independent variables called factor A and factor B. The procedure enables you to test the main effect of each of the two variables plus the effect of their interaction.

2. Two-way ANOVA is used in factorial studies in which every designated level of factor A is combined with all designated levels of factor B, producing a $\cdot$ b treatment combinations.

3. A main effect is the effect that one independent variable has on the dependent variable irrespective of the influence of the second independent variable; it is the effect of one factor literally averaged across the other factor.

4. An interaction effect means that the influence of one independent variable is modified when it is combined with a second independent variable; the effect of factor A changes at different levels of factor B. An $A \times B$ interaction is indicated when the mean differences of factor A change in size or direction (i.e., positive versus negative) across different levels of factor B. An $A \times B$ interaction is also indicated by dramatically converging or crossing lines in a graph of the treatment-combination means.

5. Like a one-way ANOVA, a two-way ANOVA involves partitioning the total data variance into a between-treatments component and a within-treatments component. In two-way ANOVA, however, there is a secondary partitioning procedure, in which the between-treatments sum of squares is broken down into sums of squares for a factor A effect, a factor B effect, and an $A \times B$ interaction effect.

6. A significant interaction means that the effect of variable A *depends* on which level of variable B is present.

REVIEW QUESTIONS

1. Define or describe: factorial design, factorial experiment, and completely randomized factorial experiment.
2. What information is provided by a two-way ANOVA that is not available from a one-way ANOVA? Describe the nature of each additional piece of information.
3. In your own words, explain what is meant by the idea that there are two levels of variance partitioning in two-way ANOVA. Be precise and complete.
4. Define or describe main effect, interaction effect, degrees of freedom for the interaction effect, and mixed factorial design.
5. One theory of psychological determinants of obesity asserts that obese people are more responsive to all stimuli than are normal-weight persons. A researcher had a hunch that some obese people selectively overreact to food-related stimuli but not to other kinds of stimuli. To test the latter hypothesis, ten obese and ten normal persons were given a letter recognition task in which common letters are flashed on a screen for $\frac{1}{20}$ second. On a random basis, five obese and five normal subjects received a food-related image along with the target letter, whereas the other five subjects in each group received a non-food-related image with each letter. The dependent variable was the number of recognition errors made in 20 trials. The raw scores from this 2×2 mixed factorial study appear in the table. Using the .05 level of significance, carry out an ANOVA, and draw appropriate conclusions about effects.

	Food Stimulus	Non-Food Stimulus
	10	4
	11	3
Obese	9	6
	8	2
	7	5
	6	5
	7	6
Normal	5	8
	4	7
	3	4

6. For the ANOVA computed in Question 5, state the null and alternative hypotheses, both verbally and symbolically, and determine the proportion of variance accounted for by each effect.
7. In a completely randomized factorial experiment, hypnotist rapport (high versus low) is crossed with the method of inducing hypnosis (strong versus seductive). Four of 16 volunteers

are randomly assigned to each of the treatment combinations. The dependent variable is responsiveness to 20 standard hypnotic suggestions. Using the .01 level of significance, conduct a two-way ANOVA on the results of this study. Draw appropriate conclusions.

	Strong Method	Seductive Method
High rapport	15	11
	12	19
	18	14
	15	16
Low rapport	13	6
	7	8
	13	12
	7	10

8. An investigator crossed three levels of ambient temperature (cold, normal, hot) with three levels of physical crowding (low, medium, high). The dependent variable was the number of aggressive verbal or physical responses in a one-hour period. The results of this study are shown in the table. Using alpha = .05, perform an ANOVA and come to appropriate conclusions.

	Crowding		
	Low	Medium	High
Cold temperature	0	3	4
	2	2	5
	0	2	5
	2	1	6
Normal temperature	0	2	1
	1	2	1
	0	0	3
	1	0	3
Hot temperature	2	6	7
	2	4	6
	3	5	8
	1	5	7

9. Use the ANOVA problem in Question 8.
   (a) Express the null and alternative hypotheses for all possible effects. Use both verbal and symbolic statements of these hypotheses.
   (b) Determine the proportion of variance accounted for by each of the significant effects.
   (c) Construct a line graph for each main effect, as well as one for the interaction.

10. A medical researcher wished to evaluate the effect of wall color and type of rock music on heart rate in middle-aged women. Five subjects were randomly assigned to each of the six treatment combinations created by crossing the three colors with two types of music. The raw scores of heart rates from this study are shown in the table. Using the .05 level of significance, perform an ANOVA and draw appropriate conclusions.

	Wall Color		
	Blue	White	Red
	60	68	70
	64	65	68
Soft rock	58	72	70
	58	75	70
	60 *300=90,000*	70 *350=122500*	72 *350*
	72	67	84
	70	73	88
Heavy metal	75	74	86
	65	66	89 *184900*
	68 *350*	70 *350*	83 *430*

11. For the results of the ANOVA described in Question 10, use Tukey's *HSD* to determine which particular mean differences are responsible for the significant interaction. Use the .05 level of significance and develop appropriate conclusions.

12. Complete the following ANOVA summary table. Evaluate the statistical significance of each effect.

Sources of Variation	Sums of Squares	Degrees of Freedom	Mean Squares	F Ratio
A	30	2		
B	20		10	
A × B	100			
Within treatments (error term)		36		
TOTAL	300			

42436
74964

13. Complete the following ANOVA summary table. Evaluate the statistical significance of each effect.

Sources of Variation	Sums of Squares	Degrees of Freedom	Mean Squares	F Ratio
A	85	5		
B		4	2	
A × B			32	
Within treatments (error term)		570	10	
TOTAL				

14. Complete the following ANOVA summary table. Evaluate the statistical significance of each effect.

Sources of Variation	Sums of Squares	Degrees of Freedom	Mean Squares	F Ratio
A	30	2		
B	20	1		
A × B				
Within treatments (error term)	42	54		
TOTAL	147			

15. Using the information provided by the summary table in Question 14, determine the percent of variance accounted for by each of the significant effects.

16. The table contains the treatment-combination means from a factorial study. Tentatively identify the A, B, and $A \times B$ effects, or lack thereof, in two ways: through numerical mean differences and graphically. Be sure to state a preliminary conclusion about the apparent presence or absence of each effect.

	B1	B2
A1	10	30
A2	20	0

17. Repeat the procedure from Question 16 for this table.

	B1	B2
A1	70	30
A2	40	0

18. Repeat the procedure from Question 16 for this table.

	B1	B2	B3
A1	45	35	58
A2	20	10	33
A3	60	50	73

CHAPTER 14 REPEATED-MEASURES ANALYSIS OF VARIANCE

From Chapter 13 you learned how to conduct and interpret a two-way analysis of variance. Equipped with that knowledge, you are now in an intellectual position to understand yet another kind of variance-testing procedure, called "repeated-measures ANOVA." You use repeated-measures ANOVA when the data are based on a **within-subjects experimental design**. In this kind of study, *the independent variable is manipulated "within subjects" by exposing each research participant to all levels of that variable.*[1] Because the subject is tested at every level of the independent variable, he or she is "repeatedly measured." Thus, it is appropriate to use a repeated-measures ANOVA when *there is only one independent variable, but each subject has contributed data to every condition in the study.* An example will clarify these ideas.

A SIMPLE REPEATED-MEASURES EXPERIMENT

Do mental images actually exist? Our immediate subjective answer is yes, because we all can think of times when we experienced a mental picture of some object as we were solving a problem or retrieving information from memory. However, introspective evidence generally is not sufficient to validate a concept in the behavioral sciences; we need objectively observable responses as well.

Some researchers have attempted to determine whether we use literal mental images by carrying out "mental rotation" experiments. In one kind of mental rotation study, subjects are first presented with a capital letter—R, for example. Then they see a second letter that is either the same as the first one or the mirror image of it. The subjects' task is to decide as fast as possible whether the second letter was "normal" or "reversed" (see Cooper & Shepard, 1973). The subject's job is not as easy as it might seem because most of the comparison letters are presented in a rotated orientation. The independent variable is *the number of degrees by which the second letter was rotated from a perfectly vertical position.* The dependent variable is the decision time.

Examples of rotated letters are shown in Figure 14.1. The rationale behind this type of manipulation is as follows: If subjects use literal mental images of letters to reach a decision in these problems, then they have to first rotate the internal mental image of the second letter through a "physical distance" until it is at the

---

[1] By comparison, in a "between-subjects" experimental design, each subject is exposed to only one level of the independent variable. A between-subjects design requires $(k - 1)N$ more research participants than a within-subjects design. Review Chapter 12 for examples of between-subjects designs.

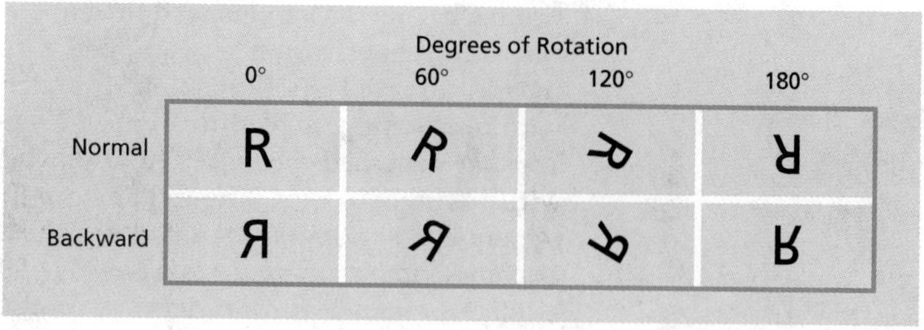

Figure 14.1 *Examples of Stimuli Used in the Mental Rotation Experiment. (Source: Figure adapted from Cooper, L.A., and Shephard, R.N. (1973). Chronometric studies of the rotation of mental images. In W.G. Chase (Ed.), Visual information processing, New York: Academic Press. Used with permission of the publisher and L.A. Cooper.)*

same orientation as the first letter. The more the comparison letter is rotated from a perfectly upright orientation, the longer the subjects' decision times should be.

Let's assume we are carrying out a mental rotation experiment involving just $N = 6$ subjects. We ask each research participant to make letter-equivalency decisions at $k = 4$ different levels of rotation: 0 degrees, 60 degrees, 120 degrees, and 180 degrees. Since each subject serves in all four rotation conditions, the independent variable is manipulated *within subjects*. Notice how this kind of manipulation differs from the other ANOVA situations you have studied, in which independent variables were manipulated *between subjects*. In those other investigations, each subject served at one and only one level of the independent variable.

As mentioned earlier, the dependent variable in the present study is the amount of time (in msec) that subjects require to decide whether the second letter matches the first one or is reversed. If the subjects use literal mental images to accomplish such comparisons, we would expect the decision time to increase systematically as a function of the degree of rotation.

Important: The essential features of an investigation that requires a repeated-measures ANOVA are that (1) there is only one independent variable, but (2) each subject serves at every level of that variable.

Despite the fact that only one independent variable is analyzed, a repeated-measures ANOVA follows the logic of a two-way ANOVA. In this kind of analysis, the subjects are treated as levels of a second "independent variable," and the subjects $\times$ treatments interaction is used as the error term (i.e., denominator) in the F ratio. I will comment further on these ideas as we go along.

Statistical Hypotheses

Under the null hypothesis, decision times are expected to be the same at all degrees of rotation:

$$H_0: \quad \mu_1 = \mu_2 = \mu_3 = \mu_4$$

Since we expect the amount of rotation to affect the decision time, we have:

H$_1$:   Not all of the means are the same.

Level of Significance

We will use a conventional level of significance: $\alpha = .05$.

Test Statistic

We will compute the omnibus F test through a repeated-measures ANOVA. In this type of analysis, three sources of variance make up the total variance: between treatments, between subjects, and the treatments $\times$ subjects interaction. The degrees of freedom associated with the sources of variance are:

$$df_{treatments} = k - 1 = 4 - 1 = 3$$
$$df_{subjects} = N - 1 = 6 - 1 = 5$$
$$df_{subjects \times treatments} = (k - 1)(N - 1) = (4 - 1)(6 - 1) = (3)(5) = 15$$

Bear in mind that the subjects $\times$ treatments mean square is considered to be an index of random error in this type of ANOVA. *The critical assumption here is that subjects do not interact with the independent variable*—that is, that the raw-score difference between any two subjects will remain approximately constant at all levels of the independent variable. Therefore, any variance associated with the subjects $\times$ treatments interaction must represent random error. I will elaborate on these points a bit later.

Critical Value of F

With alpha = .05, $df_{treatments}$ = 3, and $df_{subjects \times treatments}$ = 15, Table F in Appendix A gives an F_{crit} value of 3.29.

Decision Rule

If $F \geq 3.29$, reject H$_0$.

Computations

The results of the mental rotation experiment are listed in Table 14.1, which displays the raw scores, the treatment means, and (important for this kind of analysis) the *mean decision time for each subject*. Note well how *each person is treated statistically as a level of a second independent variable*—analogous to a "factor B" in a regular factorial design. In the lower right-hand corner of Table 14.1 you will also find the grand mean of all the data: $\overline{X}_G = 555$.

The ascending pattern of the treatment means suggests that the decision time increased systematically with the degree of mental rotation required by the stimulus, just as predicted from a "mental images" theory. Nonetheless, a question still remains: Are these mean differences statistically significant?

Table 14.1

DECISION TIME (MSEC) IN
VISUAL MEMORY AS A
FUNCTION OF DEGREE OF
ROTATION OF STIMULUS

SUBJECT	DEGREES OF ROTATION OF LETTERS				SUBJECT MEANS
	0	60	120	180	
1	650	760	780	890	770
2	550	620	680	790	660
3	400	500	520	600	505
4	350	430	450	550	445
5	400	530	510	560	500
6	350	460	480	510	450
Treatment means	450	550	570	650	
				Grand mean = 555	

Variance-Partitioning Logic Repeated-measures analysis of variance starts off like a simple one-way ANOVA, such as we would perform on data from a completely randomized research design. We first partition the total sum of squares into (1) $SS_{between\ treatments}$ and (2) $SS_{within\ treatments}$. But next the procedure follows the variance-partitioning sequence that is similar to that of a two-way ANOVA. That is, we treat the subjects in the experiment as a second independent variable. This means that we have not only $SS_{between\ treatments}$ but also $SS_{between\ subjects}$ and $SS_{subjects \times treatments}$. We obtain these last two components of variance by partitioning $SS_{within\ treatments}$; hence, $SS_{within\ treatments} = SS_{between\ subjects} + SS_{subjects \times treatments}$. Figure 14.2 summarizes the variance-partitioning operations in repeated-measures ANOVA.

Figure 14.2 *Partitioning of Variance Sequence in Repeated-Measures ANOVA*

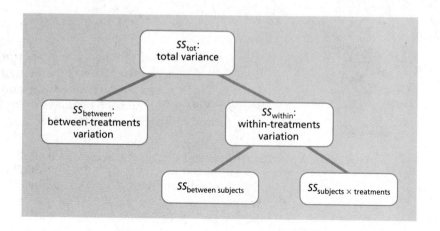

Sums of squares As noted above, the total sum of squares is attributable to three sources: $SS_{\text{between treatments}}$, $SS_{\text{between subjects}}$, and $SS_{\text{subjects} \times \text{treatments}}$.

1. Find the total sum of squares. As you now anticipate, we calculate the total sum of squares with formula (12.4). The calculations in Table 14.2 inform us that $SS_{\text{tot}} = \Sigma(X - \overline{X}_G)^2 = 473{,}400$.
2. Find the between-treatments sum of squares. We can compute this sum of squares via formula (12.5): $SS_{\text{between treatments}} = \Sigma[N(\overline{X} - \overline{X}_G)^2]$, where $N =$ number of subjects = 6.

Table 14.2

CALCULATION OF TOTAL SUM OF SQUARES FOR THE MENTAL ROTATION EXPERIMENT

ROTATION	X	$(X - \overline{X}_G)$	$(X - \overline{X}_G)^2$
0 degrees	650	95	9,025
	550	−5	25
	400	−155	24,025
	350	−205	42,025
	400	−155	24,025
	350	−205	42,025
60 degrees	760	205	42,025
	620	65	4,225
	500	−55	3,025
	430	−125	15,625
	530	−25	625
	460	−95	9,025
120 degrees	780	225	50,625
	680	125	15,625
	520	−35	1,225
	450	−105	11,025
	510	−45	2,025
	480	−75	5,625
180 degrees	890	335	112,225
	790	235	55,225
	600	45	2,025
	550	−5	25
	560	5	25
	510	−45	2,025
			473,400 = SS_{tot}

$$SS_{\text{between treatments}} = 6(450 - 555)^2 + 6(550 - 555)^2$$
$$+ 6(570 - 555)^2 + 6(650 - 555)^2$$
$$= 6(-105)^2 + 6(-5)^2 + 6(15)^2 + 6(95)^2$$
$$= 6(11,025) + 6(25) + 6(225) + 6(9,025)$$
$$= 66,150 + 150 + 1,350 + 54,150$$
$$= 121,800$$

3. Find the within-treatments sum of squares. Now we can obtain $SS_{\text{within treatments}}$ via subtraction:

$$SS_{\text{within treatments}} = SS_{\text{tot}} - SS_{\text{between treatments}}$$
$$= 473,400 - 121,800$$
$$= 351,600$$

4. Partition the within-treatments sum of squares. First, treating subjects as a second independent variable, find the between-subjects sum of squares through:

$$\overset{\textbf{any subject mean}}{\underset{\underset{\textbf{multiply by number of treatment levels}}{\uparrow}}{SS_{\text{between subjects}} = \Sigma[k(\overline{X}_{\text{subject}} - \overline{X}_{G})^2]}} \tag{14.1}$$

Thus,

$$SS_{\text{between subjects}} = 4(770 - 555)^2 + 4(660 - 555)^2 + 4(505 - 555)^2$$
$$+ 4(445 - 555)^2 + 4(500 - 555)^2 + 4(450 - 555)^2$$
$$= 4(215)^2 + 4(105)^2 + 4(-50)^2 + 4(-110)^2 + 4(-55)^2 + 4(-105)^2$$
$$= 4(46,225) + 4(11,025) + 4(2,500) + 4(12,100)$$
$$+ 4(3,025) + 4(11,025)$$
$$= 184,900 + 44,100 + 10,000 + 48,400 + 12,100 + 44,100$$
$$= 343,600$$

Then find the sum of squares for the subjects × treatments interaction:

$$SS_{\text{subjects} \times \text{treatments}} = SS_{\text{within treatments}} - SS_{\text{between subjects}} \tag{14.2}$$
$$= 351,600 - 343,600 = 8,000$$

Mean squares   The mean squares for the effect of treatments and the "error term," respectively, are:

$$MS_{\text{between treatments}} = \frac{SS_{\text{between treatments}}}{df_{\text{between treatments}}} = \frac{121,800}{3} = 40,600$$

$$MS_{\text{error}} = MS_{\text{subjects} \times \text{treatments}} = \frac{SS_{\text{subjects} \times \text{treatments}}}{df_{\text{subjects} \times \text{treatments}}}$$

$$= \frac{8,000}{15} = 533.33$$

It is not necessary for us to compute a mean square for subjects because *the "subjects factor" is extracted solely for the purpose of reducing the size of the error term in the F ratio, not for any substantive purpose.* (See the next section: "The Theory Behind Repeated-Measures ANOVA.")

F test   The F ratio in repeated-measures ANOVA is obtained in the usual way:

$$F = \frac{MS_{\text{effect}}}{MS_{\text{error}}} = \frac{MS_{\text{between treatments}}}{MS_{\text{subjects} \times \text{treatments}}} = \frac{40,600}{533.33} = 76.13$$

The complete ANOVA results appear in Table 14.3.

Decision

Since F_{crit} is only 3.29, the computed F ratio of 76.13 is obviously significant, and our decision is to reject H_0. The degree of physical rotation of the comparison stimulus reliably determines the amount of time required for subjects to complete their mental rotations.

Additional Analyses and Conclusions

Pairwise comparisons   With $\alpha = .05$, Tukey's *HSD* for this problem is $q\sqrt{MS_{\text{error}}/N} = 4.08\sqrt{\frac{533.33}{6}} = 4.08\sqrt{88.89} = 4.08 \cdot 9.43 = 38.47$. Hence, treatment means that are at least 38.47 units apart

Table 14.3

ANOVA SUMMARY TABLE
FOR THE MENTAL ROTATION
EXPERIMENT

SOURCES OF VARIATION	SUMS OF SQUARES	DEGREES OF FREEDOM	MEAN SQUARES	F RATIO
Between treatments	121,800	3	40,600	76.13*
Between subjects	343,600	5		
Subjects × treatments (error)	8,000	15	533.33	
TOTAL	473,400	23		

* F ratio is significant, $p < .05$.

are reliably different. From this result, we know that all the mean differences between the levels of letter rotation are significant except the difference between the means for 60 degrees ($\overline{X} = 550$) and 120 degrees ($\overline{X} = 570$).

Effect size   The magnitude of the degree of physical rotation effect is assessed in the same way as it was for a one-way ANOVA:

$$\eta^2 = \frac{SS_{\text{between treatments}}}{SS_{\text{tot}}} = \frac{121,800}{473,400} = .257$$

This means that the degree of rotation accounted for about 25.7% of the variance in the subjects' decision times in the present sample.[2]

It is interesting that systematic individual differences among subjects explained:

$$\eta^2 = \frac{SS_{\text{between subjects}}}{SS_{\text{tot}}} = \frac{343,600}{473,400} = .726$$

or 72.6% of the sample variance. So, naturally occurring differences among subjects accounted for almost three times as much variation as the independent variable. Had we not systematically removed these subject differences in the analysis, they would have greatly inflated the denominator of the F ratio and produced a nonsignificant outcome. But what justifies the removal of subject variance in this kind of analysis? The rationale underlying repeated-measures analysis of variance will answer this question.

THE THEORY BEHIND REPEATED-MEASURES ANOVA

According to the logic of the F test, the F statistic is the ratio:

$$\frac{\text{error variance } + \text{ effect of independent variable}}{\text{error variance}}$$

But this ratio is made up of somewhat different components in regular one-way ANOVA and in repeated-measures ANOVA.

The Situation in One-Way ANOVA

You use a standard one-way analysis of variance when you are working with data from an independent-samples research design in which *different subjects serve in the different treatment conditions*. In that type of ANOVA, the "error variance" is made up of two sources of random variation:

---

[2] Recall that eta-squared indexes the proportion of variance explained in the sample, not necessarily in the population represented by the sample.

1. "Random sampling error," represented by naturally occurring *individual differences between subjects* within a sample
2. "Experimental error," caused by confusing instructions, unreliable measurement, distractions, inattention, equipment malfunction, and so on.

A closer look at the *F* ratio reveals:

$$F = \frac{\text{random sampling error} + \text{experimental error} + \text{ effect of independent variable}}{\text{random sampling error} + \text{ experimental error}}$$

That is, in a one-way ANOVA, both the numerator and the denominator of the *F* statistic contain variance that results from (1) naturally occurring differences among subjects and (2) experimental error. In addition, the numerator contains variance that results from the treatments, if the independent variable had an effect. The implication of the complete specification of the *F* ratio in one-way ANOVA is this: Between-subjects differences contribute to both the between-treatments variance ($MS_{\text{between treatments}}$) and the within-treatments variance ($MS_{\text{within treatments}}$). Just as random between-person variation contributes to differences among raw scores within samples, it also contributes to part of the variation between sample means.

Why Subject Variance Is Removed in Repeated-Measures ANOVA

A close examination of the components of variance in repeated-measures analysis of variance discloses a slightly different picture. The key point is that, *because the same subjects are used in all treatments, the between-treatments variance ($MS_{\text{between treatments}}$) does not contain any variation resulting from random individual differences among subjects*. Rather, variation associated with naturally occurring individual differences exists only in the within-treatments variance ($MS_{\text{within treatments}}$). Thus, if we conducted a standard *F* test on data from a repeated-measures design, the ratio would look like this:

$$F = \frac{MS_{\text{between treatments}}}{MS_{\text{within treatments}}}$$

$$= \frac{\text{experimental error} + \text{ effect of independent variable}}{\text{random sampling error} + \text{ experimental error}}$$

where "random sampling error" refers to naturally occurring individual differences among subjects.

If you compare the numerator with the denominator, it should be obvious that such a ratio mixes apples with oranges. It certainly would not give us a valid indication of the effect of the independent variable; rather, the ratio would tend to give us an F statistic that is smaller than it should be.

To arrive at an accurate F ratio, we need to partition the within-treatments variation into a part that is caused by individual differences among subjects and a part that results solely from experimental error. This is exactly what we do when we treat between-subjects differences as a second independent variable and remove its sum of squares from the within-samples sum of squares. The eventual result of that operation in repeated-measures ANOVA is an appropriate F ratio that consists of:

$$F = \frac{MS_{\text{between treatments}}}{MS_{\text{subjects} \times \text{treatments}}}$$

$$= \frac{\text{experimental error} + \text{treatment effect}}{\text{experimental error}}$$

Why the Interaction Term Reflects Experimental Error

It still might not be clear to you why the subjects $\times$ treatments variance is considered random error in this type of ANOVA. The important idea to remember here is that repeated-measures ANOVA rests on the *crucial assumption that there is no reliable interaction between subjects and the experimental treatment*—in other words, that the difference between subject 2 and subject 3, let's say, is the same at each and every level of the independent variable. We can get some notion of how well this assumption holds in the present example by looking at Figure 14.3. I have plotted each subject's decision time at each level of the independent variable. Notice that the subjects' performance functions are essentially parallel, that is, there are no large changes in between-subjects differences as we look from one level of the independent variable to another. Fundamentally, there is no interaction between subjects and treatment, as attested to by the value of eta-squared for that interaction:

$$\eta^2 = \frac{SS_{\text{subjects} \times \text{treatments}}}{SS_{\text{tot}}} = \frac{8,000}{473,400} = .017$$

The eta-squared value tells us that the subjects $\times$ treatments interaction accounts for only 1.7% of the total variation in decision times. And yet the subjects' performance functions are not perfectly parallel, are they? At a couple of points in the figure, there is even some minor crossing over of functions. If there is no subjects $\times$ treatments interaction, how do we explain these small departures from the assumption? The answer is that the random

Figure 14.3 *Plot of Subjects' Decision Times Showing No Subjects × Treatments Interaction (Note: Repeated-measures ANOVA assumes that there is no interaction between subjects and treatments.)*

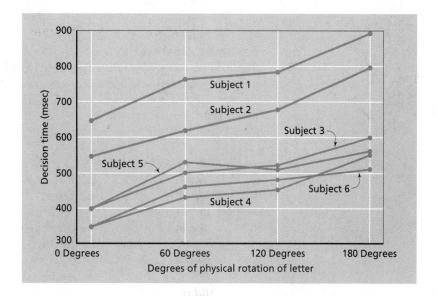

processes categorized as "experimental error"—confusing instructions, unreliable measurement, distractions, inattention, equipment malfunction, and so on—throw subjects' scores off their pace just a little at different times in the experiment. It is these small chance factors that cause the data to deviate slightly from the assumption of no interaction, which makes $MS_{\text{subjects} \times \text{treatments}}$ larger than 0. But to the extent that there is some variance associated with $MS_{\text{subjects} \times \text{treatments}}$, this variance should be considered purely random error.

The same experimental error factors also add random effects to the between-treatments differences. Thus, the F ratio of $MS_{\text{between treatments}} / MS_{\text{subjects} \times \text{treatments}}$ really is (experimental error + treatment effect)/(experimental error), and it is entirely sensible to use the subjects × treatments mean square as the error term.

ADVANTAGES AND DISADVANTAGES OF REPEATED MEASURES

Because the same subjects are used in all treatments, a repeated-measures experimental design necessarily yields correlated samples of data in the various treatments. Repeated measures ANOVA therefore has the same advantages and disadvantages as the correlated-samples t test that you studied in Chapter 11.

Advantages

Methodological advantage Compared with a completely randomized, independent-groups design, the repeated-measures design is more economical of subjects. For example, if you wanted to conduct a study with four levels of the independent variable (i.e., four treatments) with ten observations at each level, you would need the services of 40 subjects in an independent-groups experimental design. By contrast, a repeated-measures design involving the same four treatments with ten observations per treatment would require only ten subjects; each of the subjects would simply serve in each of the four treatments.

Statistical advantage The data from a repeated-measures study could be analyzed with a standard one-way analysis of variance. But the special repeated-measures ANOVA is the appropriate method for significance tests in this kind of research design. Relative to a standard ANOVA, the repeated-measures ANOVA has the same statistical benefit as its cousin, the correlated-samples t test (see Chapter 11). It tends to increase the size of the F statistic by removing individual-difference variation from the "error term," or denominator, in the F ratio. And, other things being constant, the larger the F statistic, the greater the likelihood of rejecting the null hypothesis. Repeated-measures ANOVA is more powerful than standard ANOVA *to the extent that the assumption of no subjects × treatments interaction is met.*

Look again at Table 14.3. Had we conducted a *standard* one-way ANOVA on the mental rotation data—instead of a repeated-measures ANOVA—the between-subjects sum of squares (336,400) would have been included in the within-treatments sum of squares, yielding an "error" sum of squares of 343,600 + 8,000 = 351,600. Under those circumstances, the denominator in the F ratio would have been

$$MS_{\text{w}} = \frac{SS_{\text{w}}}{df_{\text{w}}} = \frac{343,600 + 8,000}{5 + 15} = \frac{351,600}{20} = 17,580$$

instead of 533.33. And the resulting F ratio would have been $40,600/17,580 = 2.31$ (which is nonsignificant) instead of 76.13. The standard one-way ANOVA simply would not have been powerful enough to result in a rejection of H_0.

Disadvantages

Methodological disadvantage Many, if not most, questions addressed by behavioral research cannot be answered through a repeated-measures design. The main obstacle to the widespread use of the repeated-measures method is "carryover effects": Serving in one condition of an experiment changes the way the subject responds in another condition of the same experiment.

Consider research on the effectiveness of different types of psychotherapy. If a repeated-measures approach were taken to addressing therapy effectiveness, then each subject would serve in the therapy A condition at one time in the study and in therapies B and C at other times. But it is likely that receiving therapy A would influence the subject's response to the other two types of therapy, rendering them either more or less effective than they would be if experienced alone. Consequently, you would not be able to assess the usual effect of those therapies through a repeated-measures experiment. Such research questions require that you use an independent-samples design and simply resign yourself to living with the lower statistical power of an independent-samples analysis of variance. But, should you always assume that an independent-samples ANOVA will be less powerful than a repeated-measures analysis?

Potential statistical disadvantage   There is a penalty for using repeated-measures ANOVA. Relative to a comparable independent-samples analysis, it does reduce the denominator degrees of freedom by $N - 1$, which tends to diminish the power of the test. That is, the repeated-measures error term has $(k - 1)(N - 1)$ degrees of freedom—versus $k(N - 1)$ for a completely randomized ANOVA.

In most cases the positive effect of a smaller "error term" more than compensates for the loss of a few degrees of freedom. But if, for whatever reason, the assumption of no subjects $\times$ treatments interaction is untrue, then the error term in the denominator of the F ratio will contain much more variance than that due to experimental error. The resulting F ratio will be much smaller than it should be. That, coupled with the loss of $N - 1$ degrees of freedom, could actually make the repeated-measures analysis less powerful than the independent-samples analysis.

OTHER ASSUMPTIONS OF REPEATED-MEASURES ANOVA

In addition to the assumption of no interaction between subjects and treatments, repeated-measures analysis of variance rests on the following assumptions:

1. The dependent variable's scale of measurement is at the interval or ratio level.
2. The observations within each treatment are independent of one another.[3]

---

[3] This assumption simply means that, *within any treatment*, no subject's score on the dependent variable should influence any other subject's score. It's okay for the data to be correlated *between* samples in this type of ANOVA.

3. The dependent variable is normally distributed in the populations represented by the treatments.
4. The variances of the populations are equal.

A SECOND EXAMPLE

An additional repeated-measures analysis will give you some practice with what we've covered in this chapter.

The Problem

A time-and-motion analyst wants to find out whether the productivity of factory employees varies significantly as a function of their work shift. She randomly samples ten employees and records the number of units each produces in a one-hour interval during each of three work shifts: the daylight shift (7 A.M. to 3 P.M.), the swing shift (3 P.M. to 11 P.M.), and the graveyard shift (11 P.M. to 7 A.M.). Each of the employees is measured once in each of the three work shift conditions. Productivity is assessed by the number of units produced during the one-hour observation period of each shift.

Null and Alternative Hypotheses

The null hypothesis asserts no productivity differences among the three shifts:

$$H_0: \quad \mu_1 = \mu_2 = \mu_3$$

The alternative hypothesis states that there will be at least one mean difference among the shifts:

$$H_1: \quad \text{Not all the } \mu\text{'s are the same.}$$

Level of Significance

The alpha level is set at .05.

Test Statistic

A repeated-measures ANOVA will be used with the following degrees of freedom:

$$df_{\text{between treatments}} = k - 1 = 3 - 1 = 2$$

$$df_{\text{between subjects}} = N - 1 = 10 - 1 = 9$$

$$df_{\text{subjects} \times \text{treatments}} = (k - 1)(N - 1) = (3 - 1)(10 - 1) = (2)(9) = 18$$

Critical Value Needed for Significance

Table F of Appendix A shows that with 2 and 18 degrees of freedom and $\alpha = .05$, $F_{\text{crit}} = 3.55$.

Decision Rule

If $F \geq 3.55$, reject H_0.

Computations

The results of the study are shown in Table 14.4, which displays the raw scores, treatment means, subject means, and grand mean for the worker productivity scores. Note that the grand mean = $\overline{X}_G = 10$.

Find the sums of squares   We will calculate the following sum of squares in the order given: total sum of squares, between-treatments sum of squares, between-subjects sum of squares, and subjects × treatments sum of squares.

1. Compute SS_{tot}. The computations in Table 14.5 show that $SS_{tot} = \Sigma(X - \overline{X}_G)^2 = 286$.

2. Compute $SS_{between\ treatments}$

$$SS_{between\ treatments} = \Sigma[N(\overline{X} - \overline{X}_G)^2]$$

$$= 10(10 - 10)^2 + 10(11 - 10)^2 + 10(9 - 10)^2$$

$$= 10(0)^2 + 10(1)^2 + 10(-1)^2$$

$$= 0 + 10 + 10$$

$$= 20$$

Table 14.4

EMPLOYEE PRODUCTIVITY AS A FUNCTION OF WORK SHIFT		WORK SHIFT			SUBJECT MEANS
	SUBJECT	DAYLIGHT	SWING	GRAVEYARD	
	1	8	10	6	8
	2	13	16	13	14
	3	11	12	10	11
	4	7	9	8	8
	5	5	5	5	5
	6	14	15	10	13
	7	9	8	7	8
	8	12	13	11	12
	9	8	10	6	8
	10	13	12	14	13
	Treatment means	10	11	9	
				Grand mean = 10	

Table 14.5

	SHIFT	X	$(X - \bar{X}_G)$	$(X - \bar{X}_G)^2$
CALCULATION OF TOTAL SUM OF SQUARES FOR WORK-SHIFT STUDY		8	−2	4
		13	3	9
		11	1	1
		7	−3	9
	Daylight	5	−5	25
		14	4	16
		9	−1	1
		12	2	4
		8	−2	4
		13	3	9
		10	0	0
		16	6	36
		12	2	4
		9	−1	1
	Swing	5	−5	25
		15	5	25
		8	−2	4
		13	3	9
		10	0	0
		12	2	4
		6	−4	16
		13	3	9
		10	0	0
		8	−2	4
	Graveyard	5	−5	25
		10	0	0
		7	−3	9
		11	1	1
		6	−4	16
		14	4	16
				$SS_{tot} = 286$

3. Compute $SS_{\text{between subjects}}$. This sum of squares is computed with formula (14.1). Since there are so many subject means in this problem ($N = 10$), we will use a worksheet approach to this computation. Table 14.6 shows that:

$$SS_{\text{between subjects}} = \Sigma[k(\bar{X}_{\text{subject}} - \bar{X}_G)^2] = 240$$

Table 14.6

CALCULATION OF BETWEEN-
SUBJECTS SUM OF SQUARES
IN WORK-SHIFT STUDY

SUBJECT	SUBJECT MEANS	$(\bar{X}_{SUBJECT} - \bar{X}_G)$	$(\bar{X}_{SUBJECT} - \bar{X}_G)^2$	k	$k(\bar{X}_{SUBJECT} - \bar{X}_G)^2$
1	8	−2	4	3	12
2	14	4	16	3	48
3	11	1	1	3	3
4	8	−2	4	3	12
5	5	−5	25	3	75
6	13	3	9	3	27
7	8	−2	4	3	12
8	12	2	4	3	12
9	8	−2	4	3	12
10	13	3	9	3	27

$$SS_{\text{between subjects}} = 240$$

4. Compute $SS_{\text{subjects} \times \text{treatments}}$. To arrive at the subjects × treatments sum of squares, there really is no need to compute the within-treatments sum of squares, even though I did that earlier to make a theoretical point. Note that

$$SS_{\text{tot}} = SS_{\text{between treatments}} + SS_{\text{within treatments}}$$

But since

$$SS_{\text{within treatments}} = SS_{\text{between subjects}} + SS_{\text{subjects} \times \text{treatments}}$$

the model becomes:

$$SS_{\text{tot}} = SS_{\text{between treatments}} + SS_{\text{between subjects}} + SS_{\text{subjects} \times \text{treatments}}$$

Therefore, once we calculate the between-treatments and between-subjects sums of squares, the subjects × treatments sum of squares can be determined very simply through:

$$SS_{\text{subjects} \times \text{treatments}} = SS_{\text{tot}} - SS_{\text{between treatments}} - SS_{\text{between subjects}} \quad (14.3)$$

$$= 286 - 20 - 240$$

$$= 26$$

Find the mean squares In a repeated-measures analysis, it is necessary to find only two mean squares.

1. Compute the between-treatments mean square.

$$MS_{\text{between treatments}} = \frac{SS_{\text{between treatments}}}{df_{\text{between treatments}}} = \frac{20}{2} = 10.00$$

2. Compute the subjects × treatments mean square.

$$MS_{\text{subjects} \times \text{treatments}} = \frac{SS_{\text{subjects} \times \text{treatments}}}{df_{\text{subjects} \times \text{treatments}}} = \frac{26}{18} = 1.44$$

Compute the F ratio   The results of this ANOVA appear in Table 14.7.

$$F = \frac{MS_{\text{between treatments}}}{MS_{\text{subjects} \times \text{treatments}}} = \frac{10.00}{1.44} = 6.94$$

Decisions

Since the computed F of 6.94 exceeds $F_{\text{crit}} = 3.55$, the appropriate decision is to reject H_0. The productivity of the employees does vary reliably with their work shifts. We also find

$$HSD = q\sqrt{MS_{\text{error}}/N} = 3.61\sqrt{144/10} = 3.61\sqrt{0.144}$$

$$= 3.61 \cdot 0.38 = 1.37$$

To be significant, the absolute difference between any two means must be at least 1.37. Therefore, the employees are more productive on the swing shift ($\overline{X} = 11$) than on the Graveyard shift ($\overline{X} = 9$), but neither of those productivities differs from the daylight shift ($\overline{X} = 10$).

Finally, since

$$\eta^2 = \frac{SS_{\text{between treatments}}}{SS_{\text{tot}}} = \frac{20}{286} = .07$$

only 7% of variation in the sample of productivity scores is attributable to the influence of the work shift. This outcome suggests that, though statistically significant, the work shift may have little practical influence on productivity in the plant studied.

Table 14.7

ANOVA SUMMARY TABLE FOR WORK-SHIFT STUDY

SOURCES OF VARIATION	SUMS OF SQUARES	DEGREES OF FREEDOM	MEAN SQUARES	F RATIO
Between Treatments	20	2	10	6.94*
Between Subjects	240	9		
Subjects × Treatments (Error)	26	18	1.44	
TOTAL	286	29		

*F ratio is significant, $p < .05$.

Box 14.1

THE 11 STEPS OF REPEATED-MEASURES ANOVA

To perform a repeated-measures ANOVA:

1. Compute the total sum of squares: $SS_{tot} = \Sigma(X - \overline{X}_G)^2$

2. Compute the between-treatments sum of squares: $\Sigma[N(\overline{X} - \overline{X}_G)^2$, where N is the number of subjects

3. Compute the between-subjects sum of squares: $SS_{between\ subjects} = \Sigma[k(\overline{X}_{subject} - \overline{X}_G)^2]$, where k is the number of treatments

4. Compute the subjects x treatments sum of squares:

   $SS_{subjects \times treatments} = SS_{tot} - SS_{between\ treatments} - SS_{between\ subjects}$

5. Determine the total degrees of freedom: $df_{tot} = Nk - 1$

6. Determine the between-treatments degrees of freedom: $df_{between\ treatments} = k - 1$

7. Determine the between-subjects degrees of freedom: $df_{between\ subjects} = N - 1$

8. Determine the subjects $\times$ treatments degrees of freedom: $df_{subjects \times treatments} = (k - 1)(N - 1)$

9. Compute the between-treatments mean square: $MS_{between\ treatments} = SS_{between\ treatments}/df_{between\ treatments}$

10. Compute the subjects $\times$ treatments mean square: $MS_{subjects \times treatments} = SS_{subjects \times treatments}/df_{subjects \times treatments}$

11. Compute the F ratio: $F = MS_{between\ treatments}/MS_{subjects \times treatments}$

TAKING STOCK

You have learned about a type of analysis of variance specifically designed for investigations in which a single group of subjects serves in multiple treatment conditions. Although many behavioral research problems are not amenable to a repeated-measures design, using such a design can result in a more powerful test of the null hypothesis.

As we end this chapter we also complete our study of analysis of variance in general. It is likely that the knowledge gained from this chapter and the last two will prove useful to you in future courses, graduate school, and your career. ANOVA plays a very prominent role in behavioral science research.

Unfortunately, your data won't always meet, or even approximate, the assumptions of ANOVA. When you find yourself in that situation, you will need to use a different category of statistical methods called "nonparametric tests." Not coincidentally, we will examine these alternative methods in the next chapter.

KEY TERM

within-subjects experimental design

SUMMARY

1. You conduct a repeated-measures ANOVA when the data are based on a within-subjects experimental design in which the independent variable is manipulated "within subjects" by exposing each research participant to all levels of that variable. Because the subject is tested at every level of the independent variable, he or she is "repeatedly measured." It is appropriate to use a repeated-measures ANOVA when *there is only one independent variable, but each subject has contributed data to every condition in the study.*

2. A repeated-measures ANOVA is conducted according to the same logic as a two-way ANOVA in which the subjects are treated as levels of a second independent variable and the treatments × subjects interaction is the "error term" in the F ratio.

3. Compared with a standard one-way ANOVA, the main advantage of a repeated-measures ANOVA is a large increase in power, the probability of rejecting a false H_0. This increase in power is realized by removing the effect of subject differences from the error term of the F ratio, an operation that is justified only if there is no subjects × treatments interaction. When the assumption of no subjects × treatment interaction is violated, a repeated-measures ANOVA can be less powerful than a standard one-way ANOVA.

4. Repeated-measures designs also have a methodological advantage: They require fewer subjects than independent-samples designs. There is also a potential methodological disadvantage: One treatment can have carryover effects on other treatments, thereby making the results uninterpretable.

REVIEW QUESTIONS

1. Summarize the similarities and differences between two-way ANOVA and repeated-measures ANOVA. Include the effects

analyzed, nature of the error term, and number of independent variables involved.

2. When data are generated by a within-subjects experimental design, why is it advantageous to analyze those data with a repeated-measures ANOVA rather than a standard one-way ANOVA? Be specific and complete.

3. In your own words, explain the idea that the subjects × treatments interaction is to be considered an error in a repeated-measures ANOVA. Also, state the logic behind using the subjects × treatments mean square as the denominator of the F ratio.

4. In a study of the effect of pupil size on attractiveness, eight subjects rated pictures of the faces of male models on a seven-point attractiveness scale, with higher ratings meaning greater perceived attractiveness. (Prior research had established that this scale has interval properties.) The facial snapshots were selected on the basis of prior ratings that had shown them to be equally attractive, but some pictures were touched up to increase the apparent size of the models' pupils. In this study, each subject rated pictures with all three levels of pupil enlargement. The ratings appear in the table. Using alpha = .01, carry out an ANOVA on these data, and develop an appropriate conclusion.

	Pupil Size		
Subjects	Small	Average	Large
1	3 9	4	6
2	3 9	3	5
3	4 16	5	7
4	3 9	4	6
5	5 25	6	7
6	5 25	5	7
7	4 16	5	7
8	4 16	5	7
	31 125	37 177	52 342

5. Use the results of the ANOVA set up in Question 4.
   (a) State the null and alternative hypotheses, both verbally and symbolically.
   (b) Determine the proportion of total variance explained by pupil size.
   (c) Use Tukey's *HSD* to assess differences between pairs of treatment means.

6. Using the data presented in Question 4, conduct a standard one-way ANOVA to test the effect of pupil size on perceived attractiveness. Set alpha to .01 and draw an appropriate conclusion.

7. A cognitive scientist was interested in the effect of "spacing of practice" on memory. He presented five subjects with a long list of nouns to memorize—one word at a time. All the words appeared twice in the list, but the interval between occurrences of an item was systematically varied. Some words were repeated back to back with themselves (0 spacing), whereas other words were repeated after 5, 10, or 15 intervening words were presented. All subjects received all four spacing intervals. The number of words recalled in each condition appear in the table. Using the .05 level of significance, carry out an ANOVA and reach an appropriate conclusion.

Subject	Spacing Interval			
	0	5	10	15
1	12	18	20	20
2	15	20	23	25
3	13	20	22	23
4	11	16	19	20
5	9	15	16	18

8. Complete the following ANOVA summary table. Evaluate the statistical significance of the treatment effect, with $\alpha = .05$.

Sources of Variation	Sums of Squares	Degrees of Freedom	Mean Squares	F Ratio
Between treatments	22	3		
Between subjects	40			
Subjects × treatments (error term)			1.67	
TOTAL	82			

9. Complete the following ANOVA summary table. Evaluate the statistical significance of the treatment effect, with $\alpha = .05$.

Sources of Variation	Sums of Squares	Degrees of Freedom	Mean Squares	F Ratio
Between treatments	35	5		
Between subjects				
Subjects × treatments (error term)	135	45		
TOTAL	270			

10. Although procedurally different, a repeated-measures ANOVA and the correlated-samples t test are based on the same theory. In fact, the correlated-samples t test is a special version of the repeated-measures ANOVA. To demonstrate this, use repeated-measures ANOVA to reanalyze Devoe's (1990) "Silent Scream" data, which appear in Table 11.2 of Chapter 11. Using $\alpha = .05$, do you reach the same conclusion as that suggested by the t test performed on those data? Verify that, in the two-sample situation, $t^2 = F$.

11. Summarize the methodological advantage and disadvantage of repeated-measures research designs.

12. A counselor is interested in assessing self-esteem changes in college students over their four years of college. The counselor randomly selects ten students who agree to complete a questionnaire at the end of each year at the college. The students' yearly self-esteem scores appear in the table. Using the .01 level of significance, carry out an ANOVA and reach an appropriate conclusion about self-esteem changes.

Student	Freshman	Sophomore	Junior	Senior
Emily	13	16	20	22
John	23	25	28	30
Hank	17	19	20	24
Susan	20	23	24	26
Nicole	28	29	29	34
Paul	15	19	24	27
Jack	23	23	28	30
Becky	30	30	31	33
Lois	13	18	20	22
Dan	20	23	25	29

13. Use the results of the ANOVA set up in Question 12.
    (a) State the null and alternative hypotheses, both verbally and symbolically.
    (b) Determine the proportion of total variance accounted for by year in college.
    (c) Use Tukey's *HSD* to assess differences between pairs of treatment means.

14. Six depressed patients in a psychiatric facility receive five electroconvulsive shock treatments over a period of three weeks. At the end of each week, the patients' depression levels are measured with a standardized test. The weekly depression scores are shown in the table. Using the .01 level of significance, carry out an ANOVA and reach an appropriate conclusion concerning the effect of the treatment on depression scores.

		Week	
Patient	1	2	3
Harry	9	7	6
Helen	13	12	12
Henry	15	16	16
Heloise	10	9	10
Harriet	9	9	9
Harold	7	6	5

15. Assume that each of the following lines gives you information about a separate repeated-measures ANOVA. For each analysis, state how many subjects were in the study and how many treatment conditions there were.

(a) $df_{tot} = 39$, $df_{between\ treatments} = 3$

(b) $df_{tot} = 34$, $df_{between\ treatments} = 4$

(c) $df_{subjects \times treatments} = 38$, $df_{between\ subjects} = 19$

CHAPTER 15 NONPARAMETRIC TESTS

The final segment of our quantitative adventure will focus on some less well known hypothesis-testing procedures called "nonparametric tests." Normally we use these procedures when we decide that conventional tests—z tests, t tests, and ANOVA—won't produce accurate results because one or more of their assumptions cannot be met. In this chapter, we will cover (1) the nature, rationale, and limitations of nonparametric tests and (2) the logic and application of three often-used nonparametric tests.

THE NONPARAMETRIC STORY

Meaning of *Nonparametric*

The definition of *nonparametric* is most readily understood by contrasting nonparametric statistics with the more familiar parametric statistics. **Parametric tests**, such as the t and F tests, *assess the equivalence of specific parameters*. In the two-sample case, for instance, the hypothesis tested is H_0: $\mu_1 = \mu_2$. If a t ratio is significant, H_0 is rejected and the two population means represented by the sample data are deemed nonequivalent. The t test has established the inequality of *two specific parameters*, μ_1 and μ_2.

In contrast, **nonparametric tests** are *procedures designed to assess the equivalence of whole distributions*, not just specific parameters; hence, they are "non-parametric." In the two-sample case, the appropriate nonparametric null hypothesis is H_0: distribution$_1$ = distribution$_2$. Notice that this null hypothesis does not assert the equivalence of any particular parameters. Rather, it holds that two population distributions are equivalent in all respects. If the nonparametric test comes out significant, that means that the two distributions differ from each other *in at least one way*.

How might two population distributions differ? Well, obviously, they might have different means, and that disparity would cause the test statistic to be significant. But, perhaps less obvious, the distributions might also differ in variability, skewness, or shape. Any of these differences might also cause the nonparametric test statistic to be significant.

In short, nonparametric tests are designed to be sensitive to any and all differences between population distributions: They are more general tests of significance than are parametric procedures. For example, if a chi-square test (a nonparametric technique) is significant, we know that the two population distributions represented by our samples differ in some way—in central tendency, variance, shape, or some combination of these characteristics. In contrast, if a t ratio (a parametric test) is significant, we know that the significance is due to a difference between μ_1 and μ_2, period.

Neither a shape difference nor a variability difference could have been responsible for the significant *t* ratio, *as long as the assumptions of the t test are true*. Let's review these assumptions and their role in the emergence of nonparametric techniques.

What Nonparametric Tests Are Supposed to Do

 Parametric tests, such as *z*, *t*, and *F*, make the following restrictive assumptions:

1. The observations are independent and are either randomly selected from their populations or randomly assigned to treatment groups (in a true experiment).
2. The populations from which the observations were drawn are normally distributed.
3. The populations from which the observations were drawn have equal variances.

In addition, parametric statistical procedures are most accurate when the data are on an *interval or ratio scale of measurement*. They tend to be less trustworthy when applied to data that are on a nominal or ordinal scale.

Nonparametric tests were developed to be free of the assumptions of (1) normality of the population distributions, (2) equal variances, and (3) interval measurement of the dependent variable. Indeed, they are sometimes referred to as "distribution-free tests," "assumption-free tests," or "assumption-freer tests" (Kirk, 1990). However, even nonparametric tests assume that the observations in the data sample are independent of one another—that no observation within a sample in any way influences another observation in that sample.

More to the point, mathematicians created nonparametric tests expressly to serve as alternative approaches to hypothesis testing in situations where the assumptions of the parametric tests are likely to be false. So under what circumstances might you use a nonparametric technique instead of, say, a *t* test? Most often you would opt for one of these alternative tests when (1) you either know or strongly suspect that your data are not on an interval scale, or (2) your sample size is less than $N = 30$ and you have good reason to believe that the data came from a population distribution that is markedly skewed.[1]

In addition to being less constrained by assumptions than conventional tests, many nonparametric tests were designed with a

---

[1] Possible nonequivalence of population variances is generally not a concern because it causes serious problems only when samples sizes are very unequal or very small ($N < 7$).

second advantage in mind: They are said to be easier to use than their parametric counterparts because they involve simpler computations. Thus, if you wish to perform a test of significance with just a pencil and your $7.95 calculator, a nonparametric approach might be the shorter route to the goal.

CONCEPT RECAP: SCALES OF MEASUREMENT

An **interval scale** has *equal units of measurement* all along it; it provides information about a person's position on a dimension as well as information about the "distance" beteen any two persons on that dimension.

A **ratio scale** has both equal intervals and a true zero point.

An **ordinal scale** *provides only information on people's ranks* (relative positions), and no information on the distance between persons along the measurement dimension; scale units are not equal.

A **nominal scale** *merely classifies persons into discrete (either–or) categories.*

A Realistic Appraisal of Nonparametric Tests

How well do nonparametric tests do what they were designed to do? Overall, I would give their performance an "adequate to good" rating. My reasons are as follows:

Computational simplicity is overstated As you will discover, some nonparametric techniques can entail fairly tedious computational steps. In fact, the less familiar nonparametric tests seem to become more computationally awkward as one becomes more skilled at carrying out conventional statistical tests. Moreover, virtually all statistical calculations are now done by computer software; even very complicated procedures are completed almost instantaneously. Consequently, the computational ease argument no longer carries much weight.

Robustness of conventional tests is underestimated **Robustness** is defined as *the degree to which a statistical test yields accurate results even when one of its assumptions is violated.* Computer experiments and distribution-simulation tests conducted over the past 30 years or so have revealed that parametric tests generally work well even when certain of their assumptions are violated. In particular, they are robust against violation of the normality assumption when the sample size is "large" ($N \geq 30$) and against violation of the equal-variances assumption when the sample sizes are equal and all N's ≥ 7. What's more, parametric tests often work adequately even when data are on a nominal or ordinal scale (see Baker, Hardyck & Petrinovich, 1966; Myers, DiCecco, White & Borden, 1982).

In light of this information, it is rarely necessary to substitute a nonparametric technique for a conventional test. Parametric tests usually work fine even when conditions are not perfect.

Power reduction is the price   Compared with corresponding parametric tests, nonparametric procedures more often have less statistical power. That is, they are somewhat less likely than conventional procedures to reject a false null hypothesis and, therefore, somewhat more likely to result in a Type II decision error (Hays, 1988). So it is generally unwise to choose a nonparametric test over a parametric test unless there is a clear and compelling reason to do so. Is there ever such a reason?

Justifying This Chapter

In view of my critical evaluation of nonparametric tests, you are undoubtedly wondering why I think it is necessary that you study the material that follows. A fundamental knowledge of elementary nonparametric procedures will benefit you in three ways:

1. For a variety of reasons, ranging from the admirably rational to the mindlessly emotional, many behavioral scientists occasionally apply nonparametric procedures to their data. Since much of that work eventually gets published, you will see the results of nonparametric tests in your professional books and journals. It will behoove you to understand what is being reported!
2. In some research situations, the parent population is known, *in fact*, to be markedly nonnormal, and very small samples are being used. Likewise, many estimation, rating, and classification methods in the behavioral sciences unquestionably depart from interval measurement. It is in these types of situations that nonparametric methods are most legitimately implemented. You should know how to use them so that you will have statistical options when such circumstances arise in your research.
3. Some nonparametric tests are so effective and versatile that they are used regularly in the behavioral sciences. You need to include such practical tools in your statistical armamentarium because you will find so many different ways to apply them.

In regard to the last point, the chi-square statistic is the indisputable champion in the analysis of nominal-scale data. It can easily be argued that chi-square comes closest to fulfilling the original intent of those who developed nonparametric methods.

ANALYZING NOMINAL DATA: THE CHI-SQUARE STATISTIC

You will use the chi-square test when your data are in the form of frequency counts. A **frequency count** is the *number of persons (or other units of observation) that fall into a particular nominal-scale category*, such as the number of left-handers in your statistics class.

A **chi-square** (pronounced **KI-square**) **test** *compares observed frequency counts*—the actual number of left-handers in your class, let's say—*to expected frequency counts*—the number of left-handers that *should* be in your class based on the proportion of "lefties" in the general population. If the observed frequencies differ from the expected frequencies by more than an amount predicted by the theory of chance variation, then the chi-square is statistically significant. A significant chi-square statistic usually means that the distribution of outcomes across categories has been reliably altered by some variable. In other words, the population distribution represented by the sample data is different from the population distribution specified (1) by some theory, (2) by the results of prior research, or (3) by chance alone.

Chi-square tests take two general forms: a "goodness-of-fit" test and a "test of association." Inasmuch as these procedures serve different functions in applied statistics, we will consider them separately.

Chi-Square Goodness-of-Fit Test

The **goodness-of-fit test** *assesses whether observed frequency counts "fit" some preexisting "model distribution" or deviate reliably from that model.* The model can come from any of three sources:

1. *A scientific theory*—for example, a genetic theory that predicts the percentages of left-handers and right-handers in the general population.
2. *Pre-existing population research*—for example, previous findings showing that about 12% of the general population is left-handed.
3. *A "rectangular" chance distribution* that divides up all possible outcomes evenly. If handedness were *purely* a random phenomenon, about 33.33% of the population would fall into each of the left-handed, right-handed, and ambidextrous categories.

A modest example   Suppose we have reason to believe that an atypical kind of brain organization tends to endow people with a special talent related to stage and screen acting. Let's say that this kind of brain organization also contains a larger than normal left-handedness component. Our theory leads to the following prediction: The distribution of "handedness" among the top 500 stage and screen actors in the United States will differ reliably from the distribution of handedness in the general U.S. population.

To test this theory we first establish a population distribution model. Based on previous surveys, we discover the following distribution of handedness in the general population:

Left-handed: 12% (proportion = .12)

Ambidextrous: 8% (proportion = .08)

Right-handed: 80% (proportion = .80)

Next we secure a publication that profiles the top 500 stage and screen actors in the United States. The document lists each star's vital characteristics, including handedness. Handedness among the "Fortunate 500" sample is distributed in this way:

Left-handed $f_o = 102$

Ambidextrous $f_o = 57$

Right-handed $f_o = 341$

where f_o stands for **observed frequency**, *the actual number of sample outcomes that belong to a particular nominal category.*

The question is, How do these observed frequencies compare with the model distribution shown earlier? Is there a significant deviation of the population distribution represented by our sample of actors from the general population distribution that gave rise to our model? The statistical hypotheses that pertain to this question are:

H_0: actors' population distribution = model distribution

H_1: actors' population distribution ≠ model distribution

Bear in mind that, being the ideal nonparametric statistic, *chi-square tests the equivalence of entire distributions, not particular parameters.*

Finding expected frequencies Chi-square compares observed frequencies with **expected frequencies**, defined as *the frequency counts predicted by a preexisting model.* We already have the observed frequency for each category of handedness. But where are the expected frequencies? We compute them in the goodness-of-fit test via:

$$f_e = \text{(theoretical proportion)}(N) \qquad (15.1)$$

where $N = \Sigma f_o$ = the total number of observations in the sample. Using formula (15.1), we find that the expected frequencies in this investigation are:

$f_e = (.12)(500) = 60$ for left-handedness in the sample of actors

$f_e = (.08)(500) = 40$ for ambidexterity in the sample of actors

$f_e = (.80)(500) = 400$ for right-handedness in the sample of actors

Logic of chi-square   To find out whether the distribution of observed frequencies departs significantly from the distribution of expected frequencies, we will work with the standard chi-square formula:

$$\text{Chi-square} = \chi^2 = \Sigma\left[\frac{(f_o - f_e)^2}{f_e}\right] \qquad (15.2)$$

where ① and ② point to $(f_o - f_e)^2$, ③ points to f_e, and ④ points to the summation.

FORMULA GUIDE

① Subtract each expected frequency from the corresponding observed frequency.

② Square each deviation between observed and expected frequencies.

③ Divide each squared deviation by the corresponding expected frequency.

④ Sum all the results of step 3.

Note that χ^2 is the conventional symbol for the chi-square statistic. Also observe that the more the observed frequencies deviate from the expected frequencies, the larger the squared deviations become. And, other things remaining constant, the bigger the $(f_o - f_e)^2$ values become, the larger χ^2 will be. But just how large does χ^2 have to be to qualify as "significant"? The answer awaits us in the chi-square distribution.

Chi-square distribution   The significance of a computed chi-square statistic is evaluated by comparing that statistic to a sampling distribution of chance chi-squares. If alpha = .05 and the computed χ^2 falls into the most extreme 5% of the chi-square sampling distribution, then the data are declared significant.

The mathematical characteristics of a chi-square distribution are entirely a function of the degrees of freedom associated with the sample data. *Remember this*: For a goodness-of-fit chi-square, degrees of freedom = $df = (c - 1)$, where c = the number of nominal categories involved in the sample. In the present example, there are three categories of handedness, so $c = 3$ and $df = (3 - 1) = 2$.

There is a different chi-square sampling distribution for every possible degrees of freedom. The chi-square sampling distribution for $df = 2$ is shown in Figure 15.1. Notice that the distribution is skewed to the right, which is true of all chi-square distributions based on $df < 10$. Figure 15.2 reveals that chi-square sampling distributions becomes less skewed as degrees of freedom increases. Also, since χ^2 is computed from squared deviations, all chi-square values are positive, and 0 is the lowest possible value—which, incidentally, is precisely *the value of χ^2 expected under the null hypothesis.*

Finding the critical value in the χ^2 distribution   The critical χ^2 value in any problem is the chance chi-square that marks off the region of rejection in the sampling distribution. χ^2_{crit} is a function of both the level of significance and degrees of freedom. Table C in Appendix A lists the critical values of chi-square. There you will find that when $df = 2$ and $\alpha = .05$, $\chi^2_{crit} = 5.99$. *The null hypothesis should be rejected if the computed chi-square equals or exceeds χ^2_{crit}.* Thus, we will reject H_0 if our computed $\chi^2 \geq 5.99$.

Computations   Chi-square is very easy to compute if you use a chi-square worksheet, such as Table 15.1. In a computational worksheet, a logically arranged sequence of columns is used to carry out the sequence of steps specified by a statistical formula, in this case formula (15.2): $\chi^2 = \Sigma[(f_o - f_e)^2 / f_e]$. In Table 15.1, you can see that:

- Column 1 identifies the nominal categories that contain the frequency counts.

Figure 15.1 *Chi-Square Distribution with 2 Degrees of Freedom*

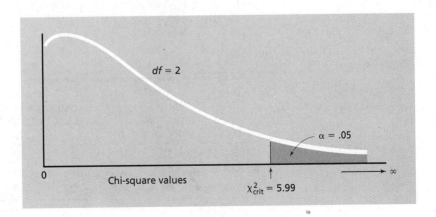

Figure 15.2 *Chi-Square Distribution Is Positively Skewed but Becomes Less Skewed As Degrees of Freedom Increases*

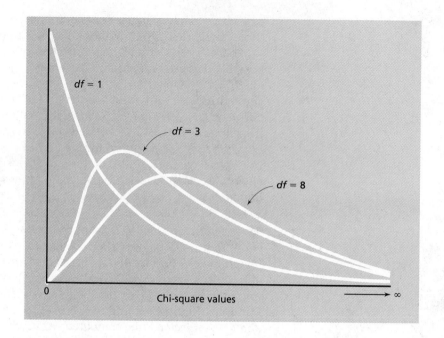

- Column 2 lists the proportions that the model predicts for each nominal category.
- Column 3 repeatedly lists N, the grand total of subjects in the study.
- Column 4 enables you to conveniently produce the expected frequencies (f_e) simply by multiplying each column 2 entry by the corresponding column 3 entry.
- Column 5 lists the observed frequency (f_o) that goes with each f_e. *Note well: It is important to properly pair up each f_o with the corresponding f_e.* Otherwise, χ^2 will be inaccurate.
- Columns 6–8 actually carry out formula (15.2): In column 6, each f_e is subtracted from each corresponding f_o. In column 7, each $(f_o - f_e)$ is squared. In column 8, each $(f_o - f_e)^2$ is divided by the corresponding f_e.

Table 15.1

GOODNESS-OF-FIT
CHI-SQUARE WORKSHEET

(1) CATEGORY	(2) THEORETICAL PROPORTION	(3) N	(4) EXPECTED FREQUENCY, f_e	(5) OBSERVED FREQUENCY, f_o	(6) $(f_o - f_e)$	(7) $(f_o - f_e)^2$	(8) $\dfrac{(f_o - f_e)^2}{f_e}$
Left-handed	.12	500	60	102	42	1764	29.400
Ambidextrous	.08	500	40	57	17	289	7.225
Right-handed	.80	500	400	341	−59	3481	8.703

$$\chi^2 = 45.328$$

The beauty of this approach is that you get χ^2 by simply summing the values in column 8. You can see that $\chi^2 = 45.33$. How does this result compare with χ^2_{crit}, as shown in Figure 15.1, and what does it mean?

Making and reporting a decision Since our computed χ^2 of 45.33 exceeds $\chi^2_{crit} = 5.99$, it is appropriate to reject H_0. The sample data do not fit the preexisting model. Accordingly, we conclude that "The distribution of handedness among eminent stage and screen stars is significantly different from the distribution of handedness in the general population, $\chi^2(2) = 45.33$, $p < .05$, with a higher incidence of left-handedness appearing in the population of stars." You should notice how the chi-square result is reported, including that the degrees of freedom is given in parentheses. This is the proper way to present the chi-square statistic in a research report.

Chi-Square Test of Association

The second version of chi-square, the **test of association**,[2] *determines whether there is a reliable association between two variables.* This second implementation uses the same general logic—and exactly the same formula—as the goodness-of-fit test. But the primary aim is not to assess the fit of frequency counts to some theoretical model. Rather, the chi-square test of association assesses whether there is a significant correlation between two variables. Also, in this kind of chi-square analysis, the sample frequency counts exist within a two-dimensional contingency table, such as Table 15.2. A **contingency table** *displays a frequency count of outcomes for each possible combination of the levels of two variables.* One variable in Table 15.2 is gender, and the other variable is hiring decision: hire or reject. Since each of the variables has two levels, Table 15.2 is a "2 × 2 contingency table."

The example depicted in Table 15.2 represents a fairly common issue in modern personnel departments: Is the frequency of hiring job applicants correlated with gender? The data come from a hypothetical company that we'll call Consolidated Engineering. They represent the number of hirings and rejections within the job-applicant pool over a two-year period. The director of personnel at Consolidated is keenly aware of the need to meet EEOC nondiscrimination guidelines in hiring practices. In the past two years, Consolidated has received 225 applications from females and has

---

[2] This second use of chi-square is also referred to as the "chi-square test of independence." These two terms are simply mirror-image expressions, since two variables are "independent" of each other if they are not "associated."

Table 15.2

CHI-SQUARE CONTINGENCY
TABLE SHOWING OBSERVED
FREQUENCIES OF HIRINGS
AND REJECTIONS BY
CONSOLIDATED ENGINEERING
AS A FUNCTION OF
APPLICANTS' GENDER

		INDEPENDENT VARIABLE: APPLICANT GENDER		
		Female	Male	Row Totals
DEPENDENT VARIABLE: EMPLOYMENT DECISION	Hired	*a* 80	*b* 107	187
	Rejected	*c* 145	*d* 230	375
	Column Totals	225	337	562 = *N*

hired 80 of those candidates; the figures for males are 337 applications and 107 hirings. To meet EEOC standards, there must not be any long-term correlation between applicant gender and hiring decisions.

How is this a test of association?   *A correlation exists within a contingency table to the extent that a particular level of one variable is associated with a particular level of a second variable more often than we would expect on the basis of chance alone.* How can we determine whether the data of a contingency table show this kind of association between particular levels of two nominal-scale variables? In the present kind of research situation, a correlation would exist to the extent that a much higher percentage of females than males fall into the hired category, for example. In that case, level 1 of the gender variable (i.e., female) would be disproportionately represented at level 1 of the employment-decision variable (i.e., hired). The latter outcome would also mean that females are disproportionately underrepresented at level 2 of the employment-decision variable (rejected). On the other hand, males would be overrepresented at level 2 of the employment-decision variable. In the language of correlation, the gender variable would tend to predict values of the employment-decision variable in the following manner: If female, the candidate will tend to be hired; if male, the candidate will tend to be rejected.

To the degree that there is such an association between gender and hiring decisions, the actual frequency of females (or males) hired will deviate from the frequency expected on the basis of chance alone. In short, such a correlation will result in big discrepancies between observed and expected frequencies, which will produce large $(f_o - f_e)^2$ quantities and, hence, a large χ^2 value.

In regard to frequencies in Table 15.2, is the hiring decision associated with females proportionately more or less often than it is associated with males? It's hard to tell by just looking at the raw frequency counts because the female and male sample sizes are so different. Even if we converted the frequencies to percents, we could only speculate on the reliability of any apparent pattern in the contingency table. However, a chi-square test of association can give us a precise answer to this question. If the chi-square test is significant, then it is very likely that there is a "contingency," or reliable relationship, between gender and hiring decisions. If χ^2 is not significant, then hiring is not "contingent" on gender; that is, no correlation exists.

For another slant on the relationship between correlation and the chi-square test of association, see Box 15.1.

IMPORTANT DISTINCTION: A chi-square test of association has a very different use than the chi-square goodness-of-fit test. The test of association usually involves *two variables* (though it can involve more than two), and a significant outcome means that the two variables tend to predict each other; that is, they are correlated.

In contrast, the goodness-of-fit test evaluates the extent to which a set of frequencies conforms to a theoretical model on a *single dimension*; that is, *only one variable is involved*. A significant outcome means that the observed frequencies deviate reliably from the model. Since only one variable is involved, the goodness-of-fit test cannot assess a correlation between two variables.

Statement of hypotheses   There are at least two ways to state the statistical hypotheses in a chi-square test of association. The more generic expression is in keeping with the fact that chi-square assesses the equivalence of population distributions:

H_0: distribution of observed frequencies = distribution of chance frequencies

H_1: distribution of observed frequencies ≠ distribution of chance frequencies

But more directly relevant to the idea of a test of association is:

H_0: There is no correlation between gender and hiring decisions

H_1: There is a correlation between gender and hiring decisions

Anatomy of a contingency table   As we proceed through this problem, I will use terms that pertain to features of contingency tables in general.

Cells: the boxes that contain the frequency counts. The number of cells in a contingency table always equals the number of levels of the first variable multiplied by the number of levels of the second variable. In this example, we have $2 \times 2 = 4$ cells. It is conventional to label each cell with a lowercase letter—cell a, cell b, cell c, and cell d—to facilitate communication about the contingency table.

Row total: the sum of the frequency counts of all cells that are in a particular row of the table. The first row total in Table 15.2 is the sum of the counts in cells a and b: $80 + 107 = 187$. The second row total equals $c + d = 145 + 230 = 375$. Row totals always appear in the right margin of a contingency table.

Column total: the sum of the frequency counts of cells that are in a particular column of the table. The first column total in Table 15.2 is $a + c = 80 + 145 = 225$. The second column total is $b + d = 107 + 230 = 337$. Column totals always appear at the bottom of a contingency table.

Box 15.1

THE PHI COEFFICIENT: CONTINGENCY TABLE, CORRELATION, AND EFFECT SIZE

A statistically significant chi-square test of association means that there is a correlation between the two variables that are represented in the contingency table. This idea is reinforced by the fact that you can directly convert a significant chi-square to the equivalent of the Pearson r (the linear correlation coefficient) *whenever you are using a 2 x 2 contingency table*. The correlation coefficient applied to a 2 x 2 contingency table is called the **phi coefficient,** and it is computed via:

$$\text{Phi coefficient} = \varphi = \sqrt{\frac{\chi^2}{N}}$$

where χ^2 is the obtained chi-square value and N is the grand total of the frequencies in the table.

φ is algebraically equivalent to the Pearson r. So a significant χ^2 tells you that there is a reliable relationship between the two nominal variables, and φ tells you how strongly the two variables are correlated. Like r^2, φ^2 is the proportion of variance in one variable that is explained by the other variable, for the sample of observations analyzed; that is, φ^2 is an index of effect size. It is computed simply through χ^2/N.

There are two caveats to heed in the application of the phi coefficient. First, neither φ nor φ^2 should routinely be computed unless χ^2 is significant. If χ^2 is significant, then φ is significant. Otherwise, φ should be considered equal to 0. Second, φ can be applied only when χ^2 is computed from a 2 x 2 table. For larger tables, you should use a slightly different index of correlation called Cramer's V statistic (Hays, 1988).

Grand total: the sum of all cell frequency counts in a contingency table. Here, grand total = $N = a + b + c + d = 80 + 107 + 145 + 230 = 562$. *The row totals should always sum to N (187 + 375 = 562); so should the column totals (225 + 337 = 562). It is wise to use this check on your arithmetic.*

Degrees of freedom and the critical value of χ^2   Remember that χ^2_{crit} is determined by the combination of the alpha level and degrees of freedom. We'll use a conventional level of significance: α = .05. When you are working with a contingency table, compute degrees of freedom via:

$$df = (r - 1)(c - 1) \tag{15.3}$$

where r = the number of rows in the contingency table and c = the number of columns in the table. In the present example, $r = 2$ and $c = 2$. Therefore, $df = (2 - 1)(2 - 1) = (1)(1) = 1$.

Consulting Table C of Appendix A, we discover that with $df = 1$, $\chi^2_{crit} = 3.84$ at the .05 level. Thus, our computed χ^2 must equal or exceed 3.84 to be considered significant. Before we can compute, we need to find the expected frequencies for the contingency table.

Expected frequencies   Table 15.3 shows both the observed and the expected frequencies for each cell of the contingency table. How are the expected frequencies determined in this type of problem? In statistical theory, the f_e values in a chi-square test of association are the *frequencies predicted by the null hypotheses—the frequencies we would get if only random variation was behind the distribution of observations in the cells of the table.* Procedurally, the expected frequency for any particular cell is found through:

Table 15.3

CHI-SQUARE CONTINGENCY TABLE SHOWING OBSERVED AND EXPECTED FREQUENCIES OF HIRING DECISIONS AT CONSOLIDATED ENGINEERING

		INDEPENDENT VARIABLE: APPLICANT GENDER		
		Female	Male	Row Totals
DEPENDENT VARIABLE: EMPLOYMENT DECISION	Hired	a $f_o = 80$ $f_e = 74.87$	b $f_o = 107$ $f_e = 112.13$	187
	Rejected	c $f_o = 145$ $f_e = 150.13$	d $f_o = 230$ $f_e = 224.87$	375
	Column Totals	225	337	562 = N

$$\text{Expected frequency} = f_e = \frac{\text{row total} \times \text{column total}}{N} \qquad (15.4)$$

To find the expected frequency for any cell, then, you multiply the total of the row in which the cell resides by the total of the column in which the cell resides; then you divide that product by the grand total.

Let's find the expected frequency for cell a:

$$f_e = \frac{\text{row total} \times \text{column total}}{N} = \frac{187 \times 225}{562}$$

$$= \frac{42,075}{562} = 74.87$$

We could compute the three remaining f_e values with the same formula, but there is a shortcut that makes finding expected frequencies much easier: Just as the observed frequencies in a row of the table must sum to the row total, so should the expected frequencies. Therefore, *the last f_e in any row or column can be found by subtraction once you have calculated all the other f_e values in that row or column.*

For example, since "f_e for cell b" + "f_e for cell a" must sum to the first row total, 187, the expected frequency for cell b must equal "first row total" – "f_e for cell a" = 187 – 74.87 = 112.13.

Likewise, the expected frequency for cell c is "first column total" – "f_e for cell a" = 225 - 74.87 = 150.13.

Finally, the expected frequency for cell d is "second column total" – "f_e for cell b" = 337 – 112.13 = 224.87.

A general principle to remember is that *you need to use formula (15.4) only as many times as the contingency table has degrees of freedom. Once $(r - 1)(c - 1)$ expected frequencies have been calculated, the remaining ones can be found through simple subtraction* because they are "not free to vary"; they are *fixed* by the marginal totals. In fact, this is precisely what is meant by the statement that a contingency table has only $(r - 1)(c - 1)$ degrees of freedom.

Rationale behind the formula for f_e   The concept behind formula (15.4) goes something like this: *If only chance is operating to distribute the frequency counts among cells, then the observations in each row of the contingency table should be distributed across the columns in simple proportion to the column totals.* For example, the proportion of total observations in the first column is 225/562 = .40036. Similarly, the proportion of total observations in the second column is 337/562 = .59964. Therefore, if only chance is operating, then 40.036% of the 187 observations in the first row are expected to fall into cell a, and 59.964% should land in cell b.

This logic works: 40.036% of 187 is $(.40036)(187) = 74.87$, which is the f_e that we calculated with formula (15.4). Similarly cell b's f_e is $(.59964)(187) = 112.13$. To repeat, the expected frequencies are generated by the theory of random variation. The question still remains: Do the observed frequencies deviate significantly from chance expectations? Chi-square will answer that for us.

 Computations The answer comes swiftly in this problem. Table 15.4 is a worksheet showing the results of the usual sequence of computational steps in a chi-square problem. The last column contains the value of $(f_o - f_e)^2/f_e$ for each cell of the contingency table. The sum of that column is $\chi^2 = \Sigma[(f_o - f_e)^2/f_e] = 0.879$. Since χ_{crit} is 3.84 and our computed χ^2 is less than that value, we fail to reject the null hypothesis. Our conclusion: "The data provide no evidence that hiring decisions are associated with gender at Consolidated Engineering, $\chi^2(1) = 0.879$, $p > .05$."

You now have a fundamental grasp of the why and how of the chi-square test of association. To broaden your understanding, we will apply your knowledge to a larger contingency table. You will find that more complicated chi-square problems require only a little more patience. The principles and procedures remain the same.

Chi-Square Test of Association: A Second Example

Statement of the problem Pseudomemories are false or distorted memories that are implanted through hypnotic suggestion. Many researchers and theorists question the nature of pseudomemories: Are they genuine changes in the accuracy of memory representations brought about through unconscious or subconscious processes? Or are they a result of a more or less conscious response bias associated with the role-playing element of hypnosis? That is, do some hypnotized persons intentionally feign memory change in order to please the hypnotist and "be a good hypnotic subject"?

Table 15.4

CHI-SQUARE WORKSHEET FOR THE DATA ON HIRING DECISIONS AS A FUNCTION OF GENDER	GENDER					$\dfrac{(f_o - f_e)^2}{f_e}$
	CELL	f_o	f_e	$(f_o - f_e)$	$(f_o - f_e)^2$	
	a	80	74.87	5.13	26.317	0.352
	b	107	112.13	−5.13	26.317	0.235
	c	145	150.13	−5.13	26.317	0.175
	d	230	224.87	5.13	26.317	0.117
						$\chi^2 = 0.879$

Murrey, Cross, and Whipple (1992) conducted a four-group study to answer these questions. All subjects viewed a videotape of a staged robbery in which the robber was wearing neither a hat nor a mask. Within a week of this experience, subjects in groups 1, 2, and 3 listened to an audiotape that was designed to heighten their memories of the robbery. Among other things, the audiotape suggested to the listener that the robber had been wearing a hat at the time of the dastardly deed. Subjects who heard the tape were characterized as "influenced" subjects. Group 4 did not hear the tape and so was "not influenced." Groups 1 and 2 listened to the suggestion-laden audiotape while they were hypnotized. Group 3 listened to the tape in their normal waking states. Finally, groups 2, 3, and 4 were offered a monetary reward for achieving high memory accuracy on a quiz about the robbery; group 1 was not offered any reward to induce greater memory accuracy.

In summary, the groups in the Murrey study were:

- Group 1: hypnotized, influenced, *not* rewarded
- Group 2: hypnotized, influenced, rewarded
- Group 3: *not* hypnotized, influenced, rewarded
- Group 4: *not* hypnotized, *not* influenced, rewarded

If pseudomemories are genuine memory changes and not a result of conscious response bias, then more hypnotized than non-hypnotized subjects should report the pseudomemories, *and* a reward should not affect how many subjects in the hypnotized groups evidence the suggested pseudomemory. But if pseudomemories are mainly a result of a conscious response bias associated with the role-playing aspect of hypnosis, then fewer subjects in group 2 than in group 1 ought to report pseudomemories. Group 2 subjects had been hypnotized but stood to gain money if they could ignore the pseudomemory suggestions and, thereby, achieve higher accuracy on the memory test.

You can inspect the results of this study by examining Table 15.5, which is a contingency table with two rows and four columns. Hence, eight data cells contain the frequency counts of the number of subjects in each group who accepted or rejected the suggested pseudomemory. Higher counts, of course, mean that more of the subjects accepted the pseudomemory.

Statistical hypotheses The competing hypotheses were:

H_0: Acceptance of pseudomemories is not associated with group membership.

H_1: Acceptance of pseudomemories is associated with group membership.

Table 15.5

		GROUP 1	GROUP 2	GROUP 3	GROUP 4	
				Not	Not Hypnotized,	
				Hypnotized,	Not	
		Hypnotized,	Hypnotized,	Not Hypnotized,	Influenced,	Row
		Influenced,	Influenced,	Influenced,	Influenced,	Totals
		No Reward	Reward	Reward	Reward	
		a	*b*	*c*	*d*	
	Accepted	12	6	7	3	28
FALSE SUGGESTION RESULT		*e*	*f*	*g*	*h*	
	Rejected	3	9	8	12	32
	Column Totals	15	15	15	15	60 = *N*

CHI-SQUARE CONTINGENCY TABLE SHOWING OBSERVED FREQUENCIES IN THE EXPERIMENT ON PSEUDOMEMORIES

Source: Table adapted from Murrey, G. J., Cross, H. J., and Whipple, J. (1992). Hypnotically created pseudomemories: Further investigation into the "memory distortion response bias question." *Journal of Abnormal Psychology, 101,* 75-77. Copyright 1992 by the American Psychological Association. Reprinted by permission.

Alpha level   The researchers used the 5% level of significance.

Test to be used   The researchers analyzed the number of subjects in each group who accepted the pseudomemory, as evidenced by their responses in the memory quiz. The procedure they used was a chi-square test of association. We'll do the same here.

Since the contingency table has two rows and four columns, the chi-square test has $(r - 1)(c - 1) = (2 - 1)(4 - 1) = (1)(3) = 3$ degrees of freedom.

Critical value   At the .05 level of significance and $df = 3$, Table C in Appendix A informs us that $\chi^2_{\text{crit}} = 7.82$.

Decision rule   If $\chi^2 \geq 7.82$, reject H_0.

 Computations   Before we can compute the chi-square statistic, we must determine what the expected frequencies are. From formula (15.4), the expected frequency for cell a is:

$$f_e = \frac{\text{row total} \times \text{column total}}{N} = \frac{28 \times 15}{60} = \frac{420}{60} = 7$$

For cell b:

$$f_e = \frac{\text{row total} \times \text{column total}}{N} = \frac{28 \times 15}{60} = \frac{420}{60} = 7$$

And for cell c:

$$f_e = \frac{\text{row total} \times \text{column total}}{N} = \frac{28 \times 15}{60} = \frac{420}{60} = 7$$

Now that we have found the expected frequencies for cells a, b, and c, we can determine the five remaining f_e values with subtraction. For cell d, $f_e = 28 - 7 - 7 - 7 = 7$; for cell e, $f_e = 15 - 7 = 8$. And so on. All of the expected frequencies are shown in Table 15.6.

Table 15.7 presents the straightforward computations involved in working out the chi-square value.

Decision The obtained χ^2 of 11.25 exceeds the χ^2_{crit} of 7.82. Consequently, we reject H_0, as did Murrey and colleagues. Acceptance of the suggested pseudomemory was significantly associated with group membership. Additional analyses conducted by Murrey et al. revealed that group 1 (hypnotized, influenced, not rewarded) was significantly more gullible than the other three groups, which did not differ significantly among themselves. This outcome agrees with the position that pseudomemories are influenced by conscious response bias.

Assumptions of Chi-Square

As mentioned earlier, nonparametric tests are not entirely free of assumptions. The chi-square statistic rests on two main assumptions:

Table 15.6

CHI-SQUARE CONTINGENCY TABLE SHOWING OBSERVED AND EXPECTED FREQUENCIES FROM THE EXPERIMENT ON PSEUDOMEMORIES			GROUP 1 Hypnotized, Influenced, No Reward	GROUP 2 Hypnotized, Influenced, Reward	GROUP 3 Not Hypnotized, Influenced, Reward	GROUP 4 Not Hypnotized, Not Influenced, Reward	Row Totals
FALSE SUGGESTION RESULT	Accepted		a $f_o = 12$ $f_e = 7$	b $f_o = 6$ $f_e = 7$	c $f_o = 7$ $f_e = 7$	d $f_o = 3$ $f_e = 7$	28
	Rejected		e $f_o = 3$ $f_e = 8$	f $f_o = 9$ $f_e = 8$	g $f_o = 8$ $f_e = 8$	h $f_o = 12$ $f_e = 8$	32
	Column Totals		15	15	15	15	$60 = N$

Table 15.7

CHI-SQUARE WORKSHEET
FOR PSEUDOMEMORY
STUDY

CELL	f_o	f_e	$(f_o - f_e)$	$(f_o - f_e)^2$	$\dfrac{(f_o - f_e)^2}{f_e}$
a	12	7	5	25	3.571
b	6	7	−1	1	0.143
c	7	7	0	0	0.000
d	3	7	−4	16	2.286
e	3	8	−5	25	3.125
f	9	8	1	1	0.125
g	8	8	0	0	0.000
h	12	8	4	16	2.000
					$\chi^2 = 11.250$

1. All observations in the sample are independent of one another. Effectively, this assumption is usually satisfied if no person is represented in more than one cell of the chi-square table. This assumption cannot be overemphasized, however, because the most common misuse of chi-square is to apply it to situations in which subjects appear in more than one treatment condition (Hays, 1988). Chi-square does not yield valid information under those circumstances.

2. All expected frequencies are equal to or greater than 5. What's more, chi-squares with only one degree of freedom should have all $f_e \geq 10$ (Kirk, 1990). However, some authorities are of the opinion that this assumption is too strict (see Faraone, 1982).

NONPARAMETRIC TESTS FOR ORDINAL DATA

Chi-square is a convenient and effective tool for analyzing nominal-scale data, but it isn't much good when you need to test the significance of ordinal-scale data. Anytime you have information that basically permits you to rank people on some dimension, but not much else, you are working with ordinal data. Fortunately, two relatively uncomplicated tests fill the bill in these kinds of hypothesis-testing situations: the Mann-Whitney U test and the Wilcoxon test for correlated samples. Both are suitable for use within the framework of two-sample research designs.

Mann-Whitney *U* Test

The Mann-Whitney U test is the ordinal-data counterpart to the independent-samples t test. This method is appropriate for testing the significance of the difference between two independent samples of observations, especially when you feel that the scale of measurement does not meet interval-scale requirements but does permit the ranking of subjects. As you probably remember, "independent samples" means that different subjects serve in the two treatment groups; put another way, each subject represents one, and only one, level of the independent variable. Ideally, subjects will have been randomly assigned to their respective treatment groups.

Here's an example of a data set for which a U test is appropriate: An industrial psychologist wishes to evaluate the impact on employee morale of two types of work-assignment plans: job rotation versus job enrichment. In the job rotation condition, employees are assigned to a different job within their department every week, until they have worked at all posts; then the "rotation" starts over again. The idea is to reduce boredom and convey the "big picture" of the department's overall function to each employee. By contrast, in the job enrichment condition, employees stay at one post all the time. However, each employee's job is broadened so that he or she has both more authority and more responsibility. This approach is also designed to reduce boredom; in addition, it is expected to boost the employee's ego.

I will not describe all the particulars, but the employees in a large department of the plant are randomly assigned to either job rotation or job enrichment for a period of one year. At the end of that time, independent interviewers, who are not aware of which condition each subject worked in, rate each employee's "positiveness of work attitude" on a scale ranging from 0 to 100. This is a type of "estimation" measurement technique.

The psychologist who headed up the study doesn't think that the attitude-estimation procedure produces data on an equal-interval scale but feels that the ratings are a valid way to rank people. Therefore, the results are in the form of ordinal data and are better analyzed with the Mann-Whitney U test than with a conventional t test. Strictly speaking, the t test requires interval data, but the U test assumes only an ordinal scale.

Statement of hypotheses   I established earlier that nonparametric techniques are supposed to assess overall differences between distributions, not differences between particular parameters. But the word *suppose* does not always coincide with actuality. As it turns out, the U test is primarily sensitive to a difference between two *medians* (Toothaker, 1986). That's okay, however, since ordinal data do permit the determination of a median, or middle score, in a distribution. And, in this investigation, the researcher certainly would be interested in finding out whether one group has a significantly higher median attitude score than the other. So the statistical hypotheses in this example are:

H_0:   population median$_1$ = population median$_2$

H_1:   population median$_1$ ≠ population median$_2$

The attitude ratings   The "positiveness of attitude" ratings appear in Table 15.8. There are N_A = 10 observations in group A (job rotation) and N_B = 7 observations in group B (job enrichment). Before we can carry out the Mann-Whitney U test, we will have to convert these raw ratings to ranks.

Ranking procedures   The procedure for transforming the raw data to ranks is to:

1. Combine all raw scores from both samples into a single set of scores.
2. Sort the set of $N_A + N_B$ raw scores in ascending order, and list them in a worksheet.
3. Next to the ordered list of raw scores, list the *ranks* of the scores, *giving the lowest raw score a rank of 1* and the highest raw score a rank of $N_A + N_B$.
4. *Important:* Resolve the ranks of tied raw scores.

Table 15.9 displays the results of these ranking operations. Column 1 lists all the raw scores in ascending order. Column 2 lists the ranks of those scores from 1 to 17. Note how the ranks of tied scores are resolved. There are two sets of tied scores: a set of 52's and a set of 68's. The formula for assigning tied ranks is:

Table 15.8

POSITIVE ATTITUDE RATINGS (100-POINT SCALE) OF EMPLOYEES ON TWO DIFFERENT WORK ASSIGNMENT PLANS	GROUP A JOB ROTATION	GROUP B JOB ENRICHMENT
	68	84
	25	52
	52	88
	68	75
	40	95
	68	93
	90	86
	45	
	60	
	77	

Table 15.9

COMBINED RANKS AND RANK
SUMS FOR GROUPS A AND B
FROM THE STUDY ON WORK
ASSIGNMENT PLANS

(1) ORDERED RAW SCORES OF BOTH GROUPS	(2) RANKS OF SCORES OF BOTH GROUPS	(3) GROUP IDENTIFICATION	(4) RANKS FOR GROUP A	(5) RANKS FOR GROUP B
25	1	A	1	
40	2	A	2	
45	3	A	3	
52	4.5	A	4.5	
52	4.5	B		4.5
60	6	A	6	
68	8	A	8	
68	8	A	8	
68	8	A	8	
75	10	B		10
77	11	A	11	
84	12	B		12
86	13	B		13
88	14	B		14
90	15	A	15	
93	16	B		16
95	17	B		17

Sums of ranks: $\Sigma R_A = 66.5$ $\Sigma R_B = 86.5$

Median rank: 7   13

Rank of scores in tied set

$$= \frac{\text{sum of rank positions held by tied scores}}{\text{number of tied scores in set}}$$

You can see that all of the tied scores in a particular set end up sharing the same rank. For example, since the set of tied 52's occupies the 4th and 5th rank positions in the sorted list, we have:

$$\text{Shared rank} = \frac{4+5}{2} = \frac{9}{2} = 4.5$$

Both 52's are assigned the rank of 4.5, as shown in the table. Likewise for the three tied 68's, which cover the 7th, 8th, and 9th rank positions,

$$\text{Shared rank} = \frac{7+8+9}{3} = \frac{24}{3} = 8$$

Column 3 of the worksheet (Table 15.9) lists the group identification symbol (A or B) for each rank. This is necessary in order to create the final two columns in the worksheet. In columns 4 and 5 the ranks are separated into their respective samples, group A and group B.

Logic of the *U* test   The idea underlying the *U* test is this: If two groups of ranks represent random outcomes from the same population, then the ranks from either of the groups should be unsystematically intermixed with the ranks of the other group. Furthermore, the median (middle) rank of group A should be similar to the median rank of group B.

But to the extent that group A's ranks tend to be concentrated in the higher or lower positions relative to the group B ranks, the group medians will tend to differ appreciably. This outcome should be interpreted to mean that the ranks of group A represent a different population than the ranks of group B.

Now reexamine columns 4 and 5 of the worksheet. Do the group A and group B ranks appear to be unsystematically intermixed? Notice that the group A's median rank of 7 is appreciably lower than group B's median rank of 13. These observations suggest that the two work-assignment procedures produced different levels of "positiveness of attitude." We will need to carry out a *U* test to determine whether the perceived trend is reliable.

U_{crit} You can find critical values for the Mann-Whitney *U* test in Table U of Appendix A. To locate the correct U_{crit}, you need to know what alpha level is being used, the size of group A ($N_A = 10$), and the size of group B ($N_B = 7$). Let $\alpha = .05$. Notice that values of N_A are given across the top of the table, and values of N_B run down the left-hand margin of the table. Using the $\alpha = .05$ portion of Table U, find the two critical values that lie at the intersection of N_A and N_B. The critical value in **boldface type** is for a nondirectional (two-tailed) test, which is the kind of test we want to perform.

U_{crit} in this example is 14. *Important*: In the *U* test, the *computed U must be equal to or less than U_{crit} for the null hypothesis to be rejected.* Remember this point because it represents a way of comparing computed and critical values that is opposite to what you are used to.

As Table U implies, it is possible to conduct a directional (one-tailed) *U* test if the research question requires that. If you elect to use a directional test, you simply must make sure that the difference in median ranks is in the direction predicted by the alternative hypothesis. A difference in the opposite direction, no matter how large, cannot be considered significant.

Computations   To carry out the *U* test, we must find the sum of the ranks for each group in the study. The ranks have already been summed for both groups in Table 15.9: $\Sigma R_A = 66.5$ and $\Sigma R_B = 86.5$. Next we compute U for each of the groups:

$$U_A = N_A N_B + \frac{N_A(N_A + 1)}{2} - \Sigma R_A \qquad (15.5)$$

$$= (10)(7) + \frac{10(10 + 1)}{2} - 66.5$$

$$= 70 + \frac{110}{2} - 66.5$$

$$= 70 + 55 - 66.5$$

$$= 58.5$$

$$U_B = N_A N_B + \frac{N_B(N_B + 1)}{2} - \Sigma R_B \qquad (15.6)$$

$$= (10)(7) + \frac{7(7 + 1)}{2} - 86.5$$

$$= 70 + \frac{56}{2} - 86.5$$

$$= 70 + 28 - 86.5$$

$$= 11.5$$

Decision time   To evaluate the significance of U, you compare the *smaller* of U_A or U_B with U_{crit}. *Note well*: The computed U is significant if it *is equal to or less than* U_{crit}. $U_B = 11.5$ is the smaller test statistic. Since it is less than $U_{crit} = 14$, we reject H_0. The two sets of ranks represent different populations. This result means that job enrichment resulted in higher morale than job rotation.

U test for larger samples   Perhaps you noticed that Table U accommodates sample sizes only up to $N = 20$. However, you can still do a U test with larger samples by using a special formula that converts U to a z score and then assesses the significance of the z score within the standard normal distribution. This approach is called a **normal approximation of the U distribution**, and it is most accurate when the number of observations in both samples is equal to or greater than 20. (It can be used with smaller samples, with somewhat reduced accuracy.) Three steps are involved in this procedure. We'll use the data from our example to illustrate each step.

1. Compute the population mean:

$$\mu = \frac{N_A N_B}{2} \qquad (15.7)$$

$$= \frac{10 \cdot 7}{2} = \frac{70}{2} = 35$$

2. Compute the population standard deviation:

$$\sigma = \sqrt{\frac{N_A N_B (N_A + N_B + 1)}{12}} \qquad (15.8)$$

$$= \sqrt{\frac{(10)(7)(10 + 7 + 1)}{12}}$$

$$= \sqrt{\frac{70(18)}{12}} = \sqrt{\frac{1260}{12}} = \sqrt{105} = 10.25$$

3. Finally, compute the z ratio. You may use either U_A or U_B as U in this formula:

$$z_U = \frac{U - \mu}{\sigma} \qquad (15.9)$$

We will use $U_B = 11.5 = U$ to get the z statistic:

$$z_U = \frac{11.5 - 35}{10.25} = \frac{-23.5}{10.25} = -2.29$$

As in a regular z test, when alpha = .05, z_{crit} = ±1.96. Since our computed z is more extreme than z_{crit}, it is in the region of rejection. Accordingly, we reject H_0, just as we did on the basis of the standard U test.

Note that if you had used U_A instead of U_B in the above z ratio equation, the result would have been z = +2.29. Because we conducted a two-tailed test, our statistical decision would be the same.

Assumptions of the test The Mann-Whitney U test assumes that:

1. The data are on an ordinal scale of measurement.
2. All the observations are independent of one another.
3. There are no tied ranks.

The third assumption is violated in almost every use of the U test. Fortunately, tied ranks do not usually distort the test appreciably unless the majority of the ranks are in tied sets. Under those circumstances, you will need to use a special formula that corrects for ties. See Hays (1988) and Kirk (1990).

Wilcoxon *T* Statistic

The last hypothesis-testing method on our humble agenda is the nonparametric counterpart of the correlated-samples *t* test. The Wilcoxon *T* statistic[3] is designed for data sets that are based on repeated-measures or matched-pairs studies. We will examine the Wilcoxon test in the context of a repeated-measures study, in which one group of subjects is pretested on the dependent variable, exposed to a special treatment, and then posttested on the dependent variable. You might recall from Chapter 11 that the raw scores in such a study are posttest–pretest difference scores.

This is the research situation of interest: A teacher of artistically gifted high school students believes that one's creativity can be boosted by spending several days in a quiet, isolated woodland retreat, where one can develop an undisturbed focus on nature's visual designs. He decides to test this philosophy in a before–after type of study. He randomly picks eight of his gifted students for this experience. As a pretest, each student is asked to spend the better part of a day creating an abstract painting. An independent art educator then rates the "creative artistry" of the students' works on a 30-point scale. The rater does not know whose painting she is rating or that this activity is part of a behavioral experiment.

In the second phase of this investigation, the teacher accompanies his eight charges on a week-long retreat in a relatively unpopulated forest. During the retreat, the students are encouraged to contemplate nature's unique visual designs. Upon their return to civilization, the students are again asked to draw abstract paintings. And, again, the independent art educator gives each painting a creative-artistry rating on the same 30-point dimension. This is the posttest.

Not only are the ratings used in the investigation likely to fall short of an interval scale, but it is also quite likely that the population distribution of such ratings is not normal (because the students are from an extreme, gifted population). Since the assumptions of a conventional correlated-samples *t* test are not satisfied, it is appropriate to analyze the posttest–pretest differences with the Wilcoxon *T* statistic. This test has exactly the same assumptions as the Mann-Whitney *U* test.

Statistical hypotheses   Since the Wilcoxon test is mainly sensitive to the difference between medians (Toothaker, 1986), the null and alternative hypotheses for this test are identical to those associated with the Mann-Whitney *U* test:

---

[3] Make a special point not to confuse this test, always symbolized with a capital *T*, with Student's *t* test, its parametric first cousin. *T* is to be used when *t* cannot be.

H_0:   population median$_1$ = population median$_2$

H_1:   population median$_1 \neq$ population median$_2$

In this type of research design, the pretest scores represent population 1, and the posttest scores represent population 2.

Results of the study   The pretest ratings, posttest ratings, and posttest–pretest difference scores for the eight students are listed in Table 15.10. Each difference score is found by subtracting the student's pretest rating from his or her posttest rating. Since the difference scores are the basic raw scores in a correlated-samples study, we will rank the single set of difference scores, *not* the two sets of original ratings.

Ranking procedures   To set up the ranks in this kind of analysis:

1. Convert all difference scores to ~~absolute values~~ (ignoring the sign of the difference).
2. Reorder the subjects *from lowest to highest absolute difference scores.*
3. Assign a rank to each difference score, beginning with a rank of 1 and going up to a rank of *N*, but *discard difference scores of 0.*
4. Place all the ranks of positive differences into one column.
5. Place all the ranks of negative differences into a different column.

The results of these steps are shown in Table 15.11. Be sure to note the following:

- The ranks of tied difference scores are determined in exactly the same way as they were in the Mann-Whitney *U* test.
- *Difference scores of 0 are not included in the sum of ranks.* They are discarded because they provide no information about the

[handwritten margin note: lowest score of ranks]

Table 15.10

RATED CREATIVE ARTISTRY (30-POINT SCALE) FOR EIGHT HIGH SCHOOL ART STUDENTS BEFORE AND AFTER AN INTENSIVE GROUP RETREAT	SUBJECT	BEFORE	AFTER	DIFFERENCE
	1	20	24	4
	2	15	12	−3
	3	23	25	2
	4	17	22	5
	5	10	7	−3
	6	9	20	11
	7	15	16	1
	8	30	30	0

Table 15.11

	(1)	(2)	(3)	(4)	(5)
		ABSOLUTE	RANK,	RANKS	RANKS
	DIFFERENCE	DIFFERENCE	DISCARDING	OF	OF
SUBJECT	SCORES	SCORES	0	+ DIFFERENCES	– DIFFERENCES
8	0	0	—		
7	1	1	1	1	
3	2	2	2	2	
2	–3	3	3.5		3.5
5	–3	3	3.5		3.5
1	4	4	5	5	
4	5	5	6	6	
6	11	11	7	7	
				$\Sigma R_+ = 21$	$\Sigma R_- = 7$

WORKSHEET FOR CALCULATING THE WILCOXON T STATISTIC IN THE STUDY OF CREATIVE ARTISTRY

relative dominance of positive versus negative difference scores.[4] Such information is central to the logic of this test (see the section on "Logic of the Wilcoxon Test" below).

- At the bottom of the table you will see the *sum of ranks of positive differences*, $\Sigma R_+ = 21$, and the *sum of ranks of negative differences*, $\Sigma R_- = 7$.

Computations   No additional computations are required for the Wilcoxon test. The T statistic is simply assigned the smaller of the two sums of ranks, ΣR_+ or ΣR_-. Another glance at Table 15.11 reminds us that the sum of ranks for negative differences is smaller. Therefore, $T = \Sigma R_- = 7$.

Determining T_{crit}   You can obtain the critical value of T needed for significance by consulting Table W of Appendix A and locating the value at the intersection of α and N. But be careful! N is not the number of difference scores but the number of *nonzero difference scores*. Remember that 0 differences are discarded in this procedure.

Given that $\alpha = .05$ and $N = 7$ ranked difference scores, the T_{crit} for a two-tailed test is 2. *Note well:* The computed T must be equal to or *less* than T_{crit} to be significant.

---

[4] If there is a large number of 0 differences, discarding them can bias the data somewhat. There are more complicated alternative ways of dealing with 0 differences. See Hays (1988) and Kirk (1990).

Decision Since the computed T is greater than T_{crit}, we cannot reject the null hypothesis. There is no reliable evidence here to support the assertion that spending time in nature improves the creative artistry of gifted high school students.

Logic of the Wilcoxon test What is the rationale behind obtaining the sums of ranks for positive and negative difference scores? The logic of this test is based on the idea that *an effective experimental treatment has consistent effects on difference scores, but chance produces inconsistent difference scores*. If the posttest scores vary only randomly from the pretest scores, then only chance is operating, and there ought to be approximately as many large negative difference scores as there are large positive differences. Thus, the null hypothesis leads to the expectation of equal large sums of ranks for the negative and positive differences. This kind of result means that the experimental treatment had no consistent effect on the dependent variable.

Keep in mind that *one of the sums of ranks must be a small number if the test is to be statistically significant*. The only way that will happen is if almost all the differences are positive or almost all are negative. In fact, the lowest possible sum of ranks ($\Sigma R = 0$) occurs when every posttest–pretest difference has the same sign, so that one column of rank sums has no entries in it. Such an outcome would mean that the experimental treatment had a consistent effect on the dependent variable, and so it makes sense to consider the outcome statistically significant.

KEY TERMS

parametric tests	contingency table
nonparametric tests	cells
robustness	row total
frequency count	column total
chi-square test	grand total
goodness-of-fit test	phi coefficient
observed frequency	normal approximation of the
expected frequency	U distribution
test of association	

SUMMARY

1. Parametric tests, such as the t and F statistics, assess the equivalence of specific parameters. Nonparametric tests, such as the

chi-square statistic, are designed to assess the equivalence of whole distributions. Nonparametric tests are used in lieu of parametric tests principally when the latter's assumptions of interval data and normal populations cannot be satisfied.

2. Nonparametric tests are not quite as necessary or useful as their developers had hoped. Most parametric tests are robust against violation of their assumptions, rendering nonparametric alternatives unnecessary under most circumstances. In addition, nonparametric tests tend to be less powerful than their parametric counterparts.

3. The chi-square (χ^2) test is one of the most useful and versatile nonparametric methods. Used to analyze nominal-scale data, it compares observed frequencies in nominal categories with frequencies expected on the basis of chance alone. When the observed versus expected differences are large, χ^2 is likely to be significant.

4. One application of chi-square is called the goodness-of-fit test, which appraises the degree to which observed frequencies deviate from expected frequencies generated by some empirical or theoretical model. A significant goodness-of-fit χ^2 means that the data do not conform to the model.

5. A second application of chi-square is called the "test of association." This kind of test assesses the existence of a correlation between two nominal-scale variables.

6. Other nonparametric tests are useful for testing hypotheses when the data are on an ordinal scale. The Mann-Whitney U test is the ordinal-scale substitute for the independent-samples t test. The Wilcoxon T statistic is the nonparametric counterpart of the correlated-samples t test.

REVIEW QUESTIONS

1. Define or describe: parametric tests, nonparametric tests, robustness, frequency count, chi-square test, goodness-of-fit test, and observed frequency.

2. In your own words, describe the main assumptions underlying parametric tests. Which of these assumptions do not affect many nonparametric tests? Tell why it may be neither necessary nor wise to use nonparametric tests instead of parametric tests even when it is likely that an assumption of the parametric test has not been met.

3. Records show that in the 1980s, 70% of the premedical majors at a liberal arts college were accepted into medical school prior to graduation. The dean of the college wishes to know whether the acceptance rate of her premed students is at the same level

in the 1990s. Since the spring of 1990, 108 of 180 premed majors have been admitted to medical school upon graduation. Carry out a goodness-of-fit test to answer the dean's question. Use the .05 level of significance.

(a) State the null and alternative hypotheses.
(b) State the critical value of the test statistic.
(c) Show your work.
(d) State your statistical decision and draw an appropriate conclusion about the research question.

4. A teacher casually mentions to a high school counselor that students with a certain hair color seem to be more physically aggressive than other students. The counselor has serious doubts about this hypothesis. But since the counselor happens to be in charge of all discipline problems in the school, he has records that will permit a test of this hypothesis. The observed frequency of discipline problems as a function of hair color appears in the table. Underneath the observed frequencies are the school population proportions of students in each hair-color category. Test the teacher's assertion at the .05 level.

(a) State the null and alternative hypotheses.
(b) State the critical value of the test statistic.
(c) Show your work.
(d) State your statistical decision and draw an appropriate conclusion about the research question.

	Hair Color					
	Other	Brown	Black	Red	Blond	
Discipline Problems	32	42	56	40	30	$N = 200$
Population Proportions	.17	.23	.28	.12	.20	

5. Using a three-choice straight alley, a biopsychologist trains rats to pick the left goal box over the center and right boxes. Then she transplants brain cells from the trained rats to some untrained rats. Next she tests the untrained rats to see whether they have a significant tendency to prefer the left box. Their choices appear in the table. Perform a goodness-of-fit test on these data, using $\alpha = .05$.

(a) State the null and alternative hypotheses.
(b) State the critical value of the test statistic.
(c) Show your work.
(d) State your statistical decision and draw an appropriate conclusion about the research question.

Left Box	Center Box	Right Box	
240	150	210	$N = 600$ rats

6. The overall grade distribution at a particular university is: A = .10, B = .20, C = .40, D = .20, and F = .10. A professor awarded the following number of grades in each category during the last academic year: 50 A's, 90 B's, 80 C's, 70 D's, and 40 F's. Do the professor's grading practices fit the school's grade distribution? Use $\alpha = .05$.
   (a) State the null and alternative hypotheses.
   (b) State the critical value of the test statistic.
   (c) Show your work.
   (d) State your statistical decision and draw an appropriate conclusion about the research question.

7. A potential sponsor would like to know whether local viewers prefer some evening news programs over others. The sponsor conducts a viewer preference survey based on a simple random sample of 1000 households. The results are given in the table. Conduct a test of significance on these data, using the .05 level of significance, and make a recommendation to the sponsor.

KTVO	KMDT	KLPF	KZTV	
220	200	300	280	$N = 1000$

8. Define or describe: expected frequency, test of association, cells, row total, column total, grand total, phi coefficient, and normal approximation of the U distribution.

9. Some cases of schizophrenia are associated with larger than normal brain ventricles, fluid-filled areas near the center of the brain. A psychiatrist thinks that certain subcategories of schizophrenia might be more likely than other subcategories to manifest this characteristic. The psychiatrist conducts brain scans on 151 hospitalized schizophrenics and finds the results shown in the table. Conduct an appropriate test of the psychiatrist's hypothesis. Set $\alpha = .05$.
   (a) State the null and alternative hypotheses.
   (b) State the critical value of the test statistic.
   (c) Show your work.
   (d) State your statistical decision and draw an appropriate conclusion about the research question.

	Paranoid	Catatonic	Disorganized	Undifferentiated
Enlarged Ventricles	10	6	18	32
Normal Ventricles	30	24	24	7

10. The table below shows the frequencies of new admissions to a metropolitan psychiatric clinic as a function of season. Test the hypothesis that the incidence of depression, as measured in this way, is correlated with season. Set alpha = .01.
    (a) State the null and alternative hypotheses.
    (b) State the critical value of the test statistic.
    (c) Show your work.
    (d) State your statistical decision and draw an appropriate conclusion about the research question.

	Spring	Summer	Fall	Winter
Depression	200	80	120	372
Other diagnosis	340	134	140	205

11. The data in the table were gathered in an investigation of possible gender differences in book-carrying behavior among college students. The research hypothesis is that men, more than women, tend to carry books down at the side rather than in front of them. Using alpha = .05, test this hypothesis.
    (a) State the null and alternative hypotheses.
    (b) State the critical value of the test statistic.
    (c) Show your work.
    (d) State your statistical decision and draw an appropriate conclusion about the research question.

Book-Carrying Styles			
	Down at the Side	In Front	Other
Women	24	70	6
Men	100	46	4

12. A social psychologist is interested in studying the effect of social status of others present on "good manners" among grocery shoppers. Two research confederates, posing as other shoppers, pause on their way out of the store to chat. They are standing about 6 feet apart and directly in line with the main exit from the store, about 8 feet from the door. Actual customers can leave the store by either (a) walking between the conversants (bad manners) or (b) walking around the conversants. The research confederates vary their attire and grooming at different times to simulate different levels of social status. The question is: Will the frequency of the socially preferred response vary as a function of the perceived social status of the confederates? The results of this study are given in the table. Conduct a test of significance on these data, using α = .05.
    (a) State the null and alternative hypotheses.
    (b) State the critical value of the test statistic.
    (c) Show your work.

(d) State your statistical decision and draw an appropriate conclusion about the research question.

	Perceived Social Status		
	High	Average	Low
Walk between	7	36	10
Walk around	49	36	50

13. A sociologist hypothesizes that scores on the Luscher Color Test will discriminate among "normals," schizophrenics, and sociopaths. The sociologist administers the test to subjects in each of those categories and classifies the patterns of test responses as anomic (antisocietal), socialized (prosocietal), or detached (uncorrelated with middle-class social norms). Response patterns as a function of the subject group appear in the table. Using alpha = .05, conduct a test of the sociologist's hypothesis.
    (a) State the null and alternative hypotheses.
    (b) State the critical value of the test statistic.
    (c) Show your work.
    (d) State your statistical decision and draw an appropriate conclusion about the research question.

	Response Patterns on Luscher Color Test		
	Anomic	Socialized	Detached
Normals	4	40	16
Schizophrenics	12	10	38
Sociopaths	42	10	8

14. A political scientist polls 30 social science professors and 120 students majoring in the social sciences concerning the importance of making Statistics 101 a required course for social science majors. Among the professors, 12 said it was very important to require the course, 17 said moderately important, and 1 said not at all important. Among the students, 27 responded very important, 43 responded somewhat important, and 50 responded not important at all. Test the hypothesis that social science students have a different view of the statistics requirement than their professors. Use the .01 level.

15. Find the degrees of freedom and the critical chi-square value for each of the following situations.
    (a) $\alpha = .01$, 7 x 6 contingency table
    (b) $\alpha = .01$, four nominal categories
    (c) $\alpha = .05$, 5 x 3 contingency table
    (d) $\alpha = .05$, 3 x 2 contingency table
    (e) $\alpha = .05$, 11 nominal categories

16. Refer to the data of Murrey and colleagues (1992) in Table 15.6. Using α = .05, perform an appropriate test *only on the cells of the table that contain the frequencies of hypnotized subjects.* On the basis of your test, draw a conclusion concerning the equivalence or nonequivalence of the two hypnotized groups on the acceptance of pseudomemories.
    (a) State the null and alternative hypotheses.
    (b) State the critical value of the test statistic.
    (c) Show your work.
    (d) State your statistical decision and draw an appropriate conclusion about the research question.

17. Subjects are randomly assigned to two meditation conditions. Subjects in one group meditate while using the correct mantras for their respective age brackets. (A *mantra* is a Sanskrit word thought to facilitate achievement of a meditative state.) The other subjects meditate with incorrect mantras. When finished, each subject rates the quality of the meditative experience on an 0 to 10 scale. The ratings are given in the table. Assuming ordinal measurement at best, test the hypothesis that the two groups differ in the quality of their meditative states. Use α = .05.
    (a) State the null and alternative hypotheses.
    (b) State the critical value of the test statistic.
    (c) Show your work.
    (d) State your statistical decision and draw an appropriate conclusion about the research question.

Correct Mantra	Incorrect Mantra
7	9
5	4
10	7
6	4
8	3
7	3
	2

18. A professor uses a seven-point scale to rate all of her students on their level of motivation in class. Her ratings appear in the table. Out of curiosity, she divides the students into those who sit in the front half of the room and those who sit in the back half of the room. It appears to her that students who sit near the front of the class generally show a higher level of motivation than those who sit farther back. Assuming the rating scale produces only ordinal data, test that hypothesis. Set alpha to .05.
    (a) State the null and alternative hypotheses.

(b) State the critical value of the test statistic.
(c) Show your work.
(d) State your statistical decision and draw an appropriate conclusion about the research question.

Front Half of Class	Back Half of Class
3	5
2	5
6	3
7	4
4	7
5	6
7	2
6	2
6	1

19. A panel of anthropologists appraises the preserved quality of human skulls from a particular historical period. Some of the skulls were unearthed in tropical areas, and the others were discovered in temperate regions. The panel's ratings of preserved quality appear in the table. Decide whether quality differs significantly between the tropical and temperate areas of origin. Set alpha = .01.
    (a) State the null and alternative hypotheses.
    (b) State the critical value of the test statistic.
    (c) Show your work.
    (d) State your statistical decision and draw an appropriate conclusion about the research question.

Tropical Find	Temperate Find
19	8
12	17
17	13
20	8
13	4
7	8
10	6
6	11
4	5
8	3

20. Use the normal approximation of the U distribution to test the significance of the following results of a study. Set $\alpha = .01$.

$N_A = 40 \quad N_B = 30 \quad U_A = 230 \quad U_B = 970$

21. Use the normal approximation of the U distribution to test the significance of the following results of a study. Set $\alpha = .05$.

$N_A = 25 \quad N_B = 25 \quad U_A = 300 \quad U_B = 325$

22. Ten adults were subjected to a standard pain stimulus and asked to estimate the level of pain experienced on a 100-point scale. Then each participant was put through the same procedure again after ingesting 4.5 ounces of alcohol. Apply the Wilcoxon T test to determine whether the alcohol significantly affected the estimated pain level. Use the .01 level of significance.
    (a) State the null and alternative hypotheses.
    (b) State the critical value of the test statistic.
    (c) Show your work.
    (d) State your statistical decision and draw an appropriate conclusion about the research question.

Subject	Normal State	Intoxicated State
1	70	50
2	62	50
3	80	80
4	51	48
5	67	42
6	92	74
7	81	73
8	61	72
9	48	38
10	75	35

23. Seven students were rated on manners both before and after their first year of attending finishing school. Assuming an ordinal scale of measurement in the scores in the table, determine whether the first year of training affected the students' grace and poise. Set alpha = .05.
    (a) State the null and alternative hypotheses.
    (b) State the critical value of the test statistic.
    (c) Show your work.
    (d) State your statistical decision and draw an appropriate conclusion about the research question.

Student	Before	After
1	3	3
2	4	6
3	2	5
4	2	6
5	5	6
6	6	7
7	1	6

24. One member of each of six pairs of identical twins was randomly assigned to a high-energy diet for a period of six months.

The co-twin in each pair ate a standard diet during the same time. Then all the twins competed in a 5-mile run. The positions in which they finished are shown in the table. Note that the data are in the form of matched pairs. Use the Wilcoxon T test to assess the effectiveness of the high-energy diet on performance in this race. Set the level of significance to .05.

(a) State the null and alternative hypotheses.
(b) State the critical value of the test statistic.
(c) Show your work.
(d) State your statistical decision and draw an appropriate conclusion about the research question.

Twin Pair	Standard Diet Twin 1	High-Energy Diet Twin 2
1	6	8
2	1	4
3	2	3
4	5	12
5	11	7
6	10	9

25. Seven hyperallergic persons were in the throes of an allergic skin reaction. In the initial stage of their treatment at a major university hospital, each subject agreed to permit an acupuncturist to treat a randomly selected side of his or her body. At the end of 24 hours, independent medical assistants rated the severity of the rash on each side of the subjects' bodies. Use the Wilcoxon test to find out whether acupuncture had an impact on the severity ratings shown in the table. Set alpha = .05.

(a) State the null and alternative hypotheses.
(b) State the critical value of the test statistic.
(c) Show your work.
(d) State your statistical decision and draw an appropriate conclusion about the research question.

Subject	Treated	Untreated
1	3	3
2	1	3
3	1	2
4	0	3
5	1	5
6	0	5
7	2	1

CHAPTER 16   BRINGING IT ALL TOGETHER

Behavioral statistics is logical, and each procedure by itself is fairly easy to understand and carry out. But to the student who is just completing a first course on the topic, the sheer number of procedures makes the practice of statistics look like a complex maze. After a while you begin to wonder just when you should use one kind of test or technique rather than another. Take heart. The maze will become increasingly friendly and manageable as you continue to use statistics in future courses and, later, in your day-to-day professional work.

Time and experience are bound to clarify your conception of statistics. Meanwhile, I hope that you will often find this chapter helpful as a guide to choosing the most appropriate kind of analysis for a particular data set. The chapter provides relatively simple flowcharts to assist in the selection of

- Methods of descriptive statistics
- Methods of prediction and estimation
- Tests of significance

The charts don't cover all possible factors that could be considered in choosing a statistical method. Nonetheless, they will give you adequate direction for most of your data-analysis problems. I hope that, by providing the big picture, they will also strengthen your grasp of material we've gone over in this book.

CHOOSING A METHOD OF DESCRIBING DATA

When you are trying to choose an appropriate method of summarizing results, the first questions to ask are: What specific kinds of descriptive techniques do I need to meet my objective in analyzing the data? Do I want to describe a relationship between two variables, graph frequencies or means, arrange my findings in a table, compute an average, or compute an index of variation? The flowchart in Figure 16.1 will help you take it from there.

Relationships Between Variables

To summarize the strength and direction of a linear association between variables X and Y, you will almost always want to apply the Pearson r, or linear correlation coefficient (Chapter 6). The absolute value of r ranges between .00 and 1.00, with larger values representing relatively stronger relationships. A negative r value indicates that the higher values of X tend to be associated with lower values of Y.

Figure 16.1 *Choosing a Method of Describing a Data Set*

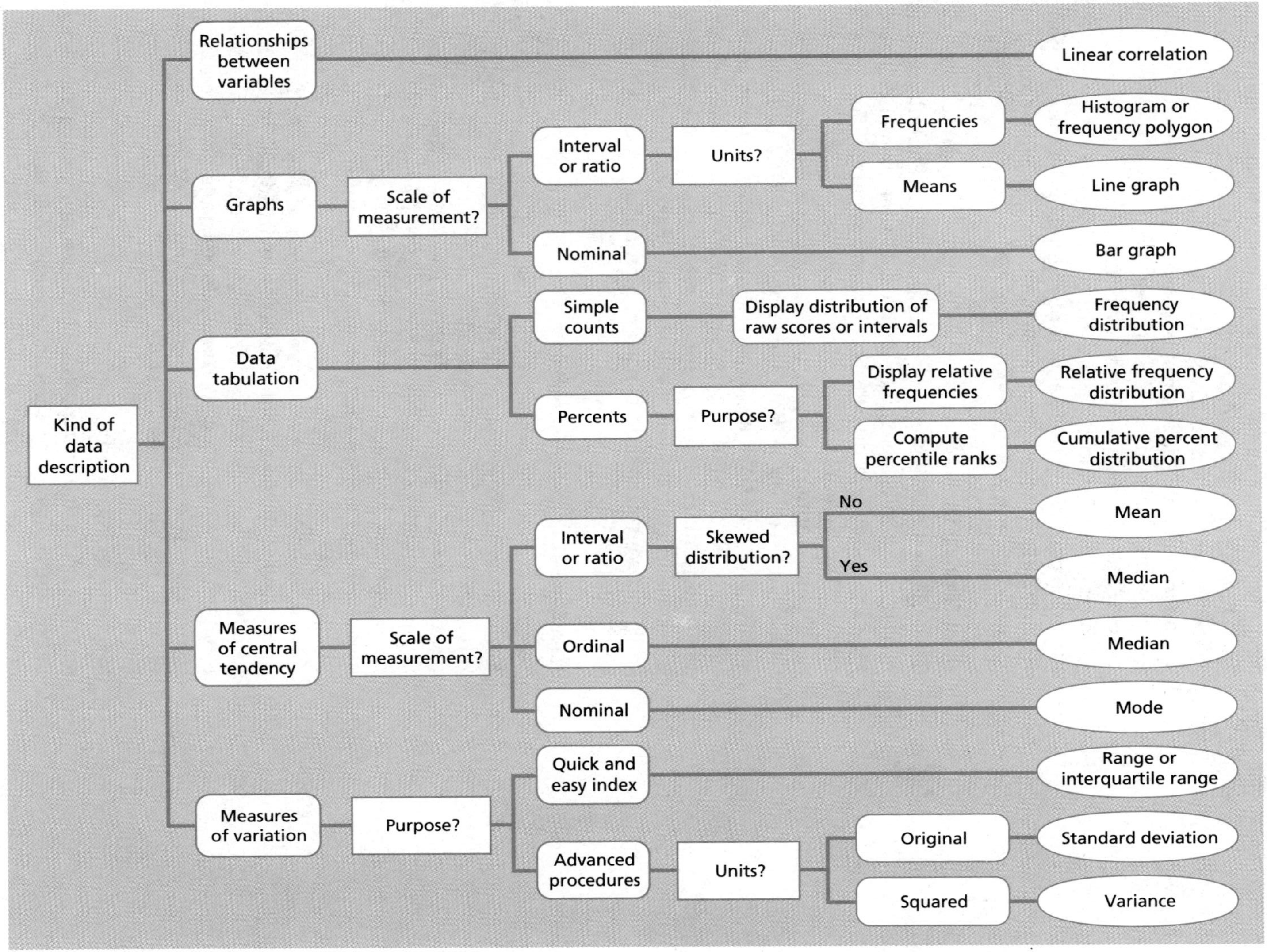

Graphs

Behavioral scientists use graphs to more efficiently communicate their findings to other behavioral scientists and the general public. Once you've decided that you want to graph your data, the next question concerns the scale of measurement to be used. If the data are on a nominal scale, a simple bar graph is an appropriate way to represent the data with a picture (Chapter 3).

Interval and ratio data present more options. A graph of frequencies can be in the form of a histogram or a frequency polygon (Chapter 3). To compare means, however, you will normally want to use a line graph (Chapter 4).

Data Tabulation

The purpose of placing scores in tables (i.e., "tabulation") is to arrange data in a systematic fashion so that the general characteristics of the data distribution become apparent. Tabulation facilitates the communication of your results.

If your intention is simply to display frequency counts of scores, you should set up a table containing either an ungrouped frequency distribution or a grouped frequency distribution (with class intervals). If you want to represent the frequencies as percents of the total number of scores, then a relative frequency distribution is the way to go. Do you want to convert the scores to percentile ranks? If so, you should construct a cumulative percent distribution (Chapter 3).

Measures of Central Tendency

Measures of central tendency are designed to represent the average of typical scores in a distribution. The particular index of "average" you choose will depend largely on the scale of measurement you may assume for your data. The mode is the only measure of central tendency that is legitimate for a set of nominal-scale data. Ordinal data can be summarized by either the median or the mode, but not the mean. Normally, you want to use the median (Chapter 4).

Scores on an interval or ratio scale are best typified by the mean, unless the distribution of data is markedly skewed. Then the median will be a more accurate measure of central tendency (Chapter 4). When it can be used, the mean is preferred over the median because it has many advanced applications—especially in the context of hypothesis testing.

Measures of Variation

Measures of variation supplement measures of central tendency by indexing the degree to which scores tend to deviate from their "average." If you simply need a quick description of the variation in a data distribution, either the simple range or the interquartile range will suffice. Once you have ordered the data from highest score to lowest, those measures of variation are easy to compute. But advanced statistical procedures—including procedures of hypothesis testing, prediction, and estimation—require either the variance or the standard deviation (Chapter 4). As a purely descriptive statistic, the standard deviation has the advantage of being in the original units of measure. In contrast, the variance is in squared units, which makes it difficult to interpret as a description of the data.

PREDICTION OR ESTIMATION

In prediction and estimation procedures, you use the known to make an educated guess about the unknown. You use known X values to project unknown Y values or known sample data to infer an unknown population mean. Figure 16.2 summarizes considerations in choosing the appropriate procedure for problems that involve prediction or estimation.

Predicting Y Values

Anytime you need to predict individual raw scores on variable Y from raw scores on variable X, consider using the linear regression formula: $Y' = bX + a$. This model produces good predictions of Y scores to the extent that X and Y have a strong *straight-line* relationship.

Parameter Estimation

Figure 16.2 *Choosing a Method of Prediction or Estimation*

In parameter estimation, you use a sample statistic to infer a population parameter. You will almost always use interval estimation,

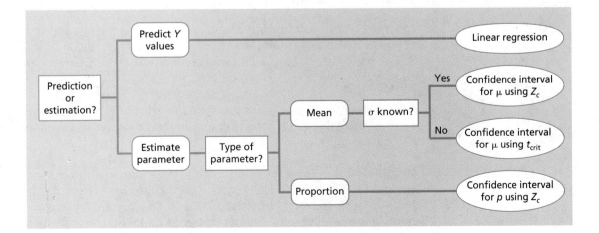

which places a confidence interval around the estimate of the parameter. The confidence interval is a range of values that is likely to include the parameter.

When you set up a confidence interval for a proportion, it is always appropriate to use the z statistic: $P \pm z\sigma_P$. When constructing a confidence interval for a mean, however, you can use z only if you know the population standard deviation or variance. If σ is unknown, then t_{crit} must be substituted for z_c: $\overline{X} \pm t_{crit}s_{\overline{X}}$.

CHOOSING A TEST OF SIGNIFICANCE

Hypothesis testing involves carrying out tests of statistical significance. The objective of doing these tests is to decide whether the results of a study should be attributed to chance or to a reliable effect of some variable. The number of samples in your study is a major determinant of the particular test to be used. See the flowchart in Figure 16.3.

One-Sample Tests

Nominal data   If your study involves the simple classification of traits or responses into nominal categories, then a chi-square goodness-of-fit test is called for (Chapter 15). This test compares the actual distribution of frequencies across categories with the distribution that is predicted by some theory or model. A significant outcome means that the empirical frequencies are not a "good fit" to the model.

Ordinal data   There is no generally accepted one-sample test for scores that are on an ordinal scale. In such a case, you can convert the data to nominal-scale frequencies and use the chi-square goodness-of-fit test.

Interval or ratio data   Interval or higher data present additional choices. If the population standard deviation is available, use a one-sample z test. If not, conduct a one-sample t test (Chapter 10).

Two-Sample Tests

Your choice of two-sample tests of significance hinges on two considerations: (1) scale of measurement and (2) whether the samples are independent or correlated.

Nominal data   Two groups of nominal data can be analyzed with the chi-square test of association. The two groups constitute two levels of the independent variable, and the nominal categories by which the groups are being compared make up the dependent

Figure 16.3 *Choosing a Test of Significance*

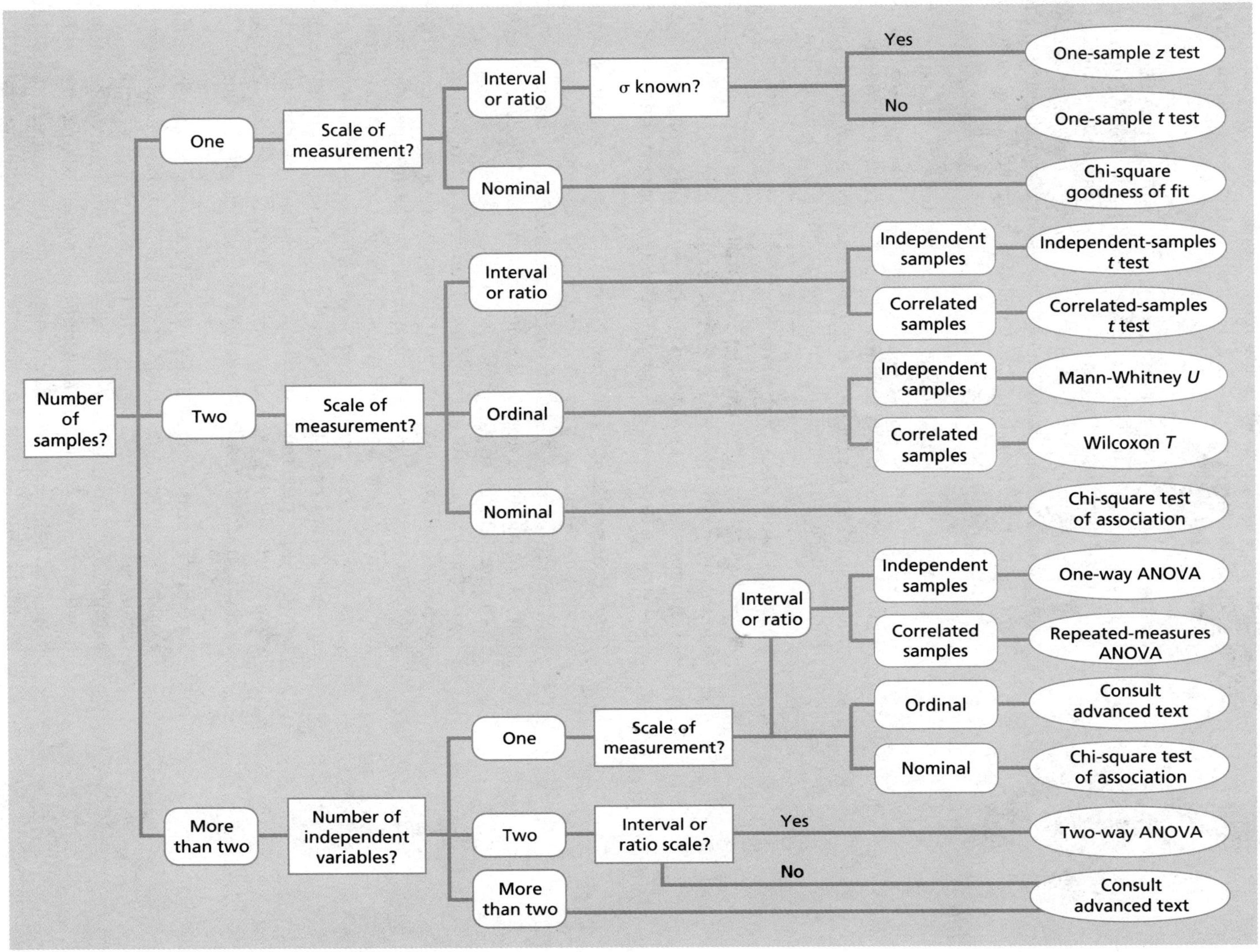

variable. If chi-square is significant, then the dependent variable is reliably associated with the independent variable. A nonsignificant chi-square means that the two variables are not correlated (Chapter 15).

Ordinal data   Independent samples of ordinal data are best analyzed via the Mann-Whitney U test (Chapter 15). A significant U statistic means that the group medians are reliably different.

If the samples are correlated, as in a before-after study, you should use the Wilcoxon T test to assess the equivalence of the two conditions (Chapter 15). This test is interpreted in the same way as the Mann-Whitney U.

Interval or ratio data   For raw scores that are measured at the interval or ratio level, a two-sample t test is the preferred significance-testing technique (Chapter 11). For matched-pairs and before-after research designs, use the correlated-samples t test. Otherwise, compute an independent-samples t ratio.

Tests for More Than Two Samples

Hypothesis testing when there are more than two samples in the same study requires you to consider three matters: (1) the number of independent variables, (2) the scale of measurement, and (3) whether the samples are independent or correlated.

One independent variable   If you have multiple samples of nominal-scale data, you should use the chi-square test of association (Chapter 15). A significant outcome means that the distribution of frequencies across the dependent-variable categories differs from one sample to another.

If the dependent variable is measured at the interval or ratio level, you will want to carry out an analysis of variance. If different subjects serve at the respective levels of the independent variable, use a simple one-way ANOVA (Chapter 12). But if the same subjects serve in all conditions, a repeated-measures ANOVA is required (Chapter 14). A significant F statistic indicates that between-samples mean differences are reliably larger than random error variation—which, in turn, tells you that the independent variable had a reliable impact on the dependent variable.

Two independent variables   The presence of two independent variables in the same study creates a "factorial research design," which enables you to study the effect of the interaction of those variables. Assuming that the data are at the interval or ratio level, a two-way ANOVA is the appropriate hypothesis-testing technique (Chapter 13). If the factorial design data do not meet at least interval-scale criteria, you have to use methods that exceed the scope of this text. Consult advanced statistics books.

APPENDIXES

APPENDIX A   STATISTICAL TABLES

Table C

CRITICAL VALUES OF THE CHI-SQUARE DISTRIBUTION

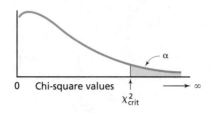

df	ALPHA LEVEL			
	.10	*.05*	*.02*	*.01*
1	2.71	3.84	5.41	6.64
2	4.61	5.99	7.82	9.21
3	6.25	7.82	9.84	11.34
4	7.78	9.49	11.67	13.28
5	9.24	11.07	13.39	15.09
6	10.65	12.59	15.03	16.81
7	12.02	14.07	16.62	18.48
8	13.36	15.51	18.17	20.09
9	14.68	16.92	19.68	21.67
10	15.99	18.31	21.16	23.21
11	17.28	19.68	22.62	24.73
12	18.55	21.03	24.05	26.22
13	19.81	22.36	25.47	27.69
14	21.06	23.69	26.87	29.14
15	22.31	25.00	28.26	30.58
16	23.54	26.30	29.63	32.00
17	24.77	27.59	31.00	33.41
18	25.99	28.87	32.35	34.81
19	27.20	30.14	33.69	36.19
20	28.41	31.41	35.02	37.57
21	29.62	32.67	36.34	38.93
22	30.81	33.92	37.66	40.29
23	32.01	35.17	38.97	41.64
24	33.20	36.42	40.27	42.98
25	34.38	37.65	41.57	44.31
26	35.56	38.89	42.86	45.64
27	36.74	40.11	44.14	46.96
28	37.92	41.34	45.42	48.28
29	39.09	42.56	46.69	49.59
30	40.26	43.77	47.96	50.89

Source: James D. Evans

Table F

CRITICAL VALUES OF THE *F* DISTRIBUTION

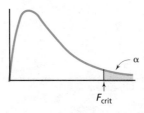

		DEGREES OF FREEDOM FOR THE NUMERATOR: *df1*											
DEGREES OF FREEDOM: DENOMINATOR	**ALPHA LEVEL**	**1**	**2**	**3**	**4**	**5**	**6**	**7**	**8**	**9**	**10**	**11**	**12**
df2	α												
1	.05	161.45	199.50	215.71	224.58	230.16	233.99	236.77	238.88	240.54	241.88	242.98	243.91
	.01	4052	5000	5403	5625	5764	5859	5928	5981	6022	6056	6083	6106
2	.05	18.51	19.00	19.16	19.25	19.30	19.33	19.35	19.37	19.38	19.40	19.40	19.41
	.01	98.50	99.00	99.17	99.25	99.30	99.33	99.36	99.37	99.39	99.40	99.41	99.42
3	.05	10.13	9.55	9.28	9.12	9.01	8.94	8.89	8.85	8.81	8.79	8.76	8.74
	.01	34.12	30.82	29.46	28.71	28.24	27.91	27.67	27.49	27.35	27.23	27.13	27.05
4	.05	7.71	6.94	6.59	6.39	6.26	6.16	6.09	6.04	6.00	5.96	5.94	5.91
	.01	21.20	18.00	16.69	15.98	15.52	15.21	14.98	14.80	14.66	14.55	14.45	14.37
5	.05	6.61	5.79	5.41	5.19	5.05	4.95	4.88	4.82	4.77	4.74	4.70	4.68
	.01	16.26	13.27	12.06	11.39	10.97	10.67	10.46	10.29	10.16	10.05	9.96	9.89
6	.05	5.99	5.14	4.76	4.53	4.39	4.28	4.21	4.15	4.10	4.06	4.03	4.00
	.01	13.75	10.92	9.78	9.15	8.75	8.47	8.26	8.10	7.98	7.87	7.79	7.72
7	.05	5.59	4.74	4.35	4.12	3.97	3.87	3.79	3.73	3.68	3.64	3.60	3.57
	.01	12.25	9.55	8.45	7.85	7.46	7.19	6.99	6.84	6.72	6.62	6.54	6.47
8	.05	5.32	4.46	4.07	3.84	3.69	3.58	3.50	3.44	3.39	3.35	3.31	3.28
	.01	11.26	8.65	7.59	7.01	6.63	6.37	6.18	6.03	5.91	5.81	5.73	5.67
9	.05	5.12	4.26	3.86	3.63	3.48	3.37	3.29	3.23	3.18	3.14	3.10	3.07
	.01	10.56	8.02	6.99	6.42	6.06	5.80	5.61	5.47	5.35	5.26	5.18	5.11
10	.05	4.96	4.10	3.71	3.48	3.33	3.22	3.14	3.07	3.02	2.98	2.94	2.91
	.01	10.04	7.56	6.55	5.99	5.64	5.39	5.20	5.06	4.94	4.85	4.77	4.71

(Table F continued)

DEGREES OF FREEDOM: DENOMINATOR df2	ALPHA LEVEL α	DEGREES OF FREEDOM FOR THE NUMERATOR: df1											
		1	2	3	4	5	6	7	8	9	10	11	12
11	.05	4.84	3.98	3.59	3.36	3.20	3.09	3.01	2.95	2.90	2.85	2.82	2.79
	.01	9.65	7.21	6.22	5.67	5.32	5.07	4.89	4.74	4.63	4.54	4.46	4.40
12	.05	4.75	3.89	3.49	3.26	3.11	3.00	2.91	2.85	2.80	2.75	2.72	2.69
	.01	9.33	6.93	5.95	5.41	5.06	4.82	4.64	4.50	4.39	4.30	4.22	4.16
13	.05	4.67	3.81	3.41	3.18	3.03	2.92	2.83	2.77	2.71	2.67	2.63	2.60
	.01	9.07	6.70	5.74	5.21	4.86	4.62	4.44	4.30	4.19	4.10	4.02	3.96
14	.05	4.60	3.74	3.34	3.11	2.96	2.85	2.76	2.70	2.65	2.60	2.57	2.53
	.01	8.86	6.51	5.56	5.04	4.69	4.46	4.28	4.14	4.03	3.94	3.86	3.80
15	.05	4.54	3.68	3.29	3.06	2.90	2.79	2.71	2.64	2.59	2.54	2.51	2.48
	.01	8.68	6.36	5.42	4.89	4.56	4.32	4.14	4.00	3.89	3.80	3.73	3.67
16	.05	4.49	3.63	3.24	3.01	2.85	2.74	2.66	2.59	2.54	2.49	2.46	2.42
	.01	8.53	6.23	5.29	4.77	4.44	4.20	4.03	3.89	3.78	3.69	3.62	3.55
17	.05	4.45	3.59	3.20	2.96	2.81	2.70	2.61	2.55	2.49	2.45	2.41	2.38
	.01	8.40	6.11	5.18	4.67	4.34	4.10	3.93	3.79	3.68	3.59	3.52	3.46
18	.05	4.41	3.55	3.16	2.93	2.77	2.66	2.58	2.51	2.46	2.41	2.37	2.34
	.01	8.29	6.01	5.09	4.58	4.25	4.01	3.84	3.71	3.60	3.51	3.43	3.37
19	.05	4.38	3.52	3.13	2.90	2.74	2.63	2.54	2.48	2.42	2.38	2.34	2.31
	.01	8.18	5.93	5.01	4.50	4.17	3.94	3.77	3.63	3.52	3.43	3.36	3.30
20	.05	4.35	3.49	3.10	2.87	2.71	2.60	2.51	2.45	2.39	2.35	2.31	2.28
	.01	8.10	5.85	4.94	4.43	4.10	3.87	3.70	3.56	3.46	3.37	3.29	3.23
21	.05	4.32	3.47	3.07	2.84	2.68	2.57	2.49	2.42	2.37	2.32	2.28	2.25
	.01	8.02	5.78	4.87	4.37	4.04	3.81	3.64	3.51	3.40	3.31	3.24	3.17
22	.05	4.30	3.44	3.05	2.82	2.66	2.55	2.46	2.40	2.34	2.30	2.26	2.23
	.01	7.95	5.72	4.82	4.31	3.99	3.76	3.59	3.45	3.35	3.26	3.18	3.12

(Table F continued)

DEGREES OF FREEDOM: DENOMINATOR	ALPHA LEVEL	DEGREES OF FREEDOM FOR THE NUMERATOR: $df1$											
		1	2	3	4	5	6	7	8	9	10	11	12
$df2$	α												
23	.05	4.28	3.42	3.03	2.80	2.64	2.53	2.44	3.37	2.32	2.27	2.24	2.20
	.01	**7.88**	**5.66**	**4.76**	**4.26**	**3.94**	**3.71**	**3.54**	**3.41**	**3.30**	**3.21**	**3.14**	**3.07**
24	.05	4.26	3.40	3.01	2.78	2.62	2.51	2.42	2.36	2.30	2.25	2.22	2.18
	.01	**7.82**	**5.61**	**4.72**	**4.22**	**3.90**	**3.67**	**3.50**	**3.36**	**3.26**	**3.17**	**3.09**	**3.03**
25	.05	4.24	3.39	2.99	2.76	2.60	2.49	2.40	2.34	2.28	2.24	2.20	2.16
	.01	**7.77**	**5.57**	**4.68**	**4.18**	**3.85**	**3.63**	**3.46**	**3.32**	**3.22**	**3.13**	**3.06**	**2.99**
26	.05	4.23	3.37	2.98	2.74	2.59	2.47	2.39	2.32	2.27	2.22	2.18	2.15
	.01	**7.72**	**5.53**	**4.64**	**4.14**	**3.82**	**3.59**	**3.42**	**3.29**	**3.18**	**3.09**	**3.02**	**2.96**
27	.05	4.21	3.35	2.96	2.73	2.57	2.46	2.37	2.31	2.25	2.20	2.17	2.13
	.01	**7.68**	**5.49**	**4.60**	**4.11**	**3.78**	**3.56**	**3.39**	**3.26**	**3.15**	**3.06**	**2.99**	**2.93**
28	.05	4.20	3.34	2.95	2.71	2.56	2.45	2.36	2.29	2.24	2.19	2.15	2.12
	.01	**7.64**	**5.45**	**4.57**	**4.07**	**3.75**	**3.53**	**3.36**	**3.23**	**3.12**	**3.03**	**2.96**	**2.90**
29	.05	4.18	3.33	2.93	2.70	2.55	2.43	2.35	2.28	2.22	2.18	2.14	2.10
	.01	**7.60**	**5.42**	**4.54**	**4.04**	**3.73**	**3.50**	**3.33**	**3.20**	**3.09**	**3.00**	**2.93**	**2.87**
30	.05	4.17	3.32	2.92	2.69	2.53	2.42	2.33	2.27	2.21	2.16	2.13	2.09
	.01	**7.56**	**5.39**	**4.51**	**4.02**	**3.70**	**3.47**	**3.30**	**3.17**	**3.07**	**2.98**	**2.91**	**2.84**
40	.05	4.08	3.23	2.84	2.61	2.45	2.34	2.25	2.18	2.12	2.08	2.04	2.00
	.01	**7.31**	**5.18**	**4.31**	**3.83**	**3.51**	**3.29**	**3.12**	**2.99**	**2.89**	**2.80**	**2.73**	**2.66**
60	.05	4.00	3.15	2.76	2.53	2.37	2.25	2.17	2.10	2.04	1.99	1.95	1.92
	.01	**7.08**	**4.98**	**4.13**	**3.65**	**3.34**	**3.12**	**2.95**	**2.82**	**2.72**	**2.63**	**2.56**	**2.50**
120	.05	3.92	3.07	2.68	2.45	2.29	2.17	2.09	2.02	1.96	1.91	1.87	1.83
	.01	**6.85**	**4.79**	**3.95**	**3.48**	**3.17**	**2.96**	**2.79**	**2.66**	**2.56**	**2.47**	**2.40**	**2.34**
200	.05	3.89	3.04	2.65	2.42	2.26	2.14	2.06	1.98	1.93	1.88	1.84	1.80
	.01	**6.76**	**4.71**	**3.88**	**3.41**	**3.11**	**2.89**	**2.73**	**2.60**	**2.50**	**2.41**	**2.34**	**2.27**
∞	.05	3.84	3.00	2.60	2.37	2.21	2.10	2.01	1.94	1.88	1.83	1.79	1.75
	.01	**6.63**	**4.61**	**3.78**	**3.32**	**3.02**	**2.80**	**2.64**	**2.51**	**2.41**	**2.32**	**2.24**	**2.18**

Note: The values in lightface type are the critical F statistics at the .05 level. The values in **boldface** type are the critical F statistics at the .01 level.

Source: James D. Evans

Table Q

CRITICAL VALUES OF THE STUDENTIZED RANGE STATISTIC (q)

ERROR DEGREES OF FREEDOM df	ALPHA LEVEL α	2	3	4	5	6	7	8	9	10	11	12
2	.05	6.08	8.33	9.80	10.90	11.70	12.40	13.00	13.50	14.00	14.40	14.70
	.01	14.00	19.00	22.30	24.70	26.60	28.20	29.50	30.70	31.70	32.60	33.40
3	.05	4.50	5.91	6.82	7.50	8.04	8.48	8.85	9.18	9.46	9.72	9.72
	.01	8.26	10.60	12.20	13.30	14.20	15.00	15.60	16.20	16.70	17.80	17.50
4	.05	3.93	5.04	5.76	6.29	6.71	7.05	7.35	7.60	7.83	8.03	8.21
	.01	6.51	8.12	9.17	9.96	10.60	11.10	11.50	11.90	12.30	12.60	12.80
5	.05	3.64	4.60	5.22	5.67	6.03	6.33	6.58	6.80	6.99	7.17	7.32
	.01	5.70	6.98	7.80	8.42	8.91	9.32	9.67	9.97	10.24	10.48	10.70
6	.05	3.46	4.34	4.90	5.30	5.63	5.90	6.12	6.32	6.49	6.65	6.79
	.01	5.24	6.33	7.03	7.56	7.97	8.32	8.61	8.87	9.10	9.30	9.48
7	.05	3.34	4.16	4.68	5.06	5.36	5.61	5.82	6.00	6.16	6.30	6.43
	.01	4.95	5.92	6.54	7.01	7.37	7.68	7.94	8.17	8.37	8.55	8.71
8	.05	3.26	4.04	4.53	4.89	5.17	5.40	5.60	5.77	5.92	6.05	6.18
	.01	4.75	5.64	6.20	6.62	6.96	7.24	7.47	7.68	7.86	8.03	8.18
9	.05	3.20	3.95	4.41	4.76	5.02	5.24	5.43	5.59	5.74	5.87	5.98
	.01	4.60	5.43	5.96	6.35	6.66	6.91	7.13	7.33	7.49	7.65	7.78
10	.05	3.15	3.88	4.33	4.65	4.91	5.12	5.30	5.46	5.60	5.72	5.83
	.01	4.48	5.27	5.77	6.14	6.43	6.67	6.87	7.05	7.21	7.36	7.49
11	.05	3.11	3.82	4.26	4.57	4.82	5.03	5.20	5.35	5.49	5.61	5.71
	.01	4.39	5.15	5.62	5.97	6.25	6.48	6.67	6.84	6.99	7.13	7.25
12	.05	3.08	3.77	4.20	4.51	4.75	4.95	5.12	5.27	5.39	5.51	5.61
	.01	4.32	5.05	5.50	5.84	6.10	6.32	6.51	6.67	6.81	6.94	7.06
13	.05	3.06	3.73	4.15	4.45	4.69	4.88	5.05	5.19	5.32	5.43	5.53
	.01	4.26	4.96	5.40	5.73	5.98	6.19	6.37	6.53	6.67	6.79	6.90

k = NUMBER OF MEANS

(Table Q continued)

CRITICAL VALUES OF THE STUDENTIZED RANGE STATISTIC (q)

ERROR DEGREES OF FREEDOM df	ALPHA LEVEL α	k = NUMBER OF MEANS										
		2	3	4	5	6	7	8	9	10	11	12
14	.05	3.03	3.70	4.11	4.41	4.64	4.83	4.99	5.13	5.25	5.36	5.46
	.01	4.21	4.89	5.32	5.63	5.88	6.08	6.26	6.41	6.54	6.66	6.77
15	.05	3.01	3.67	4.08	4.37	4.59	4.78	4.94	5.08	5.20	5.31	5.40
	.01	4.17	4.84	5.25	5.56	5.80	5.99	6.16	6.31	6.44	6.55	6.66
16	.05	3.00	3.65	4.05	4.33	4.56	4.74	4.90	5.03	5.15	5.26	5.35
	.01	4.13	4.79	5.19	5.49	5.72	5.92	6.08	6.22	6.35	6.46	6.56
17	.05	2.98	3.63	4.02	4.30	4.52	4.70	4.86	4.99	5.11	5.21	5.31
	.01	4.10	4.74	5.14	5.43	5.66	5.85	6.01	6.15	6.27	6.38	6.48
18	.05	2.97	3.61	4.00	4.28	4.49	4.67	4.82	4.96	5.07	5.17	5.27
	.01	4.07	4.70	5.09	5.38	5.60	5.79	5.94	6.08	6.20	6.31	6.41
19	.05	2.96	3.59	3.98	4.25	4.47	4.65	4.79	4.92	5.04	5.14	5.23
	.01	4.05	4.67	5.05	5.33	5.55	5.73	5.89	6.02	6.14	6.25	6.34
20	.05	2.95	3.58	3.96	4.23	4.45	4.62	4.77	4.90	5.01	5.11	5.20
	.01	4.02	4.64	5.02	5.29	5.51	5.69	5.84	5.97	6.09	6.19	6.28
24	.05	2.92	3.53	3.90	4.17	4.37	4.54	4.68	4.81	4.92	5.01	5.10
	.01	3.96	4.55	4.91	5.17	5.37	5.54	5.69	5.81	5.92	6.02	6.11
30	.05	2.89	3.49	3.85	4.10	4.30	4.46	4.60	4.72	4.82	4.92	5.00
	.01	3.89	4.45	4.80	5.05	5.24	5.40	5.54	5.65	5.76	5.85	5.93
40	.05	2.86	3.44	3.79	4.04	4.23	4.39	4.52	4.63	4.73	4.82	4.90
	.01	3.82	4.37	4.70	4.93	5.11	5.26	5.39	5.50	5.60	5.69	5.76
60	.05	2.83	3.40	3.74	3.98	4.16	4.31	4.44	4.55	4.65	4.73	4.81
	.01	3.76	4.28	4.59	4.82	4.99	5.13	5.25	5.36	5.45	5.53	5.60
120	.05	2.80	3.36	3.68	3.92	4.10	4.24	4.36	4.47	4.56	4.64	4.71
	.01	3.70	4.20	4.50	4.71	4.87	5.01	5.12	5.21	5.30	5.37	5.44
∞	.05	2.77	3.31	3.63	3.86	4.03	4.17	4.29	4.39	4.47	4.55	4.62
	.01	3.64	4.12	4.40	4.60	4.76	4.88	4.99	5.08	5.16	5.23	5.29

Note: The values in lightface type are the critical q statistics at the .05 level. The values in **boldface** type are the critical q statistics at the **.01** level.

Source: Adapted from Table 29 of Pearson, E., and Hartley, H. (1966). *Biometrika tables for statisticians*, Vol. I (3rd ed.), Cambridge, England: University Press. Used with permission of the Biometrika Trustees.

Table T

CRITICAL VALUES OF STUDENT'S *t* STATISTIC

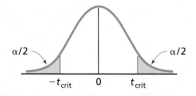

DEGREES OF FREEDOM	LEVEL OF SIGNIFICANCE: ONE-TAILED (DIRECTIONAL) TEST					
	.10	.05	.025	.01	.005	.0005
	LEVEL OF SIGNIFICANCE: TWO-TAILED (NONDIRECTIONAL) TEST					
	.20	.10	.05	.02	.01	.001
1	3.078	6.314	12.706	31.821	63.657	636.619
2	1.886	2.920	4.303	6.965	9.925	31.599
3	1.638	2.353	3.182	4.541	5.841	12.924
4	1.533	2.132	2.776	3.747	4.604	8.610
5	1.476	2.015	2.571	3.365	4.032	6.869
6	1.440	1.943	2.447	3.143	3.707	5.959
7	1.415	1.895	2.365	2.998	3.499	5.408
8	1.397	1.860	2.306	2.896	3.355	5.041
9	1.383	1.833	2.262	2.821	3.250	4.781
10	1.372	1.812	2.228	2.764	3.169	4.587
11	1.363	1.796	2.201	2.718	3.106	4.437
12	1.356	1.782	2.179	2.681	3.055	4.318
13	1.350	1.771	2.160	2.650	3.012	4.221
14	1.345	1.761	2.145	2.624	2.977	4.140
15	1.341	1.753	2.131	2.602	2.947	4.073
16	1.337	1.746	2.120	2.583	2.921	4.015
17	1.333	1.740	2.110	2.567	2.898	3.965
18	1.330	1.734	2.101	2.552	2.878	3.922
19	1.328	1.729	2.093	2.539	2.861	3.883
20	1.325	1.725	2.086	2.528	2.845	3.850
21	1.323	1.721	2.080	2.518	2.831	3.819
22	1.321	1.717	2.074	2.508	2.819	3.792
23	1.319	1.714	2.069	2.500	2.807	3.768
24	1.318	1.711	2.064	2.492	2.797	3.745
25	1.316	1.708	2.060	2.485	2.787	3.725
26	1.315	1.706	2.056	2.479	2.779	3.707
27	1.314	1.703	2.052	2.473	2.771	3.690
28	1.313	1.701	2.048	2.467	2.763	3.674
29	1.311	1.699	2.045	2.462	2.756	3.659
30	1.310	1.697	2.042	2.457	2.750	3.646
40	1.303	1.684	2.021	2.423	2.704	3.551
60	1.296	1.671	2.000	2.390	2.660	3.460
120	1.289	1.658	1.980	2.358	2.617	3.373
∞	1.282	1.645	1.960	2.327	2.576	3.291

Source: James D. Evans

Table U

CRITICAL VALUES OF THE *U* STATISTIC

The numbers listed below are the critical *U* values for α = .05. Use the lightface critical values for a one-tailed test and the **boldface critical values for a two-tailed test**. *Note:* To be significant, the *smaller computed U must be equal to or less than the critical U* (see Chapter 15). A dash (—) means that no decision is possible at the stated level of significance.

N_B \ N_A	1	2	3	4	5	6	7	8	9	10	11	12	13	14	15	16	17	18	19	20
1	—	—	—	—	—	—	—	—	—	—	—	—	—	—	—	—	—	—	0	0
2	—	—	—	—	0	0	0	1	1	1	1	2	2	2	3	3	3	4	4	4
	—	—	—	—	—	—	—	**0**	**0**	**0**	**0**	**1**	**1**	**1**	**1**	**1**	**2**	**2**	**2**	**2**
3	—	—	0	0	1	2	2	3	3	4	5	5	6	7	7	8	9	9	10	11
	—	—	—	—	**0**	**1**	**1**	**2**	**2**	**3**	**3**	**4**	**4**	**5**	**5**	**6**	**6**	**7**	**7**	**8**
4	—	—	0	1	2	3	4	5	6	7	8	9	10	11	12	14	15	16	17	18
	—	—	—	**0**	**1**	**2**	**3**	**4**	**4**	**5**	**6**	**7**	**8**	**9**	**10**	**11**	**11**	**12**	**13**	**13**
5	—	0	1	2	4	5	6	8	9	11	12	13	15	16	18	19	20	22	23	25
	—	—	**0**	**1**	**2**	**3**	**5**	**6**	**7**	**8**	**9**	**11**	**12**	**13**	**14**	**15**	**17**	**18**	**19**	**20**
6	—	0	2	3	5	7	8	10	12	14	16	17	19	21	23	25	26	28	30	32
	—	—	**1**	**2**	**3**	**5**	**6**	**8**	**10**	**11**	**13**	**14**	**16**	**17**	**19**	**21**	**22**	**24**	**25**	**27**
7	—	0	2	4	6	8	11	13	15	17	19	21	24	26	28	30	33	35	37	39
	—	—	**1**	**3**	**5**	**6**	**8**	**10**	**12**	**14**	**16**	**18**	**20**	**22**	**24**	**26**	**28**	**30**	**32**	**34**
8	—	1	3	5	8	10	13	15	18	20	23	26	28	31	33	36	39	41	44	47
	—	**0**	**2**	**4**	**6**	**8**	**10**	**13**	**15**	**17**	**19**	**22**	**24**	**26**	**29**	**31**	**34**	**36**	**38**	**41**
9	—	1	3	6	9	12	15	18	21	24	27	30	33	36	39	42	45	48	51	54
	—	**0**	**2**	**4**	**7**	**10**	**12**	**15**	**17**	**20**	**23**	**26**	**28**	**31**	**34**	**37**	**39**	**42**	**45**	**48**
10	—	1	4	7	11	14	17	20	24	27	31	34	37	41	44	48	51	55	58	62
	—	**0**	**3**	**5**	**8**	**11**	**14**	**17**	**20**	**23**	**26**	**29**	**33**	**36**	**39**	**42**	**45**	**48**	**52**	**55**
11	—	1	5	8	12	16	19	23	27	31	34	38	42	46	50	54	57	61	65	69
	—	**0**	**3**	**6**	**9**	**13**	**16**	**19**	**23**	**26**	**30**	**33**	**37**	**40**	**44**	**47**	**51**	**55**	**58**	**62**
12	—	2	5	9	13	17	21	26	30	34	38	42	47	51	55	60	64	68	72	77
	—	**1**	**4**	**7**	**11**	**14**	**18**	**22**	**26**	**29**	**33**	**37**	**41**	**45**	**49**	**53**	**57**	**61**	**65**	**69**
13	—	2	6	10	15	19	24	28	33	37	42	47	51	56	61	65	70	75	80	84
	—	**1**	**4**	**8**	**12**	**16**	**20**	**24**	**28**	**33**	**37**	**41**	**45**	**50**	**54**	**59**	**63**	**67**	**72**	**76**
14	—	2	7	11	16	21	26	31	36	41	46	51	56	61	66	71	77	82	87	92
	—	**1**	**5**	**9**	**13**	**17**	**22**	**26**	**31**	**36**	**40**	**45**	**50**	**55**	**59**	**64**	**67**	**74**	**78**	**83**
15	—	3	7	12	18	23	28	33	39	44	50	55	61	66	72	77	83	88	94	100
	—	**1**	**5**	**10**	**14**	**19**	**24**	**29**	**34**	**39**	**44**	**49**	**54**	**59**	**64**	**70**	**75**	**80**	**85**	**90**
16	—	3	8	14	19	25	30	36	42	48	54	60	65	71	77	83	89	95	101	107
	—	**1**	**6**	**11**	**15**	**21**	**26**	**31**	**37**	**42**	**47**	**53**	**59**	**64**	**70**	**75**	**81**	**86**	**92**	**98**
17	—	3	9	15	20	26	33	39	45	51	57	64	70	77	83	89	96	102	109	115
	—	**2**	**6**	**11**	**17**	**22**	**28**	**34**	**39**	**45**	**51**	**57**	**63**	**67**	**75**	**81**	**87**	**93**	**99**	**105**
18	—	4	9	16	22	28	35	41	48	55	61	68	75	82	88	95	102	109	116	123
	—	**2**	**7**	**12**	**18**	**24**	**30**	**36**	**42**	**48**	**55**	**61**	**67**	**74**	**80**	**86**	**93**	**99**	**106**	**112**
19	0	4	10	17	23	30	37	44	51	58	65	72	80	87	94	101	109	116	123	130
	—	**2**	**7**	**13**	**19**	**25**	**32**	**38**	**45**	**52**	**58**	**65**	**72**	**78**	**85**	**92**	**99**	**106**	**113**	**119**
20	0	4	11	18	25	32	39	47	54	62	69	77	84	92	100	107	115	123	130	138
	—	**2**	**8**	**13**	**20**	**27**	**34**	**41**	**48**	**55**	**62**	**69**	**76**	**83**	**90**	**98**	**105**	**112**	**119**	**127**

(Table U continued)

The numbers listed below are the critical U values for $\alpha = .01$. Use the lightface critical values for a one-tailed test and the **boldface critical values for a two-tailed test**. *Note:* To be significant, the *smaller computed U must be equal to or less than the critical U* (see Chapter 15). A dash (—) means that no decision is possible at the stated level of significance.

N_B \ N_A	1	2	3	4	5	6	7	8	9	10	11	12	13	14	15	16	17	18	19	20
1	—	—	—	—	—	—	—	—	—	—	—	—	—	—	—	—	—	—	—	—
2	—	—	—	—	—	—	—	—	—	—	—	—	0	0	0	0	0	0	1	1
	—	—	—	—	—	—	—	—	—	—	—	—	—	—	—	—	—	—	**0**	**0**
3	—	—	—	—	—	—	—	0	0	1	1	1	2	2	2	3	3	4	4	5
	—	—	—	—	—	—	—	—	**0**	**0**	**0**	**1**	**1**	**1**	**2**	**2**	**2**	**2**	**3**	**3**
4	—	—	—	—	0	1	1	2	3	3	4	5	5	6	7	7	8	9	9	10
	—	—	—	—	—	**0**	**0**	**1**	**1**	**2**	**2**	**3**	**3**	**4**	**5**	**5**	**6**	**6**	**7**	**8**
5	—	—	—	0	1	2	3	4	5	6	7	8	9	10	11	12	13	14	15	16
	—	—	—	—	**0**	**1**	**1**	**2**	**3**	**4**	**5**	**6**	**7**	**7**	**8**	**9**	**10**	**11**	**12**	**13**
6	—	—	—	1	2	3	4	6	7	8	9	11	12	13	15	16	18	19	20	22
	—	—	—	**0**	**1**	**2**	**3**	**4**	**5**	**6**	**7**	**9**	**10**	**11**	**12**	**13**	**15**	**16**	**17**	**18**
7	—	—	0	1	3	4	6	7	9	11	12	14	16	17	19	21	23	24	26	28
	—	—	—	**0**	**1**	**3**	**4**	**6**	**7**	**9**	**10**	**12**	**13**	**15**	**16**	**18**	**19**	**21**	**22**	**24**
8	—	—	0	2	4	6	7	9	11	13	15	17	20	22	24	26	28	30	32	34
	—	—	—	**1**	**2**	**4**	**6**	**7**	**9**	**11**	**13**	**15**	**17**	**18**	**20**	**22**	**24**	**26**	**28**	**30**
9	—	—	1	3	5	7	9	11	14	16	18	21	23	26	28	31	33	36	38	40
	—	—	**0**	**1**	**3**	**5**	**7**	**9**	**11**	**13**	**16**	**18**	**20**	**22**	**24**	**27**	**29**	**31**	**33**	**36**
10	—	—	1	3	6	8	11	13	16	19	22	24	27	30	33	36	38	41	44	47
	—	—	**0**	**2**	**4**	**6**	**9**	**11**	**13**	**16**	**18**	**21**	**24**	**26**	**29**	**31**	**34**	**37**	**39**	**42**
11	—	—	1	4	7	9	12	15	18	22	25	28	31	34	37	41	44	47	50	53
	—	—	**0**	**2**	**5**	**7**	**10**	**13**	**16**	**18**	**21**	**24**	**27**	**30**	**33**	**36**	**39**	**42**	**45**	**48**
12	—	—	2	5	8	11	14	17	21	24	28	31	35	38	42	46	49	53	56	60
	—	—	**1**	**3**	**6**	**9**	**12**	**15**	**18**	**21**	**24**	**27**	**31**	**34**	**37**	**41**	**44**	**47**	**51**	**54**
13	—	0	2	5	9	12	16	20	23	27	31	35	39	43	47	51	55	59	63	67
	—	—	**1**	**3**	**7**	**10**	**13**	**17**	**20**	**24**	**27**	**31**	**34**	**38**	**42**	**45**	**49**	**53**	**56**	**60**
14	—	0	2	6	10	13	17	22	26	30	34	38	43	47	51	56	60	65	69	73
	—	—	**1**	**4**	**7**	**11**	**15**	**18**	**22**	**26**	**30**	**34**	**38**	**42**	**46**	**50**	**54**	**58**	**63**	**67**
15	—	0	3	7	11	15	19	24	28	33	37	42	47	51	56	61	66	70	75	80
	—	—	**2**	**5**	**8**	**12**	**16**	**20**	**24**	**29**	**33**	**37**	**42**	**46**	**51**	**55**	**60**	**64**	**69**	**73**
16	—	0	3	7	12	16	21	26	31	36	41	46	51	56	61	66	71	76	82	87
	—	—	**2**	**5**	**9**	**13**	**18**	**22**	**27**	**31**	**36**	**41**	**45**	**50**	**55**	**60**	**65**	**70**	**74**	**79**
17	—	0	4	8	13	18	23	28	33	38	44	49	55	60	66	71	77	82	88	93
	—	—	**2**	**6**	**10**	**15**	**19**	**24**	**29**	**34**	**39**	**44**	**49**	**54**	**60**	**65**	**70**	**75**	**81**	**86**
18	—	0	4	9	14	19	24	30	36	41	47	53	59	65	70	76	82	88	94	100
	—	—	**2**	**6**	**11**	**16**	**21**	**26**	**31**	**37**	**42**	**47**	**53**	**58**	**64**	**70**	**75**	**81**	**87**	**92**
19	—	1	4	9	15	20	26	32	38	44	50	56	63	69	75	82	88	94	101	107
	—	**0**	**3**	**7**	**12**	**17**	**22**	**28**	**33**	**39**	**45**	**51**	**56**	**63**	**69**	**74**	**81**	**87**	**93**	**99**
20	—	1	5	10	16	22	28	34	40	47	53	60	67	73	80	87	93	100	107	114
	—	**0**	**3**	**8**	**13**	**18**	**24**	**30**	**36**	**42**	**48**	**54**	**60**	**67**	**73**	**79**	**86**	**92**	**99**	**105**

Source: Adapted from R. E. Kirk. (1984). *Elementary statistics* (2nd ed.). Pacific Grove, CA: Brooks/Cole. Used with permission of the publisher.

Table W

CRITICAL VALUES OF THE WILCOXON *T* STATISTIC

	LEVEL OF SIGNIFICANCE FOR A ONE-TAILED TEST					LEVEL OF SIGNIFICANCE FOR A ONE-TAILED TEST			
	.05	.025	.01	.005		.05	.025	.01	.005
	LEVEL OF SIGNIFICANCE FOR A TWO-TAILED TEST					LEVEL OF SIGNIFICANCE FOR A TWO-TAILED TEST			
N	.10	.05	.02	.01	*N*	.10	.05	.02	.01
5	0	—	—	—	28	130	116	101	91
6	2	0	—	—	29	140	126	110	100
7	3	2	0	—	30	151	137	120	109
8	5	3	1	0	31	163	147	130	118
9	8	5	3	1	32	175	159	140	128
10	10	8	5	3	33	187	170	151	138
11	13	10	7	5	34	200	182	162	148
12	17	13	9	7	35	213	195	173	159
13	21	17	12	9	36	227	208	185	171
14	25	21	15	12	37	241	221	198	182
15	30	25	19	15	38	256	235	211	194
16	35	29	23	19	39	271	249	224	207
17	41	34	27	23	40	286	264	238	220
18	47	40	32	27	41	302	279	252	233
19	53	46	37	32	42	319	294	266	247
20	60	52	43	37	43	336	310	281	261
21	67	58	49	42	44	353	327	296	276
22	75	65	55	48	45	371	343	312	291
23	83	73	62	54	46	389	361	328	307
24	91	81	69	61	47	407	378	345	322
25	100	89	76	68	48	426	396	362	339
26	110	98	84	75	49	446	415	379	355
27	119	107	92	83	50	466	434	397	373

How to use this table: Use the *smaller* computed sum of ranks as the test statistic, *T*. If the computed *T* is equal to or less than the critical value in the table, then it is significant.

Source: Adapted from R. E. Kirk. (1984). *Elementary statistics* (2nd ed.). Pacific Grove, CA: Brooks/Cole. Used with permission of the publisher.

Table Z

THE STANDARD NORMAL DISTRIBUTION

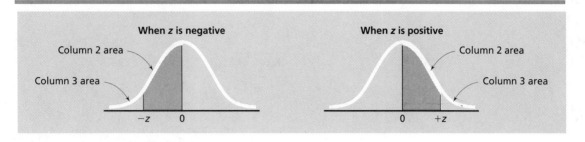

When *z* is negative — Column 2 area, Column 3 area — $-z$ 0

When *z* is positive — Column 2 area, Column 3 area — 0 $+z$

BLOCK A			BLOCK B			BLOCK C		
1	2	3	1	2	3	1	2	3
	AREA BETWEEN z AND	*AREA BEYOND*		*AREA BETWEEN z AND*	*AREA BEYOND*		*AREA BETWEEN z AND*	*AREA BEYOND*
z	*THE MEAN*	*z*	*z*	*THE MEAN*	*z*	*z*	*THE MEAN*	*z*
0.00	.0000	.5000	0.40	.1554	.3446	0.80	.2880	.2119
0.01	.0040	.4960	0.41	.1591	.3409	0.81	.2910	.2090
0.02	.0080	.4920	0.42	.1628	.3372	0.82	.2939	.2061
0.03	.0120	.4880	0.43	.1664	.3336	0.83	.2967	.2033
0.04	.0160	.4840	0.44	.1700	.3300	0.84	.2995	.2005
0.05	.0199	.4801	0.45	.1736	.3264	0.85	.3023	.1977
0.06	.0239	.4761	0.46	.1772	.3228	0.86	.3051	.1949
0.07	.0279	.4721	0.47	.1808	.3192	0.87	.3078	.1922
0.08	.0319	.4681	0.48	.1844	.3156	0.88	.3106	.1894
0.09	.0359	.4641	0.49	.1879	.3121	0.89	.3133	.1867
0.10	.0398	.4602	0.50	.1915	.3085	0.90	.3159	.1841
0.11	.0438	.4562	0.51	.1950	.3050	0.91	.3186	.1814
0.12	.0478	.4522	0.52	.1985	.3015	0.92	.3212	.1788
0.13	.0517	.4483	0.53	.2019	.2981	0.93	.3238	.1762
0.14	.0557	.4443	0.54	.2054	.2946	0.94	.3264	.1736
0.15	.0596	.4404	0.55	.2088	.2912	0.95	.3289	.1711
0.16	.0636	.4364	0.56	.2123	.2877	0.96	.3315	.1685
0.17	.0675	.4325	0.57	.2157	.2843	0.97	.3340	.1660
0.18	.0714	.4286	0.58	.2190	.2810	0.98	.3365	.1635
0.19	.0753	.4247	0.59	.2224	.2776	0.99	.3389	.1611
0.20	.0793	.4207	0.60	.2257	.2743	1.00	.3413	.1587
0.21	.0832	.4168	0.61	.2291	.2709	1.01	.3438	.1562
0.22	.0871	.4129	0.62	.2324	.2676	1.02	.3461	.1539
0.23	.0910	.4090	0.63	.2357	.2643	1.03	.3485	.1515
0.24	.0948	.4052	0.64	.2389	.2611	1.04	.3508	.1492
0.25	.0987	.4013	0.65	.2422	.2578	1.05	.3531	.1469
0.26	.1026	.3974	0.66	.2454	.2546	1.06	.3554	.1446
0.27	.1064	.3936	0.67	.2486	.2514	1.07	.3577	.1423
0.28	.1103	.3897	0.68	.2517	.2483	1.08	.3599	.1401
0.29	.1141	.3859	0.69	.2549	.2451	1.09	.3621	.1379
0.30	.1179	.3821	0.70	.2580	.2420	1.10	.3643	.1357
0.31	.1217	.3783	0.71	.2611	.2389	1.11	.3665	.1335
0.32	.1255	.3745	0.72	.2642	.2358	1.12	.3686	.1314
0.33	.1293	.3707	0.73	.2673	.2327	1.13	.3708	.1292
0.34	.1331	.3669	0.74	.2704	.2296	1.14	.3729	.1271
0.35	.1368	.3632	0.75	.2734	.2266	1.15	.3749	.1251
0.36	.1406	.3594	0.76	.2764	.2236	1.16	.3770	.1230
0.37	.1443	.3557	0.77	.2794	.2206	1.17	.3790	.1210
0.38	.1480	.3520	0.78	.2823	.2177	1.18	.3810	.1190
0.39	.1517	.3483	0.79	.2852	.2148	1.19	.3830	.1170

(Table Z continued)

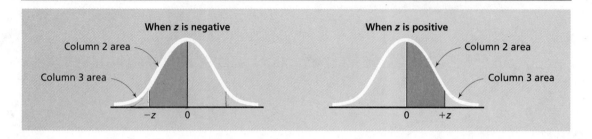

	BLOCK A			BLOCK B			BLOCK C	
1	*2*	*3*	*1*	*2*	*3*	*1*	*2*	*3*
	AREA BETWEEN z AND THE MEAN	*AREA BEYOND z*		*AREA BETWEEN z AND THE MEAN*	*AREA BEYOND z*		*AREA BETWEEN z AND THE MEAN*	*AREA BEYOND z*
z			*z*			*z*		
1.20	.3849	.1151	1.60	.4452	.0548	2.00	.4772	.0228
1.21	.3869	.1131	1.61	.4463	.0537	2.01	.4778	.0222
1.22	.3888	.1112	1.62	.4474	.0526	2.02	.4783	.0217
1.23	.3907	.1093	1.63	.4484	.0516	2.03	.4788	.0212
1.24	.3925	.1075	1.64	.4495	.0505	2.04	.4793	.0207
1.25	.3944	.1056	1.65	.4505	.0495	2.05	.4798	.0202
1.26	.3962	.1038	1.66	.4515	.0485	2.06	.4803	.0197
1.27	.3980	.1020	1.67	.4525	.0475	2.07	.4808	.0192
1.28	.3997	.1003	1.68	.4535	.0465	2.08	.4812	.0188
1.29	.4015	.0985	1.69	.4545	.0455	2.09	.4817	.0183
1.30	.4032	.0968	1.70	.4554	.0446	2.10	.4821	.0179
1.31	.4049	.0951	1.71	.4564	.0436	2.11	.4826	.0174
1.32	.4066	.0934	1.72	.4573	.0427	2.12	.4830	.0170
1.33	.4082	.0918	1.73	.4582	.0418	2.13	.4834	.0166
1.34	.4099	.0901	1.74	.4591	.0409	2.14	.4838	.0162
1.35	.4115	.0885	1.75	.4599	.0401	2.15	.4842	.0158
1.36	.4131	.0869	1.76	.4608	.0392	2.16	.4846	.0154
1.37	.4147	.0853	1.77	.4616	.0384	2.17	.4850	.0150
1.38	.4162	.0838	1.78	.4625	.0375	2.18	.4854	.0146
1.39	.4177	.0823	1.79	.4633	.0367	2.19	.4857	.0143
1.40	.4192	.0808	1.80	.4641	.0359	2.20	.4861	.0139
1.41	.4207	.0793	1.81	.4649	.0351	2.21	.4864	.0136
1.42	.4222	.0778	1.82	.4656	.0344	2.22	.4868	.0132
1.43	.4236	.0764	1.83	.4664	.0336	2.23	.4871	.0129
1.44	.4251	.0749	1.84	.4671	.0329	2.24	.4875	.0125
1.45	.4265	.0735	1.85	.4678	.0322	2.25	.4878	.0122
1.46	.4279	.0721	1.86	.4686	.0314	2.26	.4881	.0119
1.47	.4292	.0708	1.87	.4693	.0307	2.27	.4884	.0116
1.48	.4306	.0694	1.88	.4699	.0301	2.28	.4887	.0113
1.49	.4319	.0681	1.89	.4706	.0294	2.29	.4890	.0110
1.50	.4332	.0668	1.90	.4713	.0287	2.30	.4893	.0107
1.51	.4345	.0655	1.91	.4719	.0281	2.31	.4896	.0104
1.52	.4357	.0643	1.92	.4726	.0274	2.32	.4898	.0102
1.53	.4370	.0630	1.93	.4732	.0268	2.33	.4901	.0099
1.54	.4382	.0618	1.94	.4738	.0262	2.34	.4904	.0096
1.55	.4394	.0606	1.95	.4744	.0256	2.35	.4906	.0094
1.56	.4406	.0594	1.96	.4750	.0250	2.36	.4909	.0091
1.57	.4418	.0582	1.97	.4756	.0244	2.37	.4911	.0089
1.58	.4429	.0571	1.98	.4761	.0239	2.38	.4913	.0087
1.59	.4441	.0559	1.99	.4767	.0233	2.39	.4916	.0084

(Table Z continued)

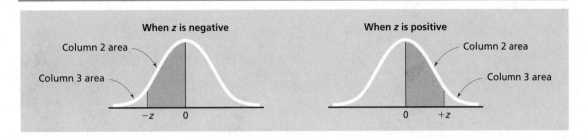

BLOCK A			BLOCK B			BLOCK C		
1	*2*	*3*	*1*	*2*	*3*	*1*	*2*	*3*
	AREA BETWEEN z AND THE MEAN	*AREA BEYOND z*		*AREA BETWEEN z AND THE MEAN*	*AREA BEYOND z*		*AREA BETWEEN z AND THE MEAN*	*AREA BEYOND z*
z			*z*			*z*		
2.40	.4918	.0082	2.72	.4967	.0033	3.04	.4988	.0012
2.41	.4920	.0080	2.73	.4968	.0032	3.05	.4989	.0011
2.42	.4922	.0078	2.74	.4969	.0031	3.06	.4989	.0011
2.43	.4925	.0075	2.75	.4970	.0030	3.07	.4989	.0011
2.44	.4927	.0073	2.76	.4971	.0029	3.08	.4990	.0010
2.45	.4929	.0071	2.77	.4972	.0028	3.09	.4990	.0010
2.46	.4931	.0069	2.78	.4973	.0027	3.10	.4990	.0010
2.47	.4932	.0068	2.79	.4974	.0026	3.11	.4991	.0009
2.48	.4934	.0066	2.80	.4974	.0026	3.12	.4991	.0009
2.49	.4936	.0064	2.81	.4975	.0025	3.13	.4991	.0009
2.50	.4938	.0062	2.82	.4976	.0024	3.14	.4992	.0008
2.51	.4940	.0060	2.83	.4977	.0023	3.15	.4992	.0008
2.52	.4941	.0059	2.84	.4977	.0023	3.16	.4992	.0008
2.53	.4943	.0057	2.85	.4978	.0022	3.17	.4992	.0008
2.54	.4945	.0055	2.86	.4979	.0021	3.18	.4993	.0007
2.55	.4946	.0054	2.87	.4979	.0021	3.19	.4993	.0007
2.56	.4948	.0052	2.88	.4980	.0020	3.20	.4993	.0007
2.57	.4949	.0051	2.89	.4981	.0019	3.21	.4993	.0007
2.58	.4951	.0049	2.90	.4981	.0019	3.22	.4994	.0006
2.59	.4952	.0048	2.91	.4982	.0018	3.23	.4994	.0006
2.60	.4953	.0047	2.92	.4982	.0018	3.24	.4994	.0006
2.61	.4955	.0045	2.93	.4983	.0017	3.25	.4994	.0006
2.62	.4956	.0044	2.94	.4984	.0016	3.30	.4995	.0005
2.63	.4957	.0043	2.95	.4984	.0016	3.35	.4996	.0004
2.64	.4959	.0041	2.96	.4985	.0015	3.40	.4997	.0003
2.65	.4960	.0040	2.97	.4985	.0015	3.45	.4997	.0003
2.66	.4961	.0039	2.98	.4986	.0014	3.50	.4998	.0002
2.67	.4962	.0038	2.99	.4986	.0014	3.60	.4998	.0002
2.68	.4963	.0037	3.00	.4987	.0013	3.70	.4999	.0001
2.69	.4964	.0036	3.01	.4987	.0013	3.80	.4999	.0001
2.70	.4965	.0035	3.02	.4987	.0013	3.90	.49995	.00005
2.71	.4966	.0034	3.03	.4988	.0012	4.00	.49997	.00003

APPENDIX B   ANSWERS TO ODD-NUMBERED REVIEW QUESTIONS

CHAPTER 1

1. Students of the behavioral sciences need to understand statistical applications in order to (a) improve their problem-solving effectiveness, (b) increase their employability and promotion potential, and (c) be able to contribute to significant advances in behavioral science knowledge.
3. Computerized statistical analysis benefits you by greatly reducing computation time and increasing accuracy. But statistical analysis does not require the use of computers; you can learn how to apply and interpret statistics without using computers at all (although it is to your advantage to learn how to do statistics with computer software). Furthermore, merely learning how to use statistical software will not, by itself, lead to a mastery of statistics. Mastery comes from actively thinking about and applying statistical theory and methods.

CHAPTER 2

1. Theoretical statistics is a branch of mathematics, and applied statistics borrows from mathematics. In everyday practice, however, researchers view statistics principally as tools for summarizing data, testing hypotheses, and making inferences and predictions about behavior.
3. A parameter is a numerical characteristic of a population—for example, the average achievement motivation score of all 7000 female executives. A statistic is a numerical characteristic of a sample—for example, the average score of 82 points made by the 400 women who participated in the study. We often use a statistic to estimate a parameter.
5. A population is the entire set of people, things, or events that the researcher wishes to study. A sample is a subset of a population. In simple random sampling, the sample is selected in such a way that every member of the population has an equal chance of being chosen. Random sampling entails selecting sampling units on the basis of chance alone.
7. A relationship is an association between two variables, whereas a hypothesis is a proposed or predicted relationship. Two examples of relationships are (a) the connection between amount of cigarette smoking and cancer rate and (b) the inverse connection between supply and demand. An example of a hypothesis is: The higher people's extroversion, the higher they will score in trivia games. Another hypothesis is: Theory X leaders will perform better than Theory Y leaders in crisis situations.
9. It is a predictive relationship because the findings came from a correlational study, not a well-controlled experiment. That is, the subjects were not randomly assigned to different levels of the smoking variable. A predictive relationship exists when you have a predictive association between two variables but no reason to infer cause and effect. A causal relationship exists when you have evidence to support the assertion that variable X causes variable Y to change. The latter kind of information usually can be derived only from a true experiment, in which subjects are randomly assigned to different treatment conditions.

11. The function of random assignment is to make different treatment groups statistically equivalent at the start of an experiment; this initial equivalence is necessary to draw cause-and-effect conclusions. That is, if the groups are equivalent at the outset of the experiment but different after the independent variable has been manipulated, then it is logical to assert that the manipulation caused the change.

13. The independent variable is method of instruction. The dependent variable is student achievement, as measured by the exam scores. This is a true experiment because the teachers (and thus their classes) are randomly assigned to one method of instruction or the other.

15. The descriptive statistics were the averages of the conventional lecture condition (89%) and the discovery learning condition (85%), respectively. Inferential statistics would be necessary to determine whether the difference in average test scores is statistically significant—that is, more than just a chance difference.

17. Inferential statistics is most likely to be used in stage 4 (finding relationships)—to determine whether a relationship between variables is reliable—and stage 6 (generalizing)—to estimate a population parameter from a sample statistic.

19. Random selection is more important because its function is to render the sample representative of the population. The purpose of parameter estimation is to draw a conclusion about a population on the basis of a sample selected from that population.

21. Statistical significance means that the basic result of a study is likely to be repeatable in additional investigations of the same type. Therefore, a significant outcome is likely to be generalizable to future studies.

23. It is hypothesized that college students who watch a lot of TV get lower grades than students who do not watch TV at all. A survey of 783 college freshmen in eight psychology classes is used to identify a group of students who watch TV at least 20 hours per week and a second group claiming abstinence from the habit. The groups' first-semester grade point averages are compared. The variables in the hypothesis are amount of TV watching and grade point average. This is not a true experiment because the students are not randomly assigned to a level of TV viewing. Consequently, the groups might differ initially in many ways, and no cause-and-effect conclusions can be drawn regarding the impact of watching TV.

25. Randomness always implies a lack of predictability between events, as in the behavior of a pair of dice or a roulette wheel. It means that outcomes are purely a function of chance. The assumption of randomness is important because both the logic of research designs and the mathematical theories underlying statistical procedures assume that either random assignment or random selection has been implemented at some point in the data-generation process. Thus, conclusions and decisions that stem from statistical analysis are valid only if the randomness assumption has been satisfied by the research procedure.

CHAPTER 3

1. Scale of measurement: levels of measurement distinguished by different mathematical properties; higher scales have all the properties of lower scales plus additional properties. Nominal scale: measurement by classifying observations into categories. Ordinal scale: measurement by ranking observations. Interval scale: measurement by locating observations along a dimension with equal intervals. Ratio scale: measurement by locating observations along a dimension with equal intervals and an absolute zero point.

3. It makes no sense to say Paula is twice as talented as Penny because mathematical ratios require a ratio scale of measurement. The talent contest results are presented as rankings on an ordinal scale.

5. Frequency distribution: a statistical table showing what responses were made and the frequency of each response. Relative frequency: the frequency of an outcome multiplied by $100/N$. Proportion: the frequency of an outcome divided by the total number of outcomes; a relative frequency. Class interval: constant range of values in a grouped frequency distribution. Mutually exclusive: class intervals are mutually exclusive if they don't overlap. Exact range: highest score – lowest score + 1.

7.

IQ	f	c%
118	1	2.00
117	1	2.00
116	0	0.00
115	1	2.00
114	0	0.00
113	1	2.00
112	0	0.00
111	0	0.00
110	1	2.00
109	0	0.00
108	1	2.00
107	1	2.00
106	1	2.00
105	1	2.00
104	1	2.00
103	3	6.00
102	3	6.00
101	5	10.00
100	7	14.00
99	5	10.00
98	4	8.00
97	3	6.00
96	1	2.00
95	1	2.00
94	1	2.00

IQ	f	%
93	1	2.00
92	1	2.00
91	0	0.00
90	1	2.00
89	0	0.00
88	0	0.00
87	1	2.00
86	0	0.00
85	1	2.00
84	0	0.00
83	1	2.00
82	1	2.00

9.

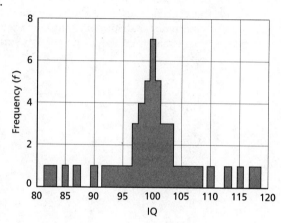

11.

CLASS INTERVAL	%
2.65–3.05	100.00
2.25–2.65	100.00
1.85–2.25	96.67
1.45–1.85	80.00
1.05–1.45	80.00
0.65–1.05	40.00
0.25–0.65	13.33

13. Range = 471; exact range = 472; interval size = 27; lowest interval is 70.5–97.5; midpoint of highest interval is 543.

15. There are too few intervals, they are not mutually exclusive, and they are of different sizes.

17.

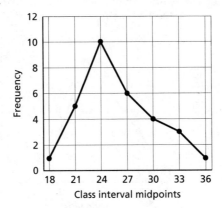

19. Bar graph: a graph that uses vertical bars to represent the frequency associated with each of several nominal categories. Statistical curve: a relative frequency polygon of a large population of values. Normal curve: a perfectly symmetrical bell-shaped statistical curve that is the chief theoretical model underlying inferential statistics. Negatively skewed: most of the observations accumulate over the high values of the measured variable. Positively skewed: most of the observations accumulate over the low values of the measured variable. Bimodal distribution: a distribution with two peaks.

21. The seven-interval distribution tells more about the typical score and the shape of the distribution. Since the top interval is empty, however, six intervals would be even better.

CLASS INTERVAL	%
17.5–25.5	20.00
9.5–17.5	60.00
1.5–9.5	20.00

CLASS INTERVAL	%
25.5–29.5	0.00
21.5–25.5	5.00
17.5–21.5	15.00
13.5–17.5	35.00
9.5–13.5	25.00
5.5–9.5	15.00
1.5–5.5	5.00

23. The distribution tends to be positively skewed because the observations tend to pile up over the lower midpoints.

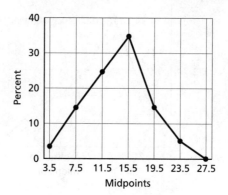

25. The distribution tends to be negatively skewed because the observations tend to pile up over the higher midpoint values. However, it would also be reasonable to argue that the distribution tends to be bimodal.

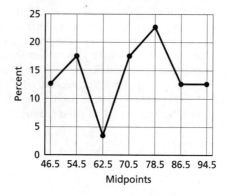

CHAPTER 4

1. Central tendency: the tendency of scores in a distribution to pile up over a particular value. Measure of central tendency: a statistical index of the average or typical score in a distribution, such as the mean, median, or mode. Mode: the most frequent score in a distribution. Median: the middle score in a distribution. Percentile: a value below which a specified percent of the distribution lies. N: the number of observations in a data set. Mean: the arithmetic average of a distribution of scores. Least squares principle: the law of statistics that underlies the mean; states that the sum of squared deviations from the mean will be smaller than the sum of squared deviations from any other value in the distribution that does not equal the mean.

3. 7.82; 7.80; 100.766; .2399

5. The mode is the only measure of central tendency that can be used with nominal-scale data, but it has no advanced uses in statistics. The

mean has a large number of uses in advanced statistical procedures, but it is sensitive to extreme scores in a markedly skewed distribution. The median is considered the most accurate descriptive index of central tendency in a markedly skewed distribution, but it has almost no advanced applications.

7. The distribution is negatively skewed: mode = 106; median = 103; mean = 101.

9. Means = 16, 48, 8. Multiplying or dividing every raw score by a constant multiplies or divides the mean by that same constant.

11. X_1: mean = 20, median = 22, mode = 22 and 24; sums of squared deviations = 200, 240, and 360 for mean, median, and mode, respectively. Calculations for mode = 22 give the same answer as those for the median, since both equal 22.

    X_2: mean = 13, median = 13, mode = 7; sums of squared deviations = 180, 180, and 540 for mean, median, and mode, respectively

    X_3: mean = 16, median = 15, mode = 11; sums of squared deviations = 280, 290, and 530 for mean, median, and mode, respectively

13. When N = 167, the rank of the median value is 84. When N = 1002, the rank of the median value is 501.5. Rank of median value = $(N + 1)/2$.

15. Pretest mean = 85.10; posttest mean = 96.15. The difference between the pretest and posttest means is −11.05, which is the same as the mean of the difference scores.

17. X_1: mean = 101; variance = 162; range = 43; interquartile range = 20.5

    X_2: mean = 101; variance = 206; range = 47; interquartile range = 25

19. Variance = 36; standard deviation = 6. Adding a constant to each raw score does not change the variance or the standard deviation.

21. Mean = 14.3; median = 13.5; mode = 13

23. Mean = 9.625; median = 9; mode = 9. The distribution tends to be positively skewed, which causes the mean to be higher than the median.

25. The percent of the distribution within two standard deviations of the mean is 93.75. Since the standard deviation is 2.50, two standard deviations = 2(2.50) = 5.00. Since the mean is 9.625, 93.75% of the scores lie between 9.625 − 5 and 9.625 + 5. A similar analysis shows that 100% of the distribution is within three standard deviations of the mean.

CHAPTER 5

1. Relative measures: transformations of raw scores to standard number scales. Proportion: the ratio of the number of target events to the total number of events in a set. Percent: proportion multiplied by 100. Measures of relative standing: mathematical transformations that locate a raw score's rank or position in a distribution. Percentile: a value below which a particular percent of the distribution lies. Percentile rank: the percent of a distribution lying below a certain value.

3. Proportion = .075; percent = 7.5

5. Percentile rank = 5% + ½ (10%), or 10

7. When X = 85, the percentile rank is 81% + ½ (6%), or 84.

9. A z score is a deviation score expressed in standard deviation units. It represents the number of standard deviations by which a particular raw score differs from the mean.

11. $z = -1.3$

13. The only frequency polygon characteristic that changed was the values along the abscissa. The shape of the distribution did not change. Converting raw scores to z scores does not alter the shape of the distribution.

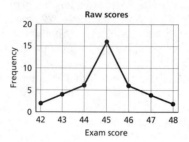
Raw scores

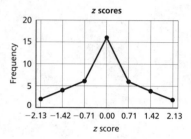
z scores

15. Two percent of the normal curve lies between $z = -2.575$ and $z = -1.96$.

17. $z = 1.28$

19. The probability that $z \le -2.575$ or $z \ge +2.575$ is .01.

21. Norman's z score is -0.84. Using formula (5.5), we find that his IQ is $(16)(-0.84) + 100$, or approximately 87.

23. 95.44% of the distribution lies between $z = -2.00$ and $z = +2.00$.

25. The z score corresponding to the bottom 20% (i.e., a lower-tail proportion of .20) is -0.84; transformed score = $(6.5)(-0.84) + 30 = -5.46 + 30 = 24.54$, or approximately 25. Someone who scored 25 or lower would be in the poor self-concept category.

CHAPTER 6

1. The most basic goal of science is to discover the laws of nature as revealed in relationships between variables. Many relationships tend to be linear. Since r is a precise quantitative gauge of the strength and direction of linear association between two variables, it is the statistic that most directly and fundamentally provides the information that science seeks.

3. Mary is correct. In regard to the Pearson r, sign is independent of size. That is, the strength of the relationship between two variables is indicated by the absolute magnitude of the correlation coefficient (or, better yet, the absolute magnitude of the squared correlation), irrespective of whether the correlation is positive or negative. A correlation of .43 is numerically higher than a correlation of .32, no matter what signs these two coefficients have.

5. (a) Negative, (b) negative, (c) positive, (d) positive, (e) negative

7. The linear correlation coefficient is a numerical index of the degree to which z scores on variable Y correspond in a systematic way to z scores on variable X.

9. $r = -6.1979/7 = -.885$

X	Y	z_X	z_Y	$z_X z_Y$
10	82	−0.74314	0.60648	−0.45070
15	72	0.21948	−0.29602	−0.06497
19	55	0.98957	−1.83026	−1.81116
17	76	0.60452	0.06498	0.03928
21	67	1.37461	−0.74726	−1.02720
6	91	−1.51323	1.41872	−2.14685
9	84	−0.93566	0.78697	−0.73634
Sums		−0.00385	0.00361	−6.19794

11. $\Sigma X = 97.00$; $\Sigma Y = 527.00$; $\Sigma X^2 = 1533.00$; $\Sigma Y^2 = 40{,}535.00$; $\Sigma XY = 6946.00$; $SS_X = 188.86$; $SS_Y = 859.43$; $CP_{XY} = -356.71$; $r = -.885$. There is no discrepancy between the results of the z-score and raw-score formulas. Any such discrepancy would be caused by rounding in the z-score method.

13. $\Sigma X = 318.00$; $\Sigma Y = 279.00$; $\Sigma X^2 = 11{,}458.00$; $\Sigma Y^2 = 9589.00$; $\Sigma XY = 8029.00$; $SS_X = 1345.60$; $SS_Y = 1804.90$; $CP_{XY} = -843.20$; $r = -.541$. Since this correlation is negative, the hypothesis of a positive correlation between extroversion and trivia scores is flatly contradicted.

15. The difference between $.70^2$ and $.50^2$ is $.49 - .25$, or $.24$. The difference between $.50^2$ and $.30^2$ is $.25 - .09$, or $.16$. This shows that $r = .70$ accounts for 24% more variance than $r = .50$, whereas $r = .50$ accounts for just 16% more variance than $r = .30$. Thus, the first difference is effectively larger. On the correlation scale, higher units are larger than lower units. However, squaring the correlation makes all units equal.

17. If one variable is causing a second variable to change, then the two variables must covary; they must be correlated. Therefore, if two variables show no correlation, they cannot have a cause-and-effect connection. Correlation is necessary for causation. The presence of a correlation, however, does not constitute proof of causality. Two "unconnected" variables can covary merely because a third factor happens to be influencing both of them in a systematic way. Consequently, a correlation is not sufficient for concluding that variable X is causing variable Y to change.

19. $\Sigma X = 7.00$; $\Sigma Y = 509.00$; $\Sigma X^2 = 7.00$; $\Sigma Y^2 = 24{,}765.00$; $\Sigma XY = 334.00$; $SS_X = 3.23$; $SS_Y = 4835.69$; $CP_{XY} = 59.92$; $r = .479$; $r^2 = .23$. Within the sample of data, leadership style accounts for 23% of the variance in type of question asked.

21. The original linear correlation is .80. Removing the odd pair of extreme scores reduces the correlation to .68. One or two extreme pairs of observations can appreciably inflate a linear correlation, so that it overestimates the typical degree of association between the variables being studied.

23. $\Sigma X = 306.00$; $\Sigma Y = 293.00$; $\Sigma X^2 = 16{,}236.00$; $\Sigma Y^2 = 14{,}447.00$; $\Sigma XY = 10{,}536.00$; $SS_X = 2859.43$; $SS_Y = 2182.86$; $CP_{XY} = -2272.29$; $r = -.91$. This is a very strong correlation, accounting for 82.7% of the variance in Y.

25. $\Sigma X = 55.00$; $\Sigma Y = 995.40$; $\Sigma X^2 = 385.00$; $\Sigma Y^2 = 99{,}187.30$; $\Sigma XY = 5389.70$; $SS_X = 82.50$; $SS_Y = 105.18$; $CP_{XY} = -85.00$; $r = -.912$. In this small sample, there is a very strong negative linear correlation between the number of children and the average IQ of the children. In this sample, family size accounts for 83.3% of variance in IQ.

CHAPTER 7

1. Linear regression analysis is a statistical procedure that uses the correlation between variables X and Y to predict future values of Y. The stronger the correlation between X and Y, the more accurately we can predict Y values from X values.

3. The higher the correlation between X and Y, the closer the *estimated* z_Y (that is, z_Y') will approximate the known z_X. When $r = 1.00$, the regression formula predicts that $z_Y = z_X$. When $r = .00$, the best prediction for every X value is that $z_Y = 0$—that is, that every Y value will equal the mean of Y (since $z_Y = 0$ is the standard score at the mean of Y). When $r < 1.00$, *the predicted Y value will be relatively closer to its mean than the corresponding X value.*

5. When $r_{XY} = .00$, the z-score regression formula predicts that every Y score will fall at the mean of the Y distribution. That is, $r_{XY} \cdot z_X$ always equals $0 \cdot z_X$, or 0, which is the mean of the z-score distribution. Therefore, the best estimate is that George's verbal aptitude score is the same as the mean score on the verbal test, or 530.

7.

z_X	z_Y'
1.440	−0.720
0.770	−0.385
0.000	0.000
−0.770	0.385
−1.440	0.720

9.

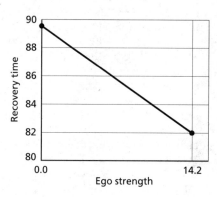

11. Mean of $X = 30.40$; mean of $Y = 68.20$; standard deviation of $X = 11.76$; standard deviation of $Y = 18.94$; $r = .862$; $b = 1.388$; $a = 26.00$

EMPLOYEE	(X) SELECTION TEST	(Y) PERF. EVALUATION	X^2	Y^2	XY
a	40	90	1600	8100	3600
b	32	74	1024	5476	2368
c	46	96	2116	9216	4416
d	50	90	2500	8100	4500
e	22	44	484	1936	968
f	10	50	100	2500	500
g	18	56	324	3136	1008
h	30	76	900	5776	2280
i	26	42	676	1764	1092
j	30	64	900	4096	1920

EMPLOYEE	(X) SELECTION TEST	PREDICTED Y
a	40	81.52
b	32	70.42
c	46	89.85
d	50	95.40
e	22	56.54
f	10	39.88
g	18	50.98
h	30	67.64
i	26	62.09
j	30	67.64

13. (a) Standard error of estimate = 10.74
    (b) Variance of prediction errors = $S_Y^2 (1 - r^2) = 358.76(1 - .862^2) = 92.32$
    (c) Coefficient of nondetermination = $S^2_{\text{pred. errors}}/S_Y^2 = 92.32/358.76 = .257$
    (d) Coefficient of nondetermination = $1 - r^2 = 1 - .862^2 = 1 - .743 = .257$

15. Slope = 0.020923; intercept = 0.547; $Y' = 0.020923X + 0.547$

17. Least squares principle: predictions generated by the linear regression formula will always yield the smallest possible sum of squared prediction errors when predictions are based on the unique values of a and b derived from the data. Standard error of estimate: a numerical index of the average amount of deviation of Y from Y' along the regression line. Homoscedasticity assumption: the assumption that the spread of actual Y values is uniform at all values of X and Y'. Bivariate

normality assumption: the assumption that the actual values of Y are normally distributed around each Y'. Cross-validation study: investigation designed to examine how well a regression equation predicts future Y values with a new sample of subjects.

19. Partitioning of variance: the results of a regression analysis are used to divide Y variance into two parts: a proportion of variance predictable from X and a proportion that is not predictable from X. Variance of prediction errors: the sum of $(Y - Y')^2$ divided by N. Coefficient of determination: the square of the correlation coefficient, which equals the proportion of Y variance explained by X. Coefficient of nondetermination: $1 - r^2$, the proportion of variance in Y not explained by X.

21. Set 2 will produce the most accurate regression formula because it is associated with the greatest percent of variance explained.

	SET 1	SET 2	SET 3
Slope	−1.20	−1.71	0.50
Intercept	52.04	47.83	86.10
Standard error	10.66	9.23	9.79
Coefficient of nondetermination	0.75	0.36	0.91
Percent of variance accounted for	25.00	64.00	9.00

23.  $Y' = -0.662X + 42.03$

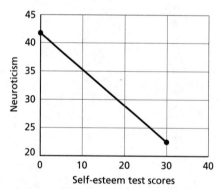

SUBJECT	(X) SELF-ESTEEM TEST	PREDICTED Y
a	27.00	24.16
b	32.00	20.85
c	40.00	15.55
d	33.00	20.18
e	24.00	26.14
f	35.00	18.86
g	20.00	28.79
h	21.00	28.13
i	36.00	18.20
j	26.00	24.82
k	30.00	22.17

SUBJECT	(X) SELF-ESTEEM TEST	PREDICTED Y
l	28.00	23.49
m	26.00	24.82
n	27.00	24.16
o	29.00	22.83

25. (a) $r^2 = .688$; percent of variance accounted for = 69

(b) $S_{est.Y} = \left(\sqrt{S_Y^2(1 - r^2)}\right)\left[\sqrt{N/(N-2)}\right] = \left(\sqrt{18.249[1 - (-.829^2)]}\right)\left(\sqrt{15/13}\right)$
$= \left(\sqrt{5.69}\right)\left(\sqrt{1.154}\right) = (2.385)(1.074) = 2.56$

(c) $Y' = -0.662(28) + 42.03 = -18.54 + 42.03 = 23.49$

(d) Lower value = $Y' - 3S_{est.Y} = 23.49 - 3(2.56) = 23.49 - 7.68 = 15.81$.
Higher value = $Y' + 3S_{est.Y} = 23.49 + 3(2.56) = 23.49 + 7.68 = 31.17$.

CHAPTER 8

1. There are several reasons that we use samples in scientific research. First, many populations are so large that it is too expensive—or even impossible—to observe every unit therein. Second, a carefully measured sample can yield more accurate data than a haphazardly observed population. Third, the process of observing might involve destruction of the observed units, which we would want to minimize.

3. (a) Random, because each professor is selected entirely on the basis of chance. (b) Nonrandom, because respondents are selected on the basis of convenience and availability, not chance. (c) Nonrandom, because respondents are selected on the basis of convenience, not chance; many of the first-year students might not be in the course that semester, and those who are taking the course that particular term might be systematically different from the general population.

5. For both $N = 1$ and $N = 2$, the mean of the sample means is equal to the population mean, 5. The smaller sample has the larger average absolute difference between the sample mean and the population mean. In accordance with the law of large numbers: on the average, larger samples yield more accurate estimates of the population mean than do smaller samples.

SAMPLE (N = 1)	$\bar{X}$	$|\bar{X} - \mu|$	SAMPLE (N = 2)	$\bar{X}$	$|\bar{X} - \mu|$
{0}	0	5	{0, 0}	0	5
{5}	5	0	{0, 5}	2.5	2.5
{10}	10	5	{5, 0}	2.5	2.5
			{0, 10}	5	0
			{0, 10}	5	0
			{5, 5}	5	0
			{5, 10}	7.5	2.5
			{10, 5}	7.5	2.5
			{10, 10}	10	5
Averages	5	3.33		5	2.22

7. $k = 16$ sample means; $N = 2$; $\mu = 5.00$; $\sigma^2 = 5.00$; $\sigma_{\bar{X}}^2 = \sigma^2/N = 5.00/2 = 2.500 = \Sigma(\bar{X} - \mu)^2/k = 40/16$. The variance of the sampling distribution equals the variance of the population divided by the sample size. The larger the sample size, the smaller the variance of the sampling distribution.

SAMPLES ($N = 2$)	$\bar{X}$	$\bar{X} - \mu$	$(\bar{X} - \mu)^2$
{2, 2}	2	−3	9
{2, 4}	3	−2	4
{4, 2}	3	−2	4
{2, 6}	4	−1	1
{6, 2}	4	−1	1
{2, 8}	5	0	0
{8, 2}	5	0	0
{4, 4}	4	−1	1
{4, 6}	5	0	0
{6, 4}	5	0	0
{4, 8}	6	1	1
{8, 4}	6	1	1
{6, 6}	6	1	1
{6, 8}	7	2	4
{8, 6}	7	2	4
{8, 8}	8	3	9
Sums	80	0	40
Averages	5	0	2.5

9. The expected value of the sample mean is the population mean: expected $\bar{X} = \mu = 6.5$.

11. Standard error of the mean $= 15/\sqrt{225} = 15/15 = 1.00$; $z = (103 - 100)/1 = 3.00$; probability (that $z \geq 3.00) = .0013$

13. Standard error of the mean $= 15/\sqrt{400} = 15/20 = 0.75$; $z = (101.47 - 100)/0.75 = 1.96$; $z = (98.53 - 100)/0.75 = -1.96$; probability (that $z < -1.96$ or $z > 1.96) = .0250 + .0250 = .05$. The key is in the wording: 98.53 is as extreme as 101.47 relative to a mean of 100.

15. Confidence coefficient: the z value in the standard normal distribution that corresponds to the desired level of confidence in interval estimation. Point estimation: use a single value (usually the sample mean) to estimate the population mean. Interval estimation: establish a range of values—called the confidence interval—that is likely to include the population mean. Confidence interval: a range of values that is likely to include the population mean. Bound on the error of estimation: the likely maximum absolute difference between the sample and population means.

17. Standard error of the mean $= \sigma_{\bar{X}} = \sigma/\sqrt{N} = 100/\sqrt{400} = 100/20 = 5.00$. The 99% confidence coefficient is $z = 2.58$. Confidence interval $= \bar{X} \pm z\sigma_{\bar{X}} = 570 \pm 2.58 \cdot 5 = 570 \pm 12.90 = [557.10$ to $582.90]$. Yes, the honors program would probably succeed. We are very confident that μ is 550 or higher.

19. Standard error of the proportion $= \sigma_P = \sqrt{pq/N} = \sqrt{(.5)(.5)/900} = \sqrt{.25/900} = \sqrt{.000278} = .0167$. The point estimate is .30. In terms of percent, the point estimate is 30%. The 99% confidence coefficient is $z = 2.58$. Confidence interval $= P \pm z\sigma_P = .30 \pm 2.58 \cdot .0167 = .30 \pm .043 = [.257$ to $.343]$. Approximately 30% of the prison population will exhibit recidivism. Taking sampling error into account, we are very confident that the actual recidivism rate in the population is not likely to be lower than 25.7% or higher than 34.3%.

21. The principal's conclusion is not necessarily accurate and is not appropriately phrased. The confidence level refers to the percent of such surveys that will yield a confidence interval containing the population parameter. In any one parameter estimation, the confidence interval will either include or exclude the population parameter. An appropriate conclusion is: I am reasonably confident that the students study at least 3.8 hours per week, but not more than 6.2 hours.

23. The conclusion is not quite appropriate in one sense but could be considered correct in another sense. The 99% confidence level refers to the percent of confidence intervals that will contain the population parameter. *In this particular analysis, the confidence interval either does or does not contain the parameter.* However, the researcher's conclusion should be considered correct at a theoretical level. That is, there is a .99 probability that 99% confidence intervals, in general, bracket the parameters of interest. A better conclusion would be that we are very confident that at least 73% of registered voters desire a more fiscally conservative government.

25. When $N = 64, \sigma_{\bar{x}} = 4/\sqrt{64} = 0.5$; when $N = 256, \sigma_{\bar{x}} = 4/\sqrt{256} = 0.25$; when $N = 1024, \sigma_{\bar{x}} = 4/\sqrt{1024} = 0.125$. Quadrupling sample size decreases the standard error of the mean by $1/2$ and increases the precision of the estimation by a factor of 2. In general, increasing sample size by a factor of x improves the precision of estimation by a factor equal to the square root of x.

CHAPTER 9

1. Statistical significance: an empirical outcome is likely to be repeatable in replications of an investigation; an outcome that is not likely to be a chance event. Null hypothesis: the hypothesis of chance; the hypothesis you test in a test of significance. Alternative hypothesis: the hypothesis that asserts that the outcome of a study will differ from chance expectations; the hypothesis that competes against the null hypothesis. Sampling distribution: a distribution of chance outcomes. Nondirectional test: a test of significance that evaluates both positive and negative differences between means or proportions. Statistically significant result: an empirical outcome that is a rare event in a distribution of chance events.

3. Null hypothesis: the population cure rate is 60%. Alternative hypothesis: the population cure rate differs from 60%. H_0: $p = .60$ and H_1: $p \neq .60$. A nondirectional hypothesis is appropriate because a difference in either direction would be meaningful. That is, it would be of

theoretical and practical interest if the new therapy impedes recovery, just as it would be of interest if the new therapy helps recovery.

5. $\sigma_P = \sqrt{pq / N} = \sqrt{(.6)(.4) / 100} = \sqrt{.24 / 100} = \sqrt{.0024} = .049$; $z = (P - p)/\sigma_P = (.73 - .60)/.049 = .13/.049 = 2.65$. Since $|2.65|$ is at least as large as 1.96, the outcome is unlikely to be a chance event. Reject H_0. The cure rate with the new therapy differs reliably from the spontaneous remission rate.

7. Level of significance: the chance probability that you use to define significant events. Critical value: the theoretical z value that is associated with the level of significance in the standard normal distribution; the value that the computed z statistic must equal or exceed to be significant. Region of rejection: the extreme portion of the standard normal distribution that contains significant values of the test statistic. Type I error: rejection of a true null hypothesis. Type II error: failure to reject a false null hypothesis. Alpha level: level of significance; the probability of a Type I error. Beta: probability of a Type II error. Power: the probability of rejecting a false hypothesis. Mean-difference size: the difference between the population mean of the null distribution and the population mean of the population from which the sample data were drawn. Small mean-difference size: an inadequate difference between the sample mean and hypothesized population mean, which lowers the power of the statistical test.

9. (a) $p < .75$; (b) $p \neq .42$; (c) $\mu > 16.34$; (d) $\mu \neq 120$; (e) $\mu < 0$

11. In a nondirectional test with $\alpha = .001$, $z_{crit} = 3.30$. Since the computed z of 2.28 is less than the critical value, you should not reject the null hypothesis.

13. You should use a directional test because only a positive difference (that is, exceeding the 7-minute standard) is of any interest in the regulation check. The attorney general doesn't care if the troopers run significantly faster times, only that they not significantly exceed 7 minutes. H_0: $\mu \leq 7$ minutes and H_1: $\mu > 7$ minutes.

15. According to the discretionary school, you should use a directional test any time you have enough information, from either theory or prior research, to clearly formulate a directional hypothesis. With this model, most tests of hypotheses in the behavioral sciences would be one-tailed. In contrast, the conservative school contends that a directional test should be used only if one of the following conditions is true: (a) just one particular direction of difference between the obtained mean and the theoretical mean has any practical value (for example, a new medicine is designed to make people better, not sicker); or (b) a researcher's theory specifically predicts either a positive or a negative outcome, and the opposite outcome would be theoretically meaningless to all researchers in that field. With this model, most tests of hypotheses in the behavioral sciences would be two tailed.

17. The sociologist probably should conduct a nondirectional test because the outcome would likely be of theoretical interest even if culturally discordant couples were actually happier than culturally concordant couples (although that outcome would contradict the sociologist's theory). Therefore, H_0: $\mu = 77$ and H_1: $\mu \neq 77$.

19.

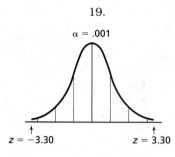
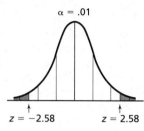
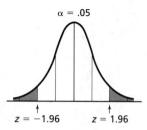

21. Probability of failing to reject $H_0 = 1 - \alpha = 1 - .10 = .90$. This decision would be correct.

23. A Type II error can occur only if the null hypothesis is false. When the null hypothesis is true, the probability of a Type II error is zero (it can't happen).

25. (1) State the problem and nature of the data: A sample of 36 culturally discordant couples has a mean score of 75 on marital happiness. Is this mean reliably different from a theoretical value of 77? (2) State the statistical hypotheses: $H_0: \mu = 77$ and $H_1: \mu \neq 77$. (3) Choose a level of significance: $\alpha = .05$. (4) Specify the test statistic: We will do a z test, where $z = (\overline{X} - \mu)/\sigma_{\overline{x}}$. (5) Determine the critical value needed for significance: $z_{crit} = 1.96$. (6) State the decision rule: If $|z| \geq 1.96$, reject H_0. (7) Compute the test statistic: From Question 17, $\sigma = 8$. The standard error of the mean $= \sigma_{\overline{x}} = \sigma/\sqrt{N} = 8/\sqrt{36} = 8/6 = 1.33$; so $z = (75 - 77)/1.33 = -2/1.33 = -1.50$. (8) Consult the decision rule and make a decision: Since the absolute value of the computed z statistic is smaller than z_{crit}, fail to reject the null hypothesis. Culturally discordant couples do not differ from the general population on marital happiness.

CHAPTER 10

1. Unbiased variance estimator: a modified kind of sample variance, obtained by dividing by $N - 1$ rather than N, that is equal to the population variance on the average. Degrees of freedom: the number of independent observations in the sample minus the number of parameters estimated in the formula. Estimated standard error: the denominator of the t ratio—calculated using s^2 rather than σ^2—that is equal to the standard error of the mean on the average.

3. To test a hypothesis about a mean, you need to form a ratio of a mean difference to the standard error of the mean; this ratio is called the test statistic and takes the form of a z score. Such a ratio enables you to locate the z score in a normal distribution of chance z scores and evaluate the probability that it has been randomly selected from the chance distribution. But calculation of the standard error requires knowledge of the population variance (or standard deviation). When that quantity is not available, you must estimate it with s^2, the unbiased estimator or σ^2. However, s^2 is a random variable, like the sample mean. Consequently, the test statistic will now have two

random variables contributing to it: the sample mean in the numerator and s^2 in the denominator. Since the standard normal distribution is based on the assumption that the sample mean is the only random variable that contributes to its values, it is no longer a valid model of chance against which to evaluate the test statistic. Under such circumstances, the t distribution is an appropriate model of chance because it takes both random variables into account. Accordingly, the new test statistic is called the t ratio, and it is evaluated against a t distribution.

5. Degrees of freedom in a t test of a mean equals $N - 1$. The variance of a t distribution is equal to (degrees of freedom)/(degrees of freedom – 2).

7. (a) 2.262; (b) 2.457; (c) 2.042 (using drop-back rule); (d) 3.373 (using drop-back rule)

9. The average unbiased variance estimate is exactly equal to the population variance ($\sigma^2 = 5.00$). This index is a good estimate of the population variance.

SAMPLE		$\overline{X}$	ΣX	ΣX^2	s^2
Sample 1	{2, 2}	2	4	8	0
Sample 2	{2, 4}	3	6	20	2
Sample 3	{4, 2}	3	6	20	2
Sample 4	{2, 6}	4	8	40	8
Sample 5	{6, 2}	4	8	40	8
Sample 6	{2, 8}	5	10	68	18
Sample 7	{8, 2}	5	10	68	18
Sample 8	{4, 4}	4	8	32	0
Sample 9	{4, 6}	5	10	52	2
Sample 10	{6, 4}	5	10	52	2
Sample 11	{4, 8}	6	12	80	8
Sample 12	{8, 4}	6	12	80	8
Sample 13	{6, 6}	6	12	72	0
Sample 14	{6, 8}	7	14	100	2
Sample 15	{8, 6}	7	14	100	2
Sample 16	{8, 8}	8	16	128	0
Averages		$\mu = 5.00$			$\sigma^2 = 5.00$

11. $N = 13$; $\Sigma X = 1270$; $\Sigma X^2 = 134{,}040$; $\overline{X} = 97.69$; $s^2 = 830.90$

13. $N = 10$; $\Sigma X = 79.10$; $\Sigma X^2 = 660.11$; $\overline{X} = 7.91$; $s^2 = 3.83$

15. $N = 20$; $\Sigma X = 5592$; $\Sigma X^2 = 1{,}653{,}270$; $\overline{X} = 279.60$; $s^2 = 4723.52$

17. (1) A psychobiologist hypothesizes that the diastolic blood pressure of Type A persons differs from average; the psychobiologist takes the blood pressure of 22 Type A men and finds that their mean diastolic blood pressure is 88. (2) H_0: $\mu = 80$ and H_1: $\mu \neq 80$. (3) $\alpha = .001$. (4) Two-tailed t test with 21 degrees of freedom. (5) $t_{crit} = 3.819$. (6) If computed $|t| \geq 3.819$, reject the null hypothesis. (7) $s_{\overline{X}} = \sqrt{16/22} = 0.85$; $t = (88 - 80)/0.85 = 9.41$. (8) Reject the null hypothesis; the diastolic blood pressure of the Type A men is significantly different from 80.

19. With degrees of freedom = $N - 2 = 19$, $t_{crit} = \pm2.093$. Since computed $t = (.40\sqrt{21 - 2})/\sqrt{1 - 40^2} = 1.90$, the correlation is not significant.

21. With degrees of freedom = $N - 2 = 8$, $t_{crit} = \pm2.306$. Since computed $t = (.60\sqrt{10 - 2})/\sqrt{1 - 60^2} = 2.12$, the correlation is not significant. The conclusion is correct.

23. $s_{\bar{x}} = \sqrt{4.7/500} = 0.097$; $t_{crit} = 2.617$; confidence interval = $28 \pm 2.617(0.097) = 28 \pm 0.254 = [27.746$ to $28.254]$. The population mean on abstract reasoning is very unlikely to be lower than 27.746 or higher than 28.254.

25. $s_{\bar{x}} = \sqrt{9.6/64} = 0.387$; $t_{crit} = 2.000$; confidence interval = $28 \pm 2.000(0.387) = 28 \pm 0.774 = [27.226$ to $28.774]$. It is unlikely that the actors' population mean on the psychopathic deviate scale is lower than 27.226 or higher than 28.774.

CHAPTER 11

1. Completely randomized two-group design: each subject is randomly assigned to one of two conditions; the independent variable is manipulated by treating the two groups differently. Sampling distribution of the mean difference: a distribution of all possible sample-mean differences from a particular population, where N_1 and N_2 are of fixed sizes for all the mean differences. Standard error of the mean difference: the standard deviation of a sampling distribution of mean differences. Pooled estimator of the population variance: a weighted average of the unbiased variance estimators in a two-group design. Static-group comparison design: a preexisting group of subjects with a particular trait is compared to another group of subjects lacking the trait.

3. Null: the mean level of stress chemical is the same in both populations; H_0: $\mu_1 = \mu_2$. Alternative: the mean level of stress chemical differs between the populations; H_1: $\mu_1 \neq \mu_2$.

5. The sampling distribution of the mean difference is made up of differences between pairs of sample means; its average value is zero. The average value of zero results from the fact that the positive mean differences perfectly cancel the negative mean differences when only random variation exists. In contrast, the sampling distribution of the mean consists of sample means and has an average value of μ.

7. I am interested in finding out which of the following new acquaintances one comes to like more: the acquaintance who is unfriendly upon being introduced but who gradually becomes friendlier, or the acquaintance who is very friendly upon being introduced but who gradually becomes unfriendly. To investigate this question, I would randomly assign half of my subjects to the "bad start/good end" condition and half to the "good start/bad end" condition. Each subject would be individually introduced to a paid actor who would carry out the script to which the subject has been randomly assigned. At the end of the subject's interaction, I would use some pretext to get him or her to evaluate the new acquaintance.

9. $\sigma_{\bar{x}} = \sqrt{490/10} = 7.00$; $\sigma_{\bar{x}_1 - \bar{x}_2} = \sqrt{(490/10) + (490/10)} = 9.90$. The result is consistent with expectations created by the text. The larger

standard error of the mean difference is in line with the assertion that there is more variability among pairs of sample means than between sample means and the population mean.

11. Null: the self-rating mean of the phobic population is equal to the self-rating mean of the nonclinical population; H_0: $\mu_1 = \mu_2$. Alternative: the self-rating mean of the phobic population is different from the self-rating mean of the nonclinical population; H_1: $\mu_1 \neq \mu_2$. With $\alpha = .05$ and degrees of freedom = 59, $t_{crit} = 2.021$ (using the drop-back rule). Decision rule: Reject H_0 if the computed $|t| \geq t_{crit}$.

13. $t = (12.5 - 9.4)/0.81 = 3.83$. Since $|t| > 2.021$, reject the null hypothesis. The social phobics have significantly lower self-ratings on public speaking competence.

15. $s_p^2 = [(45)(25.63) + (45)(40.41)]/(45 + 45) = 33.02$; $s_{\bar{X}_1 - \bar{X}_2} = \sqrt{33.02(1/46 + 1/46)} = 1.20$; $t = (35.39 - 33.67)/1.20 = 1.43$. Since the computed $t < 2.000$, do not reject the null hypothesis. The two groups were equivalent on exam performance at the start of the course.

17. Homogeneity of variance assumption: the assumption that the two populations implicated in a t test have equal variances. Correlated-samples t test: t test performed on a column of difference scores, comparing the mean of those scores to a hypothesized difference-score mean of the population. Repeated-measures design: a research design in which each subject serves in all conditions and is, thus, repeatedly measured. Difference scores: $X_1 - X_2$ values; each subject's X_2 score is subtracted from his or her X_1 score. Sample mean of the difference scores: mean of the sample of $X_1 - X_2$ values. Randomized matched-groups design: research design involving several matched pairs of subjects, in which one member of each pair is randomly assigned to the experimental condition and the other to the control condition. Error term: the denominator of the test statistic, which serves as an index of random error.

19. The data configuration in a correlated-samples research design consists of a single column of difference scores, whereas the data configuration in an independent-samples design consists of two columns of raw scores.

21. It is a correlated-samples investigation. The male–female partners are matched pairs. Performances within a pair are generally more similar than performances between pairs. Furthermore, the columns of data appear to be literally correlated. Null: the population mean difference between the female and male driving test score is zero; H_0: $\mu_D = 0$. Alternative: the population mean difference between the female and male driving test score is different from zero; H_1: $\mu_D \neq 0$. With degrees of freedom = 11, $t_{crit} = 2.201$. If $|t| \geq 2.201$, reject the null hypothesis.

23. (1) A pediatric nurse wants to find out whether the age at which children begin to walk can be influenced by special training. Eight pairs of identical twin girls under the age of 4 months serve as subjects. One member of each twin pair is randomly assigned to a special training condition and the co-twin goes to the control condition. The dependent variable is the age at which the children begin to walk. (2) H_0: $\mu_D = 0$ and H_1: $\mu_D \neq 0$. (3) $\alpha = .05$. (4) Use a correlated-samples t test with 7 degrees of freedom. (5) $t_{crit} = 2.365$. (6) If $|t| \geq 2.365$, reject the null hypothesis. (7) $s_{\bar{D}} = \sqrt{3.70/8} = 0.68$; $t = -0.63/0.68 = -0.93$. (8) Do

not reject the null hypothesis. The special training has no effect on age at which walking begins.

TWIN PAIR	SPECIAL	CONTROL	D	D²
1	12	12	0	0
2	14	16	−2	4
3	15	14	1	1
4	12	11	1	1
5	16	14	2	4
6	15	18	−3	9
7	13	16	−3	9
8	10	11	−1	1
Sum	107.00	112.00	−5.00	29.00
N	8	8	8	
Mean	13.38	14.00	−0.63	
s^2	3.98	6.57	3.70	

25. When its assumptions are met, the correlated-samples *t* test has relatively high power because it uses a smaller denominator, or error term. The smaller denominator results from computing the difference scores, which removes random variation caused by individual differences among subjects. The smaller denominator makes the *t* statistic larger, thereby increasing the chance of rejecting the null hypothesis.

CHAPTER 12

1. Completely randomized independent-samples design: multiple-condition research design in which each subject is randomly assigned to one, and only one, treatment. Analysis of variance: inferential statistical method that simultaneously assesses the significance of all differences among a set of treatment means; partitions variance into between-treatments and within-treatments components. Testwise error rate: the probability of a Type I error in a particular two-mean test of significance. Familywise error rate: the probability of at least one Type I error in the set of all significance tests carried out on a set of data. One-factor ANOVA: an analysis of variance procedure applied to levels of a single independent variable.

3. The testwise error rate is the probability of committing a Type I error in any particular significance test; the familywise error rate is the probability of committing at least one Type I error when you conduct several significance tests on the same set of data. The familywise error rate is inflated by a series of *t* tests on the same data set. The formula for the familywise error likelihood is $1 - (1 - \alpha)^C$, where *C* is the number of tests to be conducted with a data set.

5. *F* ratio: the ratio of between-treatments variance to within-treatments variance. Grand mean: the mean of all the data in a study. Mean square: a sum of squares divided by the corresponding degrees of freedom; a variance. N: the number of observations in any particular treatment. N_{tot}: the total number of observations in a study. *k*: the number of treatment groups in the study.

7. Analysis of variance summary table: a device for both displaying the results of an ANOVA and guiding the sequence of operations that produces those results. Total sum of squares: the sum of all squared deviations between raw scores and the grand mean. Within-treatments sum of squares: the sum of squared deviations between raw scores and their respective treatment means. Between-treatments sum of squares: weighted sum of squared deviations between treatment means and the grand mean, where each squared deviation is weighted by the number of observations in that treatment.

9. Worksheet: uses a sequence of columns in a table to represent the successive steps in a formula. Pairwise comparison: a test of the significance of the difference between two means at a time. Tukey's honestly significant difference: the smallest significant mean difference in a set of k means, conditional upon a constant familywise error rate. A priori comparisons: pairwise comparison tests that assess hypotheses that predated data collection; planned comparisons. Post hoc comparisons: pairwise comparison tests that are carried out only after one obtains a significant omnibus F statistic; often involve unplanned comparisons. Eta squared: the ratio of SS_B to SS_{tot}; proportion of total variation in the sample that is attributable to treatment effects.

11. With $df_B = 4$, $df_W = 15$, and $\alpha = .05$, $F_{crit} = 3.06$. Since the computed $F \geq F_{crit}$, the null hypothesis is rejected. The dose of vitamin C did affect the incidence of colds, $F(4, 15) = 3.92$, $p < .05$.

SOURCE OF VARIATION	SUMS OF SQUARES	DEGREES OF FREEDOM	MEAN SQUARES	F
Between treatments	27.20	4	6.80	3.92
Within treatments	26.00	15	1.73	
Total	53.20	19		

13. With $df_B = 2$, $df_W = 12$, and $\alpha = .05$, $F_{crit} = 3.89$. Since the computed $F < F_{crit}$, the null hypothesis is not rejected. There is no evidence in these samples that selective breeding affects emotionality, $F(2, 12) = 0.28$, $p > .05$.

SOURCE OF VARIATION	SUMS OF SQUARES	DEGREES OF FREEDOM	MEAN SQUARES	F
Between treatments	3.33	2	1.67	0.28
Within treatments	72.00	12	6.00	
Total	75.33	14		

15. With $df_B = 5$, $df_W = 24$, and $\alpha = .05$, $F_{crit} = 2.62$. Since the computed $F < F_{crit}$, the null hypothesis is not rejected. There is no evidence in these samples that Zodiak sign is related to ESP ability, $F(5, 24) = 1.51$, $p > .05$.

SOURCE OF VARIATION	SUMS OF SQUARES	DEGREES OF FREEDOM	MEAN SQUARES	F
Between treatments	124.20	5	24.84	1.51
Within treatments	394.00	24	16.42	
Total	518.20	29		

17. (a) 2.95; (b) 4.02; (c) 2.29; (d) 2.99

19. With $df_B = 3$, $df_W = 12$, and $\alpha = .05$, $F_{crit} = 3.49$. Since the computed $F \geq F_{crit}$, the null hypothesis is rejected. Self-efficacy scores vary significantly with diagnostic category, $F(3, 12) = 172.45$, $p < .05$.

SOURCE OF VARIATION	SUMS OF SQUARES	DEGREES OF FREEDOM	MEAN SQUARES	F
Between treatments	5691.00	3	1897.00	172.45
Within treatments	132.00	12	11.00	
Total	5823.00	15		

Scheffe's $S = \sqrt{(3)(3.49)(11)(\frac{1}{4} + \frac{1}{4})} = 7.59$. The only mean difference that is less than the critical S value is that between mania and paranoia. All other mean differences are significant.

MEANS			
DEPRESSION	MANIA	SCHIZOPHRENIA	PARANOIA
40.00	87.00	57.00	81.00

21. $df_{tot} = 69$; $df_W = 63$; $df_B = 6$

23.

SOURCE OF VARIATION	SUMS OF SQUARES	DEGREES OF FREEDOM	MEAN SQUARES	F
Between treatments	80.00	2	40.00	24.00
Within treatments	20.00	12	1.67	
Total	100.00	14		

25. Tukey's $HSD = 4.45\sqrt{4.00/11} = 2.68$. Mean 2 differs significantly from mean 3; there are no other significant mean differences.

CHAPTER 13

1. Factorial design: research design in which all levels of one independent variable are combined with all levels of a second independent variable. Factorial experiment: a factorial design in which subjects are randomly assigned to the various levels of at least one of the independent variables. Completely randomized factorial experiment: factorial design in which subjects are randomly assigned to all levels of both independent variables and each subject serves in only one treatment combination.

3. As in one-way ANOVA, two-way ANOVA partitions total variance into between-treatments and within-treatments variance. Two-way ANOVA further divides the between-treatments variance into three components: the main effect of factor A, the main effect of factor B, and the effect of the interaction of A and B.

5. Since all effects have the same degrees of freedom and $\alpha = .05$, F_{crit} for all effects is 4.49. There was no main effect of obesity (factor A), but there was a significant main effect of stimulus type (factor B) and a significant interaction of obesity and stimulus type. Obese subjects made many more errors when a food image was present than when a nonfood image was present, but normal-weight subjects tended to behave in the opposite way.

CELL MEANS

	FOOD STIMULUS	NON-FOOD STIMULUS
OBESE	9.00	4.00
NORMAL	5.00	6.00

SOURCE	SUMS OF SQUARES	DEGREES OF FREEDOM	MEAN SQUARES	F
Factor A	5.00	1	5.00	2.00
Factor B	20.00	1	20.00	8.00
A × B interaction	45.00	1	45.00	18.00
Error term	40.00	16	2.50	
Total	110.00	19		

7. Since all effects have the same degrees of freedom and $\alpha = .01$, F_{crit} for all effects is 9.33. There was a main effect of rapport (factor A), but there was no main effect of method of hypnosis (factor B) and no interaction of rapport and method. High rapport increases hypnotic responsiveness, regardless of the method of hypnosis used.

CELL MEANS

	STRONG METHOD	SEDUCTIVE METHOD
HIGH RAPPORT	15.00	15.00
LOW RAPPORT	10.00	9.00

SOURCE	SUMS OF SQUARES	DEGREES OF FREEDOM	MEAN SQUARES	F
Factor A	121.00	1	121.00	13.44
Factor B	1.00	1	1.00	0.11
A × B interaction	1.00	1	1.00	0.11
Error term	108.00	12	9.00	
Total	231.00	15		

9. (a) Null: the mean number of aggressive responses will not vary as a function of temperature; H_0: $\mu_{A1} = \mu_{A2} = \mu_{A3}$. Alternative: the mean number of aggressive responses will vary with temperature; H_1: not all means will be the same. Null: the mean number of aggressive responses will not vary as a function of level of crowding; H_0: $\mu_{B1} = \mu_{B2} = \mu_{B3}$. Alternative: the mean number of aggressive responses will vary with level of crowding; H_1: not all means will be the same. Null: the mean differences among the levels of the temperature variable will be the same at all levels of crowding; H_0: mean differences at B1 = mean differences at B2 = mean differences at B3. Alternative: the mean differences among the levels of the temperature variable will vary with level of crowding; H_1: not all mean differences will be the same. (b) Proportion of variance accounted for by temperature = 74.00/187.00 = .3957. Proportion of variance accounted for by crowding = 74.00/187.00 = .3957. Proportion of variance accounted for by the interaction = 16.00/187 = .0856.
(c)

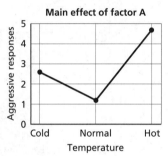

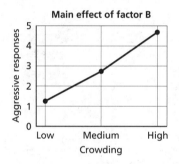

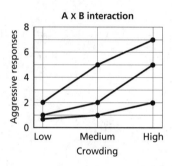

11. Tukey's $HSD = 4.37\sqrt{9.33/5} = 5.97$. Heavy metal music produces a significantly higher heart rate than soft rock when the wall color is either blue or red, but not when the walls are white. When soft rock is being played, blue walls lower heart rate, relative to other colors, but have no effect when heavy metal music is being played. Red significantly increases heart rate only when heavy metal music is playing.

13. Factor A is not significant. Factor B is not significant. The interaction is significant at the .01 level.

SOURCE	SUMS OF SQUARES	DEGREES OF FREEDOM	MEAN SQUARES	F
Factor A	85.00	5	17.00	1.70
Factor B	8.00	4	2.00	0.20
A × B interaction	640.00	20	32.00	3.20
Error term	5700.00	570	10.00	
Total	6433.00	599		

15. Proportion of variance accounted for by factor A = 30.00/147.00 = .2041. Proportion of variance accounted for by factor B = 20.00/147.00 = .1361. Proportion of variance accounted for by the A × B interaction = 55.00/147.00 = .3741.

17. The patterns of means and graphical trends suggest a main effect of factor A, a main effect of factor B, and no A × B interaction.

CELL MEANS

	B1	B2	FACTOR A MEANS
A1	70.00	30.00	50.00
A2	40.00	0.00	20.00
FACTOR B MEANS	55.00	15.00	

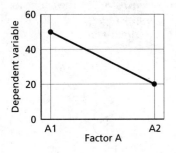 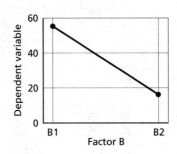 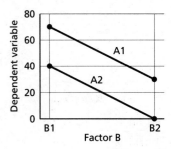

CHAPTER 14

1. Both two-way ANOVA and repeated-measures ANOVA involve calculation of SS_A, SS_B, and $SS_{A \times B}$. But repeated-measures ANOVA has only one true independent variable. The second "independent" variable is actually a subjects factor, the effect of which is removed in the analysis but not evaluated for significance. Also, unlike two-way ANOVA, repeated-measures ANOVA does not have a within-treatments error term. Rather the interaction of the independent variable with subjects is used as the error term in the denominator of the F ratio.

3. Repeated-measures ANOVA assumes that there is no interaction between the subjects factor and the independent variable. Yet there often seems to be a tiny subjects × treatments interaction even when it is reasonable to assume there is none. This small apparent interaction actually represents the random effects of experimental error influences, such as distractions, unreliable measurement, or sporadic recording errors. These influences are clearly a random-error component and, hence, $MS_{subjects \times treatments}$ is an index of random error. Since systematic individual differences have been statistically removed from the numerator of the F ratio in the ANOVA, that numerator consists of only experimental error + a possible effect of the independent variable. Since $MS_{subjects \times treatments}$ represents purely experimental error, it makes sense to use it as the denominator in the computation of F. Under these circumstances, a large F indicates a reliable impact of the independent variable.

5. (a) Null: mean attractiveness ratings will be the same at all levels of pupil size; H_0: $\mu_1 = \mu_2 = \mu_3$. Alternative: mean attractiveness ratings will vary at different levels of pupil size; H_1: not all μ values will be the same. (b) Proportion of variance accounted for = 29.25/44.00 = .6648. (c) Tukey's $HSD = 4.89\sqrt{0.101/8} = 0.55$. All the treatment means differ significantly from one another.

7. With $\alpha = .05$, $F_{crit} = 3.49$. There is a significant effect of spacing. Greater spacing is associated with higher recall scores.

SOURCE	SUMS OF SQUARES	DEGREES OF FREEDOM	MEAN SQUARES	F
Between treatments	250.15	3	83.38	217.52
Between subjects	97.00	4		
Subjects × treatments interaction	4.60	12	0.383	
Total	351.75	19		

9. The F ratio is not significant.

SOURCE	SUMS OF SQUARES	DEGREES OF FREEDOM	MEAN SQUARES	F
Between treatments	35.00	5	7.00	2.33
Between subjects	100.00	9		
Subjects × treatments interaction	135.00	45	3.00	
Total	270.00	59		

11. Advantage: relative to an independent-samples design, the repeated-measures design requires fewer subjects. Disadvantage: because of possible carryover effects, the repeated-measures design cannot be used to study certain questions.

13. (a) Null: mean self-esteem will be the same in all years of college; H_0: $\mu_1 = \mu_2 = \mu_3 = \mu_4$. Alternative: mean self-esteem will be different across years of college; H_1: not all μ values will be the same. (b) Proportion of variance accounted for = 310.67/1125.78 = .2760. (c) Tukey's $HSD = 4.91\sqrt{1.85/10} = 2.11$. All the treatment means differ significantly from one another.

15 (a) $N = 10$, $k = 4$; (b) $N = 7$, $k = 5$; (c) $N = 20$, $k = 3$

CHAPTER 15

1. Parametric tests: tests of statistical significance that assess the equivalence of specific parameters. Nonparametric tests: tests of statistical significance designed to assess the equivalence of whole distributions. Robustness: the extent to which a statistical test produces valid results even when one of its assumptions is not met by a data set. Frequency count: the number of outcomes that fall into a particular

category. Chi-square test: a test of significance that compares observed frequency counts to theoretical frequencies. Goodness-of-fit test: a type of chi-square test that assesses the degree to which a distribution of frequency counts deviates from those specified by a theoretical model. Observed frequency: the actual number of outcomes falling into a particular category.

3. H_0: distribution 1 = distribution 2 and H_1: distribution 1 $\neq$ distribution 2; with $\alpha = .05$ and $df = 1$, $\chi^2_{crit} = 3.84$; reject the null hypothesis; the 1990s percent admission rate differs from the 1980s rate.

CELL	f_o	f_e	$(f_o - f_e)$	$(f_o - f_e)^2$	$\dfrac{(f_o - f_e)^2}{f_e}$
a	108	126.00	−18.00	324.00	2.570
b	72	54.00	18.00	324.00	6.000
					$\chi^2 = 8.570$

5. H_0: distribution 1 = distribution 2 and H_1: distribution 1 $\neq$ distribution 2; with $\alpha = .05$ and $df = 2$, $\chi^2_{crit} = 5.99$; reject the null hypothesis; the rats tended to be biased toward the left box.

CELL	f_o	f_e	$(f_o - f_e)$	$(f_o - f_e)^2$	$\dfrac{(f_o - f_e)^2}{f_e}$
a	240	200.00	40.00	1600.00	8.000
b	150	200.00	−50.00	2500.00	12.500
c	210	200.00	10.00	100.00	0.500
					$\chi^2 = 21.000$

7. H_0: distribution 1 = distribution 2 and H_1: distribution 1 $\neq$ distribution 2; with $\alpha = .05$ and $df = 3$, $\chi^2_{crit} = 7.82$; reject the null hypothesis; local viewers prefer some evening news programs more than the others.

CELL	f_o	f_e	$(f_o - f_e)$	$(f_o - f_e)^2$	$\dfrac{(f_o - f_e)^2}{f_e}$
a	220	250.00	−30.00	900.00	3.600
b	200	250.00	−50.00	2500.00	10.000
c	300	250.00	50.00	2500.00	10.000
d	280	250.00	30.00	900.00	3.600
					$\chi^2 = 27.200$

9. H_0: there is no correlation between category of schizophrenia and size of brain ventricles; H_1: there is a correlation between category of schizophrenia and size of brain ventricles; with $\alpha = .05$ and $df = 3$, $\chi^2_{crit} = 7.82$; reject the null hypothesis; there is an association between category of schizophrenia and size of brain ventricles.

CELL	f_o	f_e	$(f_o - f_e)$	$(f_o - f_e)^2$	$\dfrac{(f_o - f_e)^2}{f_e}$
a	10	17.48	−7.48	55.95	3.200
b	6	13.11	−7.11	50.55	3.860
c	18	18.36	−0.36	0.13	0.010
d	32	17.05	14.95	223.50	13.110
e	30	22.52	7.48	55.95	2.480
f	24	16.89	7.11	50.55	2.990
g	24	23.64	0.36	0.13	0.005
h	7	21.95	−14.95	223.50	10.180
					$\chi^2 = 35.835$

11. H_0: there is no correlation between gender and book carrying style; H_1: there is a correlation between gender and book carrying style; with $\alpha = .05$ and $df = 2$, $\chi^2_{crit} = 5.99$; reject the null hypothesis; there is an association between gender and book carrying style.

CELL	f_o	f_e	$(f_o - f_e)$	$(f_o - f_e)^2$	$\dfrac{(f_o - f_e)^2}{f_e}$
a	24	49.60	−25.60	655.36	13.210
b	70	46.40	23.60	556.96	12.000
c	6	4.00	2.00	4.00	1.000
d	100	74.40	25.60	655.36	8.810
e	46	69.60	−23.60	556.96	8.000
f	4	6.00	−2.00	4.00	0.670
					$\chi^2 = 43.690$

13. H_0: there is no correlation between diagnostic category and Luscher profiles; H_1: there is a correlation between diagnostic category and Luscher profiles; with $\alpha = .05$ and $df = 4$, $\chi^2_{crit} = 9.49$; reject the null hypothesis; there is an association between diagnostic category and Luscher profiles.

CELL	f_o	f_e	$(f_o - f_e)$	$(f_o - f_e)^2$	$\dfrac{(f_o - f_e)^2}{f_e}$
a	4	19.33	−15.33	235.01	12.158
b	40	20.00	20.00	400.00	20.000
c	16	20.67	−4.67	21.81	1.055
d	12	19.33	−7.33	53.73	2.780
e	10	20.00	−10.00	100.00	5.000
f	38	20.67	17.33	300.33	14.530
g	42	19.33	22.67	513.93	26.587
h	10	20.00	−10.00	100.00	5.000
i	8	20.67	−12.67	160.53	7.766
					$\chi^2 = 94.876$

15. (a) $df = 30$, $\chi^2_{\text{crit}} = 50.89$; (b) $df = 3$, $\chi^2_{\text{crit}} = 11.34$; (c) $df = 8$, $\chi^2_{\text{crit}} = 15.51$; (d) $df = 2$, $\chi^2_{\text{crit}} = 5.99$; (e) $df = 10$, $\chi^2_{\text{crit}} = 18.31$

17. H_0: population median$_1$ = population median$_2$ and H_1: population median$_1 \neq$ population median$_2$. With $\alpha = .05$, $U_{\text{crit}} = 6$. $U_A = 8$; $U_B = 34$. Do not reject the null hypothesis. There is no evidence that using the correct mantra improves the effect of meditation.

ORDERED RAW SCORES BOTH GROUPS	RANKS OF SCORES BOTH GROUPS	GROUP IDENTIFICATION	RANKS FOR GROUP A	RANKS FOR GROUP B
2	1	B		1
3	2.5	B		2.5
3	2.5	B		2.5
4	4.5	B		4.5
4	4.5	B		4.5
5	6	A	6	
6	7	A	7	
7	9	A	9	
7	9	B		9
7	9	A	9	
8	11	A	11	
9	12	B		12
10	13	A	13	
		R	55	36

19. H_0: population median$_1$ = population median$_2$ and H_1: population median$_1 \neq$ population median$_2$. With $\alpha = .01$, $U_{\text{crit}} = 16$. $U_A = 32.5$; $U_B = 67.5$. Do not reject the null hypothesis. There is no evidence that the origin of the skulls is related to their preservation quality.

ORDERED RAW SCORES BOTH GROUPS	RANKS OF SCORES BOTH GROUPS	GROUP IDENTIFICATION	RANKS FOR GROUP A	RANKS FOR GROUP B
3	1	B		1
4	2.5	B		2.5
4	2.5	A	2.5	
5	4	B		4
6	5.5	B		5.5
6	5.5	A	5.5	
7	7	A	7	
8	9.5	B		9.5
8	9.5	B		9.5
8	9.5	A	9.5	
8	9.5	B		9.5
10	12	A	12	
11	13	B		13
12	14	A	14	

ORDERED RAW SCORES BOTH GROUPS	RANKS OF SCORES BOTH GROUPS	GROUP IDENTIFICATION	RANKS FOR GROUP A	RANKS FOR GROUP B
13	15.5	B		15.5
13	15.5	A	15.5	
17	17.5	B		17.5
17	17.5	A	17.5	
19	19	A	19	
20	20	A	20	
		R	122.5	87.5

21. $z_{crit} = \pm 1.96$; $\mu = 312.50$; $\sigma = 51.55$.; $z = -0.24$; do not reject H_0.
23. H_0: population $median_1$ = population $median_2$ and H_1: population $median_1 \neq$ population $median_2$. Discarding the zero differences, the effective $N = 6$. With $\alpha = .05$, $T_{crit} = 0$. $\Sigma R_+ = 21$; $\Sigma R_- = 0$; $T = 0$. Reject the null hypothesis. Attending finishing school affected the students' rated grace and poise.
25. H_0: population $median_1$ = population $median_2$ and H_1: population $median_1 \neq$ population $median_2$. Discarding the zero differences, the effective $N = 6$. With $\alpha = .05$, $T_{crit} = 0$. $\Sigma R_+ = 19.5$; $\Sigma R_- = 1.5$; $T = 1.5$. Do not reject the null hypothesis. There is no evidence that acupuncture had an impact on the rated severity of the rashes.

APPENDIX C COMPUTER APPLICATIONS OF STATISTICS

This appendix will present an overview of two generally available statistical software packages, *MINITAB* and *MYSTAT*. I will illustrate most of the statistical procedures with data sets from various chapters of this book. In each example, I will go over only the steps necessary to do a "basic essentials" analysis. Both *MINITAB* and *MYSTAT* have many extended features and capabilities that require more space than is available in this appendix. You should consult your software manual to learn how to use those advanced procedures.

My instructions and pointers will assume that the software package you are using has been installed on a hard disk, and that you are saving the data to a floppy disk in drive A: or B:. If you are running the **MINITAB** *or* **MYSTAT** *program from a floppy disk rather than a hard disk, you should study your software manual to find out how to install and start up the software.*

I will use the following conventions in this appendix:

NOTE this
instruction →
well

1. **Enter** refers to the Enter key.
2. **[Enter]** instructs you to press the Enter key after typing a software command.
3. **Esc** refers to the Esc (i.e., Escape) key.
4. **[Esc]** instructs you to press the Esc key.
5. ← refers to the Left Arrow key.
6. → refers to the Right Arrow key.
7. ↑ refers to the Up Arrow key.
8. ↓ refers to the Down Arrow key.
9. DOS refers to the Disk Operating System in your computer. When you are working in the DOS environment, you will see a "DOS prompt" that will look similar to this: C:\>. It is likely that you will be starting up your statistical software from the DOS prompt.

Also:

- In software command lines, words typed entirely in uppercase letters (e.g., SAVE) are essential instructions that must be typed just as they appear.
- In software command lines, words typed in lowercase letters are unessential plain-English qualifiers designed to make the intent of the command lines clearer to you. (For example, in "SET the data into C1," only SET and C1 are needed.) You may omit those qualifiers if you wish.
- Although not necessary, data file names will appear entirely in uppercase letters (e.g., HRSCORES).
- When files are saved or loaded, the file manipulation command will include the letter (A: or B:) of the floppy drive in which the data disk is mounted (e.g., SAVE 'B:HRSCORES').

- In MINITAB, variables are typed in a mix of uppercase and lowercase letters (e.g., 'Scores'), although they may be entered in entirely uppercase or lowercase, if you prefer.

1: GETTING STARTED

This section corresponds to Chapter 1 of this textbook. It will acquaint you with the operational structures of MINITAB and MYSTAT, basic data entry and editing commands, and file manipulation commands.

MINITAB

To start MINITAB from the DOS prompt, make sure that you are in the directory that contains the MINITAB program—usually C:\MINITAB. From the C:\MINITAB> prompt, type MINITAB and press [**Enter**]. Or, if you are running the program from within the Windows[1] environment, go to File Manager, click on the MINITAB folder, and then double-click on the MINITAB.EXE icon. You will see the greeting/copyright screen shown in Figure C1.1. Press **Enter** to move on.

Figure C1.1 *MINITAB Greeting Screen*

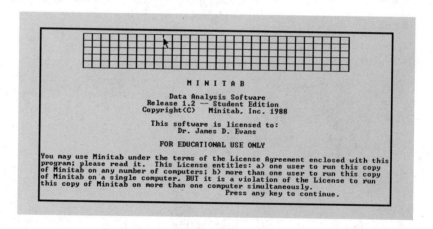

```
                    M I N I T A B
                 Data Analysis Software
               Release 1.2 -- Student Edition
               Copyright(C)  Minitab, Inc. 1988

                This software is licensed to:
                     Dr. James D. Evans

                   FOR EDUCATIONAL USE ONLY

  You may use Minitab under the terms of the License Agreement enclosed with this
  program; please read it.  This License entitles: a) one user to run this copy
  of Minitab on any number of computers; b) more than one user to run this copy
  of Minitab on a single computer, BUT it is a violation of the License to run
  this copy of Minitab on more than one computer simultaneously.
                                    Press any key to continue.
```

Structure of MINITAB Statistics programs usually have three operational modes:

1. An "input screen" for naming variables and inputting data
2. A "command screen" for issuing statistical analysis instructions
3. An "output screen" for displaying the results of the analysis

---

[1] Windows is a trademark of Microsoft Corporation.

In MINITAB these three modes are combined into a single screen, as shown in Figure C1.2. The screen does not clear or change appearance as you go from entering data to issuing software commands to displaying the results.

Figure C1.2 *MINITAB's All-Purpose Input, Command, and Output Screen*

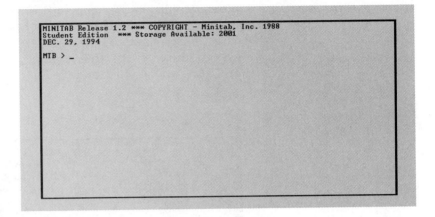

```
MINITAB Release 1.2 *** COPYRIGHT - Minitab, Inc. 1988
Student Edition  *** Storage Available: 2001
DEC. 29, 1994

MTB > _
```

Naming data sets   To practice data entry and editing, we will use the following scores made by 30 first-line managers on a test of human relations skills:

17	23	21	28	23	28	26	25	22	34
23	27	29	21	22	33	33	26	24	25
28	31	20	23	25	35	31	23	24	30

Although it is not necessary to name data sets in MINITAB, it is to your advantage to do so. Use the **NAME C1 'VarName' [Enter]** command, where 'VarName' is any label you wish to assign the variable represented by your data and C1 represents *column 1* in a data table. (If you had two sets of data in your study, you would also name C2, because column 2 would contain the second set of data.) *Note that the name of the data set must be bracketed by apostrophes*; otherwise, you'll get an error message. Also, variable names must not contain more than eight characters. See Figure C1.3. In this example, the command is NAME C1 'Scores'.

Data input   Assign data to a column (i.e., variable) using the **SET C1 [Enter]** command, as shown in Figure C1.3. Individual scores must be separated from one another by *spaces, commas, or carriage returns.*

When you reach the edge of your input screen, hit **Enter** to proceed with data entry on the next line. When you are finished inputting all the data in a column, press **Enter** and type **END [Enter]**.

You can use the **INFO [Enter]** command to display the column label(s), column name(s), and number of observations in the data set(s).

Saving your data   The command **SAVE 'B:FILENAME' [Enter]** will store the data file on the floppy disk in drive B:. 'FILENAME' can be any label that consists of eight or fewer letters and/or digits. (Don't leave any spaces.) Use **SAVE 'A:FILENAME' [Enter]** if your disk is in drive A:. In this example, I have named the file 'HRSCORES', and saved it in drive B:.

Figure C1.3 *Entering and Saving Data in MINITAB*

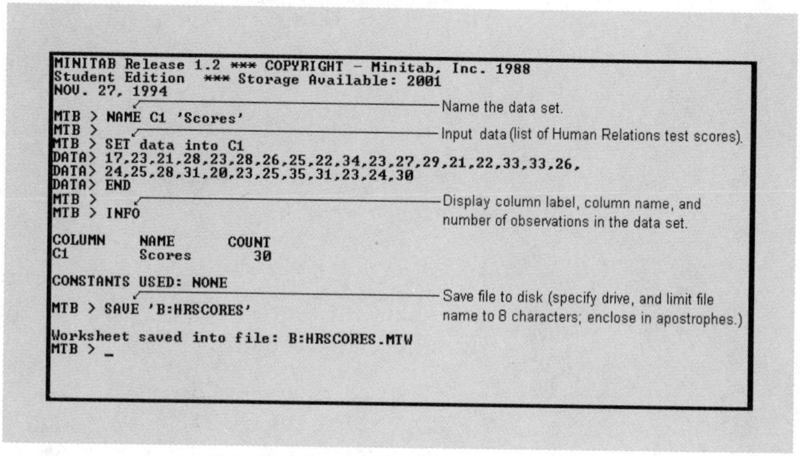

```
MINITAB Release 1.2 *** COPYRIGHT — Minitab, Inc. 1988
Student Edition  *** Storage Available: 2001
NOV. 27, 1994
MTB > NAME C1 'Scores'————————————————— Name the data set.
MTB >
MTB > SET data into C1 ————————————————— Input data (list of Human Relations test scores).
DATA> 17,23,21,28,23,28,26,25,22,34,23,27,29,21,22,33,33,26,
DATA> 24,25,28,31,20,23,25,35,31,23,24,30
DATA> END
MTB >
MTB > INFO ————————————————————————————— Display column label, column name, and
                                          number of observations in the data set.
COLUMN     NAME      COUNT
C1         Scores       30

CONSTANTS USED: NONE
                      ————————————————————— Save file to disk (specify drive, and limit file
MTB > SAVE 'B:HRSCORES'                       name to 8 characters; enclose in apostrophes.)

Worksheet saved into file: B:HRSCORES.MTW
MTB > _
```

Displaying data   As Figure C1.4 illustrates, you can display and check data through **PRINT 'VarName' [Enter]**. Thus, PRINT 'Scores' [Enter] lists the 30 raw scores entered into C1.

Editing data   The command **LET C#(R) = X [Enter]** enables you to edit data and correct data-entry errors. C# stands for a particular column of data—C1 in this example. (R) is the row that contains the raw score to be changed, and X is the new value of the score. In the present case, the value in row 3 of column 1 is changed from 21 to 15 through LET C1(3) = 15 [Enter]. Then, for demonstration purposes, the same data point is changed back to its original value of 21 via LET C(3) = 21 [Enter].

Loading data files from disk   You can retrieve data from a disk file by using **RETRIEVE 'B:FILENAME' [Enter]**. As displayed in Figure C1.5, RETRIEVE 'B:HRSCORES' [Enter] fetches the data stored earlier on drive B:. If you ever forget the name of the file

Figure C1.4 *Displaying and Editing Data in MINITAB*

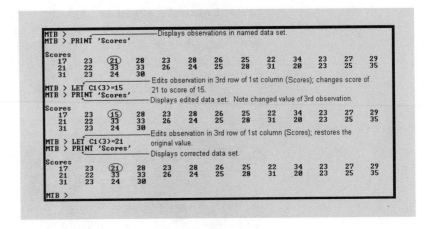

Figure C1.5 *Sending Output to a Printer and Quitting MINITAB*

you're looking for, the command **DIR B:** [Enter] will show you all of the files on the disk in drive B:.

Printing You can send all screen output to a printer through the **PAPER** [Enter] command. To turn off this feature, issue the **NOPAPER** [Enter] command. **NEWPAGE** [Enter] ejects the current page from the printer.

Quitting To leave the MINITAB program, type **STOP** [Enter].

MYSTAT

To start MYSTAT from the DOS prompt, make sure that you are in the directory that contains the MYSTAT program, usually C:\MYSTAT. From the C:\MYSTAT> prompt, type MYSTAT and press [**Enter**]. Or, if you are running the program from within the Windows environment, go to File Manager, click on the MYSTAT folder, and then double-click on the MYSTAT.EXE icon. You will see the greeting/copyright screen shown in Figure C1.6. Press **Enter** to move on.

Figure C1.6 *MYSTAT Greeting/Copyright Screen*

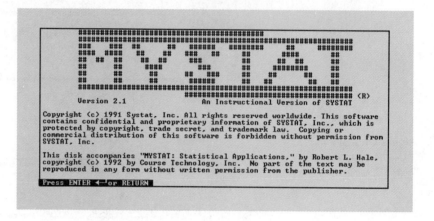

Structure of MYSTAT MYSTAT uses three different screen modes:

1. The "Command Menu," in which you issue software commands
2. The "Editor," in which you enter data
3. The "Output Screen," which displays the results of a statistical analysis.

You will encounter the Output Screen in later sections of this appendix. The Command Menu is shown in Figure C1.7. In the Command Menu mode you see the various MYSTAT commands that are available. You can execute any command by typing its name at the command prompt (>) and pressing the **Enter** key. Typing **HELP** [**Enter**] will result in a screenful of command descriptions.

Typing **EDIT** [**Enter**] at the command prompt will transport you to the Editor, illustrated in Figure C1.8. The Editor has two areas: the "Edit Window," where you enter and edit data, and the "Command Line," where you can enter some commands, such as SAVE B:FILENAME. To move back and forth between the Edit Window and the Command Line, press the **Esc** key.

Figure C1.7 *MYSTAT*
Command Menu

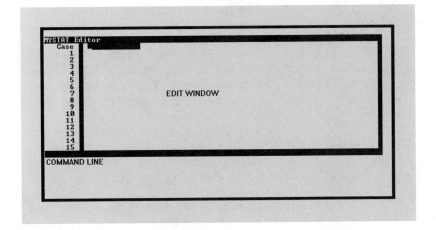

Figure C1.8 *MYSTAT*
Edit Window

Naming data sets   To demonstrate how to work with the Editor, I'll use the following data. They are the average directory-assistance response times (in seconds) of 85 telephone operators.

6.2	6.2	6.1	6.1	6.0	6.0	6.0	6.0	5.9	5.9	5.9
5.9	5.9	5.8	5.8	5.8	5.8	5.8	5.8	5.7	5.7	5.7
5.7	5.7	5.7	5.7	5.6	5.6	5.6	5.6	5.6	5.6	5.6
5.6	5.6	5.6	5.5	5.5	5.5	5.5	5.5	5.5	5.5	5.5
5.5	5.5	5.5	5.5	5.8	5.8	5.8	5.8	5.8	5.8	5.8
5.8	5.3	5.3	5.3	5.3	5.3	5.2	5.2	5.2	5.2	5.2
5.1	5.1	5.1	5.0	5.0	5.0	4.9	4.9	4.8	4.8	4.7
4.7	4.5	4.5	4.5	4.4	4.3	4.2	4.1			

Before you can start inputting these data, you must get the blinking cursor (underline mark) into the first row of the Edit Window and type in the name of the variable represented by the data set.

Use this format: 'VARNAME, where 'VARNAME is any label you wish to apply to your data set. *Important: Be sure to precede the variable name with an apostrophe*, and restrict the name to eight letters with no spaces. I named the reaction time scores 'RT, as shown in Figure C1.9

Press [**Enter**] after you finish typing the variable name. This will move the cursor into column 2. To get back into column 1, press the ← key. To move the cursor to the next row down, press the ↓ key. In general, you can move among the rows and columns in the Edit Window by using the four arrow keys.

Once you're in row 1, begin typing in the raw scores. After you type a score, press the ↓ key to move the cursor to the next row, as illustrated in Figure C1.9.

Editing data   While you are in the Edit Window, you can change and correct the data entries at any time simply by using the arrow keys to move to the target datum, typing the desired score, and pressing either ↓ or [**Enter**]. The new value will replace the old one.

Saving your data   After you input the last raw score, press the **Esc** key to move from the Edit Window to the Editor's Command Line. *Important: You must now save your data file* with the command **SAVE B:FILENAME** [**Enter**], where B: is the floppy drive where the data disk is mounted (use drive A: if you prefer) and FILENAME is any label you wish to apply to your data file. File names must consist of eight or fewer letters and/or digits. (Don't leave any spaces.) See Figure C1.10 for an example of how to use the SAVE command.

Moving between screen modes   Once you've saved a data file, you will need to return to the Command Menu screen to conduct analyses. At the Editor's Command Line, type **QUIT** and press

Figure C1.9 *Entering Data in MYSTAT*

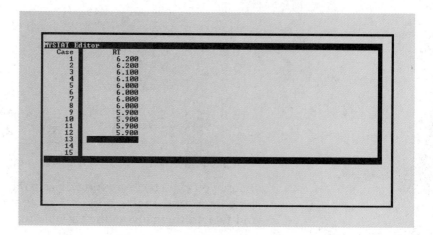

Figure C1.10 *Saving a Data File in MYSTAT*

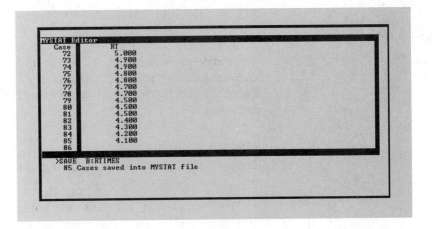

```
MYSTAT Editor
  Case        RT
    72       5.000
    73       4.900
    74       4.900
    75       4.800
    76       4.800
    77       4.700
    78       4.700
    79       4.500
    80       4.500
    81       4.500
    82       4.400
    83       4.300
    84       4.200
    85       4.100
    86
>SAVE  B:RTIMES
    85 Cases saved into MYSTAT file
```

[**Enter**]. Using the QUIT command is always the way to return to the Command Menu from the Editor.

To return to the Editor from the Command Menu, type **EDIT** [**Enter**].

Displaying data Typing **NAMES** [**Enter**] while in the Command Menu will display the name(s) of the variable(s) (i.e., columns) in your data set. The variable name(s) will appear in MYSTAT's third screen mode, the "Output Screen." *Hint*: If data-analysis commands don't work in the Command Menu, using the NAMES command will often correct the problem.

Typing **LIST** [**Enter**] while in the Command Menu will list all the variable names and data in a file. *Hint*: If LIST does not produce the expected result, issuing the **USE B:FILENAME** [**Enter**] command will usually remedy the situation. Of course, you should substitute the actual name of your data file for "FILENAME," and substitute A: for B: if your disk is in drive A:.

Printing data To send data and results of analyses to a printer, type **OUTPUT @** [**Enter**] while in the Command Menu. **OUTPUT *** [**Enter**] restores the normal state of things, such that output goes only to the screen.

Quitting MYSTAT To leave MYSTAT, type QUIT [Enter] from the Command Menu.

2: BASIC CONCEPTS AND IDEAS

In this section you will use statistical software to analyze the data from Table 2.1 of Chapter 2. That table lists the scores on the tendency to convict of two groups of subjects: 15 who advocate the use of capital punishment and 10 who do not believe in capital punishment. The question of interest is whether the death-penalty advocates are more likely than the death-penalty opponents to convict a defendant accused of a noncapital crime (burglary). If so, the "death-qualified" subjects should have a higher average (i.e., mean) score on the tendency to convict than the "non-death-qualified" subjects.

MINITAB

Figure C2.1 illustrates the following operations within MINITAB:

- Using the SET command to enter the conviction ratings of the "death-qualified" subjects into column 1 (i.e., C1) of the data array, and to enter the conviction ratings of the "non-death-qualified" subjects into column 2 (i.e., C2) of the array
- Using the NAME command to label C1 "DeathQ" and C1 "NDeathQ"
- Executing the INFO command to display the column names and the number of observations in each sample
- Issuing the SAVE command to store the columns of data into a disk file

The numerous pointers and annotations in the figure should help identify each of these commands and its effect.

Figure C2.2 shows, once again, that the PRINT command lists the data that have been inputted. But, this time, the data are displayed in a vertical, or tabulated, format. The PRINT command will always list raw scores in the tabulated format when there is more than one column of data.

The most important point made by Figure C2.2 is that the average score in a column of data is calculated through **MEAN**

Figure C2.1 *Inputting and Saving Data in MINITAB*

```
MINITAB Release 1.2 *** COPYRIGHT - Minitab, Inc. 1988
Student Edition  *** Storage Available: 2001
NOV. 2, 1994
                                              —— Input data for Sample 1 (Death Qualified)
MTB > SET data into C1
DATA> 9,6,8,5,7,7,9,5,8,10,6,7,4,7,7
DATA> END
MTB > SET data into C2                        —— Input data for Sample 2 (Non-Death Qualified)
DATA> 3,6,2,4,5,4,4,4,5,3
DATA> END
MTB >                                         —— Name the samples (limit of 8 characters)
MTB > NAME C1 'DeathQ' C2 'NDeathQ'
MTB >                                         —— Display columns, column names, and
MTB > INFO                                       number of observations in each column.

COLUMN    NAME        COUNT
C1        DeathQ       15
C2        NDeathQ      10

CONSTANTS USED: NONE
                                              —— Save file to disk (specify drive and limit file
MTB > SAVE 'B:DBELIEFS'                          name to 8 characters; enclose in apostrophes)

Worksheet saved into file: B:DBELIEFS.MTW
MTB >
```

Figure C2.2 *Computing the "Average" Score with MINITAB's MEAN Command*

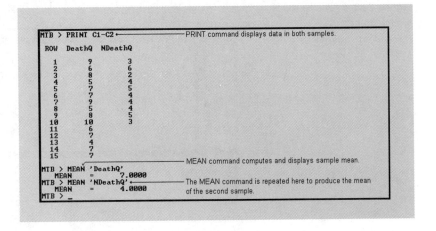

The following table represents the data displayed in the figure:

```
MTB > PRINT C1-C2 ◄──────────── PRINT command displays data in both samples.

ROW   DeathQ   NDeathQ
 1       9        3
 2       6        6
 3       8        2
 4       5        4
 5       7        5
 6       7        4
 7       9        4
 8       5        4
 9       8        5
10      10        3
11       6
12       7
13       4
14       7
15       7
                            ─── MEAN command computes and displays sample mean.
MTB > MEAN 'DeathQ'
    MEAN    =      7.0000
MTB > MEAN 'NDeathQ' ◄──────── The MEAN command is repeated here to produce the mean
    MEAN    =      4.0000        of the second sample.
MTB > _
```

'VarName' [Enter] , where 'VarName' is whatever label you have assigned to the column for which you wish to calculate an average. Note that **MEAN C# [Enter]** will work just as well, where # is the number of the column for which an average score is needed. In the example given in Figure C2.2, MEAN C1 [**Enter**] is functionally the same as MEAN 'DeathQ' [**Enter**].

The MEAN command answers the question of the relationship between attitude toward capital punishment and the tendency to convict the hypothetical burglary suspect. Figure C2.2 shows that the death-qualified subjects had a higher average tendency to convict rating (7.000) than the non-death-qualified subjects (4.000).

MYSTAT

As you use MYSTAT to work with the data on attitudes toward capital punishment (from Table 2.1), you will be introduced to (1) declaring an independent variable, (2) entering text (alpha) data, (3) coding levels of an independent variable, (4) executing the STATS command, and (5) qualifying a MYSTAT command with an "option" command.

Declaring an independent variable in MYSTAT In the first section of this appendix, you discovered that you had to declare (i.e., name) a variable before inputting the data associated with that variable. That example involved only one factor, a dependent variable. But the typical behavioral science study has both an independent variable—most often represented by two or more groups of subjects—and a dependent variable represented by the subjects' raw scores. To do calculations with the MYSTAT program, both variables must be declared in the Edit Window. Figure C2.3 shows how this is accomplished. The independent variable is declared by

Figure C2.3 *Declaring Independent and Dependent Variables in MYSTAT*

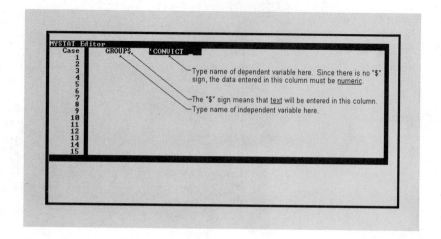

naming one column (press **Enter**), and the dependent variable is declared by naming a second column (press **Enter**).

Entering test (alpha) data Data are often in the form of numbers, called numerical data. But a dichotomous variable, such as belief in capital punishment (advocate vs. oppose), can be represented by words or letters rather than numbers. Variables represented by words are called "alpha variables," and their data are called "alpha data." In the present example, the death-qualified subjects (the first level of the independent variable) will be identified with the letter "D," and the non-death-qualified subjects (the second level of the independent variable) will be identified with the letters "ND." Two rules govern the use of alpha variables:

1. Alpha variables must be declared by appending the $ symbol to the variable name (the total number of characters must still be eight or fewer)
2. *Each value* of the alpha variable must be preceded by an apostrophe.

Hence, to declare the independent and dependent variables displayed in Figure C2.3, you must start at the top row of the first column in the Edit Window and type: 'GROUP$ [**Enter**]. This sequence will register the independent variable as an "alpha variable" and move the cursor to the top of column 2. Now type: 'CONVICT [**Enter**]. The latter keystrokes will declare the dependent variable as a "numerical variable," the default type of variable that is assumed when the variable name has no $ suffix.

Coding levels of the independent variable Having declared the independent and dependent variables, you are now ready to enter data, as shown in Figures C2.4 and C2.5. Note these two points in regard to inputting values of the independent variable:

Figure C2.4 *Coding an Independent Variable in MYSTAT*

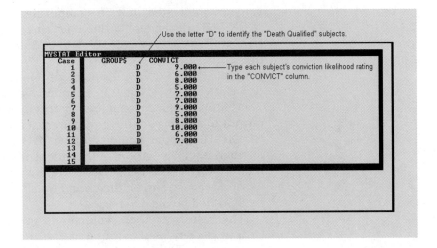

Figure C2.5 *Completing Group Coding and Data Input, and Saving the File*

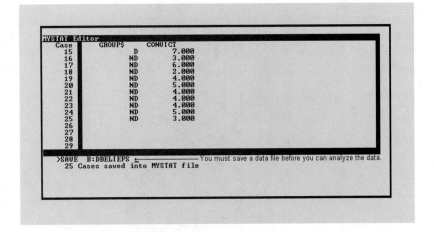

1. The letter D is used to code each death-qualified subject (i.e., subjects in the first sample), whereas the letters ND are for each non-death-qualified subject (i.e., subjects in the second sample).
2. When you type in the independent variable codes, you must precede each with an apostrophe—for example, 'D and 'ND. The leading apostrophe is always necessary when inputting "alpha" data (but should not be included with numeric data).

When you finish entering all the codes and data, press **Esc** to move to the Editor Command Line and save the data file to disk with the command SAVE B:DBELIEFS [**Enter**].

Executing the STATS command with an option command At the Editor Command Line, type QUIT and press the **Enter** key to return to the Command Menu. The **STATS VARNAME [Enter]** command will produce a number of summary statistics, but right now you are interested only in computing an average score on the "CONVICT" dependent variable *for each group of subjects.* So type STATS CONVICT /MEAN BY GROUP$ and press **Enter,** as shown in Figure C2.6.

The STATS CONVICT statement tells the software to calculate descriptive statistics on the dependent variable named "CONVICT." The /MEAN option command instructs the software to limit the output to a display of the group averages. Finally, observe that the BY GROUP$ option command tells MYSTAT to compute the mean for each level (i.e., group) of the independent variable. This manipulation works because the independent variable was declared and its values coded earlier. *Important: Note that option commands must be preceded by a slash (/).*

The results of these commands are displayed in MYSTAT's Output Screen, as illustrated in Figure C2.7. The death-qualified sample clearly had a higher conviction-rating mean.

Figure C2.6 *Using the STATS Command with the MEAN Option Command in MYSTAT*

```
  ┌───────────────────────────────────────────────────────────────────┐
  │  DEMO      EDIT      MENU      PLOT       STATS       MODEL         │
  │  HELP                NAMES     BOX        TABULATE    CATEGORY      │
  │  SYSTAT    USE       LIST      HISTOGRAM  TTEST       ANOVA         │
  │            SAVE      FORMAT    STEM       PEARSON     COVARIATE     │
  │            PUT       NOTE      TPLOT                  ESTIMATE      │
  │            SUBMIT                                                   │
  │  QUIT      OUTPUT    SORT      CHARSET    SIGN                      │
  │                      RANK                 WILCOXON                  │
  │                      WEIGHT               FRIEDMAN                  │
  ├───────────────────────────────────────────────────────────────────┤
  │ >STATS CONVICT /MEAN BY GROUP$                                     │
  │                                                                    │
  │                                                                    │
  │                                                                    │
  │                                                                    │
  │   If you are a new user, type DEMO and then press the [Enter] key. │
  └───────────────────────────────────────────────────────────────────┘
```

Figure C2.7 *Output Produced by MYSTAT's STATS /MEAN Command*

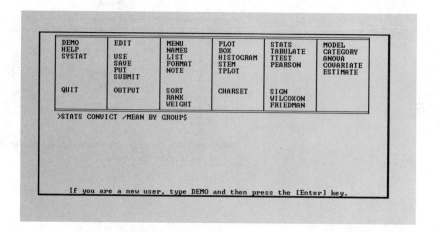

```
┌────────────────────────────────────────────────────────────────────┐
│ THE FOLLOWING RESULTS ARE FOR:                                     │
│          GROUP$    = D                                             │
│ TOTAL OBSERVATIONS:    15                                          │
│                   CONVICT                                          │
│   N OF CASES             15 ◄────────Sample size                   │
│   MEAN                7.000 ◄────────Mean conviction rating of Death Qualified sample │
│                                                                    │
│ THE FOLLOWING RESULTS ARE FOR:                                     │
│          GROUP$    = ND                                            │
│ TOTAL OBSERVATIONS:    10                                          │
│                   CONVICT                                          │
│   N OF CASES             10 ◄────────Sample size                   │
│   MEAN                4.000 ◄────────Mean conviction rating of Non-Death Qualified sample │
│                                                                    │
│ Press ENTER ◄─┘ or RETURN                                         │
└────────────────────────────────────────────────────────────────────┘
```

3: FREQUENCY DISTRIBUTIONS AND GRAPHS

One of the great benefits of statistical software is its ability to easily and accurately produce tables and graphs from raw scores. Let's find out how this is done.

MINITAB

To explore table and graph construction in MINITAB, we'll use the human relations test scores that you saved to disk in Section 1 of this appendix. (If you skipped that exercise, you may wish to complete it now.) Start up MINITAB, make sure the data-file floppy disk is in drive B: (or A:, if you prefer), and type **DIR B: [Enter]** (or **DIR A: [Enter]**). As displayed in Figure C3.1, this DIR command shows you what files are available on the disk.

Type RETRIEVE 'B:HRSCORES' **[Enter]** to load the data file, and issue the INFO command to verify that the name of the data set is 'Scores' and that there are $N = 30$ observations in the sample.

Figure C3.1 *Retrieving a Data File from Disk in MINITAB*

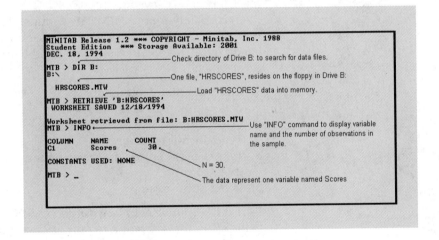

Constructing a frequency distribution and graph You can have MINITAB construct a frequency distribution by typing **HISTOGRAM C# [Enter]** or **HISTOGRAM 'VarName' [Enter]**, where # is the number of the column that contains the data and 'VarName' is a label that you have assigned to the data set.

Figure C3.2 displays the result of the HISTOGRAM instruction. Notice that the command yields a combination of (1) a grouped frequency distribution based on interval midpoints and (2) a histogram in which the frequency of each class interval is represented by the number of asterisks (*) in the corresponding row of the graph.

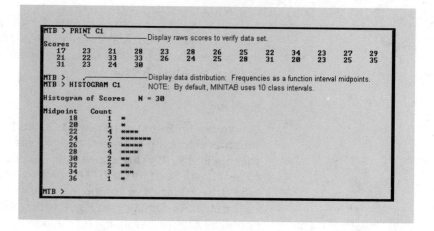

Using subcommands in MINITAB You might have noticed that the HISTOGRAM command defaults to ten class intervals. But in Chapter 3 of the textbook, these same data were organized into a frequency distribution with seven intervals. How can you get MINITAB to use seven intervals?

Fortunately, many **MINITAB** commands have "subcommands" that you can use to override default settings or specify particular conditions. To append a subcommand to a main command, follow these steps:

1. Input the main command.
2. Follow the main command with a semicolon.
3. Press **Enter**.
4. Type the subcommand.
5. Type a period after the subcommand. (*Important:* Don't forget the period!)
6. Press **Enter** again.

Figure C3.3 illustrates how to use the **INCREMENT = K** subcommand. K is the interval size you are assigning. By setting K to 3, you can force the HISTOGRAM routine to use seven class intervals (see Chapter 3). Notice how well the results correspond to Table 3.2 of Chapter 3.

MYSTAT

To learn how to create tables and graphs in MYSTAT, we'll work with the telephone operators' response times data set that was saved to disk in the first section of this appendix. (If you skipped that exercise, you might wish to do it now.) Table 3.4 of Chapter 3 displayed the response times.

Retrieving files from disk in MYSTAT Figure C3.4 shows that you can load a MYSTAT data file from disk by typing the instruction

Figure C3.3 *Using the INCREMENT Subcommand in MINITAB*

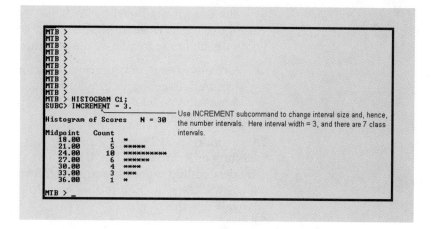

Figure C3.4 *Retrieving a Data File from Disk in MYSTAT*

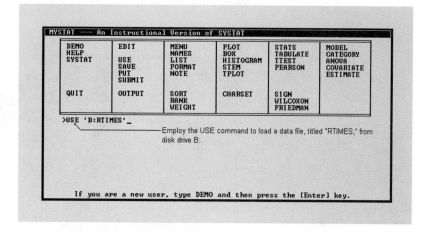

USE B:FILENAME [Enter] from the Command Menu Screen. In this example, the response times data are loaded from drive B: via USE B:RTIMES **[Enter]**. The successful loading operation is then verified by the message shown on MYSTAT's Output Screen, which reminds you that "RT" is the variable name associated with these scores (see Figure C3.5). You might also like to issue a LIST RT **[Enter]** instruction to inspect the data on the Output Screen.

Using MYSTAT's TABULATE command If you are in the Command Menu, you can produce an ungrouped frequency distribution by entering **TABULATE VARNAME [Enter]**. Since the variable name in this example is RT, you would type TABULATE RT **[Enter]**. This main command produces a horizontally arranged distribution. Since you normally will want a vertical listing of the distribution, you must append the **/LIST** option command to TABULATE VARNAME, as illustrated in Figure C3.5.

The output produced by TABULATE RT /LIST **[Enter]** appears in Figure C3.6. Observe that the TABULATE command gives you not only a simple frequency distribution, but also:

- A relative frequency (percent) distribution
- A cumulative frequency distribution
- A cumulative relative frequency (cumulative percent) distribution

Also, notice that the raw scores in this distribution are in an ascending order from the top of the table downward. This is opposite the descending order of tabulated information presented by many statistics texts, including this one. Nonetheless, the information conveyed is the same.

Drawing a histogram in MYSTAT To depict a distribution in a histogram, you simply type **HISTOGRAM VARNAME /BARS = K**

Figure C3.5 *Using the TABULATE Command in MYSTAT*

COUNT	CUM COUNT	PCT	CUM PCT	RT
1	1	1.2	1.2	4.100
1	2	1.2	2.4	4.200
1	3	1.2	3.5	4.300
1	4	1.2	4.7	4.400
3	7	3.5	8.2	4.500
2	9	2.4	10.6	4.700
2	11	2.4	12.9	4.800
2	13	2.4	15.3	4.900
3	16	3.5	18.8	5.000
3	19	3.5	22.4	5.100
5	24	5.9	28.2	5.200
5	29	5.9	34.1	5.300
8	37	9.4	43.5	5.400
12	49	14.1	57.6	5.500
10	59	11.8	69.4	5.600
7	66	8.2	77.6	5.700
6	72	7.1	84.7	5.800
5	77	5.9	90.6	5.900
4	81	4.7	95.3	6.000
2	83	2.4	97.6	6.100
2	85	2.4	100.0	6.200

Press ENTER ←┘ or RETURN

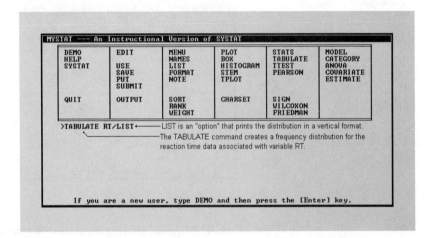

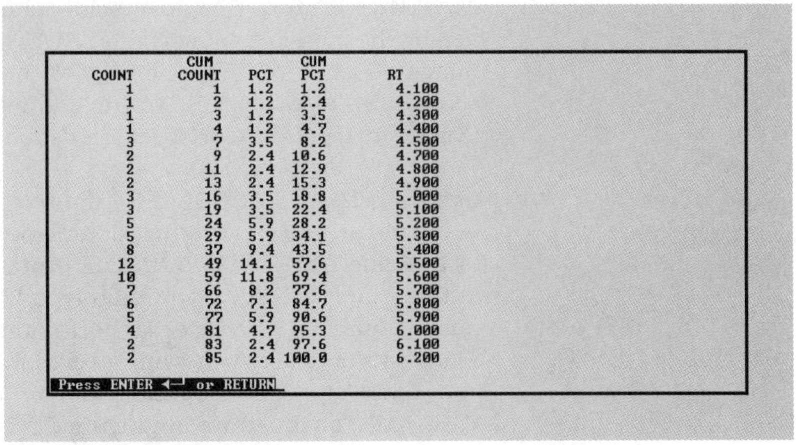

Figure C3.6 *Frequency Distribution Created with the TABULATE Command*

[**Enter**], where VARNAME is whatever you have named the variable of interest and K is the number of bars you want to have in the histogram. In this example, you should enter HISTOGRAM RT / BARS = 11, as shown in Figure C3.7. Figure C3.8 displays the result of this instruction.

Spreadsheet Graphing

You can construct very sophisticated graphs easily and quickly with spreadsheet software. I'll demonstrate spreadsheet graphing by using Lotus 1-2-3 for Windows to graph the frequency distribution in Table 3.3.

Entering data into the spreadsheet   Using Table 3.3 as a reference, type the class intervals into cells A2 through A8, as shown in Figure C3.9. *Important*: Enter the class intervals *as text* by

Figure C3.7 *Using the HISTOGRAM Command in MYSTAT*

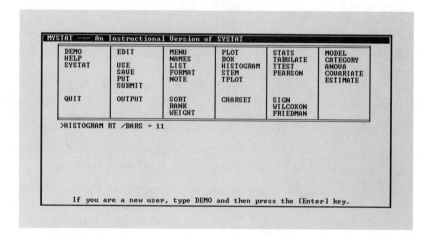

Figure C3.8 *A Histogram Created by MYSTAT*

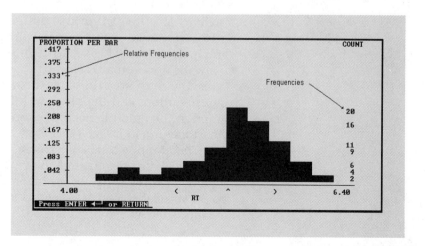

Figure C3.9. *Entering Frequency Distribution Data in a Spreadsheet*

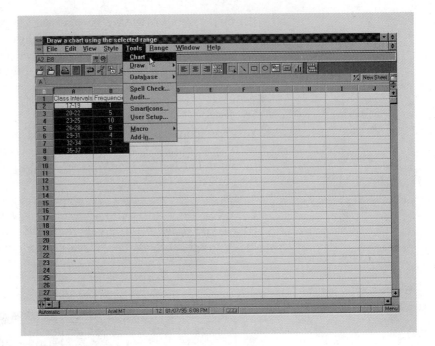

preceding each one with an apostrophe: '17-19, for example. Next type the frequency of each interval in cells B2 through B8. Enter those data as numbers, not text.

Creating a frequency polygon To draw a frequency polygon for the distribution, follow these steps:

1. Holding down the left mouse button, drag the mouse pointer across all cells that contain class intervals and frequencies, so that they become highlighted in black (Figure C3.9).
2. Use the mouse to pull down the Tools menu, and click on Chart (Figure C3.9).
3. Holding down the left mouse button, drag the mouse pointer across the area of the screen where you want to draw the graph, as shown in Figure C3.10. (An outline of a rectangle will mark the area you are selecting.)
4. When the chart area is exactly the way you want it, release the mouse button, and the program will automatically create the graph. If the graph is not a frequency polygon, pull down the Chart menu and select Type. Next choose "Line" and click on OK. The graph will now become a frequency polygon, and it should look like the one in Figure C3.11.
5. Finally, in three separate operations, *double-click* on the Title box, *Y*-axis label, and *X*-axis label to change them to appropriate labels, as shown in Figure C3.12.

Figure C3.10 *Dragging the Mouse Pointer to Create a Graph in a Spreadsheet*

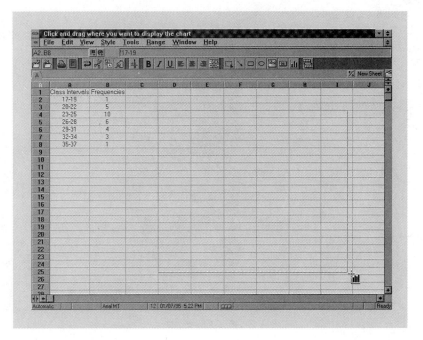

Figure C3.11 *Preliminary Graph Resulting from the Chart Option in a Spreadsheet*

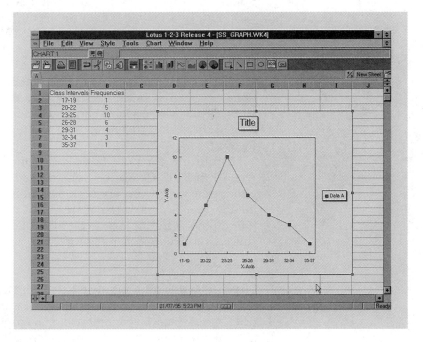

Figure C3.12 *Spreadsheet Graph with Appropriate Title and Axis Labels*

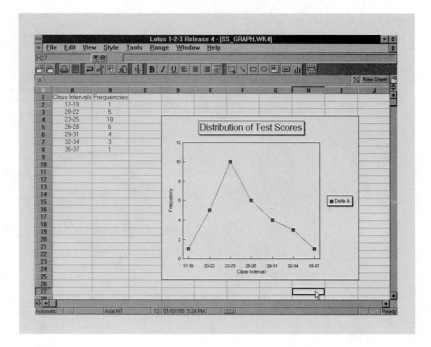

4: SUMMARY MEASURES

Summary measures refer to measures of central tendency and variation. How do you obtain the mean, median, range, and standard deviation from MINITAB and MYSTAT? It couldn't be easier.

MINITAB

The observations of interest in this section are the 30 human relations test scores saved earlier (in Section 1 of this appendix) as HRSCORES. These scores can also be found in Table 4.1. Load the data from disk and check their accuracy, as demonstrated in Figure C4.1

To quickly compute basic descriptive statistics in MINITAB, type **DESCRIBE C# [Enter]**, where # is the number of the column of data for which you are seeking summary measures. Since the data for the 'Scores' variable are stored in column 1, the appropriate command in this case is DESCRIBE C1 **[Enter]**. The results of this command appear in Figure C4.2.

MYSTAT

The data in this example are taken from Table 4.4 of Chapter 4. The observations are the reaction times of student drivers under two drug states: placebo condition versus amphetamine condition. You should input and save these raw scores, as shown in Figure C4.3.

Figure C4.1 *Retrieving the HRSCORES Data File from Disk*

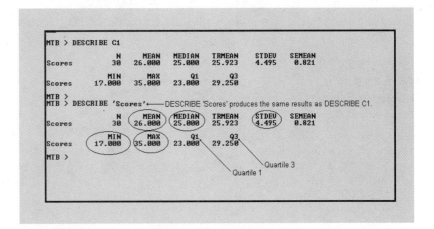

```
MINITAB Release 1.2 *** COPYRIGHT - Minitab, Inc. 1988
Student Edition  *** Storage Available: 2001
DEC. 21, 1994

MTB > RETRIEVE 'B:HRSCORES'
 WORKSHEET SAVED 12/18/1994

Worksheet retrieved from file: B:HRSCORES.MTW
MTB >
MTB > INFO

COLUMN    NAME      COUNT
C1        Scores      30

CONSTANTS USED: NONE

MTB > PRINT 'Scores'

Scores
   17    23    21    28    23    28    26    25    22    34    23    27    29
   21    22    33    33    26    24    25    28    31    20    23    25    35
   31    23    24    30

MTB >
```

Figure C4.2 *Using the DESCRIBE Command in MINITAB*

```
MTB > DESCRIBE C1

              N      MEAN    MEDIAN    TRMEAN     STDEV    SEMEAN
Scores       30    26.000    25.000    25.923     4.495     0.821

            MIN       MAX        Q1        Q3
Scores   17.000    35.000    23.000    29.250
MTB >
MTB > DESCRIBE 'Scores'←——DESCRIBE 'Scores' produces the same results as DESCRIBE C1.

              N      MEAN    MEDIAN    TRMEAN     STDEV    SEMEAN
Scores       30    26.000    25.000    25.923     4.495     0.821

            MIN       MAX        Q1        Q3
Scores   17.000    35.000    23.000    29.250
MTB >
```

Quartile 3

Quartile 1

Figure C4.3 *Entering and Saving Drug-Effects Data in MYSTAT*

```
MYSTAT Editor
  Case    PLACEBO      AMPHETAM
     7     21.000        21.000
     8     20.000        21.000
     9     20.000        20.000
    10     20.000        20.000
    11     20.000        20.000
    12     20.000        20.000
    13     20.000        19.000
    14     19.000        19.000
    15     19.000        18.000
    16     19.000        18.000
    17     19.000        17.000
    18     18.000        17.000
    19     18.000        16.000
    20     17.000        15.000
    21

>SAVE B:DRUG
   20 Cases saved into MYSTAT file
```

Now use QUIT to return to the Command Menu, and execute the STATS command on the PLACEBO and AMPHETAM variables, as shown in Figure C4.4. This instruction generates the measures of central tendency and variation exhibited in Figure C4.5.

Figure C4.4 *Using the STATS Command in MYSTAT*

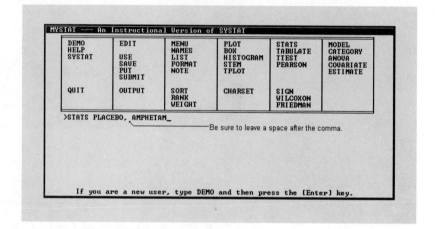

Figure C4.5 *Output Produced by the STATS Command*

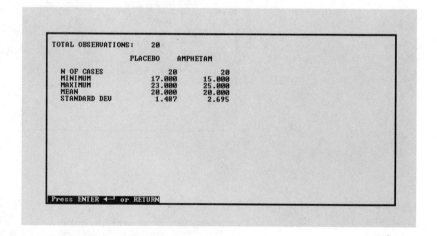

5: RELATIVE MEASURES AND THE NORMAL CURVE

Relative measures refer to percents, percentiles, percentile ranks, *z* scores, and various kinds of transformed scores. Of all these measures, percentiles and percentile ranks are often considered the most tedious to develop because computing them often requires that you construct a cumulative percent distribution. Both MINITAB and MYSTAT have simple commands for generating such distributions.

MINITAB

Table 5.2 of Chapter 5 contains an ungrouped cumulative percent distribution of 40 final exam scores from a humanities class. Figures C5.1 and C5.2 illustrate how to replicate that distribution in the MINITAB environment. Note how to carry out the **TALLY C#; [Enter]** command. Observe that the resulting cumulative percent distribution perfectly duplicates the one that appears in Table 5.2 of the text, except that MINITAB lists the scores from lowest to highest, rather than from highest to lowest. Also, notice how the **ALL. [Enter]** subcommand is appended to the TALLY instruction. Without the ALL subcommand, TALLY yields only a simple frequency distribution.

Figure C5.1 *Entering and Saving Humanities Test Scores in MINITAB*

```
MINITAB Release 1.2 *** COPYRIGHT - Minitab, Inc. 1988
Student Edition  *** Storage Available: 2001
DEC. 21, 1994

MTB > SET into C1
DATA> 48,48,47,47,47,47,46,46,46,46,46,46,
DATA> 45,45,45,45,45,45,45,45,45,45,45,45,45,45,45,45,
DATA> 44,44,44,44,44,44,43,43,43,43,42,42
DATA> END
MTB > SAVE AS 'B:HUMEXAM'

Worksheet saved into file: B:HUMEXAM.MTW
MTB >
```

Figure C5.2 *Using the TALLY Command to Produce a Cumulative Percent Distribution in MINITAB*

```
DEC. 21, 1994

MTB > SET into C1
DATA> 48,48,47,47,47,47,46,46,46,46,46,46,
DATA> 45,45,45,45,45,45,45,45,45,45,45,45,45,45,45,45,
DATA> 44,44,44,44,44,44,43,43,43,43,42,42
DATA> END
MTB > SAVE AS 'B:HUMEXAM'

Worksheet saved into file: B:HUMEXAM.MTW
MTB > TALLY C1;
SUBC> ALL.
```

C1	COUNT	CUMCNT	PERCENT	CUMPCT
42	2	2	5.00	5.00
43	4	6	10.00	15.00
44	6	12	15.00	30.00
45	16	28	40.00	70.00
46	6	34	15.00	85.00
47	4	38	10.00	95.00
48	2	40	5.00	100.00
N=	40			

Calculate percentile ranks from cumulative percents in this column.

```
MTB >
```

MYSTAT

The data in this example are taken from Table 5.1 of Chapter 5. The observations are $N = 200$ final exam scores from a statistics class. You should input and save data from Table 5.1, as illustrated, in part, in Figure C5.3. To generate a cumulative percent distribution for these scores, execute the TABULATE command with the /LIST option command, as depicted in Figure C5.4. The resulting distribution appears in Figure C5.5. Compare this cumulative percent distribution with that in Table 5.1 of Chapter 5.

Figure C5.3 *Entering and Saving Statistics Exam Scores in MYSTAT*

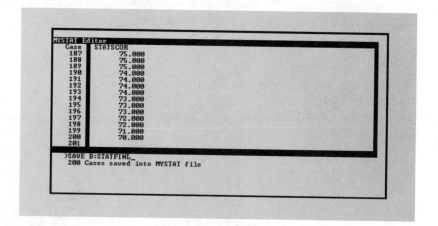

Figure C5.4 *Using the TABULATE Command (with the LIST Option Command) in MYSTAT*

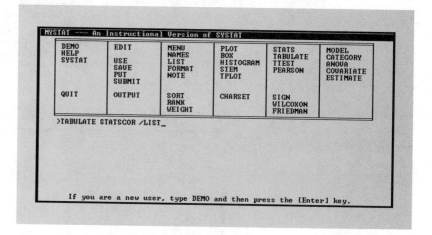

Figure C5.5 *Output Produced by the TABULATE Command*

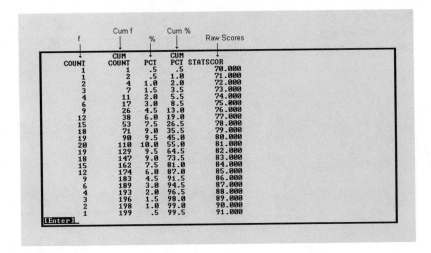

	Cum f	%	Cum %	Raw Scores

COUNT	CUM COUNT	PCT	CUM PCT	STATSCOR
1	1	.5	.5	70.000
1	2	.5	1.0	71.000
2	4	1.0	2.0	72.000
3	7	1.5	3.5	73.000
4	11	2.0	5.5	74.000
6	17	3.0	8.5	75.000
9	26	4.5	13.0	76.000
12	38	6.0	19.0	77.000
15	53	7.5	26.5	78.000
18	71	9.0	35.5	79.000
19	90	9.5	45.0	80.000
20	110	10.0	55.0	81.000
19	129	9.5	64.5	82.000
18	147	9.0	73.5	83.000
15	162	7.5	81.0	84.000
12	174	6.0	87.0	85.000
9	183	4.5	91.5	86.000
6	189	3.0	94.5	87.000
4	193	2.0	96.5	88.000
3	196	1.5	98.0	89.000
2	198	1.0	99.0	90.000
1	199	.5	99.5	91.000

[Enter]

6: LINEAR CORRELATION

MINITAB

Our example for the topic of linear correlation involves the ten pairs of scores from Table 6.2, which shows selection test scores and performance evaluation ratings for the ten employees of XYZ Company. The selection test score is variable X (the predictor), and the performance evaluation rating is variable Y (the criterion).

Constructing a scatterplot in MINITAB Figures C6.1 and C6.2 demonstrate the use of the **PLOT** instruction, which creates a scatter diagram for the paired scores. The syntax is **PLOT C2 versus C1 [Enter]** , where column 2 (i.e., C2) is the Y variable that you want to plot against the X variable in column 1 (i.e., C1).

Figure C6.1 *Using the PLOT Command in MINITAB*

```
MINITAB Release 1.2 *** COPYRIGHT - Minitab, Inc. 1988
Student Edition  *** Storage Available: 2001
DEC. 23, 1994

MTB > SET data into C1
DATA> 40,32,46,50,22,10,18,30,26,30
DATA> END
MTB > NAME C1 'SelTest'
MTB > SET data into C2
DATA> 90,74,96,90,44,50,56,76,42,64
DATA> END
MTB > NAME C2 'PerfEval'
MTB >                        ────PLOT command creates a scatter plot of Y vs. X.
MTB > PLOT C2 versus C1
```

Figure C6.2 *Output Generated by the PLOT Command in MINITAB*

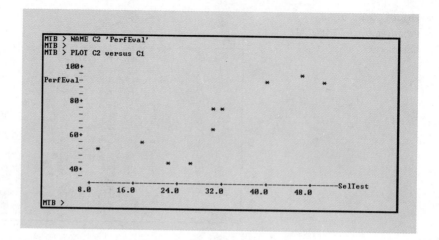

Computing the Pearson *r* in MINITAB To calculate the linear correlation between selection test scores and performance evaluation ratings, use the **CORRELATE** command, as displayed in Figure C6.3. It doesn't matter whether C2 or C1 is specified first.

Figure C6.3 *Using the CORRELATE Command in MINITAB*

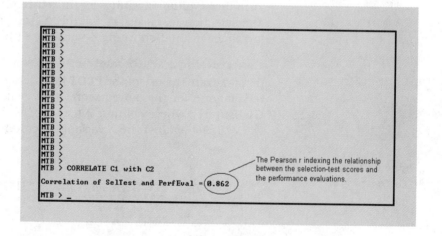

MYSTAT

The data in this example are taken from Table 6.3 and concern the relationship between neuroticism (variable *Y, the criterion*) and self-esteem (variable *X, the predictor*). Figure C6.4 shows how the variable names and 15 pairs of scores are entered. It doesn't matter which column *X* and *Y* are assigned to, as long as you remember which variable you consider to be the criterion (*Y*) variable.

Figure C6.4 *Entering and Saving Pairs of X and Y Scores in MYSTAT*

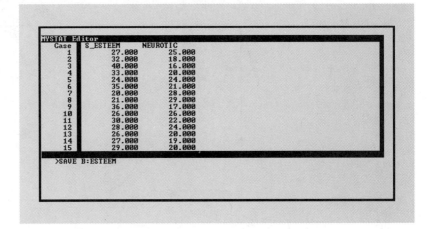

Plotting a scatter diagram in MYSTAT   Figures C6.5 and C6.6 illustrate how to use the PLOT instruction to produce a scatterplot of the neuroticism scores against levels of self-esteem. The important point to remember is that the criterion variable should be specified first after you type PLOT. Thus, the general syntax is: **PLOT VAR_Y * VAR_ X [Enter]**. Observe that *a space must be left on each side of the asterisk.*

Figure C6.5 *Using the PLOT Command in MYSTAT*

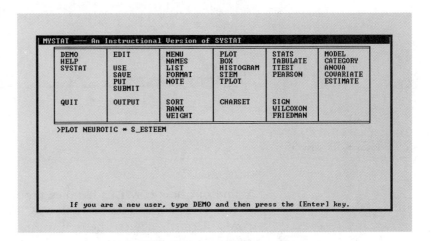

Computing the Pearson *r* in MYSTAT   You can calculate the linear correlation by typing **PEARSON VAR_Y, VAR_X [Enter]**. In the present example, the instruction PEARSON NEUROTIC, S_ESTEEM [Enter] produced the correlation coefficient that appears in Figure C6.7. The PEARSON command always yields a "correlation matrix," which displays the correlation of each variable with every other variable, including itself.

Figure C6.6 *Output Produced by MYSTAT's PLOT Command*

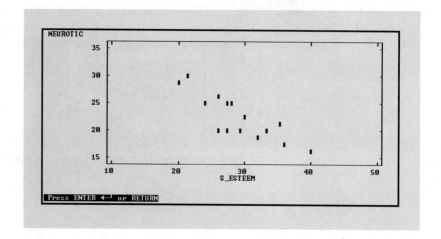

Figure C6.7 *Linear Correlation Coefficient Produced by MYSTAT's PEARSON Command*

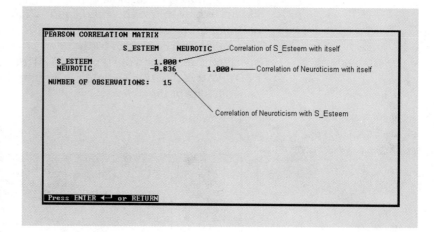

7: LINEAR REGRESSION

MINITAB

To find out how MINITAB handles problems in linear regression, we will analyze the data in Table 7.3. The data are nine pairs of scores. Performance on the first exam in a psychology course is variable X (the predictor), and final average in the course is variable Y (the criterion). We will use the software to set up a regression equation that can be used to predict final averages from scores on the first exam.

Figure C7.1 displays the execution of the REGRESS command. The general syntax is **REGRESS 'VarY' on K predictors in 'VarX'** [**Enter**]. 'VarY' is the criterion variable ('FinalAve' in the present example), and 'VarX' is the predictor variable ('Exam1' here). K = 1 because there is only one predictor variable in the type of linear regression discussed in this text. The output produced by the REGRESS command appears in Figure C7.2.

MYSTAT

The data in this example are taken from Table 7.7 and concern the relationship between level of depression (variable *Y, the criterion*) and number of therapy sessions (variable *X, the predictor*). Figure C7.3 shows how the variable names and 17 pairs of scores are entered. It doesn't matter which columns *X* and *Y* are assigned to, as long as you remember which variable you consider to be the criterion (*Y*) variable.

Figure C7.1 *Using the REGRESS Command in MINITAB*

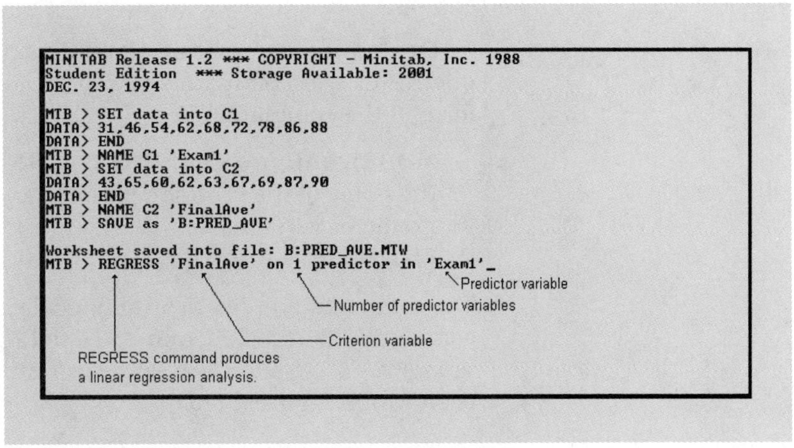

Figure C7.2 *Output Generated by the REGRESS Command in MINITAB*

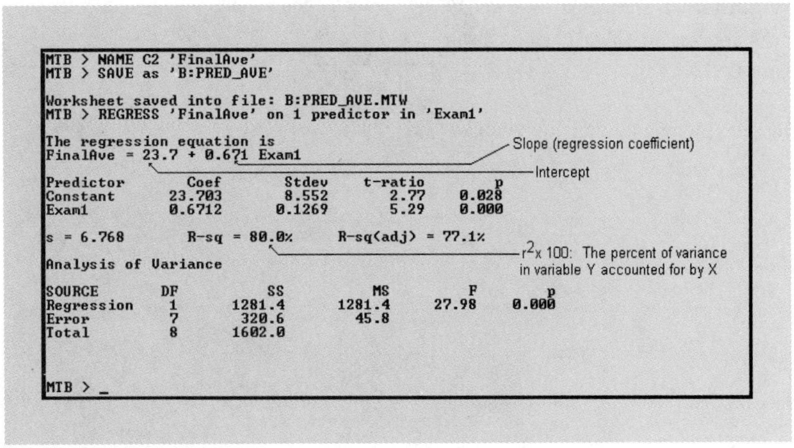

Figure C7.3 *Entering and Saving Pairs of X and Y Scores in MYSTAT*

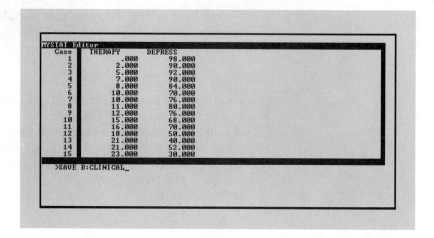

```
MYSTAT Editor
  Case     THERAPY       DEPRESS
    1          .000        98.000
    2         2.000        90.000
    3         5.000        92.000
    4         7.000        90.000
    5         8.000        84.000
    6        10.000        70.000
    7        10.000        76.000
    8        11.000        80.000
    9        12.000        76.000
   10        15.000        68.000
   11        16.000        70.000
   12        18.000        50.000
   13        21.000        40.000
   14        21.000        52.000
   15        23.000        30.000
>SAVE B:CLINICAL_
```

Setting up a regression equation in MYSTAT requires a sequence of two commands: MODEL and ESTIMATE. The general form of the commands is:

1. **MODEL VAR_Y = CONSTANT + VAR_X [Enter]** (where VAR_Y is the name of the criterion variable and VAR_X is the name of the predictor variable)
2. **ESTIMATE [Enter]**

The command line clears automatically after you input the MODEL command. The MODEL and ESTIMATE instructions are illustrated in Figures C7.4 and C7.5, and the output for the present example is shown in Figure C7.6.

Figure C7.4 *Entering the Regression MODEL in MYSTAT*

```
MYSTAT --- An Instructional Version of SYSTAT

  DEMO       EDIT       MENU        PLOT        STATS       MODEL
  HELP                  NAMES       BOX         TABULATE    CATEGORY
  SYSTAT     USE        LIST        HISTOGRAM   TTEST       ANOVA
             SAVE       FORMAT      STEM        PEARSON     COVARIATE
             PUT        NOTE        TPLOT                   ESTIMATE
             SUBMIT

  QUIT       OUTPUT     SORT        CHARSET     SIGN
                        RANK                    WILCOXON
                        WEIGHT                  FRIEDMAN

>MODEL DEPRESS = CONSTANT + THERAPY_




                 If you are a new user, type DEMO and then press the [Enter] key.
```

Figure C7.5 *Using the ESTIMATE Command to Conduct Linear Regression in MYSTAT*

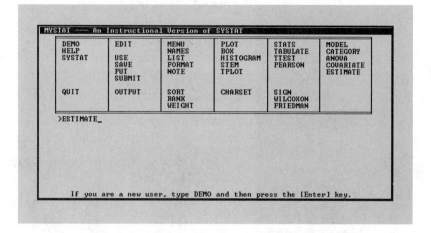

Figure C7.6 *Linear Regression Output Produced by MYSTAT's ESTIMATE Command*

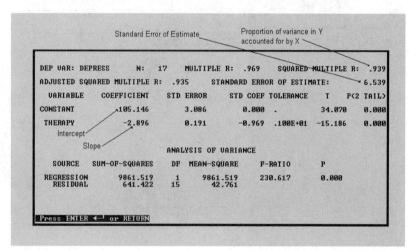

8: SAMPLING DISTRIBUTIONS

MINITAB

MINITAB has a slick little command for quickly and easily constructing a confidence interval around a sample mean, using the equivalent of $\overline{X} + z_c\sigma_{\overline{X}}$ [formula (8.6)]. To use this command, you must either know or be able to make a reasonable assumption about the value of the population standard deviation—referred to as *sigma*.

To demonstrate the command, we'll use a hypothetical set of $N = 25$ scores on the verbal scale of the Scholastic Aptitude Test (SAT). Figure C8.1 displays those data as entered into MINITAB. The objective is to compute the mean of the sample and set up a 95% confidence interval around that mean. We will assume that $sigma = \sigma = 100$.

As shown in Figure C8.1, the syntax of the confidence-interval instruction is **ZINTERVAL assuming sigma = K using the data in C# [Enter]**, where K is the actual or assumed value of the population standard deviation and # is the number of the column that contains the sample data. The 95% confidence interval for the SAT scores appears at the bottom of Figure C8.1.

The ZINTERVAL command defaults to a 95% confidence level. If you want to use a different confidence level, you should include the K% argument in the instruction, like this:

ZINTERVAL with K% confidence assuming sigma = K using the data in C# [Enter]

where K% is the desired confidence level (90% or 99%, for example).

Figure C8.1 *Using the ZINTERVAL Command in MINITAB*

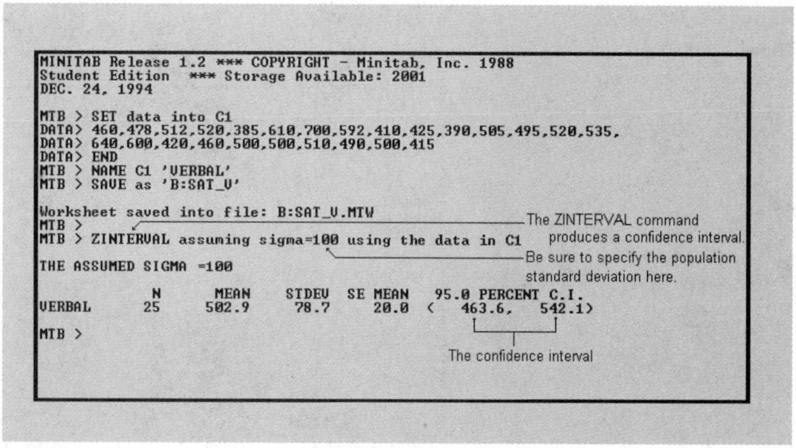

MYSTAT

Unfortunately, MYSTAT does not have a built-in routine to construct a confidence interval around a mean. Nonetheless, you can use **STATS VARNAME /MEAN [Enter]** to compute a sample mean. If you know or can make a reasonable assumption about the value of the population standard deviation, you can compute the boundaries of the confidence interval through $\bar{X} \pm z_c \sigma_{\bar{X}}$ [formula (8.6)].

9: LOGIC OF HYPOTHESIS TESTING

MINITAB

This section will familiarize you with MINITAB's ZTEST instruction, which tests the null hypothesis that a population mean (μ) is equal to a particular value. You use it with a set of sample data. The test compares the sample mean with the population mean specified under the null hypothesis, using the equivalent of $z = (\overline{X} - \mu)/\sigma_{\overline{X}}$. If the computed z ratio is significant, you reject H_0 and conclude that the mean of the population differs from the hypothesized μ.

To demonstrate the command, we'll use a hypothetical set of $N = 64$ scores on the Beck Depression Inventory. Figure C9.1 displays those data as entered into MINITAB. The objective is to compute the mean of the sample and compare it with a hypothesized population mean of $\mu = 25$. This problem is taken from the example in Chapter 9 in which a social worker assesses the effectiveness of enhanced ambient lighting therapy on the mood of 64 seasonally depressed patients. We will assume that $sd = \sigma = 6$. Be aware that, to work with the ZTEST command, you must know the population standard deviation or be able to make a reasonable assumption about its value.

As shown in Figure C9.1, the syntax of the instruction is

ZTEST of population mean=K1 assuming population sd=K2 on data in 'VarName' [Enter]

where K1 is the hypothesized population mean, K2 is the actual or assumed value of the population standard deviation, and 'VarName' is the name of the data set. The result of the ZTEST instruction for the present sample is shown in Figure C9.1. Note the P VALUE in the output. This is the probability that the computed z ratio is a chance event. *If the P VALUE is equal to or less than the alpha level you chose, then the z ratio is significant.*

Figure C9.1 *Using the ZTEST Command in MINITAB*

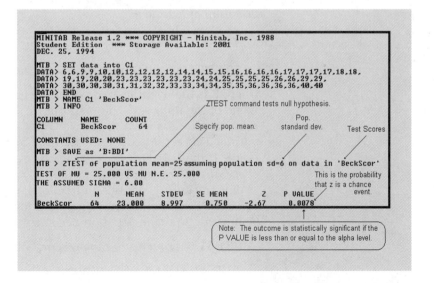

MYSTAT

Unfortunately, MYSTAT does not have a built-in routine to conduct a *z* test of a mean. Nonetheless, you can use **STATS VARNAME / MEAN [Enter]** to compute a sample mean. If you know or can make a reasonable assumption about the value of the population standard deviation, then you can compute the *z* ratio via $z = (\overline{X} - \mu)/\sigma_{\overline{X}}$. If the computed *z* is equal to or greater than the z_{crit} at the stated alpha level, then it is significant.

10: ONE-SAMPLE *t* STATISTIC

MINITAB

This section will acquaint you with MINITAB's TTEST instruction, which tests the null hypothesis that a population mean (μ) is equal to a particular value. You use it with a set of sample data when a *z* test is not feasible because the population standard deviation is unknown. The test compares the sample mean with the population mean specified under the null hypothesis, using the equivalent of $t = (\overline{X} - \mu)/S_{\overline{X}}$. If the computed *t* ratio is significant, you reject H_0 and conclude that the mean of the population differs from the hypothesized μ.

To demonstrate the TTEST instruction, we'll use the T-maze data from Table 10.1. Figure C10.1 displays those data as entered into MINITAB. The objective is to compute the mean of the sample and compare it with a hypothesized population mean of $\mu = 60$ "safe" turns in the T-maze. Since the test in this example is non-directional, the computed *t* statistic is significant if its probability is equal to or less than alpha, regardless of the direction of the deviation of $\overline{X}$ from μ.

As shown in Figure C10.1, the syntax of the TTEST instruction is

TTEST of population mean=K on 'VarName' [Enter]

where K is the hypothesized population mean and 'VarName' is the name of the data set. The result of the TTEST instruction for the T-maze data is shown in Figure C10.1. Note the P VALUE in the output. This is the probability that the computed *t* ratio is a chance event. *If the P VALUE is equal to or less than the alpha level you chose, then the t ratio is significant.*

If you want to conduct a directional (i.e., one-tailed) test using the TTEST command, you must include the ALTERNATIVE sub-command. So the instruction becomes

TTEST of population mean=K on 'VarName'; [Enter]

ALTERNATIVE CODE. [Enter]

where CODE is either 1 or -1, depending on whether your alternative hypothesis predicts a positive (1) or a negative (–1) difference between $\overline{X}$ and μ.

Figure C10.1 *Using the TTEST Command in MINITAB*

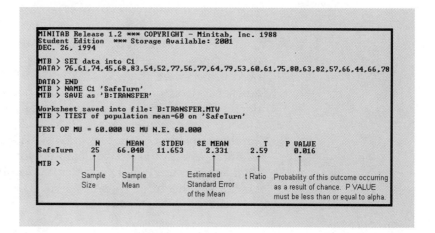

```
MINITAB Release 1.2 *** COPYRIGHT - Minitab, Inc. 1988
Student Edition  *** Storage Available: 2001
DEC. 26, 1994

MTB > SET data into C1
DATA> 76,61,74,45,68,83,54,52,77,56,77,64,79,53,60,61,75,80,63,82,57,66,44,66,78

DATA> END
MTB > NAME C1 'SafeTurn'
MTB > SAVE as 'B:TRANSFER'

Worksheet saved into file: B:TRANSFER.MTW
MTB > TTEST of population mean=60 on 'SafeTurn'

TEST OF MU = 60.000 US MU N.E. 60.000

             N      MEAN    STDEV   SE MEAN      T   P UALUE
SafeTurn    25    66.040   11.653    2.331    2.59    0.016

MTB >
```

| | Sample Size | Sample Mean | Estimated Standard Error of the Mean | t Ratio | Probability of this outcome occurring as a result of chance. P VALUE must be less than or equal to alpha. |

MYSTAT

This section will introduce you to MYSTAT's **TTEST** instruction, which tests the null hypothesis that a population mean (μ) is equal to a particular value. You use it with a set of sample data when a z test is not feasible because the population standard deviation is unknown. The test compares the sample mean with the population mean specified under the null hypothesis, using the equivalent of $t = (\overline{X} - \mu)/S_{\overline{X}}$. If the computed t ratio is significant, you reject H_0 and conclude that the mean of the population differs from the hypothesized μ.

To demonstrate the TTEST instruction, we'll use the weight gains data from Table 10.3. Figure C10.2 displays those data as entered with the MYSTAT Editor. The objective is to compute the mean weight gain of the sample and compare it with a hypothesized population mean of $\mu = 0$. Since the test in this example is a directional test, the computed t statistic is significant if its probability is equal to or less than alpha *and* the sample mean is greater than the hypothesized population mean. Therefore the outcome cannot be considered significant if $\overline{X}$ is less than μ, regardless of how large the mean difference is.

To execute the TTEST command:

1. Input the VARNAME and sample data into column 1 in the Editor (Figure C10.2).
2. Name the second column POP_MEAN (Figure C10.2).
3. Use MYSTAT's LET command to set all values of POP_MEAN to 0 (Figures C10.2 and C10.3).

4. SAVE the data file to disk under the name HORMONE (Figure C10.3).
5. Type QUIT [**Enter**] to return to the Command Menu.
6. Type TTEST WTGAIN, POP_MEAN [**Enter**] (Figure C10.4).

The result of these actions appears in MYSTAT's Output Screen, displayed in Figure C10.5.

Notice that Figure C10.5 shows PROB = .178, which means that the probability of getting the t ratio by chance alone is .178. You need to be aware that MYSTAT's TTEST routine does not distinguish between one-tailed (i.e., directional) and two-tailed (i.e., nondirectional) tests. *When performing a one-tailed test*, as was done in this example, *you need to divide the PROB value by 2* to obtain the correct chance probability. Therefore, the correct PROB is .0854. The t ratio is significant only if PROB is equal to or less than alpha. Since PROB is greater than alpha (.05) in this instance, t is not significant.

Figure C10.2 *Using MYSTAT's LET Command to Set a Column of Values to a Constant*

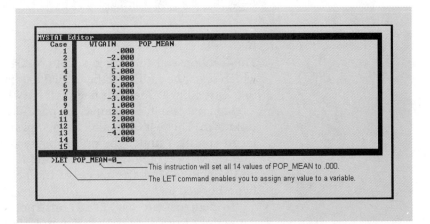

Figure C10.3 *Completing Data Entry and Saving the Data to Disk in MYSTAT*

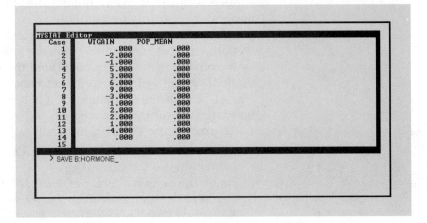

Figure C10.4 *Executing MYSTAT's TTEST Command*

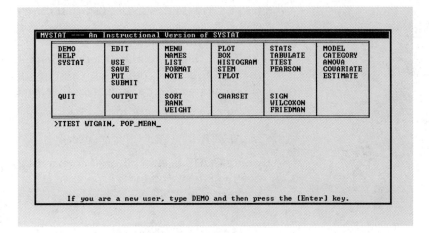

Figure C10.5 *Output of MYSTAT's TTEST Command*

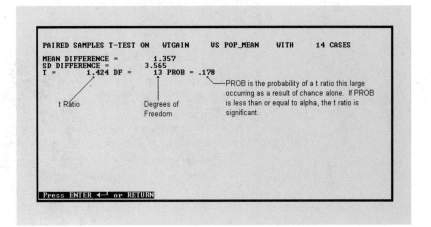

11: TWO-SAMPLE *t* TESTS

MINITAB

To conduct an independent-samples *t* test in MINITAB, you use the TWOSAMPLE instruction. This instruction and its effects are illustrated in Figure C11.1. The example compares the effectiveness of two approaches to teaching high school mathematics: programmed learning versus standard lecture. The data are final exam scores in the math course. The syntax of the TWOSAMPLE command is:

TWOSAMPLE with K% confidence for data in C1 and C2 [Enter]

where K% is the confidence level that corresponds to a particular alpha level. For example, K% is 95% when the alpha level is 5%. The result of the TWOSAMPLE instruction for the teaching-methods data is shown in Figure C11.1. Note the *P* value in the output. This is the probability that the computed *t* ratio is a chance event. *If the P value is equal to or less than the alpha level you chose, then the t ratio is significant.*

If you want to conduct a directional (i.e., one-tailed) test using the TWOSAMPLE command, you must include the ALTERNATIVE subcommand. So the instruction becomes

TWOSAMPLE with K% confidence for data in C1 and C2; [Enter]

ALTERNATIVE CODE. [Enter]

where CODE is either 1 or –1, depending on whether your alternative hypothesis predicts a positive (1) or a negative (–1) difference between μ_1 and μ_2.

Figure C11.1 *Using the TWOSAMPLE Command in MINITAB*

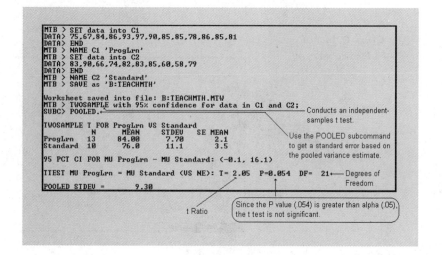

MYSTAT

To conduct an independent-samples *t* test in MYSTAT, you use the TTEST instruction. The example I'll use to illustrate this command compares the effectiveness of two approaches to teaching high school mathematics: programmed learning versus standard lecture. The data are final exam scores in the math course. They are distributed in the following way:

Programmed learning: 75, 67, 84, 86, 93, 97, 90, 85, 85, 78, 86, 85, 81

Standard lecture: 83, 90, 66, 74, 82, 83, 85, 60, 58, 79

Integer coding To conduct an independent-samples *t* test in MYSTAT, you must represent the two samples as two levels of an independent variable (let's call the variable GROUP). After naming and inputting the dependent variable in the usual way, you enter each subject's level of the independent variable as an "integer code." In this example, each subject in the programmed instruction group is represented by the integer 0, and each subject in the standard lecture group is represented by the integer 1. Figure C11.2 illustrates integer coding. Notice how each subject has two data entries in the Edit Window, one under the EXAMSCOR variable and another under the GROUP variable.

Figure C11.2 *Using Integer Coding in MYSTAT*

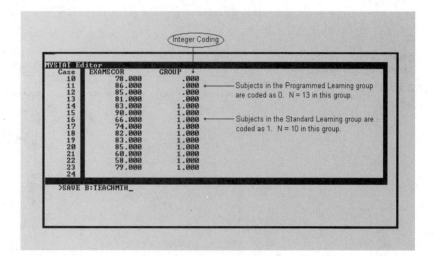

Doing the independent-samples test   Figure C11.3 illustrates how to execute the TTEST command in the independent-samples situation, and Figure C11.4 shows the output of the instruction. If you can assume that the population variances are equal (and normally you can), you should refer to the results for the pooled-variances *t* test.

Figure C11.3 *Executing MYSTAT's TTEST Command for Independent Samples*

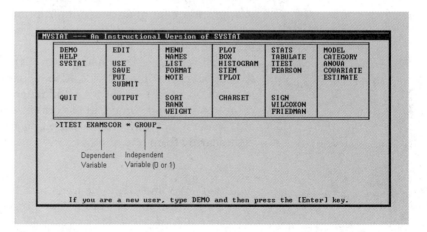

Notice that Figure C11.4 shows PROB = .054, which means that the probability of getting the *t* ratio by chance alone is .054. You need to be aware that MYSTAT's TTEST routine does not distinguish between one-tailed (i.e., directional) and two-tailed (i.e., nondirectional) tests. *When performing a one-tailed test, you need to divide the PROB value by 2* to obtain the correct chance probability. The *t* ratio is significant only if PROB is equal to or less than alpha. Since PROB is greater than alpha (.05) in this instance, *t* is not significant. Thus, you conclude that the population means do not differ.

Figure C11.4 *The Output of MYSTAT's TTEST Command for Independent Samples*

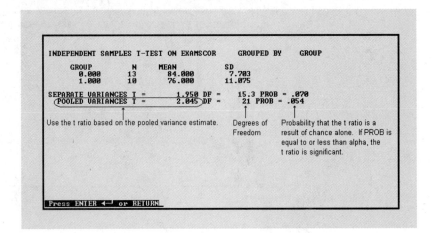

12: ANALYSIS OF VARIANCE

MINITAB

Figures C12.1 and C12.2 demonstrate how to carry out an analysis of variance with the **AOVONEWAY** command. This instruction is appropriate for analyzing data from multiple-samples studies in which there is only one independent variable. The data are taken from Table 12.1.

You'll recognize the basic parts of the ANOVA summary table that appears in Figure C12.2. Also, observe that the AOVONEWAY instruction produces a 95% confidence interval for each sample mean, so that you can conduct pairwise comparisons (see Chapter 12). Means that have nonoverlapping confidence intervals differ significantly from one another.

MYSTAT

Figures C12.3 and C12.4 demonstrate how to carry out an analysis of variance in MYSTAT. This procedure is appropriate for

Figure C12.1 *Entering Data for the AOVONEWAY Command in MINITAB*

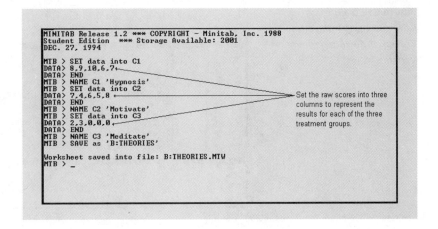

Figure C12.2 *Output from the AOVONEWAY Command in MINITAB*

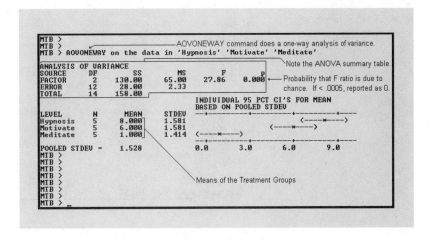

analyzing data from multiple-samples studies in which there is only one independent variable. The data are taken from Table 12.10.

Notice that you must use integer coding to represent the independent variable: 1 = control group, 2 = um-huh group, 3 = huh-uh group, and 4 = smile group.

After you enter the data, code the groups, and save the file, use QUIT to return to the Command Menu. Then type this sequence of instructions to execute the ANOVA:

1. CATEGORY GROUP = 4 [**Enter**]   ←**Note that the GROUP variable has four levels**

2. ANOVA Y [**Enter**]

3. ESTIMATE [**Enter**]

Figure C12.3 *Entering Data and Using Integer Coding of the Independent Variable in MYSTAT*

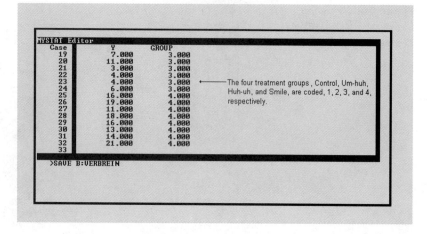

Figure C12.4 *The Output of MYSTAT's Analysis of Variance Procedure*

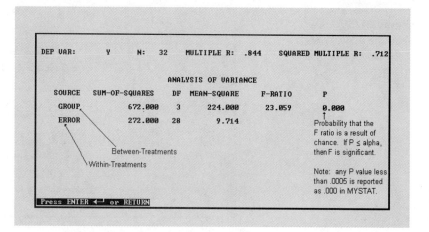

13: TWO-WAY ANALYSIS OF VARIANCE

MINITAB

Figures C13.1 and C13.2 demonstrate how to carry out a two-way analysis of variance via the TWOWAY command. This instruction is appropriate for analyzing data from multiple-samples studies in which there are two independent variables. The data are taken from Table 13.1.

To use the TWOWAY command properly, you must use "integer coding." Notice that all dependent-variable raw scores are SET into C1 (Figure C13.1). You represent the first independent variable (factor *A*) with the integers 1 (for placebo) and 2 (for Valium) in C2. Likewise, you represent the second independent variable (factor *B*) with the integers 1 (for no noise) and 2 (for noise) in C3. *Important*:

Figure C13.1 *Entering Data for the TWOWAY Command in MINITAB*

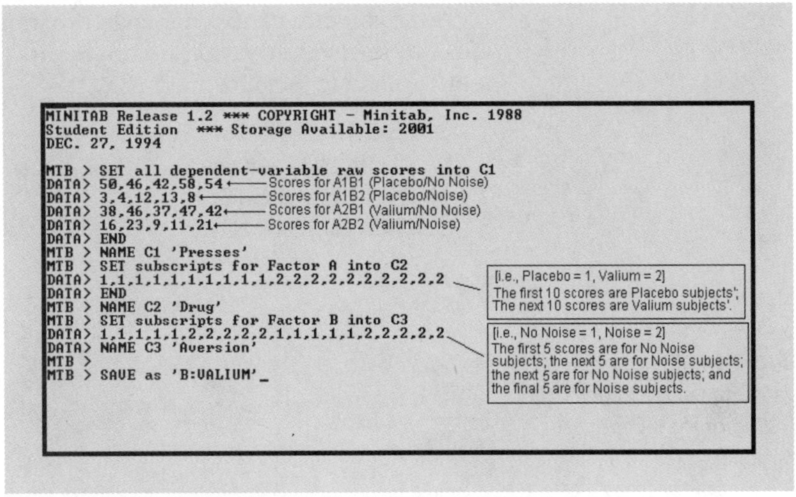

```
MINITAB Release 1.2 *** COPYRIGHT - Minitab, Inc. 1988
Student Edition  *** Storage Available: 2001
DEC. 27, 1994

MTB > SET all dependent-variable raw scores into C1
DATA> 50,46,42,58,54 ←——— Scores for A1B1 (Placebo/No Noise)
DATA> 3,4,12,13,8 ←——— Scores for A1B2 (Placebo/Noise)
DATA> 38,46,37,47,42 ←——— Scores for A2B1 (Valium/No Noise)
DATA> 16,23,9,11,21 ←——— Scores for A2B2 (Valium/Noise)
DATA> END
MTB > NAME C1 'Presses'
MTB > SET subscripts for Factor A into C2
DATA> 1,1,1,1,1,1,1,1,1,1,2,2,2,2,2,2,2,2,2,2 ——— [i.e., Placebo = 1, Valium = 2]
DATA> END                                          The first 10 scores are Placebo subjects';
MTB > NAME C2 'Drug'                               The next 10 scores are Valium subjects'.
MTB > SET subscripts for Factor B into C3
DATA> 1,1,1,1,1,2,2,2,2,2,1,1,1,1,1,2,2,2,2,2 ——— [i.e., No Noise = 1, Noise = 2]
DATA> NAME C3 'Aversion'                            The first 5 scores are for No Noise
MTB >                                               subjects; the next 5 are for Noise subjects;
MTB > SAVE as 'B:VALIUM' _                          the next 5 are for No Noise subjects; and
                                                    the final 5 are for Noise subjects.
```

Figure C13.2 *Output from the TWOWAY Command in MINITAB*

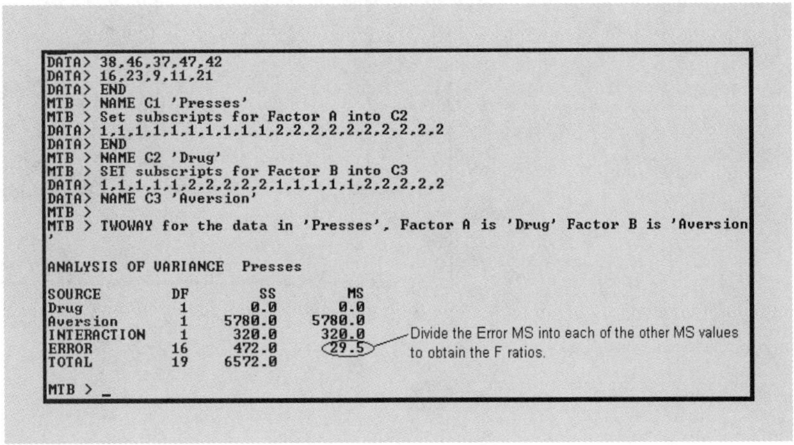

```
DATA> 38,46,37,47,42
DATA> 16,23,9,11,21
DATA> END
MTB > NAME C1 'Presses'
MTB > Set subscripts for Factor A into C2
DATA> 1,1,1,1,1,1,1,1,1,1,2,2,2,2,2,2,2,2,2,2
DATA> END
MTB > NAME C2 'Drug'
MTB > SET subscripts for Factor B into C3
DATA> 1,1,1,1,1,2,2,2,2,2,1,1,1,1,1,2,2,2,2,2
DATA> NAME C3 'Aversion'
MTB >
MTB > TWOWAY for the data in 'Presses', Factor A is 'Drug' Factor B is 'Aversion'

ANALYSIS OF VARIANCE  Presses

SOURCE        DF      SS      MS
Drug           1     0.0     0.0
Aversion       1  5780.0  5780.0
INTERACTION    1   320.0   320.0 ——— Divide the Error MS into each of the other MS values
ERROR         16   472.0   (29.5)     to obtain the F ratios.
TOTAL         19  6572.0

MTB > _
```

Each subject is represented in the same position in each of the three columns of data. For example, the first subject has the first raw score (50) in C1, the first factor *A* code (1) in C2, and the first factor *B* code (1) in C3.

MYSTAT

Figures C13.3 and C13.4 demonstrate how to carry out a two-way analysis of variance in MYSTAT. This procedure is appropriate for analyzing data from multiple-samples studies in which there are two independent variables. The data are taken from Table 13.7.

Notice that you must use integer coding to represent the independent variables. Also observe that a separate coding column is required for each factor.

After you enter the data, code the groups, and save the file, use QUIT to return to the Command Menu. Then type this sequence of instructions to execute the ANOVA:

1. CATEGORY FACTOR_A = 2   ←Note that FACTOR_A
FACTOR_B = 3 [**Enter**]   has two levels and
2. ANOVA ERRORS [**Enter**]   FACTOR_B has three.
3. ESTIMATE [**Enter**]

Figure C13.3 *Entering Data and Using Integer Coding of the Independent Variables in MYSTAT*

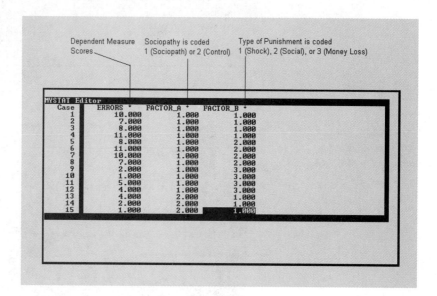

Figure C13.4 *The Output of MYSTAT's Two-Way Analysis of Variance Procedure*

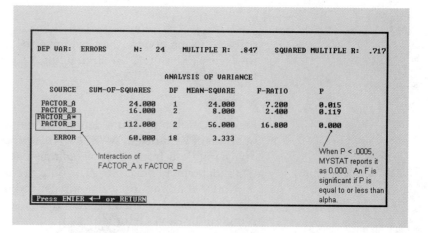

14: REPEATED-MEASURES ANALYSIS OF VARIANCE

MINITAB

Figures C14.1 and C14.2 demonstrate how to carry out a repeated-measures analysis of variance with the TWOWAY command. The data are taken from Table 14.1.

To use the TWOWAY command properly, you must use "integer coding." Notice that all dependent-variable raw scores are SET into C1 (Figure C14.1). You represent the independent variable with the integers 1, 2, 3, and 4 (for different degrees of rotation) in C2. Likewise, you represent the six subjects with the integers 1–6 in C3. *Important*: Each subject is represented in the same position in each of the three columns of data. For example, the first subject has the first raw score (650) in C1, the first rotation code (1) in C2, and the first subject code (1) in C3. Because this is a repeated-measures study, the first subject is also represented in the 7th, 13th, and 19th positions in all three columns.

Figure C14.1 *Entering Data for the Repeated-Measures ANOVA in MINITAB*

```
MINITAB Release 1.2 *** COPYRIGHT - Minitab, Inc. 1988
Student Edition   *** Storage Available: 2001
DEC. 27, 1994

MTB > SET all dependent-variable raw scores into C1
DATA> 650,550,400,350,400,350
DATA> 760,620,500,430,530,460
DATA> 780,680,520,450,510,480
DATA> 890,790,600,550,560,510
DATA> END
MTB > NAME C1 'DecTime'
MTB > SET independent-variable subscripts into C2
DATA> 1,1,1,1,1,1,2,2,2,2,2,2,3,3,3,3,3,3,4,4,4,4,4,4
DATA> END
MTB > NAME C2 'Rotation'
MTB > SET subjects subscripts into C3
DATA> 1,2,3,4,5,6,1,2,3,4,5,6,1,2,3,4,5,6,1,2,3,4,5,6
DATA> END
MTB > NAME C3 'Subjects'
MTB > SAVE as 'B:IMAGERY'

Worksheet saved into file: B:IMAGERY.MTW
MTB > _
```

Figure C14.2 *Output from a Repeated-Measures ANOVA in MINITAB*

```
MTB > NAME C1 'DecTime'
MTB > SET independent-variable subscripts into C2
DATA> 1,1,1,1,1,1,2,2,2,2,2,2,3,3,3,3,3,3,4,4,4,4,4,4
DATA> END
MTB > NAME C2 'Rotation'
MTB > SET subjects subscripts into C3
DATA> 1,2,3,4,5,6,1,2,3,4,5,6,1,2,3,4,5,6,1,2,3,4,5,6
DATA> END
MTB > NAME C3 'Subjects'
MTB > SAVE as 'B:IMAGERY'

Worksheet saved into file: B:IMAGERY.MTW
MTB >
MTB >
MTB > TWOWAY for the data in 'DecTime', Factor is 'Rotation' remove 'Subjects'

ANALYSIS OF VARIANCE  DecTime

SOURCE      DF      SS       MS
Rotation     3   121800    40600
Subjects     5   343600    68720 ←——Ignore this MS in Repeated-Measures ANOVA
ERROR       15     8000      533 ←——This is the MS for the Interaction of the
TOTAL       23   473400                Independent Variable and Subjects. Divide this
                                        into the MS for Rotation to get F.
MTB > _
```

MYSTAT

Figures C14.3 and C14.4 demonstrate how to carry out a repeated-measures analysis of variance in MYSTAT. The data are taken from Table 14.4.

Notice that you must use integer coding to represent both the independent variable and the subjects factor. Also observe that a separate coding column is required for each factor.

After you enter the data, code the groups, and save the file, use QUIT to return to the Command Menu. Then type this sequence of instructions to execute the ANOVA:

1. MODEL = CONSTANT + INDVAR + SUBJECT [**Enter**]
2. ESTIMATE [**Enter**]

Note that, in MYSTAT, you actually use a regression command to do repeated-measures ANOVA.

Figure C14.3 *Entering Data and Using Integer Coding for Repeated-Measures ANOVA in MYSTAT*

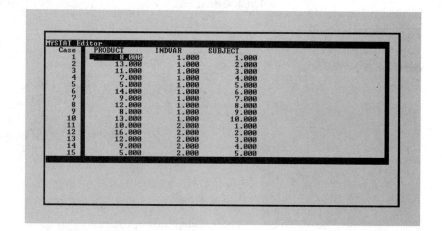

Figure C14.4 *The Output of MYSTAT's Repeated-Measures Analysis of Variance Procedure*

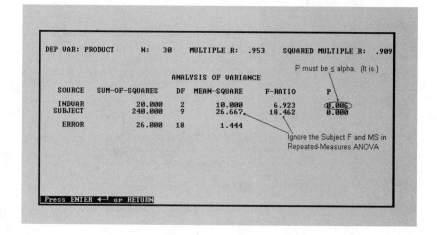

15: NONPARAMETRIC TESTS

MINITAB

Doing a chi-square test   Figures C15.1 and C15.2 demonstrate how to conduct a chi-square test of association with MINITAB's CHISQUARE command. The data are taken from Table 15.1.

To use the CHISQUARE command, you first SET the cell frequency counts of the contingency table into C1 and C2 (Figure C15.1). Then you issue the following instruction:

CHISQUARE of table in 'Col1Name' and 'Col2Name' [Enter]

In this example, 'Col1Name' is 'Female', and 'Col2Name' is 'Male'.

As shown in Figure C15.2, the CHISQUARE instruction produces both a complete contingency table and the results of the significance test.

Figure C15.1 *Entering Data for the CHISQUARE Procedure in MINITAB*

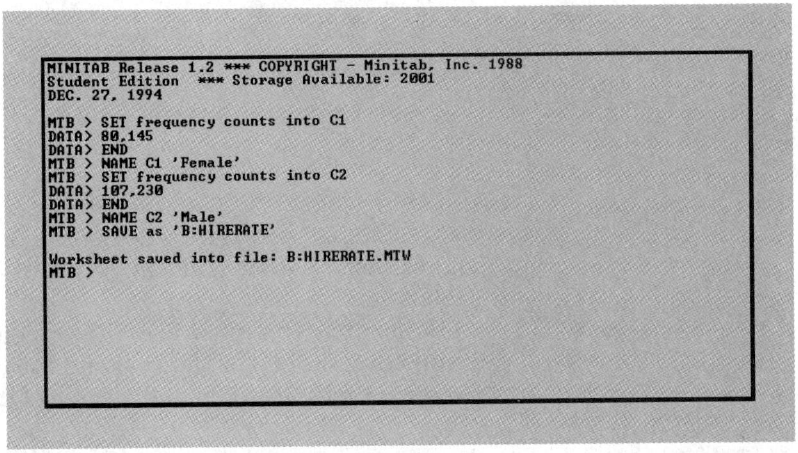

```
MINITAB Release 1.2 *** COPYRIGHT - Minitab, Inc. 1988
Student Edition  *** Storage Available: 2001
DEC. 27, 1994

MTB > SET frequency counts into C1
DATA> 80,145
DATA> END
MTB > NAME C1 'Female'
MTB > SET frequency counts into C2
DATA> 107,230
DATA> END
MTB > NAME C2 'Male'
MTB > SAVE as 'B:HIRERATE'

Worksheet saved into file: B:HIRERATE.MTW
MTB >
```

Figure C15.2 *Output from the CHISQUARE Command in MINITAB*

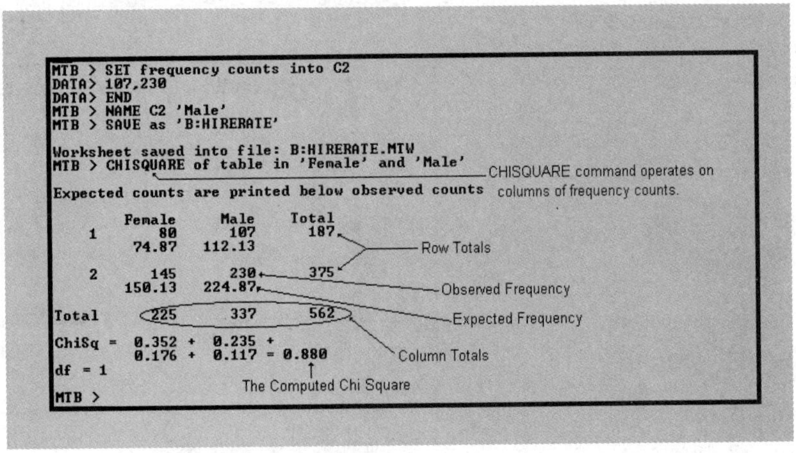

```
MTB > SET frequency counts into C2
DATA> 107,230
DATA> END
MTB > NAME C2 'Male'
MTB > SAVE as 'B:HIRERATE'

Worksheet saved into file: B:HIRERATE.MTW
MTB > CHISQUARE of table in 'Female' and 'Male'    CHISQUARE command operates on
                                                   columns of frequency counts.
Expected counts are printed below observed counts

          Female    Male    Total
    1        80      107     187            Row Totals
          74.87    112.13

    2       145      230     375            Observed Frequency
         150.13    224.87
                                            Expected Frequency
Total      225      337     562

ChiSq =  0.352 +  0.235 +
         0.176 +  0.117 = 0.880            Column Totals
df = 1
                          The Computed Chi Square
MTB >
```

Conducting the Mann-Whitney *U* test in MINITAB Figure C15.3 illustrates how to do a *U* test in MINITAB. Note that the program converts the raw scores to ranks. All you have to do is enter the groups' respective raw scores into C1 and C2 and type the command.

Figure C15.3 *Using the Mann-Whitney command in MINITAB*

```
DEC. 28, 1994

MTB > SET data into C1
DATA> 68,25,52,68,40,68,90,45,60,77
DATA> END
MTB > NAME C1 'Group_A'
MTB > SET data into C2
DATA> 84,52,88,75,95,93,86
DATA> END
MTB > NAME C2 'Group_B'
MTB > SAVE as 'B:JOBATID'

Worksheet saved into file: B:JOBATID.MTW
MTB > MANN-WHITNEY on 'Group_A' and 'Group_B'
```
———— MANN-WHITNEY command does U test.
```
Mann-Whitney Confidence Interval and Test

Group_A    N = 10     MEDIAN =    64.00    ← Medians of    ⟍ Smaller of Us
Group_B    N =  7     MEDIAN =    86.00       Groups
POINT ESTIMATE FOR ETA1-ETA2 IS     -23.50
95.5  PCT C.I. FOR ETA1-ETA2 IS ( -43.01,   -7.00)
W =      66.5
TEST OF ETA1 = ETA2 VS. ETA1 N.E. ETA2 IS SIGNIFICANT AT  0.0248

MTB > _                                    p must be ≤ alpha.(It is.)
```

MYSTAT

Figures C15.4 and C15.5 demonstrate how to carry out a chi-square test of association in MYSTAT. The data are taken from Table 15.5.

Notice that you must use integer coding to represent both the treatment groups (1, 2, 3, or 4) and the dependent variable, where 1 = accept false memory and 2 = reject false memory.

Figure C15.4 *Using Integer Coding for a Chi-Square Test in MYSTAT*

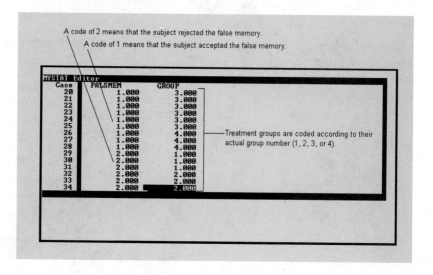

A code of 2 means that the subject rejected the false memory.
A code of 1 means that the subject accepted the false memory.

```
MYSTAT Editor
Case    FALSMEM      GROUP
 20      1.000       3.000
 21      1.000       3.000
 22      1.000       3.000
 23      1.000       3.000
 24      1.000       3.000
 25      1.000       3.000
 26      1.000       4.000
 27      1.000       4.000
 28      1.000       4.000
 29      2.000       1.000
 30      2.000       1.000
 31      2.000       1.000
 32      2.000       2.000
 33      2.000       2.000
 34      2.000       2.000
```
Treatment groups are coded according to their actual group number (1, 2, 3, or 4).

After you code the groups and their responses and save the file, use QUIT to return to the Command Menu. Then type the following instruction to execute the chi-square:

TABULATE FALSEMEM * GROUP [**Enter**]

The output shown in Figure C15.5 includes the contingency table, chi-square value, probability of the chi-square under the null hypothesis, and various correlation coefficients based on the chi-square. In the output, you should refer to the PEARSON CHI-SQUARE, not the LIKELIHOOD RATIO CHI-SQUARE.

Figure C15.5 *The Output of MYSTAT's Chi-Square Procedure*

```
TABLE OF  FALSMEM   (ROWS) BY   GROUP   (COLUMNS)
FREQUENCIES
              1.000    2.000    3.000    4.000    TOTAL

   1.000        12        6        7        3       28

   2.000         3        9        8       12       32

TOTAL           15       15       15       15       60   The Computed Chi Square


TEST STATISTIC                         VALUE        DF        PROB
     PEARSON CHI-SQUARE               11.250         3        .010
     LIKELIHOOD RATIO CHI-SQUARE      11.969         3        .007

COEFFICIENT                            VALUE     ASYMPTOTIC STD ERROR
     PHI                              .4330
     CRAMER V                         .4330
     CONTINGENCY                      .3974
     GOODMAN-KRUSKAL GAMMA            .5462          .14455
     KENDALL TAU-B                    .3546          .10349
     STUART TAU-C                     .4333          .12664
     SPEARMAN RHO                     .3884          .11172
     SOMERS D     (COLUMN DEPENDENT)  .4353          .12698
[Enter]
```

REFERENCES

Argyris, C. (1980). *Inner contradictions of rigorous research.* New York: Academic Press.

Bakan, D. (1966). The test of significance in psychological research. *Psychological Bulletin, 66,* 423–437.

Baker, B. O., Hardyck, C. D., & Petrinovich, L. F. (1966). Weak measurements vs. strong statistics: An empirical critique of S. S. Stevens' proscriptions on statistics. *Educational and Psychological Measurement, 26,* 291–309.

Barber, T. X. (1976). *Hypnosis: A scientific approach.* New York: Psychological Dimensions.

Cohen, J. (1977). *Statistical analysis for the behavioral sciences.* New York: Academic Press.

Cooper, L. A., & Shepard, R. N. (1973). Chronometric studies of the rotation of mental images. In W. G. Chase (Ed.), *Visual information processing.* New York: Academic Press.

Devoe, C.J. (1990). *Effects of the film "The Silent Scream" on attitudes toward abortion.* Unpublished doctoral dissertation. The Professional School of Psychological Studies, San Diego.

Dollard, J., Doob, L., Miller, N., Mowrer, O. H., & Sears, R. R. (1939). *Frustration and aggression.* New Haven, CT: Yale University Press.

Evans, J. D. (1985). *Invitation to psychological research.* New York: Holt, Rinehart & Winston.

Evans, J. D., & Peeler, L. (1979). Personalized comments on returned tests improve test performance in introductory psychology. *Teaching of Psychology, 6,* 57.

Faraone, S. (1982). Chi-square in small samples. *American Psychologist, 37,* 107.

Fisher, R. A. (1966). *The design of experiments* (8th ed.). New York: Harper & Row.

Haas, G. L., & Sweeney, J. A. (1992). Premorbid and onset features of first-episode schizophrenia. *Schizophrenic Bulletin, 18,* 373–386.

Hare, R. D. (1970). *Psychopathy: Theory and research.* New York: Wiley.

Hays, W. L. (1988). *Statistics* (4th ed.). New York: Holt, Rinehart & Winston.

Howard, K. I., Kopta, S. M., Krause, M. S., & Orlinsky, D.E. (1986). The dose-effect relationship in psychotherapy. *American Psychologist, 41,* 159–164.

Kerlinger, F. N. (1979). *Behavioral research: A conceptual approach.* New York: Holt, Rinehart & Winston.

Kimmel, H. D. (1957). Three criteria for the use of one-tailed tests. *Psychological Bulletin, 54,* 351–353.

Kirk, R .E. (1990). *Statistics: An introduction* (3rd ed.). New York: Holt, Rinehart & Winston.

Kirk, R. E. (1995). *Experimental design: Procedures for the behavioral sciences* (3rd ed.). Pacific Grove, CA: Brooks/Cole.

Lehman, D. R., Lempert, R. O., & Nisbett, R. E. (1988). The effects of graduate training on reasoning: Formal discipline and thinking about everyday-life events. *American Psychologist, 43,* 431–442.

Lickey, M. E., & Gordon, B. (1991). *Medicine and mental illness.* New York: Freeman.

McGuire, W.J. (1973). The yin and yang of progress in psychology: Seven koan. *Journal of Personality and Social Psychology, 26,* 446–456.

Moore, D. S. (1991). *Statistics: Concepts and controversies.* New York: Freeman.

Murrey, G. J., Cross, H. J., & Whipple, J. (1992). Hypnotically created pseudomemories: Further investigation in the "memory-distortion or response-bias" question. *Journal of Abnormal Psychology, 101,* 75–77.

Myers, D. G. (1990). *Social psychology* (3rd ed.). New York: McGraw-Hill.

Myers, J. C., DiCecco, J. V., White, J. B., & Borden, V. M. (1982). Repeated measurements on dichotomous variables: Q and F tests. *Psychological Bulletin, 92,* 517–525.

Newnham, J. (1988). *Color ranking as a measure of self-esteem.* Unpublished master's thesis. Lindenwood College, St. Charles, MO.

Occupational Outlook Handbook. (1990). Washington, D.C.: Bureau of Labor Statistics.

Occupational Outlook Handbook. (1994). Washington, D.C.: Bureau of Labor Statistics.

Rapee, R. M., & Lim, L. (1992). Discrepancy between self- and observer ratings of performance in social phobics. *Journal of Abnormal Psychology, 101,* 728–731.

Rocco, Y. (1987). *The relationship between employment status, leisure, and self-esteem for women in mid-life.* Unpublished master's thesis. Lindenwood College, St. Charles, MO.

Safer, D. J., & Krager, J. M. (1988). A survey of medication treatment for hyperactive/inattentive students. *Journal of the American Medical Association, 260,* 2256–2259.

Sheehan, P. W., Green, V., & Truesdale, P. (1992). Influence of rapport on hypnotically induced pseudomemory. *Journal of Abnormal Psychology, 101,* 690–700.

Splinter, J. P. (1989). *Functional priorities within marriage: Moral values, spouse, children, and occupation.* Unpublished master's thesis. Lindenwood College, St. Charles, MO.

Stinson, F. S., DeBakey, S. F., Grant, B. F., & Dawson, D. A. (1992). Association of alcohol problems with risk for AIDS in the 1988 National Health Survey. *Alcohol Health & Research World, 16,* 245–251.

Toothaker, L. E. (1986). *Introductory statistics for the behavioral sciences.* New York: McGraw-Hill.

Vox pop. (1993, July 5). *TIME,* p. 18.

Walizer, M. H., & Weiner, P. L. (1978). *Research methods and analysis: Searching for relationships.* New York: Harper & Row.

Westoff, L. A. (1979). Women in search of equality. *Focus, 6,* 1–19.

INDEX

TO THE OWNER OF THIS BOOK:

I hope that you have found *Straightforward Statistics for the Behavioral Sciences* useful. So that this book can be improved in a future edition, would you take the time to complete this sheet and return it? I'd really like to hear what you think.

School and address: _____

Department: _____

Instructor's name: _____

1. What I like most about this book is: _____

_____

_____

2. What I like least about this book is: _____

_____

_____

3. My general reaction to this book is: _____

_____

4. Were all of the chapters of the book assigned for you to read? _____

   If not, which ones weren't? _____

_____

5. In the space below, or on a separate sheet of paper, please write specific suggestions for improving this book and anything else you'd care to share about your experience in using the book.

_____

_____

_____

_____

_____

Optional:

Your name: _____ Date: _____

May Brooks/Cole quote you, either in promotion for *Straightforward Statistics for the Behavioral Sciences* or in future publishing ventures?

Yes: _____ No: _____

Sincerely,

James D. Evans

FOLD HERE

FOLD HERE

FORMULA NUMBER	DESCRIPTION	FORMULA
	One-Sample t Statistic	
10.1	Unbiased estimator of the population variance	$s^2 = \dfrac{\Sigma(X - \overline{X})^2}{N - 1}$
10.2	Unbiased estimator of the population variance	$s^2 = \dfrac{SS}{N - 1}$
10.3	Variance of the t distribution	$\dfrac{\text{degrees of freedom}}{\text{degrees of freedom} - 2}$
10.4	t statistic, or t ratio, for a sample mean	$t = \dfrac{\overline{X} - \mu}{s_{\overline{X}}}$
10.5	Estimated standard error of the mean	$s_{\overline{X}} = \sqrt{\dfrac{s^2}{N}}$
10.6	Estimated standard error of the mean	$s_{\overline{X}} = \dfrac{s}{\sqrt{N}}$
10.7	t statistic, or t ratio, for a correlation coefficient	$t = \dfrac{r\sqrt{N - 2}}{\sqrt{1 - r^2}}$
10.8	Confidence interval of a mean (based on a t statistic)	$\overline{X} \pm t_{\text{crit}} s_{\overline{X}}$
	Two-Sample t tests	
11.1	Standard error of the mean difference	$\sigma_{\overline{X}_1 - \overline{X}_2} = \sqrt{\dfrac{\sigma^2}{N_1} + \dfrac{\sigma^2}{N_2}}$
11.2	Standard error of the mean difference	$\sigma_{\overline{X}_1 - \overline{X}_2} = \sqrt{\sigma^2\left(\dfrac{1}{N_1} + \dfrac{1}{N_2}\right)}$
11.3	Estimated standard error of the mean difference	$s_{\overline{X}_1 - \overline{X}_2} = \sqrt{s_p^2\left(\dfrac{1}{N_1} + \dfrac{1}{N_2}\right)}$
11.4	Pooled estimate of the population variance	$s_p^2 = \dfrac{(N_1 - 1)s_1^2 + (N_2 - 1)s_2^2}{(N_1 - 1) + (N_2 - 1)}$
11.5	Independent-samples t ratio	$t = \dfrac{\overline{X}_1 - \overline{X}_2}{s_{\overline{X}_1 - \overline{X}_2}}$
11.6	Independent-samples t ratio	$t = (\overline{X}_1 - \overline{X}_2)/\sqrt{s_p^2\left(\dfrac{1}{N_1} + \dfrac{1}{N_2}\right)}$
11.7	Degrees of freedom for independent-samples t ratio	$df = (N_1 - 1) + (N_2 - 1)$
11.8	Estimated standard error of the sampling distribution of the mean of D	Estimated standard error of $\overline{D} = s_{\overline{D}} = \sqrt{\dfrac{s^2}{N}}$
11.9	Correlated-samples t ratio	$t = \dfrac{\overline{D}}{S_{\overline{D}}}$
	Analysis of Variance	
12.1	F ratio	$F = \dfrac{MS_B}{MS_W}$
12.2	Grand mean	$\overline{X}_G = \dfrac{\Sigma X}{N_{\text{tot}}}$
12.3	Grand mean, when all N's are equal	$\overline{X}_G = \dfrac{\Sigma \overline{X}}{k}$
12.4	Total sum of squares	$SS_{\text{tot}} = \Sigma(X - \overline{X}_G)^2$
12.5	Between-treatments sum of squares	$SS_B = \Sigma\left[N(\overline{X} - \overline{X}_G)^2\right]$
12.6	Within-treatments sum of squares	$SS_W = \Sigma\left[\Sigma(X - \overline{X})^2\right]$